A Unified Introduction to Linear Algebra:

Models, Methods, and Theory

A Unified Introduction to Linear Algebra:

Models, Methods, and Theory

Alan Tucker
*State University of New York
at Stony Brook*

Assisted by Donald Small
Colby College

Macmillan Publishing Company
New York

Collier Macmillan Publishers
London

Macmillan Publishing Company
866 Third Avenue, New York, New York 10022

Collier Macmillan Canada, Inc.

Library of Congress Cataloging-in-Publication Data

Tucker, Alan,
  A unified introduction to linear algebra.

  Bibliography: p.
  Includes index.
  1. Algebras, Linear.   I. Title.
QA184.T83 1987   512′.5   86-23670
ISBN   0-02-421580-5

Printing: 2 3 4 5 6 7 8   Year: 8 9 0 1 2 3 4 5 6 7

ISBN   0-02-421580-5

It is a pleasure and honor to dedicate my work on this book to my father.

A. C. T.

I would like to dedicate my efforts on this book to my parents, two great teachers.

D. B. S.

Preface

This book attempts to give students a unified introduction to the models, methods, and theory of modern linear algebra. Linear models are now used at least as widely as calculus-based models. The world today is commonly thought to consist of large, complex systems with many input and output variables. Linear models are the primary tool for analyzing these systems. A course based on this book (or one like it) should prove to be the most useful college mathematics course most students ever take. With this goal in mind, the material is presented with an eye toward making it easy to remember, not just for the next hour test but for a lifetime of diverse uses.

Linear algebra is an ideal subject for a lower-level college course in mathematics, because the theory, numerical techniques, and applications are interwoven so beautifully. The theory of linear algebra is powerful, yet easily accessible. Best of all, theory in linear algebra is likable. It simplifies and clarifies the workings of linear models and related computations. This is what mathematics is really about, making things simple and clear. It provides important answers that go beyond results we could obtain by brute computation. For too many students, mathematics is either a collection of techniques, as in calculus, or a collection of formal theory with limited applications, as in most courses after calculus (including traditional linear algebra courses). This book tries to rectify this artificial dichotomy.

Again, the applications of linear algebra are powerful, easily understood, and very diverse. This book introduces students to economic input–output models, population growth models, Markov chains, linear programming, computer graphics, regression and other statistical techniques, numerical methods for approximate solutions to most calculus problems, linear codes, and much more. These different applications reinforce each other and associated theory. Indeed, without these motivating applications, several of the more theoretical topics could not be covered in an introductory textbook.

The field of linear numerical analysis is very young, having been dependent on digital computers for its development. This field has wrought major changes in what linear algebra theory should be taught in an introductory course. The standout example of such a modern linear algebra text is G. Strang's *Linear Algebra and Its Applications*. Once the theory was needed as an alternative to numerical computation, which was hopelessly difficult. Now theory helps direct and interpret the numerical computation, which computers do for us.

Overview of the Text This book develops linear algebra around matrices. Vector spaces in the abstract are not considered, only vector spaces associated with matrices. This book puts problem solving and an intuitive treatment of theory first, with a proof-oriented approach intended to come in a second course, the same way that calculus is taught.

The book's organization is straightforward: Chapter 1 has introductory linear models; Chapter 2 has the basics of matrix algebra; Chapter 3 develops different ways to solve a system of equations; Chapter 4 has applications, and Chapter 5 has vector-space theory associated with matrices and related topics such as pseudoinverses and orthogonalization. Many linear algebra textbooks start immediately with Gaussian elimination, before any matrix algebra. Here we first pose problems in Chapter 1, then develop a mathematical language for representing and recasting the problems in Chapter 2, and then look at ways to solve the problems in Chapter 3—four different solution methods are presented with an analysis of strengths and weaknesses of each.

In most applications of linear algebra, the most difficult aspect is understanding matrix expressions, such as $Ue^{D}U^{-1}$. Students from a traditional linear algebra course have little preparation for understanding such expressions. This book constantly forces students to interpret the meaning of matrix expressions, not just perform rote computations. Matrix notation is used as much as possible, rather than constantly writing out systems of equations. The sections are generally too long to be covered completely in class; most have several examples (based on familiar models) that are designed to be read by students on their own without explanation by the instructor. The goal is for students to be able to read and understand uses of matrix algebra for themselves.

The material is unified pedagogically by the repeated use of a few linear models to illustrate all new concepts and techniques. These models give the student mental pictures to "visualize" new ideas during this course and help remember the ideas after the course is over.

Although this book is often informal ("proving theorems" by example) and sticks mainly to matrices rather than general linear transformations, it covers several topics normally left to a more advanced course, such as matrix norms, matrix decompositions, and approximation by orthogonal polynomials. These advanced topics find immediate, concrete applications. In addition, they are finite-dimensional versions of important theory in functional analysis; for example, the eigenvalue decomposition of a matrix into simple matrices is a special case of the spectral representation of linear operators.

Discrete Versus Continuous Mathematics Today there is a major curriculum debate in the mathematics community between computer science–oriented discrete mathematics and classical calculus-based mathematics. Linear algebra, especially as viewed in this book, is right in the middle of this debate. (Linear algebra and matrices have always been in the middle of such debates. Matrices were a core topic in the best-known first-year college mathematics text before 1950, Hall and Knight's *College Algebra;* and much of Kemeny, Snell, and Thompson's *Introduction to Finite Mathematics* involved new applications of linear algebra: Markov chains and linear programming.)

This book attempts to present a healthy interplay between mathematics and computer science, that is, between continuous and discrete modes of thinking. The complementary roles of continuous and discrete thinking are typified by the different uses of the euclidean norm (1_2-norm) and sum norm (1_1-norm) in this book. An important example of computer science thinking in this book is matrix representations, such as the **LU** decomposition. They are viewed as a way to preprocess the data in a matrix in order to be ready to solve quickly certain types of matrix problems.

We note that computer science even gives insights into the teaching of any linear algebra course. A computer scientist's distinction between high-level languages (such as PASCAL) and low-level languages (such as assembly language) applies to linear algebra proofs: A high-level proof involves matrix notation, such as $\mathbf{B}^T\mathbf{A}^T = (\mathbf{AB})^T$, while a low-level proof involves individual entries a_{ij}, such as $c_{ij} = \Sigma a_{ik}b_{kj}$.

Suggested Course Syllabus This book contains more than can be covered in the typical first-semester sophomore course for which it is intended. Most of Chapters 1, 2, and 3 and the first four sections of Chapter 5 should normally be covered. A freshman course would skip Chapter 5. In addition, selected sections of Chapter 4 can be chosen based on available time and the class's interests. For the student, the essence of any course should be the homework. This book has a large number of exercises at all levels of difficulty: computational exercises, applications, and proofs of much of the basic theory (with extensive hints for harder proofs). For more information about course outlines, plus suggested homework sets, sample exams, and additional solutions of exercises, see the accompanying Instructor's Manual.

At the end of the book is a list of various programming languages and software packages available for performing matrix operations. It is recommended that students have access to computers with ready matrix software in the first week.

Acknowledgments The first people to thank for help with this book are relatives. My father, A. W. Tucker, ignited and nurtured my born-again interest in linear algebra. Notes from the linear algebra course of my brother, Tom Tucker at Colgate, formed the foundation of early work on this book. My family—wife, Mandy, and daughters, Lisa and Katie—provided a supportive atmosphere that eased the long hours of writing; more concretely, Lisa's calculus project on cubic splines became the appendix to Section 4.7.

After the first draft was finished and initial reviews indicated that much rewriting was needed, the book might have died had not Don Small offered to help with the revision work. He added new examples, reworked several sections, and generally made the text better.

Many people have read various versions of this book and offered helpful suggestions, including Don Albers, Richard Alo, Tony Ralston, and Margaret Wright. Brian Kohn, Tom Hagstrom, and Roger Lutzenheiser class-tested preliminary versions of this book. Finally, thanks go to Gary W. Ostedt, Robert F. Clark, Elaine W. Wetterau, and the staff of the editorial and production departments at Macmillan.

A. T.

Contents

Preface

Chapter 3 **Solving Systems of Linear Equations** 157

Chapter 4 **A Sampling of Linear Models** 255

Interlude: **Abstract Linear Transformations and Vector Spaces** 387

Chapter 5 **Theory of Systems of Linear Equations and Eigenvalue/ Eigenvector Problems** 393

A Unified Introduction to Linear Algebra:

Models, Methods, and Theory

Introductory Models

Section 1.1 Mathematical Models

This book is concerned with ways to organize and analyze complex systems. From the time of the ancient Greeks until the middle of the twentieth century, scientists concentrated on problems involving a small number of variables. The calculus of Newton and Leibnitz dealt with functions of a single variable, later generalized to several variables. Although the functions studied in calculus can display very complex behavior, the amount of input data is usually quite small.

Today, scientists face problems involving large amounts of data. Consider the following examples of modern complex systems.

1. A mathematical model of the U.S. economy that considers the interactive effects of supplies and demands of various goods. The model may involve thousands of variables and equations.
2. The task of routing long-distance telephone calls. Every second, thousands of calls must be instantly routed from various origins to destinations through many intermediate switching stations. The system doing the routing procedure must look for circuitous indirect routes when more direct pathways are saturated.
3. Statistical studies of factors implicated in the spread of some new disease. Hundreds of causative agents must be analyzed for interactive effects; whereas neither effect A alone nor effect B alone may make a person susceptible to the illness, effects A and B together make a person very susceptible.

1

4. A mathematical model simulating the airflow around a jet aircraft whose design requires thousands of parameters to describe.

Although this book does not contain 100-, much less 1000-variable problems (5 variables are usually sufficient for realistic examples), it is concerned with mathematical techniques that can readily be applied to such large problems.

In this opening chapter we present some basic models that are used throughout the book to illustrate concepts of matrix theory and associated computations. In Chapter 2 we develop the basic tools of matrix algebra. In Chapter 3 we present various ways to solve a system of linear equations. In Chapter 4 we use matrix algebra to analyze a collection of models in greater depth. In Chapter 5 we discuss the theory of solutions to systems of equations.

Intuitively, the more information we put into a model, the more accurate should be the analysis obtained. The problem is, how do we handle all this information? How do we construct sensible models using hundreds of inputs when we do not really know the underlying mechanism by which the input variables affect the model? In the ''old'' days when scientists studied simpler systems, very accurate mathematical models were obtained, say, to describe a spherical body's rate of fall based on three critical parameters: the time elapsed since the body was dropped, the body's density, and the density of air. When hundreds of interdependent variables are involved, there is little chance of obtaining a precise mathematical model.

If we do not understand well the system we are modeling, then the structure of our mathematical model should be simple. But the model must still be useful—tell us things about the system that we could not otherwise easily find out. We shall see as we work through this book, that a linear mathematical model is often the best choice. Before defining a linear model, let us state what we mean by a mathematical model in general.

A **mathematical model** is a mathematical formulation of some class of real-world problems. The formulation may be a set of equations, it may be the minimization of some function, or it may involve integrals. The model may embody various constraints on its variables or it may be a combination of other mathematical models. Part of the modeling process involves **input values** that vary from one instance of the problem to another. These values are coefficients and constants for equations in the model.

Let us consider five simple mathematical models. The first is a physics model derived with calculus. The other four are standard high school algebra problems.

Example 1. **Falling Objects**

In physics, the height H of an object dropped off a building is modeled by the formula

$$H = -16T^2 + H_0 \tag{1}$$

where T is time (in seconds) elapsed since the object was dropped,

Figure 1.1 Equation of falling object $H = -16T^2 + 100$.

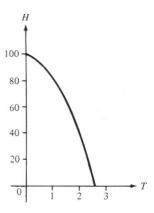

and H_0 is the height (in feet) of the building. (The numbers 16 and H_0 are the input values of the model.) One can derive this formula with calculus (if drag from air resistance is ignored).

Suppose that the building is 100 feet tall. Figure 1.1 gives the graph of the equation $H = -16T^2 + 100$. When 2 seconds has elapsed, the object's current height can be computed to be

$$H = -16(2)^2 + 100 = -64 + 100 = 36$$

To determine the time until the object hits the ground, we set $H = 0$ and solve for T:

$$0 = -16T^2 + 100 \rightarrow 16T^2 = 100$$
$$\rightarrow T^2 = \tfrac{100}{16} = \tfrac{25}{4}$$
$$\rightarrow T = \sqrt{\tfrac{25}{4}} = \tfrac{5}{2}$$ ∎

Example 2. **Elementary Algebra: A Problem of Relative Ages**

Consider the following word problem. Michael is three times the age of his sister, but in 6 years he will be only twice his sister's age. How old is Michael now?

We want to model the information in the word problem with algebraic equations. If M is Michael's current age and S is his sister's age, we have

$$M = 3S \tag{2}$$
$$M + 6 = 2(S + 6) \tag{3}$$

This model is simply an algebraic restatement of information given in words. We were told that certain quantities were equal; for example, Michael's current age equals three times his sister's current age, and we expressed these equalities in symbolic (algebraic) form.

There are two variables in this problem, but the relation between the two variables M and S is so simple—one is triple the other—that one can easily rewrite the second equation (3) in terms of one of the variables. If we write (3) in terms of S by substituting $3S$ for M, we obtain

$$(3S) + 6 = 2(S + 6) \tag{3'}$$

Solving equation (3'), we obtain $S = 6$, so $M = 3S = 18$.   ■

We now consider a slightly more complicated age problem.

Example 3. **Another Problem of Relative Ages**

Alice is currently twice as old as her brother Bill. If twice the sum of their current ages is equal to the product of their ages 4 years ago, how old is Bill?

If A represents Alice's current age and B Bill's current age, the given information can be modeled by the system of two equations

$$A = 2B \tag{4}$$
$$2(A + B) = (A - 4)(B - 4)$$

Expanding and simplifying the second equation, we obtain

$$A = 2B \tag{5}$$
$$AB - 6A - 6B + 16 = 0$$

Note that this model involves a term of "degree 2," the product AB in the second equation.

Substituting $2B$ for A in the second equation yields

$$(2B)B - 6(2B) - 6B + 16 = 0 \quad \text{or} \quad B^2 - 9B + 8 = 0 \tag{6}$$

Factoring the right equation in (6) yields

$$(B - 1)(B - 8) = 0 \tag{7}$$

and thus $B = 1$ or $B = 8$.

The solution $B = 1$ is not possible in terms of the problem statement (if Bill is currently 1 year old, what was he 4 years ago?). Thus the answer is: Bill is 8 (and Alice is 16).   ■

Although both ($B = 1$, $A = 2$) and ($B = 8$, $A = 16$) satisfy algebraic system (4), only the second pair of values makes sense in the original real-world problem. This illustrates an important aspect of modeling that is frequently assumed in this book: namely, interpreting a mathematical solution

Figure 1.2 Modeling process.

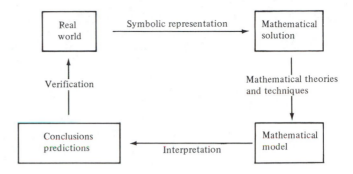

in terms of the original real-world problem to see that the solution makes sense. Figure 1.2 provides a picture of the major steps in the modeling process.

Example 4. Speed of a Canoe

Consider another classic word problem. When Mary paddles a canoe up a river (against the river's flow), the canoe goes 3 miles per hour, and when she paddles downstream the canoe goes 11 miles per hour. How fast would the canoe be going if she were paddling on a still lake?

If C is the canoe's speed in still water and R is the speed of the river, then algebra books model the problem with the two equations

$$
\begin{aligned}
C + R &= 11 \\
C - R &= 3
\end{aligned}
\tag{8}
$$

This model expresses the upstream and downstream speeds as functions of C and R by using the intuitive physical principle that the net upstream speed is the canoe speed minus the river speed, and that the net downstream speed is the canoe speed plus the river speed.

Figure 1.3 has a graph of the two equations in (8). We see that they intersect at the point $C = 7$, $R = 4$. This solution can also be obtained by solving (8) algebraically. ∎

Figure 1.3 Graphical solution of canoe problem.

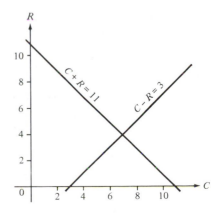

Examples 2 and 4 are called **linear models** *because the mathematical equations in the model are linear*, that is, equations of lines. Examples 1 and 3 are nonlinear models because they contain nonlinear equations. A linear equation involves sums and differences of variables or multiples of variables, such as

$$y = 2x - 6, \qquad x = \frac{y}{2} + 3 \qquad (9a)$$

$$4x - 2y = 12 \qquad (9b)$$

Note that the three preceding equations are all equivalent.

The left sides in (8) and (9b) are called **linear combinations** of variables: sums and differences of variables or of multiples of variables. A linear combination is the simplest type of expression of variables that one can build.

The widespread use of linear models results primarily from the ease of computing linear expressions as well as the existence of a powerful theory for analyzing linear models. Even when a problem is nonlinear, a linear model will often be used as an approximation. For example, for small values of x, $\sin x$ is often approximated by x. However, the reader should not expect that every solution has a satisfactory linear model, as the next example shows.

Example 5. **Speed of Canoe with Sail**

Suppose that the canoe has a sail mounted at its front. The canoe is on a lake with a wind of W miles per hour. We assume that downwind speed (moving with the wind) is again $C + W$, the sum of the canoe's speed (paddled by Mary) plus the wind's speed. However, boats with sails can move upwind by an aerodynamic principle, the same principle that holds up airplanes. So we try a linear combination of the C and W of the form $C + kW$, where k is a constant to be determined. If U is the upwind speed and D the downwind speed, we obtain the equations

$$
\begin{aligned}
\text{Downwind:} \quad & C + W = D \\
\text{Upwind:} \quad & C + kW = U
\end{aligned}
\qquad (10)
$$

Let us consider a numerical example. Suppose that our downwind speed D is 7 and our upwind speed U is 5. We try our model—the equations in (10)—with k values of $k = .75$ and $k = 1$.

$$
k = .75: \quad
\begin{aligned}
C + W &= 7 \\
C + .75W &= 5
\end{aligned}
\qquad (11)
$$

$$
k = 1: \quad
\begin{aligned}
C + W &= 7 \\
C + W &= 5
\end{aligned}
\qquad (12)
$$

Figure 1.4 Equations for canoe with sail.

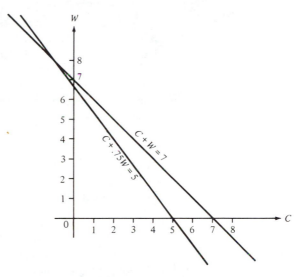

Obviously, the two equations in (12) can never be satisfied simultaneously. In (11) with $k = .75$, there are also difficulties. Subtracting the second equation in (11) from the first, we obtain $.25W = 2$ or $W = 8$. Since $C + W = 7$, then $C = 7 - W = 7 - 8 = -1$. A negative canoe speed is impossible (we assume that Mary is not cheating by paddling backwards).

A good way to see what is happening is with a graph. In Figure 1.4, we have plotted the lines in (11). Because these two lines do not meet inside the positive quadrant, there is no feasible solution. If we change our estimate for the constant k slightly from $.75$ to $\frac{2}{3}$, then we obtain a solution $W = 6$, $C = 1$, which might be close to the true value of W and C.

Instead of assigning specific values to k, U and D, let us solve the two equations of (10) for C and W in terms of these general parameters. If we subtract the upwind equation from the downwind equation, we have

$$C + W = D$$
$$\underline{C + kW = U}$$
$$(1 - k)W = D - U \qquad \text{or} \qquad W = \frac{D - U}{1 - k} \qquad (13)$$

If we substitute the formula for W found in (13) into the downwind equation, we obtain

$$C + \frac{D - U}{1 - k} = D$$

or

$$C = D - \frac{D - U}{1 - k} = \frac{(1 - k)D - (D - U)}{1 - k} = \frac{U - kD}{1 - k} \qquad (14)$$

Observe that whereas originally D and U were expressed as linear combinations of C and W, we have now expressed C and W as linear combinations of D and U. However, the critical factor here is k. Recall that the parameter k is supposed to allow us to express the upwind speed with a sail as a linear combination of C and W. In (13) and (14), C and W depend on k in a nonlinear fashion: When k approaches 1, the denominator $1 - k$ in (13) and (14) approaches 0, so the values of C and W blow up.

This sensitivity to small changes in k near 1 makes this model inherently poor. More generally, it appears that a linear combination such as $C + kW$ of C and W is unable to model properly the upwind speed of a canoe with sail. ∎

Example 5 illustrates the possible inadequacy of a linear model. It also points out that one must always be careful about the accuracy of coefficients in linear equations, because the solution depends on them in a nonlinear fashion. A system of equations is called **ill-conditioned** if a large change in the answer can be produced by a small error in the value of a coefficient (or by a roundoff error in computation, such as writing $\frac{1}{3}$ as .33). When k was near 1, the system of equations in the canoe-with-sail model was very ill-conditioned. In Section 3.5 we learn how to calculate the *condition number* of a system of equations, which tells how poorly conditioned a system is.

Section 1.1 Exercises

Summary of Exercises
These exercises examine the five models presented in this section and ask the reader to make mathematical models of similar problems. Exercises 1–3 are based on Example 1; Exercises 4–8 on Example 2; Exercises 9–15 on Example 4; Exercises 16–19 on Example 5. All the exercises require only first-year high school algebra.

1. In Example 1 about a falling object (dropped from the top of a 100-foot building), what height will the object have after 1 second?

2. If the object in Example 1 were dropped from a 400-foot building, how high would it be after 2 seconds? When would it hit the ground? What is the relation between the time of impact in this problem and the time of impact in the original 100-foot problem in Example 1? Guess the time of impact if the object were dropped from the top of a 1600-foot building, and verify that this guess is right.

3. If the object in Example 1 were dropped from a building of height H_0, solve equation (1) for the time when the object hits the ground (your answer will involve H_0 and the constant 16). Make a graph plotting the time when the object hits the ground as a function of building height.

4. Suppose that John is twice Mary's age but 4 years ago he was three times her age. Express this information as a pair of linear equations

involving J (John's current age) and M (Mary's current age). Solve for J and M.

5. Store A expects to sell three times as many books as store B this month. But next month store B's sales are scheduled to double while store A's sales stay the same. Over the 2-month period, the two stores together are expected to sell a total of 4500 books. How many books would each store sell this month?

6. Suppose that Mary and Nancy are sisters and the sum of their ages equals their older brother Bill's age, that Nancy is 4 years older than Mary, and that Bill is 6 years older than Nancy. Express this information in three linear equations in B (Bill's age), M (Mary's age), and N (Nancy's age). Express B and N in terms of M, and then solve for the ages of the three children.

7. A rectangle is twice as high as it is wide. The sum of the height and width of this rectangle is equal to one-half the area of the rectangle. Find the height and width.

8. A company has a budget of $280,000 for computing equipment. Three types of equipment are available: microcomputers at $2000 each, terminals at $500 each, and word processors at $5000 each. There should be five times as many terminals as microcomputers and two times as many microcomputers as word processors. How many machines of each type should be purchased?

9. Suppose that the canoe in Example 4 went 5 miles per hour upstream and 9 miles per hour downstream. Solve the resulting system of linear equations algebraically to determine the speed of the canoe (in still water) and the speed of the river.

10. Suppose that the canoe in Example 4 goes U miles per hour upstream and D miles per hour downstream. Find general formulas in terms of U and D for the speed of the canoe (in still water) and the speed of the river.

11. A company has $36,000 to hire a mathematician and his or her secretary. Out of respect for the mathematician's training, the mathematician will be paid $8000 more than the secretary. How much will each be paid?

12. Cook A cooks 2 steaks and 6 hamburgers in half an hour. Cook B cooks 4 steaks and 3 hamburgers in half an hour. If there is a demand for 16 steaks and 21 hamburgers, how many half-hour periods should cook A work and how many half-hour periods should cook B work to fill this demand?

13. We have two oil refineries. Refinery A produces 20 gallons of heating

oil and 8 gallons of diesel oil out of each barrel of petroleum it refines. Refinery B produces 6 gallons of heating oil and 15 gallons of diesel oil out of each barrel of petroleum it refines. There is a demand of 500 gallons of heating oil and 750 gallons of diesel oil. How many barrels of petroleum should be refined at each refinery to equal this demand?

14. The sum of John's weight and Sally's weight is 20 pounds more than four times the difference between their two weights (John is the heavier). Twice Sally's weight is 40 pounds more than John's weight. Write down these two facts in two linear equations. Simplify the first equation and solve the two equations to find the weights of John and Sally.

15. Two ferries travel across a lake 50 miles wide. One ferry goes 5 miles per hour slower than the other. If the slower ferry leaves 1 hour before the faster and arrives at the opposite shore at the same time as the other ferry, what is the speed of the slower ferry?

16. In Example 5, re-solve the downwind–upwind equations of (10) when $k = .9$ with $D = 7$ and $U = 5$. Solve again with $k = .9$ but now $D = 6.5$ and $U = 5$. Does this small change in D result in an equally small change in C?

17. In Example 5, suppose that $D = 7$ and $U = 5$, as in equations (10) and (11). Then formula (14) for C becomes

$$C = \frac{5 - 7k}{1 - k}$$

Plot C as a function of k in the interval $0 \le k \le 1$. Is C equal to $+$ infinity or equal to $-$ infinity when $k = 1$? Explain your answer.

18. In the downwind–upwind system of equations in Example 5, suppose that Mary were not paddling, so that $C = 0$. If $k = .75$ and U (upwind speed) $= 6$, what is W and what is D? In this case, what must be the relation between U and D? (That is, write U as a function of D.)

19. In the system of equations for our canoe-with-sail model,

$$C + kW = 8$$
$$C + W = 12$$

pick the unknown k so that $C = 0$ when this system is solved. What is W?

20. A refinery produces 8 gallons of gasoline and 6 gallons of heating oil from each barrel of petroleum it refines. There is a demand for 400 gallons of gasoline and 200 gallons of heating oil. Can you set up a linear equation to determine how many barrels of petroleum should be

refined? Explain your answer and tell how many barrels you would refine if you were the manager of the refinery.

21. Suppose we estimate that a child's IQ is the average of the parents' IQs and that the income the child has when grown to the age of 40 is 200 times the father's IQ plus 100 times the mother's IQ. If the child has an IQ of 120 and earns $48,000 at age 40, what are the parents' IQs? Does this model give a reasonable answer?

22. Suppose that factory *A* produces 12 tables and 6 chairs an hour while factory *B* produces 8 tables and 4 chairs an hour. How many hours should each factory work to produce 48 tables and 24 chairs? How many different solutions are there to this problem?

23. We estimate that Jack can do 3 chemistry problems and 6 math problems in an hour, while Paula can do 4 chemistry problems and 7 math problems in an hour. There are 11 chemistry problems and 17 math problems. Set up and solve a system of two linear equations to determine how long Jack and Paula should work to do these problems. What is the matter with the solution? Propose a solution that makes more sense.

Section 1.2 Systems of Linear Equations

In Section 1.1 we gave a quadratic formula $H = -16T^2 + H_0$ to model mathematically the height of an object dropped from a building H_0 feet tall. From this formula we could determine how long it took for an object to hit the ground (see Example 1 of Section 1.1). The model involved a single nonlinear equation with one variable to be determined. Nonlinear equations in one variable are not difficult to derive and solve.

When many variables need to be determined, then almost surely the mathematical model will be a system of linear equations. There are four basic reasons for using linear models.

1. There is a rich theory for analyzing and solving systems of linear equations. There is limited theory and no general solution techniques for systems of nonlinear equations.
2. Systems of nonlinear equations involving several variables exhibit very complex behavior and we rarely understand real-world phenomena well enough to use such complex models.
3. Small changes in coefficients in nonlinear systems can cause huge changes in the behavior of the systems, yet precise values for these coefficients are rarely known.
4. All nonlinear phenomena are approximately linear over small intervals; that is, a complicated curve can be approximated by a collection of many short line segments (see Figure 1.5).

Figure 1.5 Line segments approximating a curve.

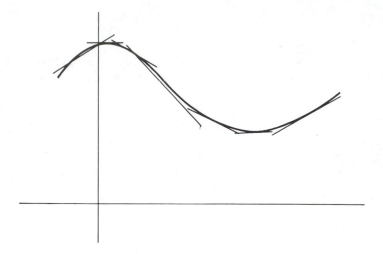

For these reasons we have the following general principle that is true
for numerical problems in all fields: physics, statistics, economics, for instance.

General Mathematical Principle for Multivariable Problems. *Any
problem involving several unknowns is normally solved by recasting
the problem as a system of linear equations.*

The two canoe problems, with and without a sail, in Section 1.1 were
simple examples of systems of linear equations. Here are two more examples. These examples, together with those in Section 1.3, are used dozens
of times throughout the book to motivate and illustrate theory and numerical
methods. It is *very important* for the reader to gain familiarity with these
examples through working some of the numerical exercises at the end of
each section.

Example 1. Oil Refinery Model

A company runs three oil refineries. Each refinery produces three petroleum-based products: heating oil, diesel oil, and gasoline. Suppose
that from 1 barrel of petroleum, the first refinery produces 20 gallons
of heating oil, 10 gallons of diesel oil, and 5 gallons of gasoline. The
second and third refineries produce different amounts of these three
products as described in the following table.

	Refinery 1	Refinery 2	Refinery 3	
Heating oil:	20	4	4	
Diesel oil:	10	14	5	(1)
Gasoline:	5	5	12	
	x_1	x_2	x_3	

Let x_i be the number of barrels of petroleum used by the ith refinery. Then the total amount of each product produced by the refineries is given by the linear expressions

$$
\begin{array}{lrrr}
\text{Heating oil:} & 20x_1 + & 4x_2 + & 4x_3 \\
\text{Diesel oil:} & 10x_1 + & 14x_2 + & 5x_3 \qquad (2) \\
\text{Gasoline:} & 5x_1 + & 5x_2 + & 12x_3
\end{array}
$$

Suppose that the demand is 500 units of heating oil, 850 units of diesel oil, and 1000 units of gasoline. What values x_1, x_2, x_3 are needed to produce these amounts? We require the x_i to satisfy the following system of linear equations.

$$
\begin{array}{rrrcr}
20x_1 + & 4x_2 + & 4x_3 & = & 500 \\
10x_1 + & 14x_2 + & 5x_3 & = & 850 \qquad (3) \\
5x_1 + & 5x_2 + & 12x_3 & = & 1000
\end{array}
$$

Later in this book we shall learn several ways to solve systems of linear equations. For now, let us use the tried-and-true method of trial and error. As an initial guess, try $x_1 = 25$, $x_2 = 25$, $x_3 = 25$. Using these values in (3), we get

$$
\begin{array}{rrrcr}
20(25) + & 4(25) + & 4(25) & = & 700 \\
10(25) + & 14(25) + & 5(25) & = & 725 \qquad (4) \\
5(25) + & 5(25) + & 12(25) & = & 550
\end{array}
$$

(It is helpful to have a programmable calculator or microcomputer for calculations with systems of linear equations.)

We need to alter the x_i values to make the first expression (heating oil) smaller and the last expression (gasoline) larger. Since x_1 makes the largest contribution to the first expression and x_3 makes the largest contribution to the last expression, we decrease x_1 and increase x_3. Suppose that we try $x_1 = 10$, $x_2 = 25$, $x_3 = 70$.

$$
\begin{array}{rrrcr}
20(10) + & 4(25) + & 4(70) & = & 580 \\
10(10) + & 14(25) + & 5(70) & = & 800 \qquad (5) \\
5(10) + & 5(25) + & 12(70) & = & 1015
\end{array}
$$

Although these x values are much better than the original ones, let us try to do better. We need to decrease the first expression and increase the second expression (without changing the third expression). To decrease the first expression, we should decrease x_1. Similarly, to increase the second expression, we increase x_2. Trying the following values, we obtain the result that production levels

$$x_1 = 5, \qquad x_2 = 35, \qquad x_3 = 70$$

yield

heating oil $= 520,$ diesel oil $= 890,$ gasoline $= 1040$

This is getting quite close to our production goals. The overproduction of 20 to 40 gallons in each product might be a reasonable safety margin that is actually desirable.

To get closer, we should decrease x_2 and x_3 a little.

$$x_1 = 5, \qquad x_2 = 33, \qquad x_3 = 68$$

yield

heating oil $= 504,$ diesel oil $= 852,$ gasoline $= 1006$

This is an excellent fit. We have been a bit lucky. To do better, we would probably have to use fractional values. (In the Exercises the reader may need more tries to get this close.) ■

We next consider a slightly more complicated supply–demand model. This model has the balancing advantage that trial-and-error calculations to estimate a solution are easier. The reader is warned that it takes a little while to get a feel for all the numbers in this model.

Example 2. A Model of General Economic Supply–Demand

We present a linear model due to W. Leontief, a Nobel Prize–winning economist. The model seeks to balance supply and demand throughout a whole economy. For each industry, there will be one supply–demand equation. In practical applications, Leontief economic models can have hundreds or thousands of specific industries. We consider an example with four industries.

The left-hand side of each equation is the supply, the amount produced by the ith industry. Call this quantity x_i; it is measured in dollars. On the right-hand side, we have the demand for the product of the ith industry. There are two parts to the demand. The first part is demand for the output by other industries (to create other products requires some of this product as input). The second part is consumer demand for the product.

For a concrete instance, let us consider an economy of four general industries: energy, construction, transportation, and steel. Suppose that the supply–demand equations are

Supply	Industrial Demands				Consumer Demand
	Energy	Construct.	Transport.	Steel	
Energy: $x_1 =$	$.4x_1 +$	$.2x_2 +$	$.2x_3 +$	$.2x_4 +$	100
Construct.: $x_2 =$	$.3x_1 +$	$.3x_2 +$	$.2x_3 +$	$.1x_4 +$	50
Transport.: $x_3 =$	$.1x_1 +$	$.1x_2 +$		$.2x_4 +$	100
Steel: $x_4 =$	$+$	$.1x_2 +$	$.1x_3$	$+$	0

$$(6)$$

The first equation, for energy, has the supply of energy x_1 on the left. The terms on the right of this equation are the various demands that this supply must meet. The first term on the right, $.4x_1$, is the input of energy required to produce our x_1 dollars of energy (.4 units of energy input for one unit of energy output). Also, the second term of $.2x_2$ is the input of energy needed to make x_2 dollars of construction. Similarly, terms $.2x_3$ and $.2x_4$ are energy inputs required for transportation and steel production. The final term of 100 is the fixed consumer demand.

Each column gives the set of input demands of an industry. For example, the third column tells us that to produce the x_3 dollars of transportation requires as input $.2x_3$ dollars of energy, $.2x_3$ dollars of construction, and $.1x_3$ dollars of steel. In the previous refinery model, the demand for each product was a single constant quantity. In the Leontief model, there are many unknown demands that each industry's output must satisfy. There is an ultimate consumer demand for each output, but to meet this demand industries generate input demands on each other. Thus the demands are highly interrelated: Demand for energy depends on the production levels of other industries, and these production levels depend in turn on the demand for their outputs by other industries, and so on.

When the levels of industrial output satisfy these supply–demand equations, economists say that the economy is *in equilibrium*.

As in the refinery model, let us try to solve this system of equations by trial-and-error. As a first guess, let us set the production levels at twice the consumer demand (the doubling tries to account for the interindustry demands). So $x_1 = 200$, $x_2 = 100$, $x_3 = 200$, and $x_4 = 0$; these are our supplies. Given these production levels, we can compute the demands from (6).

Supply		Demand	
Energy:	200	$.4(200) + .2(100) + .2(200) + .2(0) + 100 =$	240
Construct.:	100	$.3(200) + .3(100) + .2(200) + .1(0) + 50 =$	180
Transport.:	200	$.1(200) + .1(100) + + .2(0) + 100 =$	130
Steel:	0	$ + .1(100) + .1(200) =$	30

$$(7)$$

For our next approximations, let us try supply levels halfway between the supply and demand values in (7). That is, $x_1 = \frac{1}{2}(200 + 240) = 220$, and similarly, $x_2 = 140$, $x_3 = 165$, and $x_4 = 15$.

Supply	Demand
Energy: 220	$.4(220) + .2(140) + .2(165) + .2(15) + 100 = 252$
Construct.: 140	$.3(220) + .3(140) + .2(165) + .1(15) + 50 = 192.5$
Transport.: 165	$.1(220) + .1(140) + + .2(15) + 100 = 139$
Steel: 15	$ + .1(140) + .1(165) = 30.5$

$$(8)$$

The second approximation is only moderately better. The interaction effects between different industries are hard to predict. Adjusting production levels was much easier in the refinery problem, where the demand for each product was constant.

Let us stop trying to be clever and just use the simple-minded approach of setting production levels (i.e., supply levels) equal to the previous demand levels. So from (8), we try

Supply	Demand
Energy: 252	$.4(252) + .2(192) + .2(139) + .2(30) + 100 = 273$
Construct.: 192	$.3(252) + .3(192) + .2(139) + .1(30) + 50 = 214$
Transport.: 139	$.1(252) + .1(192) + + .2(30) + 100 = 150$
Steel: 30	$ + .1(192) + .1(139) = 33$

$$(9)$$

The demand values here have been rounded to whole numbers. The supplies and demands are getting a little closer together in (9). Repeating the process of setting the new supply levels equal to the previous demand levels (i.e., the demands on the right side in (9)) yields

Supply	Demand
Energy: 273	$.4(273) + .2(214) + .2(150) + .2(33) + 100 = 289$
Construct.: 214	$.3(273) + .3(214) + .2(150) + .1(33) + 50 = 229$
Transport.: 150	$.1(273) + .1(214) + + .2(33) + 100 = 155$
Steel: 33	$ + .1(214) + .1(150) = 36$

$$(10)$$

Repeating this process again, we have

Supply		Demand
Energy:	289	.4(289) + .2(229) + .2(155) + .2(36) + 100 = 300
Construct.:	229	.3(289) + .3(229) + .2(155) + .1(36) + 50 = 240
Transport.:	155	.1(289) + .1(229) + + .2(36) + 100 = 159
Steel:	36	+ .1(229) + .1(155) = 38

$$(11)$$

Observe that in successive rounds (9), (10), (11), supplies are rising. This is because as we produce more, we need more input which requires us to produce still more, and so on. It may be that this iteration will go on forever, and no equilibrium exists. On the other hand, the gap between supplies and demands is decreasing.

Leontief proposed a constraint on the input costs that we shall show (in Section 3.4) guarantees that an equilibrium exists. The constraint is

Input Constraint. Every industry is profitable: Every industry must require less than \$1 of inputs to produce \$1 of output.

In mathematical terms, this means that the sum of the coefficients in each column must be less than 1. Our data in (6) satisfy this constraint, so an equilibrium does exist for this four-industry economy. Moreover, the iteration process of repeatedly setting production levels equal to the previous demands will converge to this equilibrium. The reader should check that the following numbers are equilibrium values (rounded to the nearest integer).

Equilibrium: energy = 325,   construction = 265,

transportation = 168,   steel = 43   ■

Note that any system of linear equations can be rewritten in the form of supply–demand equations with x_i appearing alone on the left side of the ith equation, as in the Leontief supply–demand model (6). It is standard practice to solve large systems of linear equations by some sort of iterative method. The nature of the supply–demand equations suggested the iterative scheme we used here, letting the demands from one round be the production levels of the next round.

Section 1.2 **Exercises**

Summary of Exercises
Exercises 1–5 are based on the refinery model. Exercises 6–8 are other problems involving a system of three linear equations in three unknowns. Exercises 9–12 are based on the Leontief economic model. Exercises 13 and 14 involve converting the refinery problem into a system of Leontief-type equations.

Exercises 1–3 refer to the refinery model in Example 1.

1. Suppose that refinery 1 processes 15 barrels of petroleum, refinery 2 processes 20 barrels, and refinery 3 processes 60 barrels. With this production schedule, for which product does production deviate the most from the set of demands 500, 850, 1000?

2. Suppose that the demand for heating oil grows to 800 gallons, while other demands stay the same. Find production levels of the three refineries to meet approximately this new set of demands (by "approximately" we mean with no product off by more than 30 gallons).

3. Suppose that refinery 3 is improved so that each barrel of petroleum yields 8 gallons of heating oil, 10 gallons of diesel oil, and 20 gallons of gasoline. Find production levels of the three refineries to meet approximately the demands (by "approximately" we mean with no product off by more than 30 gallons).

4. Consider the following refinery model. There are three refineries 1, 2, and 3 and from each barrel of crude petroleum, the different refineries produce the following amounts (measured in gallons) of heating oil, diesel oil, and gasoline.

	Refinery 1	Refinery 2	Refinery 3
Heating oil	6	3	2
Diesel oil	4	6	3
Gasoline	3	2	6

Suppose that we have the following demand:

> 280 gallons of heating oil,
>
> 350 gallons of diesel oil, and
>
> 350 gallons of gasoline.

(a) Write a system of equations whose solution would determine production levels to yield the desired amounts of heating oil, diesel oil, and gasoline. As in Example 1, let x_i be the number of barrels processed by the ith refinery.

(b) Find an approximate solution to this system of equations with no product off by more than 30 gallons from its demand.

5. Repeat the refinery model in Exercise 4 with new demand levels of 500 gallons heating oil, 300 gallons diesel oil, and 600 gallons gasoline. Try to find an approximate solution (within 30 gallons) with this set of demands. Something is going wrong and there is no valid set of pro-

duction levels to attain this set of demands. What is invalid about the solution to this refinery problem?

Extra Credit: Try to explain in words why this set of demands is unattained while the demands in Exercise 4 were attainable.

6. The staff dietician at the California Institute of Trigonometry has to make up a meal with 600 calories, 20 grams of protein, and 200 milligrams of vitamin C. There are three food types to choose from: rubbery jello, dried fish sticks, and mystery meat. They have the following nutritional content per ounce.

	Jello	Fish Sticks	Mystery Meat
Calories	10	50	200
Protein	1	3	.2
Vitamin C	30	10	0

(a) Make a mathematical model of the dietician's problem with a system of three linear equations.
(b) Find an approximate solution (accurate to within 10%).

7. A furniture manufacturer makes tables, chairs, and sofas. In one month, the company has available 300 units of wood, 350 units of labor, and 225 units of upholstery. The manufacturer wants a production schedule for the month that uses all of these resources. The different products require the following amounts of the resources.

	Table	Chair	Sofa
Wood	4	1	3
Labor	3	2	5
Upholstery	2	0	4

(a) Make a mathematical model of this production problem.
(b) Find an approximate solution (accurate to within 10%).

8. A company has a budget of $280,000 for computing equipment. Three types of equipment are available: microcomputers at $2000 a piece, terminals at $500 a piece, and word processors at $5000 a piece. There should be five times as many terminals as microcomputers and two times as many microcomputers as word processors. Set this problem up as a system of three linear equations. Determine approximately how many machines of each type there should be by solving by trial-and-error.

Note: Check your answer by expressing the numbers of terminals and microcomputers in terms of the number of word processors and solving the remaining single equation in one unknown.

Exercises 9–11 are based on the Leontief model in Example 2.

9. If we produced 300 units of energy, 250 units of construction, 160 units of transportation, and 40 units of steel, what would be the largest deviation between supply and demand among the four commodities?

10. Start the iteration procedure followed in (9), (10), and (11) with an initial set of supplies equal to the consumer demands, that is, $x_1 = 100$, $x_2 = 50$, $x_3 = 100$, $x_4 = 0$. Compute the right sides of the equations in (6) with this set of x_i's and let the resulting numbers be the new values for x_1, x_2, x_3, x_4; compute the right sides again with these new x_is; and so on. Do this iteration five times. Do the successive sets of x_is appear to be converging toward the equilibrium values given at the end of Example 2?

11. This exercise explores the effect on all industries of changes in one industry. Quadrupling the price of petroleum had a widespread effect on all industrial sectors in the 1970s. But smaller changes in one seemingly unimportant industry can also result in important changes in many other industries.
    (a) Change the system of equations in the Leontief model in (6) by decreasing the coefficient of x_1 (energy) in the construction equation from .3 to .2 (this is the result of new energy efficiencies in construction equipment). We want to know how this change affects our economy. Iterate five times, as in Exercise 10, with this altered system using as starting x_i's the equilibrium values for the original model: $x_1 = 325$, $x_2 = 265$, $x_3 = 168$, $x_4 = 43$.
    (b) Repeat part (a), but now decrease the coefficient of x_4 (steel) in the transportation equation from .2 to .1.
    (c) Repeat part (a), but now increase the coefficient of x_2 (construction) in the energy equation from .2 to .3.

12. Consider the Leontief system

$$x_1 = .4x_1 + .3x_2 + .3x_3 + 100$$
$$x_2 = .3x_1 + .4x_2 + .3x_3 + 100$$
$$x_3 = .3x_1 + .3x_2 + .4x_3 + 100$$

Here the column sums are 1, violating the Leontief input constraint given in the text. Show that this system cannot have a solution.

Hint: Add the three equations together.

Extra Credit: Try to explain in economic terms why no solution exists.

13. **(a)** Rewrite the refinery model's equations in (3) to look like Leontief equations as follows. Divide the first equation by 20 (the coefficient of x_1 in that equation), divide the second equation by 14, and divide the third equation by 12. Next move the x_2 and x_3 terms in the new first equation over to the right side, leaving just x_1 on the left (the new equation should be $x_1 = -.2x_2 - .2x_3 + 25$); similarly, in the second equation leave just x_2 on the left and in the third equation leave just x_3 on the left. Note that the Leontief input constraint about column sums is not satisfied by this system.

    (b) Use the iteration method introduced in equations (9), (10), and (11) (see also Exercise 10) to get an approximate solution to the refinery problem (do five iterations, starting with the "consumer demands" of 25, 50, 100).

14. **(a)** Rewrite the refinery model's equations in (3) to look somewhat like Leontief equations as follows. In the first equation, move the x_2 and x_3 terms to the right side and also move 19 of the 20 units of the x_1 term to the right, leaving just x_1 on the left (the equation is now $x_1 = -19x_1 - 4x_2 - 4x_3 + 500$). Similarly, in the second equation leave just x_2 on the left; move everything else to the right side. In the third equation leave just x_3 on the left. Note that Leontief's input constraint about column sums is far from satisfied by this system.

    (b) Try using the iteration method introduced in equations (9), (10), and (11) (see also Exercise 10) to get an approximate solution to the refinery problem (do five iterations starting with the guess of $x_1 = 25$, $x_2 = 50$, $x_3 = 100$). Does the iteration process seem to be converging?

Section 1.3 Markov Chains and Dynamic Models

The refinery and economic supply–demand models of Section 1.2 were *static* in the sense that we solved them once and that was it. There was one set of production levels required, not a sequence of levels that would be needed to describe an economy changing over time. A model that tries to predict the behavior of a system over a period of time is called a *dynamic* model. In this section we examine two dynamic linear models.

The first dynamic model we consider involves probability. This model, called a Markov chain, will arise over and over in this book, so it is important to understand the model well. The concepts of probability we need for this model are simple and intuitive.

A **Markov chain** is a probabilistic model that describes the random movement over time of some activity. At each period of time, the activity is in one of several possible states. States might be amounts won in gambling, different weather conditions (e.g., sunny, snowy), or numbers of jobs waiting to be processed by a computer. The model specifies probabilities that tell how the activity changes states from period to period. Consider a simple two-state Markov chain.

Example 1. **Markov Chain for Weather**

Suppose that we have two states of the weather: sunny or cloudy. If it is sunny today, the probability is $\frac{3}{4}$ that it will be sunny tomorrow, and $\frac{1}{4}$ that it will be cloudy tomorrow. If it is cloudy today, then the probability is $\frac{1}{2}$ that it will be sunny tomorrow and $\frac{1}{2}$ that it will be cloudy tomorrow. It is convenient to display these probabilities in an array.

		Today	
		Sunny	**Cloudy**
Tomorrow	**Sunny**	$\frac{3}{4}$	$\frac{1}{2}$
	Cloudy	$\frac{1}{4}$	$\frac{1}{2}$

The probabilities in this array are called **transition probabilities**, and the array is called a **transition matrix**. The probabilities in each column of the transition matrix must add up to 1. A convenient way to display the information in a Markov chain is with a transition diagram. The diagram for the weather Markov chain is drawn in Figure 1.6. There is a node for each state and arrows between states. Beside the arrow from state A to state B we write the transition probability of going from state A to state B.

The transition probabilities of a Markov chain tell us the chances of being in different states one period later. We need a formula from probability theory to be able to calculate the probabilities of where an activity will be after several periods. As in weather forecasting, *it is predictions many periods into the future that are most interesting*.

To state this probability formula, we need to introduce some notation. Let $p_1, p_2, \ldots, p_n$ be the probabilities of being in state 1, state 2, . . ., state n. This set of probabilities is called a **probability distribution**. Let a_{ij} be the transition probability of going from state j to state i. In the weather Markov chain, if state 1 is sunny and state 2 is cloudy, the transition probabilities are

$$\begin{bmatrix} a_{11} = \frac{3}{4} & a_{12} = \frac{1}{2} \\ a_{21} = \frac{1}{4} & a_{22} = \frac{1}{2} \end{bmatrix} \qquad \begin{array}{l} \text{Weather} \\ \text{Markov chain} \end{array}$$

Figure 1.6

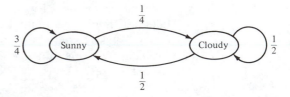

Formula for Distribution of Next States in Markov Chain. If $p_1, p_2, \ldots, p_n$ is the current probability distribution of the activity, then the probability distribution $p'_1, p'_2, \ldots, p'_n$ for the next period is given by

$$
\begin{aligned}
p'_1 &= a_{11}p_1 + a_{12}p_2 + a_{13}p_3 + \cdots + a_{1n}p_n \\
p'_2 &= a_{21}p_1 + a_{22}p_2 + a_{23}p_3 + \cdots + a_{2n}p_n \\
p'_3 &= a_{31}p_1 + a_{32}p_2 + a_{33}p_3 + \cdots + a_{3n}p_n \\
&\vdots \qquad \vdots \qquad \vdots \qquad \vdots \qquad \qquad \vdots \\
p'_n &= a_{n1}p_1 + a_{n2}p_2 + a_{n3}p_3 + \cdots + a_{nn}p_n
\end{aligned}
\tag{1}
$$

To illustrate this formula with the weather Markov chain, let $p_1 = \frac{3}{4}$ and $p_2 = \frac{1}{4}$. Then tomorrow's distribution p'_1, p'_2 is

$$
\begin{aligned}
p'_1 &= a_{11}p_1 + a_{12}p_2 = \tfrac{3}{4} \cdot \tfrac{3}{4} + \tfrac{1}{2} \cdot \tfrac{1}{4} = \tfrac{11}{16} \\
p'_2 &= a_{21}p_1 + a_{22}p_2 = \tfrac{1}{4} \cdot \tfrac{3}{4} + \tfrac{1}{2} \cdot \tfrac{1}{4} = \tfrac{5}{16}
\end{aligned}
\tag{2}
$$

We explain the formulas in (2) intuitively as follows. We can be in state 1 (sunny) tomorrow either because we are in state 1 today and then stay in state 1—this is the probability $a_{11}p_1$— or because we are in state 2 today and then switch to state 1—this is the probability $a_{12}p_2$. [To compute the probability of a sequence of two events, such as (i) the probability p_2 of now being in state 2, and (ii) the probability a_{12} of switching from state 2 to state 1, we multiply these two probabilities together, to get $a_{12}p_2$.]

Let us next consider a larger Markov chain that models the action of a popular video arcade game called Frogger.

Example 2. Frogger Markov Chain

We model the behavior of a frog jumping around on a four-lane highway. The possible states range from 1 = left side of highway, to 6 = right side of highway. See Figure 1.7a. Suppose that the following array gives the transition probabilities that the frog, if now in state j, will be in state i one minute later. The transition diagram for this Markov chain is given in Figure 1.7b.

		State in Current Period						
		1	2	3	4	5	6	
State	1	.5	.25	0	0	0	0	
in	2	.5	.5	.25	0	0	0	
Next	3	0	.25	.5	.25	0	0	(3)
Period	4	0	0	.25	.5	.25	0	
	5	0	0	0	.25	.5	.5	
	6	0	0	0	0	.25	.5	

Figure 1.7 (a) Six different states
for frogs on four-lane highway.
(b) Transition diagram for Markov
chain in Example 2; there is a node
for each state. The number beside
the edge from state j to state i is p_{ij}
(the probability of going from state
j to state i).

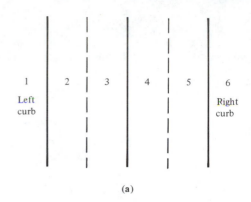

(a)

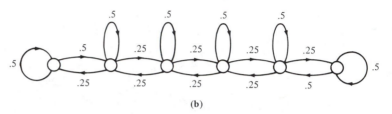

(b)

The formulas for next-period probabilities with the frog Markov
chain are

$$
\begin{aligned}
p_1' &= .50p_1 + .25p_2 \\
p_2' &= .50p_1 + .50p_2 + .25p_3 \\
p_3' &= \qquad\;\; .25p_2 + .50p_3 + .25p_4 \\
p_4' &= \qquad\qquad\qquad .25p_3 + .50p_4 + .25p_5 \\
p_5' &= \qquad\qquad\qquad\qquad\qquad .25p_4 + .50p_5 + .50p_6 \\
p_6' &= \qquad\qquad\qquad\qquad\qquad\qquad\qquad .25p_5 + .50p_6
\end{aligned}
\tag{4}
$$

Suppose that the frog starts in state 1 (left side of highway). Then
its probability distribution after 1 minute is given by the probabilities
in the first column of (4), since initially $p_1 = 1$ and other $p_i = 0$:

$$
p_1 = .5, \qquad p_2 = .5, \qquad \text{other } p_i = 0
\tag{5}
$$

Let us use (4) to compute the probability distribution for the frog after
2 minutes (remember that only p_1 and p_2 are nonzero):

$$
\begin{aligned}
p_1' &= .5p_1 + .25p_2 = .5 \times .5 + .25 \times .5 = .375 \\
p_2' &= .5p_1 + \;\; .5p_2 = .5 \times .5 + \;\; .5 \times .5 = .5 \\
p_3' &= 0 \;\; + .25p_2 = \;\;\;\; 0 \;\; + .25 \times .5 = .125
\end{aligned}
\tag{6}
$$

Other $p_i' = 0$. Note that the sum of the probabilities in (6) equals 1,
as it should.

We can continue iterating with formula (4) to find the distribution
after 3 minutes, after 4 minutes, and so on (it helps to let a computer

Table 1.1

Minutes	State 1	2	3	4	5	6
0	1	0	0	0	0	0
1	.50	.50	0	0	0	0
2	.375	.50	.125	0	0	0
3	.312	.469	.188	.031	0	0
4	.273	.438	.219	.063	.007	0
5	.246	.410	.234	.088	.020	.002
6	.226	.387	.242	.122	.046	.011
10	.176	.320	.241	.150	.080	.030
15	.144	.272	.226	.172	.128	.056
20	.126	.244	.217	.183	.157	.073
25	.116	.226	.210	.190	.173	.084
100	.1	.2	.2	.2	.2	.1
200	.1	.2	.2	.2	.2	.1
1000	.1	.2	.2	.2	.2	.1

do this). Table 1.1 gives these probabilities, assuming that the frog started in state 1 (left side of highway).

Observe that the probabilities converge to the distribution $p_1 = .1$, $p_2 = .2$, $p_3 = .2$, $p_4 = .2$, $p_5 = .2$, $p_6 = .1$ (and then stay the same forever). Very interesting!

Would this long-term distribution evolve from any starting distribution? Do all Markov chains exhibit this type of long-term distribution? Can the long-term distribution be computed more simply than iterating the equations in (4) 100 times? (*Answer*: Definitely yes.) ∎

Markov chains are a very useful type of linear model—linear because formula (1) for next-state probabilities is a system of linear equations. Part of their usefulness is due to results in matrix algebra that provide simple answers to all the questions just posed and many more.

Next we look at a simpler dynamic model, an ecological model that traces the sizes of populations of rabbits and foxes. The simplicity of the model allows us to experiment more, changing the values of the coefficients to exhibit a variety of different long-term trends.

Example 3. Growth Model for Rabbits and Foxes Model

Consider the following model for the monthly growth of populations of foxes and rabbits. If R and F are the numbers of rabbits and foxes this month, let R' and F' be the numbers next month given in the equations

$$R' = R + bR - eF$$
$$F' = F - dF + e'R$$
(7)

where b is the birthrate of rabbits, d the death rate of foxes, and e and e' are eating rates. The term $-eF$ in the rabbit equation is negative and the term $+e'R$ in the fox equation is positive because foxes eat rabbits. The results R', F' after 1 month can become new values for R and F to project the populations 2 months hence, and so on, as we did in the frog Markov chain.

Normally, the (positive) constants would be estimated for us by ecologists. But let us make up some reasonable-sounding values for these constants and see what sort of behavior this model predicts. Suppose that we try

$$R' = R + .2R - .3F$$
$$F' = F - .1F + .1R$$
(8)

and start with $R = 100$, $F = 100$. Then using (8) repeatedly to compute the populations in successive months, we get

0 months:	100 rabbits,	100 foxes
1 month:	90 rabbits,	100 foxes
2 months:	78 rabbits,	99 foxes
3 months:	64 rabbits,	97 foxes
4 months:	48 rabbits,	94 foxes
5 months:	29 rabbits,	89 foxes
6 months:	8 rabbits,	83 foxes
7 months:	-15 rabbits,	76 foxes

(9)

A negative number means that the rabbits became extinct.

If there are no rabbits, then the fox equation in (8) becomes

$$F' = F - .1F \rightarrow F' = .9F$$

and the foxes will eventually die out, too. This behavior of foxes killing off the rabbits and then starving to death is reasonable.

Let us try new starting values for R and F that will allow the rabbits to increase in size. The term $+.2R - .3F$ in the rabbit equation is the amount the rabbit population changes from this month to the next. For this term to be positive we require $.2R$ to be more than $.3F$. Suppose that we choose $R = 100$, $F = 50$:

0 months:	100 rabbits, 50 foxes
1 month:	105 rabbits, 55 foxes

2 months:	109 rabbits, 60 foxes	
3 months:	113 rabbits, 65 foxes	
4 months:	116 rabbits, 70 foxes	
5 months:	119 rabbits, 75 foxes	
6 months:	120 rabbits, 79 foxes	
7 months:	121 rabbits, 83 foxes	(10)
8 months:	120 rabbits, 87 foxes	
9 months:	118 rabbits, 90 foxes	
10 months:	115 rabbits, 93 foxes	
11 months:	110 rabbits, 95 foxes	
12 months:	103 rabbits, 97 foxes	
13 months:	94 rabbits, 97 foxes	
14 months:	84 rabbits, 97 foxes	
15 months:	72 rabbits, 96 foxes	

While the rabbits increased initially, so did the foxes (since they fed off the rabbits). After 7 months there were enough foxes so that they were eating rabbits faster than new rabbits were being born and the rabbit population began to decline. After 13 months (when there are fewer rabbits than foxes), the foxes begin to decline, also. Now we are in the same situation as before, in (9).

It appears that the equations in our model (8) make it inevitable that the foxes will grow to a level where they eat rabbits faster than rabbits are born, causing the rabbits to decline to extinction. Then the foxes become extinct, too. The reader is asked in the Exercises to try other values for b, d, e, and e' in this model and to explore the resulting behavior.

Akin to the stable probability distribution in the frog Markov chain, let us try to determine values of the coefficients in our model that will permit the rabbit and fox populations to stabilize, that is, to remain the same forever. This means that $R' = R$ and $F' = F$.

We return to the original general model

$$R' = R + bR - eF$$
$$F' = F - dF + e'R$$

(11)

When $R' = R$ and $F' = F$, we have

$$R = R + bR - eF$$
$$F = F - dF + e'R$$

or

$$bR - eF = 0$$
$$e'R - dF = 0$$

(12)

Note that in (12), the order of the terms $+e'R$ and $-dF$ in the second equation was reversed. Let us solve this pair of linear equations for R and F. Obviously, $R = F = 0$ is a solution. But we want another solution. We use the standard method for eliminating one of the variables in (12): Multiply the first equation by d and the second by e and then subtract the second from the first.

$$
\begin{aligned}
dbR - deF &= 0 \\
-(ee'R - edF &= 0) \\
\hline
(bd - ee')R &= 0
\end{aligned}
$$

If $bd - ee' = 0$, then R (and F) need not be 0. Note that

$$bd - ee' = 0 \leftrightarrow bd = ee' \leftrightarrow \frac{e}{b} = \frac{d}{e'} \tag{13}$$

Suppose that $e/b = d/e'$. Then one can show that R and F are solutions to (11) if and only if

$$\frac{R}{F} = \frac{e}{b} \quad \left(= \frac{d}{e'} \right) \tag{14}$$

For example, the system

$$
\begin{aligned}
R' &= R + .1R - .15F \\
F' &= F - .15F + .1R
\end{aligned} \tag{15}
$$

has stable values $R = 15$, $F = 10$ or $R = 6$, $F = 4$. In fact, any pair (R, F) is stable in (15) if

$$\text{Stable } R, F \text{ values for (15):} \qquad \frac{R}{F} = \frac{.15}{.1} \quad \text{or} \quad R = \frac{3}{2}F \tag{16}$$

Further, if we start with values for R and F that are not stable, then over successive months the rabbit and fox populations always move toward one of these stable pairs of values, just like the Markov chain. In this model the "law of nature" is that there should be a 3 : 2 ratio of rabbits to foxes. Figure 1.8 shows sample curves along which unstable values move in approaching a stable value. For example, if we start iterating (15) with $R = 50$, $F = 40$, we have

0 months:	50	rabbits,	40	foxes
1 month:	49	rabbits,	39	foxes
2 months:	48	rabbits,	38	foxes
3 months:	47	rabbits,	37	foxes
⋮	⋮		⋮	

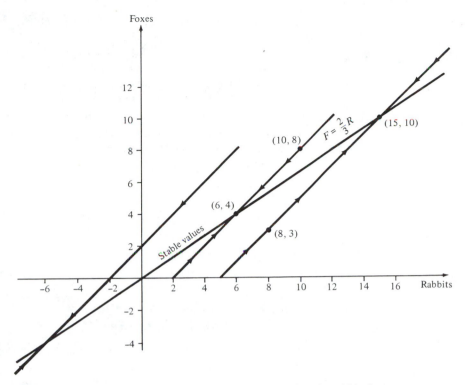

Figure 1.8 Stable values and trajectories to stable values in rabbit–fox growth model.

$$
\begin{array}{lllll}
10 \text{ months:} & 42 & \text{rabbits,} & 32 & \text{foxes} \\
\vdots & & \vdots & & \vdots \\
20 \text{ months:} & 37 & \text{rabbits,} & 27 & \text{foxes} \\
\vdots & & \vdots & & \vdots \\
30 \text{ months:} & 34 & \text{rabbits,} & 24 & \text{foxes} \\
\vdots & & \vdots & & \vdots \\
40 \text{ months:} & 32.5 & \text{rabbits,} & 22.5 & \text{foxes} \\
\vdots & & \vdots & & \vdots \\
50 \text{ months:} & 31.5 & \text{rabbits,} & 21.5 & \text{foxes} \\
\vdots & & \vdots & & \vdots \\
75 \text{ months:} & 30.4 & \text{rabbits,} & 20.4 & \text{foxes} \\
\vdots & & \vdots & & \vdots \\
100 \text{ months:} & 30.1 & \text{rabbits,} & 20.1 & \text{foxes}
\end{array}
\tag{17}
$$

Clearly, the populations are approaching the stable sizes of 30 rabbits and 20 foxes.

The starting pair (100, 80) goes to (60, 40); (80, 30) goes to (150, 100). If we write the starting values in the form $R = r$, $F = r - k$, then the limiting stable values will always turn out to be $R = 3k$, $F = 2k$. If the starting values have $R < F$, the limiting values will both be negative, with rabbits becoming negative (extinct) first. If $R = F$, both populations approach 0.

Why does this happen? How much of this behavior would occur if we used other values for the constants b, d, e, e' that satisfied the condition $e/b = d/e'$?   ∎

We conclude this section by introducing a nonlinear model for rabbit and fox populations. This nonlinear model can simulate a cyclic behavior that occurs frequently in nature.

Example 4. Nonlinear Model for Rabbits and Foxes

Let us consider the following nonlinear model for the monthly growth of populations of foxes and rabbits. Again, if R and F are the numbers of rabbits and foxes this month, then R' and F' are the numbers next month. Our system of equations is

$$R' = R + bR - eRF$$
$$F' = F - dF + e'RF$$
(18)

We now use the terms $-eRF$ and $+e'RF$ for the effect of foxes eating rabbits because the chances of a fox catching a rabbit depend on the abundance of both species.

Let us suppose that $b = d = .1$ and $e = e' = .01$.

$$R' = R + .1R - .01RF$$
$$F' = F - .1F + .01RF$$
(19)

If initially we let $R = 30$ and $F = 30$, then using (19) we obtain the table whose values are plotted in Figure 1.9.

Number of Months	Rabbits	Foxes
1	30	30
2	32	28
3	34	26
5	40	23
10	58	17
15	87	15

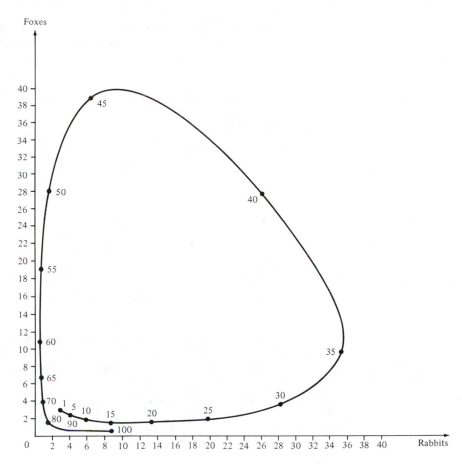

Figure 1.9 Cyclic pattern of population sizes in nonlinear rabbit–fox growth model.

20	132	15
25	196	20
30	282	36
35	354	96
40	261	278
45	62	389
50	15	280
60	6	107
70	7	40
80	15	15
90	35	7
100	88	4

(20)

When we started with a small number of both rabbits and foxes, the fox population declined further for a few months (from the *-dF* term) while the rabbit population started growing (from the $+bR$ term); the $-eRF$ and $+e'RF$ terms have little effect because the constants e and e' are so small. Soon the fox population grows and the rabbit population stops growing and starts to decline as the foxes eat the rabbits (the terms $-eRF$ and $+e'RF$ come into effect when R and F are large). As the rabbit population declines, so the fox population soon declines because of less food. We eventually find the sizes of the two populations back at the levels at which we started. The cycle time in this model is about 80 months.

This cycle of sudden growth followed by sudden decline characterizes the behavior of periodic pests, such as gypsy moths. The moths are dormant for many years and then have sudden, major outbreaks when few of their natural enemies are present (for gypsy moths, the enemy is a parasitic wasp). Eventually, the large numbers of the pest stimulate the appearance of its predator. The moths are killed off by the wasps and then the wasps die for lack of food. The dormant period begins again.

Linear models cannot produce this type of behavior.   ■

Note that after one cycle (80 months) the population sizes in (20) are 15 rabbits and 15 foxes, half the starting sizes of 30. This slippage in size is a fault in the model, due to the fact that we rounded time into units of months. If time were measured in days or seconds, the slippage would be less. Exercise 21 describes how to convert this model into units of days or seconds.

Section 1.3 Exercises

Summary of Exercises
Exercises 1–11 involve Markov chain models. Exercises 12–19 examine the model in Example 3 and similar linear growth models. Exercises 20 and 21 examine the behavior of the nonlinear model in Example 4. Exercises 11, 20, and 21 require computer programs.

1. Using the weather Markov chain in Example 1, simulate the weather over 10 days by flipping a coin to determine the chances of sunny or cloudy weather the next day according to the Markov chain's transition probabilities. If currently sunny, flip once and a head means sunny the next day and a tail means cloudy the next day. If currently cloudy, flip twice and when either flip is a head it is sunny the next day and when both flips are tails it is cloudy the next day. To start, assume that the previous day was sunny. What fraction of the 10 days was sunny?

2. In the weather Markov chain, starting with the probability distribution $(1, 0)$ (a sunny day), compute and plot (in p_1, p_2 coordinates) the distribution over five successive days. Repeat the process starting with the probability distribution $(0, 1)$. Can you guess the value of the stable distribution to which your points are converging?

3. In the frog Markov chain, what is the probability distribution in the next period if the current distribution is
   (a) $p_3 = 1$, all other $p_i = 0$?
   (b) $p_2 = .5$, $p_3 = .5$, all other $p_i = 0$?
   (c) $p_2 = .25$, $p_3 = .25$, $p_4 = .5$, all other $p_i = 0$?
   (d) $p_1 = .1$, $p_2 = .2$, $p_3 = .2$, $p_4 = .2$, $p_5 = .2$, $p_6 = .1$?

4. The printing press in a newspaper has the following pattern of break-downs. If it is working today, tomorrow it has 90% chance of working (and 10% chance of breaking down). If the press is broken today, it has a 60% chance of working tomorrow (and 40% chance by being broken again).
   (a) Make a Markov chain for this problem; give the matrix of transition probabilities and draw the transition diagram.
   (b) If there is a 50–50 chance of the press working today, what are the chances that it is working tomorrow?
   (c) If the press is working today, what are the chances that it is working in 2 days' time?

5. If the local professional basketball team, the Sneakers, wins today's game, they have a $\frac{2}{3}$ chance of winning their next game. If they lose this game, they have a $\frac{1}{2}$ chance of winning their next game.
   (a) Make a Markov chain for this problem; give the matrix of transition probabilities and draw the transition diagram.
   (b) If there is a 50–50 chance of the Sneakers winning today's game, what are the chances that they win their next game?
   (c) If they won today, what are the chances of winning the game after the next?

6. If the stock market went up today, historical data show that it has a 60% chance of going up tomorrow, a 20% chance of staying the same, and a 20% chance of going down. If the market was unchanged today, it has a 20% chance of being unchanged tomorrow, a 40% chance of going up, and a 40% chance of going down. If the market goes down today, it has a 20% chance of going up tomorrow, a 20% chance of being unchanged, and a 60% chance of going down.
   (a) Make a Markov chain for this problem; give the matrix of transition probabilties and the transition diagram.
   (b) If there is a 30% chance that the market goes up today, a 10% chance that it is unchanged, and a 60% chance that it goes down, what is the probability distribution for the market tomorrow?

7. The following model for learning a concept over a set of lessons identifies four states of learning: I = Ignorance, E = Exploratory Thinking, S = Superficial Understanding, and M = Mastery. If now in state I, after one lesson you have $\frac{1}{2}$ probability of still being in I and $\frac{1}{2}$ probability of being in E. If now in state E, you have $\frac{1}{4}$ probability of being in I, $\frac{1}{2}$ in E, and $\frac{1}{4}$ in S. If now in state S, you have $\frac{1}{4}$ probability of

being in E, $\frac{1}{2}$ in S, and $\frac{1}{4}$ in M. If in M, you always stay in M (with probability 1).

(a) Make a Markov chain model of this learning model.

(b) If you start in state I, what is your probability distribution after two lessons? After three lessons?

8. (a) Make a Markov chain model for a rat wandering through the following maze if at the end of each period, the rat is equally likely to leave its current room through any of the doorways. The states of the Markov chain are the rooms.

(b) If the rat starts in room 1, what is the probability that it is in room 4 two periods later?

9. Make a Markov chain model of a poker game where the states are the number of dollars a player has. With probability .3 a player wins 1 dollar in a period, with probability .4 a player loses 1 dollar, and with probability .3 a player stays the same. The game ends if the player loses all his or her money *or* if the player has 6 dollars (when the game ends, the Markov chain stays in its current state forever). The Markov chain should have seven states, corresponding to the seven different amounts of money: 0, 1, 2, 3, 4, 5, or 6 dollars. If you now have $2, what is your probability distribution in the next round? In the round after that?

10. Three tanks A, B, C are engaged in a battle. Tank A, when it fires, hits its target with hit probability $\frac{1}{2}$. B hits its target with hit probability $\frac{1}{3}$, and C with hit probability $\frac{1}{6}$. Initially (in the first period), B and C fire at A and A fires at B. Once one tank is hit, the remaining tanks aim at each other. The battle ends when there is one or no tank left. Make a Markov chain model of this battle.

Assistance in Computing Probabilities: Let the states of the Markov chain be the eight different possible subsets of tanks currently in action: ABC, AB, AC, BC, A, B, C, None. When in states A or B or C or None, the probability of staying in the current state is 1—this simulates the battle being over. One can never get to state AB. (Why?) So one only needs to determine the transition probabilities from states ABC, AC, and BC. From states AC and BC, the transition probabilities are products of the probability that each remaining tank hits or misses its target. For example, the probability of going from state AC to state A is the product of the probability that A hits C—$\frac{1}{2}$—times the probability that C misses A—$\frac{5}{6}$. So this probability is $(\frac{1}{2})(\frac{5}{6}) = \frac{5}{12}$. It takes some knowledge of probability to compute the transition probabilities from state ABC. From ABC there is a $\frac{5}{18}$ chance of remaining in state ABC (all tanks miss), a $\frac{5}{18}$ chance of going to state AC (A hits B but B and C miss A), a $\frac{4}{18}$ chance of going to state BC (at least one of B or C hits A and A misses B), and a $\frac{4}{18}$ chance of going to state C (at least one of B or C hits A and A hits B).

11. Use a computer program to follow the Markov chains in the following examples and exercises for 50 periods by iterating the next-period formula (1) as done in Example 2.
    (a) Example 1, starting in state Sunny.
    (b) Example 1, starting in state Cloudy.
    (c) Example 2, starting in state 4.
    (d) Example 2, starting with $p_1 = p_6 = .5$, other $p_i = 0$.
    (e) Exercise 4, starting in state Broken.
    (f) Exercise 5, starting in state Win.
    (g) Exercise 6, starting in state Market Unchanged.
    (h) Exercise 7, starting in state I.
    (i) Exercise 8, starting in state Room 1.
    (j) Exercise 9, starting in state $2.
    (k) Exercise 9, starting in state $3.
    (l) Exercise 9, starting in state $4.
    (m) Exercise 10, starting in state ABC.

12. For the rabbit–fox model in equations (8), use hand calculations to verify the population sizes for months 1, 2, and 3 given in table (9). To get the sizes after 1 month, set $R = 100$, $F = 100$ (the starting sizes) and evaluate the right sides of the equations in (8). Next take the values you obtained for R' and F' and let these be the new R and F. Repeat this process three times.

13. For the rabbit–fox model in equations (8), suppose that the initial population sizes are $R = 50$, $F = 50$.
    (a) Calculate by hand the population sizes after 1 month, after 2 months, and after 3 months.
    (b) Use a computer or calculator to compute the population sizes over 8 months.

14. Consider the following rabbit–fox models and an initial population size of $R = 100$, $F = 100$. In each case, compute the population sizes after 1 month, after 2 months, and after 3 months.
    (a) $R' = R + .3R - .2F$ (b) $R' = R + .3R - .2F$
         $F' = F - .2F + .1R$ $F' = F - .1F + .2R$

15. Consider the following goat–sheep models, where the two species compete for common grazing land. In each case, compute the population sizes after 1 month, after 2 months, and after 3 months if the initial population is 50 goats and 100 sheep.

(a) $G' = G + .2G - .3S$ **(b)** $G' = G + .2G - .1S$
      $S' = S + .2S - .2G$ $S' = S + .4S - .2G$

16. This exercise concerns the rabbit–fox model in equations (15). For given initial population sizes, calculate the population sizes after 1 month, after 2 months, and after 3 months. Also plot the trajectory of population sizes from the starting values to the stable sizes (as in Figure 1.8). The initial sizes are
    (a) $R = 30, F = 24$ **(b)** $R = 8, F = 3$ **(c)** $R = 8, F = 10$
    (d) $R = 10, F = 10$

17. Consider the rabbit–fox model

$$R' = R + .1R - .15F$$
$$F' = F - .3F + .2R$$

What is the equation of the line of stable population sizes? For given initial population sizes, calculate the population sizes after 1 month, after 2 months, and after 3 months. Also predict the stable sizes to which these populations are converging. Compare your numbers with the calculations in Exercise 16. The initial sizes are
    (a) $R = 30, F = 24$ **(b)** $R = 8, F = 3$ **(c)** $R = 8, F = 10$
    (d) $R = 10, F = 10$

18. Consider the rabbit–fox model

$$R' = R + .1R - .2F$$
$$F' = F - .4F + .2R$$

On a graph plot the following:
    (a) The line of stable population sizes.
    (b) The trajectory of population sizes starting from (10, 15).
    (c) The trajectory of population sizes starting from (10, 30).
    (d) The trajectory of population sizes starting from (20, 10).

19. Consider the rabbit–fox model

$$R' = R + 2R - 3F$$
$$F' = F - 3F + 2R$$

On a graph plot the following:
    (a) The line of stable population sizes.
    (b) The trajectory of population sizes starting from (10, 15).
    (c) The trajectory of population sizes starting from (1, 2).
    (d) How do the trajectories of this model differ from those in Figure 1.8?

20. Use a computer program to follow the behavior of the nonlinear rabbit–fox model in (19) over a period of 100 months (as in Figure 1.9) with the following starting values:
(a) $R = 6$, $F = 6$ (b) 100, 100   (c) 10, 10

21. In Example 4, if the change in R is $.1R - .01RF$ in 1 month, then in 1 day we would expect $\frac{1}{30}$ of such a change [i.e., $(\frac{1}{300})R - (\frac{1}{3000})RF$]; similarly for the change in foxes. Write out the full set of equations for this model with time measured in days. Starting with $R = 3$, $F = 3$, follow the populations as before for 3000 days ($= 100$ months). (Use a computer program.) How do your results compare with those in table (20)?

Projects

22. Use a computer program to follow the populations for many periods in the models in Exercises 14 and 15. Try a couple of different starting population sizes. In each case describe in words the long-term trends of the populations.

23. Make a thorough analysis of long-term trends for the rabbit–fox model

$$R = R + bR - eF$$
$$F = F - dF + e'R$$

for different values of the positive parameters b, d, e, e'. That is, list all possible long-term trends and give conditions on the parameters that tell when each trend occurs. For example, one trend is that both populations become extinct, with rabbits dying out first. Determine the conditions experimentally by trying many different specific parameter values and in each case computing the population sizes over many months.

Section 1.4 **Linear Programming and Models Without Exact Solutions**

In this section we examine two very important variations on the problem of solving n linear equations in n unknowns. We illustrate these variations with the refinery problem from Section 1.2.

Example 1. **Refinery Problem Revisited with One Refinery Broken**

The original refinery problem had three refineries and three products: heating oil, diesel oil, and gasoline. We wanted the production levels, x_1, x_2, x_3, of the refineries to meet demands of 500 gallons of heating oil, 850 gallons of diesel oil, and 1000 gallons of gasoline. The equations we got were

$$
\begin{array}{llrcrcrcr}
\text{Heating oil:} & 20x_1 & + & 4x_2 & + & 4x_3 & = & 500 \\
\text{Diesel oil:} & 10x_1 & + & 14x_2 & + & 5x_3 & = & 850 & \quad (1) \\
\text{Gasoline:} & 5x_1 & + & 5x_2 & + & 12x_3 & = & 1000
\end{array}
$$

Suppose that the third refinery breaks down and we have to try to meet the demands with two refineries. Our system of equations is now

Two-Refinery Production Problem

$$
\begin{array}{llrcrcr}
\text{Heating oil:} & 20x_1 & + & 4x_2 & = & 500 \\
\text{Diesel oil:} & 10x_1 & + & 14x_2 & = & 850 & \quad (2) \\
\text{Gasoline:} & 5x_1 & + & 5x_2 & = & 1000
\end{array}
$$

A system like (2) with more equations than unknowns is called *overdetermined* and does not normally have a solution. All one can ask for is an approximate solution. We want a "solution" of x_1, x_2 values that makes the total production of the two remaining refineries *as close as possible* to the demands. In Chapter 5 we define precisely the term "as close as possible" and then show how to solve this problem.   ∎

If we take the situation in system (2) to a greater extreme, with, say, 10 or 50 equations but just two unknowns, then we have a famous estimation problem in statistics.

Example 2. **Predicting Grades in College**

A guidance counselor at Scrooge High School wants to develop a simple formula for predicting a Scrooge graduate's GPA (grade-point average) at the local state college as a function of the student's GPA at Scrooge High. The formula would be a linear model of the form

$$
\text{college GPA} = q \times (\text{Scrooge GPA}) + r \qquad (3a)
$$

or

$$
C = qS + r \qquad (3b)
$$

where C stands for college GPA, S stands for Scrooge GPA, and r, q are constants to be determined based on the performances of past Scrooge graduates. Suppose that the data from eight students are

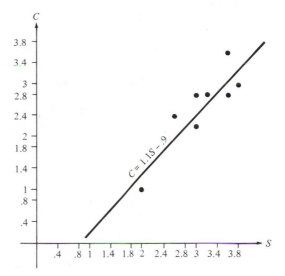

Figure 1.10 Estimated equation for relation between Scrooge GPA and college GPA.

			Predicted C
Student	S (Scrooge GPA)	C (College GPA)	$1.1 \times S - .9$
A	3.0	2.2	2.4
B	3.6	3.6	3.06
C	2.6	2.4	1.96
D	3.2	2.8	2.62
E	2.0	1.0	1.3
F	3.0	2.8	2.4
G	3.8	3.0	3.28
H	3.6	2.8	3.06

(4)

One should pick the constants q and r so that the predicted college GPA given by the expression $qS + r$ will be *as close as possible* to the actual college GPA for these students. Using a method discussed in Chapter 4, we set $q = 1.1$ and $r = -.9$. The predicted college GPAs with this formula are given in the last column of the table. Figure 1.10 has a plot of the C and S values from (4) along with the suggested line $C = 1.1S - .9$. ∎

This is the same sort of problem that we faced in finding an approximate solution to the refinery problem in Example 1. The statistical name for this type of estimation problem is **regression**.

Suppose that the counselor wanted to break down the Scrooge grade average into various components $G_{M/S}$, G_E, and $G_{H/L}$, representing the student's grades in the three subject areas of math/science, English, and history/languages. The counselor would give these three components separate weightings in a formula like

$$G_C = q_1 G_{M/S} + q_2 G_E + q_3 G_{H/L} + r \tag{5}$$

The solution of regression problems is discussed in Chapters 4 and 5.

Next let us consider the situation where we have more unknowns than equations. Again we use the refinery model.

Example 3. **Refinery Model Revisited Without Diesel Oil**

Suppose now that there is no demand for diesel oil and the three refineries just produce heating oil and gasoline. So the system of equations to be satisfied is

Two-Product Refinery Problem

$$
\begin{array}{lrcl}
\text{Heating oil:} & 20x_1 + 4x_2 + 4x_3 & = & 500 \\
\text{Gasoline:} & 5x_1 + 5x_2 + 12x_3 & = & 1000
\end{array} \tag{6}
$$

This system of two equations in three unknowns is called *underdetermined* in the sense that there are not enough constraints to determine each unknown uniquely. The solution we found in Section 1.2 for the refinery problem with all three equations is clearly valid with two equations: $x_1 = 5$, $x_2 = 33$, $x_3 = 68$. But many other solutions are possible. In particular, we could shut down one of the refineries, say refinery 3, as in Example 1. In mathematical terms, we seek a solution to (6) with $x_3 = 0$. Dropping the x_3 terms from (6), we have

$$
\begin{array}{lrcl}
\text{Heating oil:} & 20x_1 + 4x_2 & = & 500 \\
\text{Gasoline:} & 5x_1 + 5x_2 & = & 1000
\end{array} \tag{7}
$$

This system of equations is easily solved by high school algebra: Subtract four times the second equation from the first to eliminate x_1 and obtain $-16x_2 = -3500$ or $x_2 = \frac{3500}{16} = 218\frac{3}{4}$. Now the gasoline equation becomes $5x_1 + 5(218\frac{3}{4}) = 1000$; dividing this equation by 5 and solving for x_1 yields $x_1 = -18\frac{3}{4}$. Unfortunately, a negative value for x_1 is nonsense.

Next try shutting down the first refinery by setting $x_1 = 0$. We have

$$
\begin{array}{lrcl}
\text{Heating oil:} & 4x_2 + 4x_3 & = & 500 \\
\text{Gasoline:} & 5x_2 + 12x_3 & = & 1000
\end{array} \tag{8}
$$

Solving (8) for x_2 and x_3 yields the solution $x_2 = \frac{500}{7} \approx 71$, $x_3 = \frac{375}{7} \approx 54$. We could set $x_2 = 0$ and get another solution. There are many more solutions in which all the refineries are running.

Which solution would we use in practice? The answer is, the solution that is most efficient. That is, the solution that is the cheapest.

Each refinery will have a cost of operation. Suppose that the costs to refine a barrel of oil (the units for the x_i's) are

Refinery Operation Costs

Refinery 1	$30 per barrel
Refinery 2	$25 per barrel
Refinery 3	$20 per barrel

(9)

Then we want a solution to the two-product production problem (6) (with no x_i negative) for which the total refining cost of $30x_1 + 25x_2 + 20x_3$ is minimized. The complete mathematical statement of this problem is

Optimal Refinery Production Problem

$$\text{Minimize } 30x_1 + 25x_2 + 20x_3$$

subject to the constraints

$$20x_1 + 4x_2 + 4x_3 = 500$$
$$5x_1 + 5x_2 + 12x_3 = 1000$$
$$x_1 \geq 0, \quad x_2 \geq 0, \quad x_3 \geq 0$$

(10)

∎

The problem of optimizing (minimizing or maximizing) a linear expression subject to constraints that are linear equations or inequalities is called **linear programming**. Linear programming is the most important mathematical tool in management science. There are thousands of different real-world problems that can be posed as linear programming problems. When scientists at Bell Laboratories recently proposed a new, more efficient way to solve linear programming problems, the announcement was a front-page story in major newspapers.

The optimal refinery production problem (10) involved a set of linear equations as constraints together with the inequalities $x_1 \geq 0$, $x_2 \geq 0$, $x_3 \geq 0$. As we shall see shortly, it is easier to solve linear programs in which all the constraints are inequalities. Exercise 13 shows how to convert the equations in (10) into linear inequalities. This conversion is discussed further in Section 4.6.

As an example of how linear programs with inequalities are solved, we look at a simple two-variable maximization problem.

Example 4. A Linear Program: Optimal Production of Two Crops

Suppose that a farmer has 200 acres on which he can plant any combination of two crops, corn and wheat. Corn requires 4 worker-days of labor and $20 of capital for each acre planted, while wheat requires 1 worker-day of labor and $10 of capital for each acre planted. Suppose also that corn produces $60 of revenue per acre and wheat produces $40 of revenue per acre. If the farmer has $2200 of capital and 320 worker-days of labor available for the year, what is the most profitable planting strategy?

If

$$C = \text{number of acres of corn}$$
$$W = \text{number of acres of wheat}$$

the constraints on land, labor, and capital are given by the following system of linear inequalities:

$$
\begin{array}{lrl}
\text{Land:} & C + & W \le 200 \\
\text{Labor:} & 4C + & W \le 320 \\
\text{Capital:} & 20C + & 10W \le 2200
\end{array}
\tag{11}
$$

also,

$$W \ge 0, \qquad C \ge 0$$

Subject to these constraints, we want to determine C and W so as to maximize the total revenue.

$$\text{Maximize } 60C + 40W \tag{12}$$

The expression to be maximized is called the *objective function*.

When only two variables are involved, one can plot the inequality constraints and display the region of (x_1, x_2)-points that simultaneously satisfy all the constraints in (11). This region is called the **feasible region** of the linear program, and its points **feasible points**. See the shaded area in Figure 1.11. Recall that to find the points satisfying an inequality such as $C + W \le 200$, we plot the line $C + W = 200$ and then shade the line and all points on the lower left side of the line.

Once we have plotted the feasible region for (11), it remains to find out which feasible point maximizes $60C + 40W$. The following geometric insight greatly simplifies the solution of linear programs (this theorem is proved in Section 4.6).

Theorem. A linear objective function assumes its maximum and minimum values on the boundary of the feasible region. In fact, the optimal value is achieved at a corner point of this boundary.

Now we can solve our linear program. The theorem tells us where to look for the optimal (C, W)-value—at the corners of the feasible region. To find a corner that lies at the intersection of two constraint lines, we solve for a (C, W) point that lies on both lines—the same old problem of solving two equations in two unknowns. Once we determine the coordinates of a corner of the feasible region, we can

Figure 1.11 Feasible region for two-crop linear program.

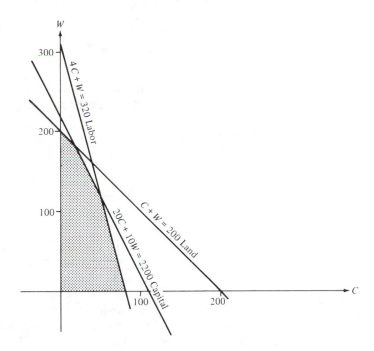

evaluate the expression $60C + 40W$ at that corner. The corner that maximizes this expression is the answer to our optimal production problem. In the current problem we can see from Figure 1.11 which intersections of constraint lines form corners of the feasible region.

Table 1.2 lists the coordinates of the corners and the associated objective function values. So the optimal production schedule is to plant 20 acres of corn and 180 acres of wheat. In Section 4.6 we present a more general, systematic approach to find the maximizing corner in a linear program.

Table 1.2

Corner Coordinates	Intersecting Constraints	Objective Function
(0, 0)	$C \geq 0$ and $W \geq 0$	0
(0, 200)	$C \geq 0$ and Land	8000
(20, 180)	**Land and Capital**	**8400*****
(50, 120)	Capital and Labor	7800
(80, 0)	Labor and $W > 0$	4800

∎

Before leaving this example, we note that the farmer's constraints (Land, Labor, and Capital) determined the feasible region, and the "marketplace" (prices for corn and wheat) determined the objective function. If the market prices for corn or wheat change, or equivalently, the farmer receives a subsidy for one of the crops, the optimal solution may change.

Table 1.3

Corner Coordinates	Intersecting Constraints	Objective Function
(0, 0)	$C \geq 0$ and $W \geq 0$	0
(0, 200)	$C \geq 0$ and Land	8000
(20, 180)	Land and Capital	9000
(50, 120)	**Capital and Labor**	**9300*****
(80, 0)	Labor and $W > 0$	7200

For many years, the Federal Farm Program has offered crop subsidies in order to influence both the types of crops grown and the total number of acres planted.

To illustrate how a crop subsidy can cause land to be taken out of production, let us suppose in Example 4 that the farmer receives a subsidy for corn that increases the revenue from \$60 to \$90 per acre. Then the objective function is now $90C + 40W$. The values of this new objective function at the corner points of the feasible region are shown in Table 1.3. The new optimal strategy is to plant 50 acres of corn and 120 acres of wheat, for a total of 170 acres. The subsidy results in the farmer removing 30 acres from production.

Section 1.4 Exercises

Summary of Exercises
Exercises 1–5 involve overdetermined systems and regression. Exercises 6–12 involve linear programming. Exercise 13 tells how to convert a system of equations into a system of inequalities (this conversion is discussed further in Chapter 4).

1. Use a trial-and-error approach to estimate as closely as possible an approximate solution to the refinery problem in Example 1.

2. (a) Repeat Exercise 1, but now refinery 2 rather than refinery 3 is missing.
   (b) Repeat Exercise 1, but now refinery 1 rather than refinery 3 is missing.
   (c) If you had to close down one refinery, which refinery would you pick in order to meet the demand as closely as possible with the remaining two refineries?

3. Consider the following system of equations, which might represent supply–demand equations for chairs, tables, and sofas from two factories:

	Factory 1		Factory 2		Demand
Chairs:	$10x_1$	+	$6x_2$	=	200
Tables:	$7x_1$	+	$7x_2$	=	150
Sofas:	$5x_1$	+	$4x_2$	=	100

Find an approximate solution to this system of equations by trial and error.

4. For the following sets of x-y points, estimate a line to fit the points as closely as possible.
   (a) (1, 1), (2, 3), (3, 2), (4, 6), (5, 5)
   (b) (1, 6), (2, 4), (3, 3), (3, 2), (4, −1), (4, 0)

5. Suppose that the estimate for GPA in college in Example 2 had been

$$G_C = .6G_{M/S} + .3G_E + .3G_{H/L} - 1$$

where $G_{M/S}$, G_E, and $G_{H/L}$ are the GPAs in mathematics/science, English, and humanities/languages. Based on this predictor, on which courses should students work hardest (if students want to improve their expected college GPA)?

6. Find a solution to the refinery problem in Example 3 in which the values of x_2 and x_3 are the same.

7. Change the labor constraint in the crop linear program of Example 4 to be $4C + 2W = 320$. Now what would be the optimal solution?

8. Suppose that a Ford Motor Company factory requires 7 units of metal, 20 units of labor, 3 units of paint, and 8 units of plastic to build a car, while it requires 10 units of metal, 24 units of labor, 3 units of paint, and 4 units of plastic to build a truck. A car sells for $6000 and a truck for $8000. The following resources are available: 2000 units of metal, 5000 units of labor, 1000 units of paint, and 1500 units of plastic.
   (a) State the problem of maximizing the value of the vehicles produced with these resources as a linear program.
   (b) Plot the feasible region of this linear program.
   (c) Solve this linear program by the method in Example 4, by determining the coordinates of the corners of the feasible region and finding which corner maximizes the objective function.

   Hint: By looking at the objective function, you should be able to tell which corners are good candidates for the maximum.

9. Suppose that a meal must contain *at least* 500 units of vitamin A, 1000 units of vitamin C, 200 units of iron, and 50 units of protein. A dietician has the following two foods from which to choose:

Meat: One unit of meat has 20 units of vitamin A, 30 units of vitamin C, 10 units of iron, and 15 units of protein.

Fruit: One unit of fruit has 50 units of vitamin A, 100 units of vitamin C, 1 unit of iron, and 2 units of protein.

Meat costs 50 cents a unit and fruit costs 40 cents a unit.

(a) State the problem of minimizing the cost of a meal that meets all the minimum nutritional requirements as a linear program (now you want to minimize the objective function).

(b) Plot the boundary of the feasible region for this linear program.

(c) Solve this linear program by the method in Example 4 [see Exercise 8, part (c)].

10. Consider the two-refinery problem in Example 1.

$$\begin{array}{lrcll}
\text{Heating oil:} & 20x_1 + 4x_2 &=& 500 & \\
\text{Diesel oil:} & 10x_1 + 14x_2 &=& 850 & (2) \\
\text{Gasoline:} & 5x_1 + 5x_2 &=& 1000 &
\end{array}$$

Suppose that it costs \$30 to refine a barrel in refinery 1 and \$25 a barrel in refinery 2. What is the production schedule (i.e., values of x_1, x_2) that minimize the cost while producing *at least* the amounts demanded of each product (i.e., at least 500 gallons of heating oil, etc.)?

Hint: Solve by the method in Exercise 8, part (c).

11. Consider the following two linear programs.

(i) Maximize $3x_1 + 3x_2$	(ii) Minimize $10x_1 + 8x_2$
subject to	subject to
$x_1 \geq 0, \quad x_2 \geq 0$	$x_1 \geq 0, \quad x_2 \geq 0$
$x_1 + 2x_2 \leq 10$	$x_1 + 2x_2 \geq 3$
$2x_1 + x_2 \leq 8$	$2x_1 + x_2 \geq 3$

Solve them and show that the optimum values of these two objective functions are the same.

12. Set up the following problem as a linear program, but *do not* solve. There are two truck warehouses and two stores that sell trucks. The following table gives the cost of transporting a truck from one of the warehouses to one of the stores.

	Store 1	Store 2
Warehouse 1	\$40	\$50
Warehouse 2	\$60	\$40

Warehouse 1 has 100 trucks and warehouse 2 has 80. Store 1 needs at least 50 trucks and store 2 needs at least 100 trucks. Find the cheapest way to meet the stores' demand.

Hint: Let the variables be x_{11}, x_{12}, x_{21}, and x_{22}, where x_{ij} is the amount shipped from warehouse i to store j.

13. To convert a system of equations in which each variable must be ≥ 0, into a system of inequalities with each variable ≥ 0, we perform the following steps.

    (i) Pick a variable in the first equation and solve that equation for the chosen variable, that is, so that the chosen variable is alone on one side of the equation. For example, in the system of equations

    $$2x + 4y + 6z = 8$$
    $$x + 3y + 2z = 6$$
    $$x \geq 0, y \geq 0, z \geq 0$$

    if we pick x in the first equation, then we rewrite it as

    $$(*) \qquad x = -2y - 3z + 4$$

    (ii) Replace the chosen variable in the other equations by substituting in its place the right-hand side in (*). So the second equation in this problem becomes

    $$(-2y - 3z + 4) + 3y + 2z = 6$$

    or

    $$y - z = 2$$

    (iii) Since the chosen variable is ≥ 0, the right side of (*) must be ≥ 0, that is,

    $$-2y - 3z + 4 \geq 0$$

    or

    $$4 \geq 2y + 3z \qquad \text{(equivalently, } 2y + 3z \leq 4)$$

    Now the original three-variable problem has been reduced to a two-variable problem with one equation converted into an inequality.

    $$2y + 3z \leq 4$$
    $$y - z = 2$$
    $$y \geq 0, \quad z \geq 0$$

(iv) Repeat the entire procedure for a variable in the second equation. If we pick y, solve the second equation for y in terms of z, substitute this expression involving z in place of y in the first inequality; also make this expression in z be ≥ 0.

(a) Complete step iv.

(b) Convert the refinery linear program at the end of Example 3 into a linear program with inequalities, and solve the linear program.

(c) Convert the following system of equations for nonnegative variables into a system of inequalities.

$$x_1 + 3x_2 + 2x_3 = 10$$
$$2x_1 + 5x_2 - 4x_3 = 15$$
$$x_1 \geq 0, \quad x_2 \geq 0, \quad x_3 \geq 0$$

Section 1.5 Arrays of Data and Linear Filtering

In the previous sections we have encountered arrays of numbers that were the coefficients of systems of equations. But not all arrays of numbers are sets of coefficients. There are many problems in which arrays of numbers are input data to be analyzed. In statistics, we study huge data sets that come in sequences, two-dimensional arrays, and more complex structures. In the field of information processing and pattern recognition, certain information must be extracted from the data, be it a coded message or a picture. Both of these fields make heavy use of linear models to process arrays of data.

The examples in this section illustrate the use of linear models in pattern recognition and encoding of information. First we consider an example in which the data to be processed are letters, not numbers.

Example 1. Linear Models for Encoding Alphabetic Messages

A common approach to coding an alphabetic message is to treat each letter in the message as a number between 1 and 26: A $\rightarrow$ 1, B $\rightarrow$ 2, C $\rightarrow$ 3, . . ., Z $\rightarrow$ 26. The simplest way to encode a message is to convert each letter (number) to a different letter (number) using some simple arithmetic formula. For example, we could add 7 to each number (this shifts the corresponding letter seven places to the right in the alphabet) or multiply each number by 11. However, these operations will sometimes convert a number between 1 and 26 to a number greater than 26.

To ensure that the result of some calculation is a number between 1 and 26, we usually assume that all arithmetic is done mod 26. As an example,

$$7 \times 13 = 91 = 13 \quad (\text{mod } 26) \qquad \text{since } 91 = 3 \times 26 + 13$$

Encoding schemes based on adding a constant to each number (letter) or multiplying by a constant are easy for an outsider to break because once one letter is guessed, the constant can be determined. For example, if X_L is the original letter and X_C the coded letter, the encoding

$$X_C = 7X_L \tag{1}$$

transforms THE into JDI since in numbers T = 20, H = 8, and E = 5, so

$$7 \cdot T = 7 \cdot 20 = 140 \equiv (\text{mod } 26) \quad 10 = J$$
$$7 \cdot H = 7 \cdot 8 = 56 \equiv (\text{mod } 26) \quad 4 = D$$
$$7 \cdot E = 7 \cdot 5 = 35 \equiv (\text{mod } 26) \quad 9 = I$$

Now if we guess that I is the encoding of E, it is easy to compute that the constant in (1) is 7. Even a code with multiplication and addition, such as $X_C = 9X_L + 21$, is easy to break.

A better scheme, in which frequent words and letters are scrambled, is to encode numbers in pairs using two linear equations of the following form. Let L_1, L_2 be a pair of original letters (represented as numbers between 1 and 26), and C_1, C_2 be the coded letters (also represented as numbers) into which L_1, L_2 are transformed.

$$C_1 \equiv aL_1 + bL_2 \quad (\text{mod } 26) \tag{2}$$
$$C_2 \equiv cL_1 + dL_2 \quad (\text{mod } 26)$$

For example, in the scheme

$$C_1 \equiv 9L_1 + 17L_2 \quad (\text{mod } 26) \tag{3}$$
$$C_2 \equiv 7L_1 + 2L_2 \quad (\text{mod } 26)$$

the pair E, C, represented numerically as 5, 3, would be encoded as

$$C_1 \equiv 9 \times 5 + 17 \times 3 = 96 \equiv 18 \quad (\text{mod } 26) = R \tag{4}$$
$$C_2 \equiv 7 \times 5 + 2 \times 3 = 41 \equiv 15 \quad (\text{mod } 26) = O$$

To use (2) for a whole message, we divide the string of m letters (ignoring punctuation) into $m/2$ successive pairs. Observe that if the fifth letter in the message were an E, the fifth letter in the encoded sequence would vary depending on what the sixth letter was [with which E is paired in (2)]. There are four constants [a, b, c, d in (2)] in this encoding scheme, and hence $26^4 = 456,976$ different possible schemes. Moreover, there are no meaningful patterns of frequently used letters to help a codebreaker.

If the codebreaker had access to a large computer, we could counter this by grouping letters in sets of 5 and replace (2) by a scheme involving five equations of linear combinations of five letters. Now there would be 25 constants yielding $26^{25} \simeq 10^{35}$ schemes—and we can go to sleep knowing that our code is secure.

Although it is important to keep this code from being broken, it is also important that a code not be too hard to decode by a receiver. If the receiver knows the constants in (2), he or she still has to reverse the encoding process by solving a pair of equations in two unknowns. For example, if (3) were being used and the pair R, O (= 18, 15) generated in (4) were received, the decoder would have to solve the system of equations

$$9L_1 + 17L_2 \equiv 18 \quad (\text{mod } 26) \tag{5}$$
$$7L_1 + 2L_2 \equiv 15 \quad (\text{mod } 26)$$

For a more complex scheme of five equations in five unknowns, the decoding problem gets harder, especially since arithmetic is mod 26. Fortunately, we shall show in Chapter 3 that there exist simple formulas for decoding so that the original pair L_1, L_2 (or 5-tuple) can be computed as a linear combination of the coded pair C_1, C_2. For example, the decoding equations for (3) are

$$L_1 \equiv 18C_1 + 3C_2 \quad (\text{mod } 26) \tag{6}$$
$$L_2 \equiv 15C_1 + 3C_2 \quad (\text{mod } 26)$$ ∎

The next two examples involve data analysis.

Example 2. **The Mean of a Data Set**

The most basic piece of statistical information about a set of data is the **mean**, or average, of the data. The mean is obtained by summing all data values and dividing by the number of data. For example, the mean of the sequence of numbers 2, 3, 13, 3, 7, 1, 9, 3, 4, 5 is $(2 + 3 + 13 + 3 + 7 + 1 + 9 + 3 + 4 + 5)/10 = 5$.

The mean is a linear combination of these 10 data values in which each value is multiplied by $\frac{1}{10}$ (recall from Section 1.1 that a linear combination is an expression of the form $c_1x_1 + c_2x_2 + \cdots + c_nx_n$). ∎

Example 3. **Smoothing a Time Series**

In many situations one receives a sequence of numbers recorded over time that form a pattern, but randomness in nature or in recording and transmitting the numbers has obscured the pattern. Such sequences are called **time series**. Consider the series of readings in Data Plot 1 taken

Time	Data	
1	23	————————————
2	27	—————————————
3	21	———————————
4	32	———————————————
5	29	——————————————
6	26	—————————————
7	30	———————————————
8	29	——————————————
9	26	—————————————
10	26	—————————————
11	27	—————————————
12	19	—————————
13	24	————————————
14	22	———————————
15	25	————————————
16	20	—————————
17	16	———————
18	27	—————————————
19	19	—————————
20	15	———————
21	13	——————
22	25	————————————
23	18	————————
24	22	———————————
25	21	———————————
26	24	————————————
27	25	————————————
28	28	——————————————
29	23	————————————
30	27	—————————————

Data Plot 1

over a period of time. Suppose that this time series gives the numbers of people applying for welfare aid in some city in successive months (the numbers are presented in units of 100). The values might equally well have represented levels of X-rays measured in a spacecraft or the numbers of new houses started in the United States in successive months.

To help picture the data, we plot the numbers in a graph, with time measured on the vertical axis and the data values on the horizontal axis. (The axes are omitted in the graph.)

We want to try to find a long-term pattern in this time series by smoothing the data—that is, reducing the jumps in data from one period to the next. In engineering, the task of smoothing a noisy electronic signal is called **filtering** (the term is now also used in nonengineering settings).

The simplest way to filter a time series is by replacing the ith value d_i by the average of d_i and the two adjacent values d_{i-1} and d_{i+1}. That is, we form a new time series whose ith value d_i' is given by

$$d_i' = \frac{d_{i-1} + d_i + d_{i+1}}{3} \tag{7}$$

For example, we replace d_{15} by

$$d_{15}' = \frac{d_{14} + d_{15} + d_{16}}{3} = \frac{22 + 25 + 20}{3} = 22$$

(When the value of d_i' is fractional, we shall round to the nearest integer.) The formula in (7) is not defined for the first and last values ($i = 1$ and $i = 27$); instead, let us set $d_1' = (d_1 + d_2)/2$ and $d_{27}' = (d_{26} + d_{27})/2$. The complete system of linear equations (the linear model) for filtering is then

$$
\begin{aligned}
d_1' &= \tfrac{1}{2}d_1 &&+ \tfrac{1}{2}d_2 \\
d_2' &= \tfrac{1}{3}d_1 &&+ \tfrac{1}{3}d_2 &&+ \tfrac{1}{3}d_3 \\
d_3' &= \tfrac{1}{3}d_2 &&+ \tfrac{1}{3}d_3 &&+ \tfrac{1}{3}d_4 \\
&\vdots &&\vdots &&\vdots &&\vdots \\
d_{n-1}' &= \tfrac{1}{3}d_{n-2} &&+ \tfrac{1}{3}d_{n-1} &&+ \tfrac{1}{3}d_n \\
d_n' &= \tfrac{1}{2}d_{n-1} &&+ \tfrac{1}{2}d_n
\end{aligned}
\tag{8}
$$

Our new time series looks as shown in Data Plot 2. This time series is much smoother. There is a clear trend of increasing values, then decreasing, then increasing, and finishing relatively level.

To smooth these data further, we could apply the smoothing transformation (8) again to this new time series. Instead, let us smooth the original data (in Data Plot 1) by applying a weighted average of five values in which d_i is weighted more and d_{i-2} and d_{i+2} are weighted less:

$$d_i'' = \frac{d_{i-2} + 2d_{i-1} + 3d_i + 2d_{i+1} + d_{i+2}}{9} \tag{9}$$

For example, now d_{15} is replaced by

$$d_{15}'' = \frac{d_{13} + 2d_{14} + 3d_{15} + 2d_{16} + d_{17}}{9}$$

$$= \frac{24 + 2 \times 22 + 3 \times 25 + 2 \times 20 + 16}{9} = \frac{199}{9} \simeq 22$$

Time	Data
1	25
2	24
3	27
4	27
5	29
6	28
7	28
8	28
9	27
10	26
11	24
12	23
13	22
14	24
15	22
16	20
17	21
18	21
19	20
20	16
21	18
22	19
23	22
24	20
25	22
26	23
27	26
28	25
29	26
30	25

Data Plot 2

For d_1'', drop the missing terms from (8) to obtain $d_1'' = (3d_1 + 2d_2 + d_3)/6$, and similarly for d_2'', d_{29}'', d_{30}''. The new time series obtained when transformation (8) is applied to the original data is as shown in Data Plot 3. Observe how smooth this transformed time series is as compared to the original one. The Exercises have other examples of time series for which filtering reveals important trends.

We can show that applying the three-value average (7) to Data Plot 2; or, equivalently, applying the three value average twice to the original data, produces the same time series as in Data Plot 3. In general, successively performing two (or more) linear filterings is equivalent to performing another (more complicated) filtering; see the Exercises for examples. ∎

Time	Data	
1	24	_____
2	25	_____
3	26	_____
4	28	_____
5	28	_____
6	29	_____
7	28	_____
8	28	_____
9	27	_____
10	26	_____
11	25	_____
12	23	_____
13	23	_____
14	23	_____
15	22	_____
16	21	_____
17	21	_____
18	21	_____
19	19	_____
20	18	_____
21	17	_____
22	19	_____
23	20	_____
24	21	_____
25	22	_____
26	24	_____
27	25	_____
28	26	_____
29	26	_____
30	26	_____

Data Plot 3

████████████

Example 4. **Linear Filtering in Pattern Recognition**

When the TV camera that serves as the eyes of a robot transmits a picture to the robot's computer, the picture is sent as a two-dimensional array of numbers that indicate the darkness of each point in the picture. A computer program to perform pattern recognition must determine what the robot is seeing by analyzing this digital representation of the picture. Light reflecting off an object or confusing patterns in the background can make a simple object quite difficult to recognize. Transformations to filter the data and increase the level of contrast are an essential part of any pattern recognition program.

Suppose that there are nine shades of darkness of a point, represented by integers 0 through 8, 0 for white (least dark) and 8 for black:

$$0 = \quad, \quad 1 = \text{▦}, \quad 2 = \text{▦}, \quad 3 = \text{▦}, \quad 4 = \text{▦},$$
$$5 = \text{▦}, \quad 6 = \text{▦}, \quad 7 = \text{▦}, \quad 8 = \text{■}$$

A picture represented by the 12×12 array of darkness values shown in Data Plot 4 has been received from a robot's TV camera (of course, a 12×12 array would only be a small section of the full TV image).

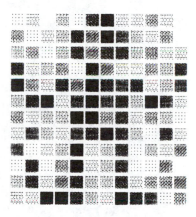

Data Plot 4

Let us perform a linear filtering on this array of darkness values. We replace each value by a weighted average of the value and the eight values surrounding it, as denoted by 1's and 4's in Figure 1.12.

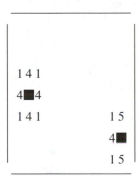

We use a weighting in which the old value gets a weight of 16, the four neighboring values in the same row or column (denoted by 4's in Figure 1.12) get a weight of 4, and the diagonal neighboring values (denoted by 1's in Figure 1.12) get a weight of 1; we divide this weighted sum by 36. (At the edges of the array, we increase the weights on the border values to 5, 20, 5, as shown in Figure 1.12.) The transformed array is shown in Data Plot 5.

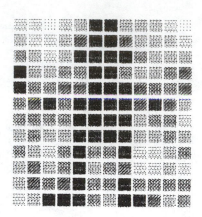

Data Plot 5

We might now start to perceive a person in the figure. But to see the person clearly we need greater contrast. To increase the contrast between light and dark, we apply the following linear function with roundoff: $f(x) = 3x - 9$; a value below 0 is rounded to 0 and a value above 8 is rounded to 8. The table for this contrast function is

Old Value	0	1	2	3	4	5	6	7	8
New Value	0	0	0	0	3	6	8	8	8

With this contrast function, Data Plot 5 becomes Data Plot 6. Now the person is fairly visible, perhaps with an object by the left foot. Greater contrast would help a little more. A good computer program to recognize patterns should now be able to "see" that the object pictured is a human being.

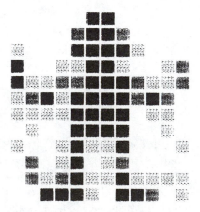

Data Plot 6

"Adaptive" filtering schemes do a small amount of filtering and then look for "borders" between light and dark regions. The region around a border is then subjected to a contrast transformation to accentuate the border, while nonborder regions are filtered as above. ∎

Section 1.5 **Exercises**

Summary of Exercises

Exercises 1–5 are associated with Example 1 about codes. Exercises 6–12 are associated with Example 3 on filtering time series; Exercises 10–12 involve algebraic composition of transformations—they require more maturity. Exercises 13–15 are associated with Example 4 about two-dimensional arrays forming "pictures."

1. Evaluate the following expressions mod 26. *All answers must be positive numbers between 1 and 26.*
   (a) 7×7 **(b)** 12×5 **(c)** -5×5 **(d)** -11×19
   (e) $12 \times (14 + 17)$

2. Use the encoding $C \equiv 5L + 7$ (mod 26) to encode the letters in the following words.
   (a) BE   **(b)** AT   **(c)** APE

3. Use the encoding in equations (3) to encode the following pairs of letters.
   (a) BE   **(b)** AT   **(c)** CC

4. Use the decoding in equations (6) to decode the following pairs of letters.
   (a) BG   **(b)** CC   **(c)** RD

5. Determine the value(s) of x that satisfies (satisfy) the following equations mod 26. Which equations have unique solutions?
   (a) $9x \equiv 11$ **(b)** $7x \equiv 13$ **(c)** $14x \equiv 3$ **(d)** $11x \equiv 9$

6. Apply the filtering transformation in formula (7) to the first 15 numbers in the time series in Data Plot 2. (You should get the same results as in Data Plot 3.)

7. Apply the following filtering transformations to Data Plot 1 (explain how you alter these transformations for the first and last values).

   (a) $d'_i = \dfrac{d_{i-2} + d_i + d_{i+2}}{3}$

(b) $d'_i = \dfrac{d_{i-2} + d_{i-1} + d_i + d_{i+1} + d_{i+2}}{5}$

(c) $d'_i = \dfrac{d_{i-3} + d_{i-1} + d_{i+1} + d_{i+3}}{4}$

8. Consider the time series 2, 10, 4, 12, 6, 14, 8, 16, 10, 18, 12, 20. Apply transformations (a), (b), and (c) from Exercise 7. Which of the transformations smooth this time series well, and which do a poor job?

9. Consider the time series 1, 4, 2, 5, 8, 6, 3, 10, 3, 12, 10, 9, 8, 12, 18, 13, 21, 16, 16. Apply transformations (a), (b), and (c) from Exercise 7. Which of the transformations smooth this time series well, and which do a poor job?

10. Show algebraically that if the transformation in formula (7) were applied to a time series and applied again to the resulting time series, then the cumulative result would be the same as the transformation in formula (9).

11. **(a)** Suppose that the transformation in Exercise 7, part (a) is applied twice (as described in Exercise 10). Give a formula for the cumulative transformation.
    (b) Suppose that the transformation in Exercise 7, part (a) is applied to a time series and then the resulting time series is filtered by the transformation in Exercise 7, part (b). Give a formula for the cumulative transformation.

12. Suppose that $d'_i = a_1 d_{i-1} + a_2 d_i + a_3 d_{i+1}$, $d''_i = b_1 d_{i-1} + b_2 d_i + b_3 d_{i+1}$. Give a formula for the transformation obtained by performing the first and then the second transformation on a time series.

13. Apply the contrast function (given just before Data Plot 6) to Data Plot 6.

14. Apply the following filtering transformations to the upper 8-by-8 corner of Data Plot 4 (just the first 8 rows and first 8 columns), and then apply the contrast function given in the text. The transformations are described by a 3-by-3 square of weights, as in Figure 1.12. Explain what you did at the borders.

     (a) 1   1   1   **(b)** 0   1   0   **(c)** 1   2   1   **(d)** 1   2   1
            1   1   1   1   4   1   2   8   2   2   0   2
            1   1   1   0   1   0   1   2   1   1   2   1

How effectively does each transformation help reveal the human being in the picture?

15. Consider the following "pictures," given in terms of numbers rather than darkness levels. Apply the transformation in Example 4 followed by the contrast function. Give your answer in darkness levels. What is the letter or number in each picture?

(a)					(b)					(c)				
5	7	1	2	7	8	7	2	5	8	3	3	5	1	2
8	4	1	5	1	7	6	1	3	7	4	7	3	3	8
7	3	5	2	1	4	8	3	7	6	3	8	4	1	7
7	7	5	7	8	8	5	8	5	1	2	6	3	3	8
4	8	6	7	5	7	4	8	7	3	6	6	1	6	6
8	4	3	4	8	6	8	2	8	5	3	8	2	4	7
7	8	7	4	7	7	4	4	6	7	1	3	1	5	1
5	7	6	8	6	8	7	1	1	8	5	1	6	2	3

Matrices

2

Section 2.1 Examples of Matrices

An essential tool in all mathematical modeling is good notation. This is especially true for models with large systems of equations or arrays of data. Two characteristics of good notation are

1. To provide a way to express complex operations simply.
2. To help a reader concentrate on the central features of a model without being overwhelmed by numbers.

Most data can be naturally organized into tables. Sometimes the table consists of a single list, as in a list of scores of students on a test. Sometimes the table has the form of a rectangular array with several columns and rows, as in a teacher's record of the scores of students on all tests in a course; here we have one column for each test and one row for each student. The mathematical name for a rectangular array of numbers is a matrix. The most common type of matrix in mathematical applications is the array of the coefficients in some system of linear equations.

In this section we introduce basic matrix notation. Matrix notation takes most people a little time to learn. But in a short while the reader will find it impossible to talk about linear models without using matrix notation.

A **matrix** is a rectangular array of numbers. We speak of an *m*-by-*n* matrix when the matrix has *m* rows and *n* columns, and we use capital boldface letters, such as **A**, to denote matrices. (The common handwritten

way to indicate a matrix is with a wavy line under the letter, such as $\underset{\sim}{A}$.)
We use the notation a_{ij} to denote the number in matrix **A** occurring in row
i and column j. This is similar to the computer programming notation A(I,J).
Examples of matrices are

$$\mathbf{A} = \begin{bmatrix} 1 & 2 & 3 \\ 4 & 5 & 6 \end{bmatrix} \quad \text{and} \quad \mathbf{M} = \begin{bmatrix} 4 & 3 & 8 \\ 9 & 5 & 1 \\ 2 & 7 & 6 \end{bmatrix} \tag{1}$$

An ordered list of n numbers is called a **vector** or an n-vector. We use
lowercase boldface letters, such as **v**, to denote vectors; v_i is our name for
the ith entry in vector **v**. Examples of vectors are

$$\mathbf{v} = [1, 2, 3, 4] \quad \text{and} \quad \mathbf{c} = \begin{bmatrix} 7 \\ 8 \\ 9 \end{bmatrix}$$

Sometimes we write a vector as a row of numbers, sometimes as a column,
but a vector is formally just an ordered list.

An n-vector is just a 1-by-n matrix or an n-by-1 matrix. Conversely,
an m-by-n matrix **A** can be thought of as a set of m row vectors (each of
length n) or as a set of n column vectors (each of length m). We use the
following notation:

$\mathbf{a}_i^R$ denotes the ith row vector in **A**.

$\mathbf{a}_j^C$ denotes the jth column vector in **A**.

We omit the R (or C) superscript when it is clear from the discussion that
we are talking about rows (or columns).

For example, in the matrix **A** in (1),

$$a_{12} = 2, \qquad a_{22} = 5, \qquad \mathbf{a}_1^R = [1, 2, 3], \qquad \mathbf{a}_2^C = \begin{bmatrix} 2 \\ 5 \end{bmatrix}$$

Summarizing our matrix notation, we can write a general matrix **A** in
the following ways:

$$\mathbf{A} = \begin{bmatrix} a_{11} & a_{12} & a_{13} & \cdots & a_{1j} & \cdots & a_{1n} \\ a_{21} & a_{22} & a_{23} & \cdots & a_{2j} & \cdots & a_{2n} \\ a_{31} & a_{32} & a_{33} & \cdots & a_{3j} & \cdots & a_{3n} \\ \vdots & \vdots & \vdots & \cdots & \vdots & \cdots & \vdots \\ a_{i1} & a_{i2} & a_{i3} & \cdots & a_{ij} & \cdots & a_{in} \\ \vdots & \vdots & \vdots & \cdots & \vdots & \cdots & \vdots \\ a_{m1} & a_{m2} & a_{m3} & \cdots & a_{mj} & \cdots & a_{mn} \end{bmatrix} = \begin{bmatrix} \mathbf{a}_1^R \\ \mathbf{a}_2^R \\ \mathbf{a}_3^R \\ \vdots \\ \mathbf{a}_i^R \\ \vdots \\ \mathbf{a}_m^R \end{bmatrix}$$

or

$$= [\mathbf{a}_1^C \quad \mathbf{a}_2^C \quad \mathbf{a}_3^C \quad \cdots \quad \mathbf{a}_j^C \quad \cdots \quad \mathbf{a}_n^C]$$

The following examples will show how vectors and matrices arise naturally in linear models introduced in Chapter 1.

Example 1. **Matrix Notation for Oil Refinery Model**

In Section 1.2 we introduced a system of linear equations modeling the production of three products—heating oil, diesel oil, gasoline—by three refineries. The system of equations was

$$
\begin{array}{llrrrr}
\text{Heating oil:} & 20x_1 + & 4x_2 + & 4x_3 = & 500 & \\
\text{Diesel oil:} & 10x_1 + & 14x_2 + & 5x_3 = & 850 & \quad (2) \\
\text{Gasoline:} & 5x_1 + & 5x_2 + & 12x_3 = & 1000 &
\end{array}
$$

We can make a matrix **A** of the coefficients on the left sides in (2).

$$
\mathbf{A} = \begin{bmatrix} 20 & 4 & 4 \\ 10 & 14 & 5 \\ 5 & 5 & 12 \end{bmatrix} \tag{3}
$$

Each column of **A** is a vector of outputs by a refinery. For example, from 1 barrel of oil, refinery 2 produces an output vector

$$
\mathbf{a}_2^C = \begin{bmatrix} 4 \\ 14 \\ 5 \end{bmatrix}
$$

Each row on the left side is a vector of amounts produced of some product. The vector for gasoline is $\mathbf{a}_3^R = [5, 5, 12]$. The right-side numbers in (2) form a demand vector. ∎

Example 2. **Matrix Notation for Leontief Economic Model**

The Leontief model for economic equilibrium in Example 2 of Section 1.2 contained the following system of supply–demand equations:

Supply	Industrial Demand	Consumer Demand

Energy: $x_1 = .4x_1 + .2x_2 + .2x_3 + .2x_4 + 100$

Construct.: $x_2 = .3x_1 + .3x_2 + .2x_3 + .1x_4 + 50$

Transport.: $x_3 = .1x_1 + .1x_2 + + .2x_4 + 100$

Steel: $x_4 = + .1x_2 + .1x_3 + 0$

$$(4)$$

It is natural to form a matrix $\mathbf{D}$ of the coefficients of industrial demands on the right-hand side of (4). The set of consumer demands in the last column in (4) form a vector $\mathbf{c}$.

$$\mathbf{D} = \begin{bmatrix} .4 & .2 & .2 & .2 \\ .3 & .3 & .2 & .1 \\ .1 & .1 & 0 & .2 \\ 0 & .1 & .1 & 0 \end{bmatrix} \quad \text{and} \quad \mathbf{c} = \begin{bmatrix} 100 \\ 50 \\ 100 \\ 0 \end{bmatrix} \quad (5)$$

Recall that the second row $\mathbf{d}_2^R = [.3, .3, .2, .1]$ tells how much of product 2 (construction) is needed to produce 1 dollar's worth of the other products; for example, it takes $d_{23} = .2$ dollar of product 2 to make 1 dollar of product 3 (transportation). Similarly, the third column

$$\mathbf{d}_3^C = \begin{bmatrix} .2 \\ .2 \\ 0 \\ .1 \end{bmatrix}$$

tells the inputs required to make 1 dollar of product 3.   ■

Example 3. **Matrix Notation for a Markov Chain**

The transition probabilities of the Markov chain in Example 2 of Section 1.3 (about a frog wandering around in a highway) form a transition matrix $\mathbf{A}$:

$$\mathbf{A} = \begin{bmatrix} .50 & .25 & 0 & 0 & 0 & 0 \\ .50 & .50 & .25 & 0 & 0 & 0 \\ 0 & .25 & .50 & .25 & 0 & 0 \\ 0 & 0 & .25 & .50 & .25 & 0 \\ 0 & 0 & 0 & .25 & .50 & .50 \\ 0 & 0 & 0 & 0 & .25 & .50 \end{bmatrix} \quad (6)$$

The columns are associated with current states and the rows with states in the next period. Entry a_{ij} is the transition probability of going from state j to state i—the probability that if the frog is now in state j, then 1 minute later it will be in state i. The third column $\mathbf{a}_3^C$ gives the probability distribution among states next period if we are currently in state 3.

In general, we have a vector **p** of the current probability distribution $\mathbf{p} = [p_1, p_2, p_3, p_4, p_5, p_6]$; that is, p_i is the probability that the frog is currently in state i. From **p** and the transition probabilities in **A**, we obtained a system of equations that allowed us to compute the probability distribution **p**′ for the next period [see equations (4) of Section 1.3]. By repeating the process of computing the next-period probabilities k times, we could compute the probability distribution $\mathbf{p}^{(k)}$ after k periods, (see Table 1.1). ■

Example 4. A Vector as a Point in Space

A common use of vector notation is to represent points in space. In two-dimensional space, we use a 2-vector; in n-dimensional space, we use an n-vector. The point in three-dimensional x-y-z space with coordinates $x = 2$, $y = 7$, $z = 1$ is written as the vector [2, 7, 1]. Often a two- or three-dimensional vector is represented by an arrow going from the origin to the point with these coordinates, as shown in Figure 2.1.

The collection of all 3-vectors is all of three-dimensional space; all n-vectors are n-dimensional space. Much of the algebraic theory about collections of vectors is equivalent to the geometric theory of the corresponding spaces of points. In Section 5.2 we discuss some properties of the collection of vectors that satisfy a given system of linear equations.

It is often helpful to think of an n-vector as a point in n-space. For example, we can talk naturally about the distance between two vectors. We can use x-y coordinates to plot the behavior of a linear model involving two-dimensional vectors. ■

The thoughtful reader may rightly ask: "You have shown me one-dimensional arrays and two-dimensional arrays, so what about higher-dimensional arrays?" We see higher-dimensional arrays from time to time in computer programs. If programmers work with three-dimensional arrays, do not mathematicians? The answer surprisingly is basically "no" (although

Figure 2.1 Vector as an arrow.

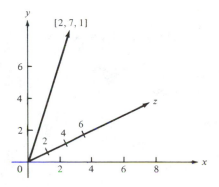

tensors are a higher-dimensional extension of matrices). Historically, matrices have been closely associated with systems of linear equations, as in Examples 1 and 2. The operations performed on matrices are defined with an eye on the associated systems of equations. The absence of a natural higher-dimensional version of a system of linear equations is the major reason why mathematicians have only been concerned with one- and two-dimensional arrays.

Matrix elements need not be numbers, as this next example illustrates.

Example 5. **Encoding Messages with Matrices**

In Example 1 of Section 1.5 we introduced some linear models for encoding messages by converting letters to numbers between 1 and 26. In this example we show how to scramble a message without transforming the letters. We place the message into a matrix and perform simple scrambling functions on this matrix.

Suppose that our message is

ALLIED SOLDIERS SHOULD REMAIN ON ALERT

The message has 33 letters. We use this list of the letters (ignoring spaces) to fill the entries in the first row, then the second row, and so on in a matrix **M**. We want **M** to be a square matrix. To accommodate 33 letters, we need a 6-by-6 matrix (with 36 entries). We add three E's (or any nonsense letters) at the end of the message to fill out the matrix.

$$\mathbf{M} = \begin{bmatrix} A & L & L & I & E & D \\ S & O & L & D & I & E \\ R & S & S & H & O & U \\ L & D & R & E & M & A \\ I & N & O & N & A & L \\ E & R & T & E & E & E \end{bmatrix} \tag{7}$$

Consider the following simple operations on a square matrix.

1. Interchange two rows (or two columns).
2. Interchange ith row and ith column.
3. Rearrange entries in a row (or column), such as reversing the order of entries or cyclicly permuting.

A suitable sequence of 10 operations, chosen from these three types, will produce an array of letters that will be impossible to unscramble without knowing what operations were performed. As an example, suppose that we interchange the first row and column; then interchange the new third row and column; and then interchange the new sixth row and column. The result is

$$\mathbf{M}^* = \begin{bmatrix} A & S & L & L & I & D \\ L & O & S & D & I & R \\ R & L & S & R & O & U \\ I & D & H & E & M & E \\ E & N & O & N & A & E \\ E & E & T & A & L & E \end{bmatrix} \tag{8}$$

Already, the message is unintelligible. ∎

Now we define some simple operations on matrices. The most basic operation is to multiply a vector or matrix by a constant c. This operation is called **scalar multiplication**. A **scalar** is a single number, as opposed to a vector or matrix. Scalar multiplication is performed by multiplying each entry in the vector or matrix by the constant c. For example,

$$\mathbf{M} = \begin{bmatrix} 2 & 4 & 5 & 1 \\ 3 & 9 & 2 & 5 \\ 1 & 6 & 6 & 2 \end{bmatrix}, \quad \text{then } 3\mathbf{M} = \begin{bmatrix} 6 & 12 & 15 & 3 \\ 9 & 27 & 6 & 15 \\ 3 & 18 & 18 & 6 \end{bmatrix}$$

Addition of vectors and matrices is straightforward—add the corresponding entries together. There is one minor problem, however. Two vectors being added together must have the same length, and two matrices being added must have the same number of rows and same number of columns. For example, if

$$\mathbf{A} = \begin{bmatrix} 1 & 5 \\ 2 & 3 \\ -7 & 0 \end{bmatrix} \quad \text{and} \quad \mathbf{B} = \begin{bmatrix} 3 & 1 \\ 0 & 4 \\ 1 & 2 \end{bmatrix}, \quad \text{then } \mathbf{A} + \mathbf{B} = \begin{bmatrix} 4 & 6 \\ 2 & 7 \\ -6 & 2 \end{bmatrix}$$

Example 6. Matrices of Test Scores

Suppose that we are recording the test scores of four students in three subjects. To preserve confidentiality, we will call the students A, B, C, and D, and the subjects 1, 2, and 3. The students have two hour exams and a final exam in each course, each graded out of 10 points. For each of the three tests we form a matrix of test scores with rows for students and columns for subjects. Call the matrices $\mathbf{S}_1$, $\mathbf{S}_2$, and $\mathbf{S}_3$ ($\mathbf{S}_3$ is the matrix of final exam scores).

$$\mathbf{S}_1 = \begin{array}{c} \\ A \\ B \\ C \\ D \end{array}\begin{array}{ccc} 1 & 2 & 3 \\ \end{array}\begin{bmatrix} 6 & 8 & 9 \\ 8 & 5 & 8 \\ 8 & 7 & 8 \\ 4 & 6 & 6 \end{bmatrix}, \quad \mathbf{S}_2 = \begin{bmatrix} 5 & 9 & 8 \\ 6 & 7 & 9 \\ 7 & 8 & 8 \\ 5 & 6 & 7 \end{bmatrix}, \quad \mathbf{S}_3 = \begin{bmatrix} 6 & 7 & 9 \\ 8 & 6 & 9 \\ 8 & 7 & 8 \\ 6 & 5 & 6 \end{bmatrix} \tag{9}$$

Then the matrix **T** of total scores of each student in each course (without any weighting to make the final more important) is

$$\mathbf{T} = \mathbf{S}_1 + \mathbf{S}_2 + \mathbf{S}_3$$

Summing the corresponding entries in $\mathbf{S}_1$, $\mathbf{S}_2$, and $\mathbf{S}_3$, we obtain the matrix **T**:

$$\mathbf{T} = \begin{bmatrix} 6 & 8 & 9 \\ 8 & 5 & 8 \\ 8 & 7 & 8 \\ 4 & 6 & 6 \end{bmatrix} + \begin{bmatrix} 5 & 9 & 8 \\ 6 & 7 & 9 \\ 7 & 8 & 8 \\ 5 & 6 & 7 \end{bmatrix} + \begin{bmatrix} 6 & 7 & 9 \\ 8 & 6 & 9 \\ 8 & 7 & 8 \\ 6 & 5 & 6 \end{bmatrix} \tag{10}$$

$$= \begin{bmatrix} 17 & 24 & 26 \\ 22 & 18 & 26 \\ 23 & 22 & 24 \\ 15 & 17 & 19 \end{bmatrix}$$

Suppose that the final should be weighted twice as much as each hour test. Each test had a total of 10 points, and we want the course score also to be out of 10 points. That is, the course score is a weighted average of the tests. Then the matrix **C** of weighted averages of course scores has the form

$$\mathbf{C} = \tfrac{1}{4}\mathbf{S}_1 + \tfrac{1}{4}\mathbf{S}_2 + \tfrac{1}{2}\mathbf{S}_3 \tag{11}$$

We compute **C** by computing the linear combination in (11) for each entry. For example, the entry c_{12}, student A's weighted average in course 2, is

$$c_{12} = \tfrac{1}{4}8 + \tfrac{1}{4}9 + \tfrac{1}{2}7 = 7\tfrac{3}{4}$$

A computer program to compute all the c_{ij} entries would look as follows:

```
    FOR I = 1 TO 4
        FOR J = 1 TO 3
            C(I, J) = .25*S1(I, J) + .25*S2(I, J) + .5*S3(I, J)
        NEXT J
    NEXT I
```

Using this program, we obtain **C** (fractions $\geq$ .5 have been rounded up; that is, 3.6 is written as 4):

$$C = \begin{array}{c} \\ A \\ B \\ C \\ D \end{array} \begin{array}{ccc} 1 & 2 & 3 \\ \begin{bmatrix} 6 & 8 & 9 \\ 8 & 6 & 9 \\ 8 & 7 & 8 \\ 5 & 6 & 6 \end{bmatrix} \end{array} \tag{12}$$

■

Example 6 implicitly shows why we rarely have to check whether a set of matrices that we want to add together has the same numbers of rows and columns. We would not want to add the matrices together unless the entries in the matrices matched up in some natural way.

Section 2.1 Exercises

Summary of Exercises
The exercises in this section are straightforward variants on the examples in this section.

1. Given the matrix

$$A = \begin{bmatrix} 1 & 2 & 3 & 4 \\ 2 & 4 & 6 & 8 \\ 3 & 5 & 7 & 9 \end{bmatrix}$$

   write out the following row and column vectors, and entries.
   (a) $\mathbf{a}_1^R$ (b) $\mathbf{a}_2^C$ (c) $\mathbf{a}_3^C$ (d) a_{22} (e) a_{31}

2. In the matrix of letters

$$A = \begin{bmatrix} E & R & S & T & A \\ N & P & O & C & W \\ H & B & U & I & L \\ M & G & Y & F & K \end{bmatrix}$$

   spell out the words given by the following sequence of entries.
   (a) $a_{31}, a_{11}, a_{35}, a_{22}$ (b) $a_{35}, a_{34}, a_{21}, a_{11}, a_{15}, a_{12}$
   (c) $a_{11}, a_{35}, a_{32}, a_{23}, a_{25}$ (d) $a_{25}, a_{15}, a_{14}, a_{24}, a_{31}, a_{23}, a_{33}, a_{14}$

3. Consider the following Markov chain model involving the states of mind of Professor Mindthumper. The states are Alert (A), Hazy (H), and Stupor (S). If in state A or H today, then tomorrow the professor has a $\frac{1}{3}$ chance of being in each of the three states. If in state S today, tomorrow with probability 1 the professor is still in state S.
   (a) Write the transition matrix $\mathbf{A}$ for this Markov chain.
   (b) Write out entry a_{23} and column $\mathbf{a}_2^C$.
   (c) Which pairs of rows and pairs of columns in this Markov chain are the same?

4. In the transition matrix **A** for the frog Markov chain in Example 3, what does entry a_{34} represent?

5. (a) In the matrix **A** for the refinery model in Example 1, state in words what the numbers in entries a_{23} and entries a_{31} represent. What do the numbers in the third column of **A** represent?
   (b) Suppose that refinery 3 is modernized and its output for each barrel of oil is doubled. What is the new matrix of coefficients?
   (c) In Example 1 of Section 1.4 we discussed the situation where refinery 3 broke down and was out of service. In this case, what is the matrix of coefficients?

6. Make a matrix for data of Scrooge high school GPAs and college GPAs for the set of students in Example 2 of Section 1.4.

7. Write out the matrix of coefficients in the inequality constraints of the linear program in Example 4 of Section 1.4.

8. Plot the following vectors as points in the *x-y* plane.
   (a) [1, 0]   (b) [2, 4]   (c) [2, −1]

9. Plot the following points on an *x-y-z* grid of the sort given in Example 4.
   (a) [1, 0, 0]   (b) [1, 1, 1]   (c) [2, 4, 1]   (d) [2, −1, 3]

10. Scramble the matrix **M** in Example 5 by performing the following sequences of changes.
    (a) Interchange row 2 and column 2; interchange row 3 and 5; interchange columns 1 and 4.
    (b) Reverse the order of the letters in row 2; do the same in column 3; interchange row 1 and column 1; do the same for row 4 and column 4.
    (c) Reverse the order of each row; then reverse the order of each column.

11. In Example 5, why is it unclear how one should define the process of interchanging row i and with column j (for $i \neq j$)?

    Hint: What will entry (1, 2) be when we try interchanging row 1 and column 2?

12. Let

$$\mathbf{A} = \begin{bmatrix} 1 & 2 & 3 & 4 \\ 2 & 4 & 6 & 8 \\ 3 & 5 & 7 & 9 \end{bmatrix} \quad \text{and} \quad \mathbf{B} = \begin{bmatrix} -1 & 0 & 2 & 1 \\ 2 & -1 & -1 & 0 \\ 2 & 0 & 0 & 2 \end{bmatrix}$$

   Determine

(a) 3**A**   (b) 2**B**   (c) − 3**B**   (d) **A** + **B**   (e) 2**A** + 3**B**
(f) 3**A** − 2**B**

13. Let all matrices in this exercise be 4-by-4. Let **I** denote the matrix with 1's in the main diagonal and 0's elsewhere. Let **J** denote the matrix with each entry equal to 1, and let **A** be the matrix

$$\mathbf{A} = \begin{bmatrix} 1 & 0 & 1 & 0 \\ 0 & 1 & 0 & 1 \\ 1 & 0 & 1 & 0 \\ 0 & 1 & 0 & 1 \end{bmatrix}$$

Express the following matrices as linear combinations of **I**, **J**, and **A**.

(a) $\begin{bmatrix} 6 & 2 & 2 & 2 \\ 2 & 6 & 2 & 2 \\ 2 & 2 & 6 & 2 \\ 2 & 2 & 2 & 6 \end{bmatrix}$ (b) $\begin{bmatrix} 0 & 1 & 0 & 1 \\ 1 & 0 & 1 & 0 \\ 0 & 1 & 0 & 1 \\ 1 & 0 & 1 & 0 \end{bmatrix}$ (c) $\begin{bmatrix} 5 & 3 & 1 & 3 \\ 3 & 5 & 3 & 1 \\ 1 & 3 & 5 & 3 \\ 3 & 1 & 3 & 5 \end{bmatrix}$

14. Show that any vector $\mathbf{x} = [x_1, x_2]$ that is a multiple of [2, 1] (i.e., $\mathbf{x} = c[2, 1]$ for some c) satisfies the system of equations

$$x_1 - 2x_2 = 0$$
$$-2x_1 + 4x_2 = 0$$

15. Suppose in Example 6 that the final exam counted three times as much as an hour exam, so that the weights on the three tests should be $\frac{1}{5}$, $\frac{1}{5}$, $\frac{3}{5}$, respectively. Recompute the course score matrix **C** with these weights.

16. Write a computer program to add two matrices **A** and **B**, where both are m-by-n. Assume m, n given and that the entries of the matrices are stored in arrays A(I,J) and B(I,J).

17. Write a computer program to read in scalars r and s and then compute the linear combination $r\mathbf{A} + s\mathbf{B}$ of the m-by-n matrices **A** and **B**. Assume m, n given and that the entries of the matrices are stored in arrays A(I,J) and B(I,J).

Section 2.2 Matrix Multiplication

In Section 2.1 we introduced vectors and matrices and showed how to add them and multiply them by a scalar. These two operations were obvious and straightforward, and accordingly they are not powerful tools. In this section we discuss multiplication of vectors and matrices. This operation is more complicated but also more useful. It provides a simple notation for expressing systems of linear equations and associated calculations.

Consider the following typical situation, which requires vector multiplication. We have a vector **p** of prices for a set of three vegetables. Suppose that $p = [.80, 1.00, .50]$, where the ith entry is the price of the ith vegetable. We are also given a vector **d** of the weekly demand in a household for these three vegetables. Suppose that $\mathbf{d} = [5, 3, 4]$. We shall define vector-times-vector multiplication so that $\mathbf{p} \cdot \mathbf{d}$ equals the cost of the household's weekly demand for these three vegetables. In this case,

$$\begin{aligned} \mathbf{p} \cdot \mathbf{d} &= [.80, 1.00, .50] \cdot [5, 3, 4] \\ &= .80 \times 5 + 1.00 \times 3 + .50 \times 4 \\ &= 4.00 + 3.00 + 2.00 = 9.00 \end{aligned}$$

Vector Multiplication

Let **a** and **b** be two n-vectors, where $\mathbf{a} = [a_1, a_2, \ldots, a_n]$ and $\mathbf{b} = [b_1, b_2, \ldots, b_n]$. Then the product $\mathbf{a} \cdot \mathbf{b}$, called the **scalar product** of **a** and **b**, is a single number (a scalar) equal to the sum of the products $a_i b_i$. That is,

$$\mathbf{a} \cdot \mathbf{b} = \sum_{i=1}^{n} a_i b_i$$

Vector multiplication $\mathbf{a} \cdot \mathbf{b}$ *makes sense only when* **a** *and* **b** *have the same length.*

The scalar product is also sometimes called the *inner product* or *dot product* (the latter term coming from the dot used in writing the product). An important geometric interpretation of scalar products is discussed in Chapter 5.

Example 1. **Calculating Time to Process Computer Jobs**

A Superduper computer requires 3 minutes to do a type 1 job (say, a statistics problem), 4 minutes to do a type 2 job, and 2 minutes to do a type 3 job. The computer has 6 type 1 jobs, 8 type 2 jobs, and 10 type 3 jobs. How long will the computer take to perform all these jobs?

If $\mathbf{t} = [3, 4, 2]$ is the vector of the times to do the various jobs and $\mathbf{n} = [6, 8, 10]$ is the vector of the numbers of each type of job, the total time required will be the value of the scalar product $\mathbf{t} \cdot \mathbf{n}$.

$$\begin{aligned} \text{Total time} = \mathbf{t} \cdot \mathbf{n} &= [3, 4, 2] \cdot [6, 8, 10] \\ &= 3 \times 6 + 4 \times 8 + 2 \times 10 \\ &= 18 + 32 + 20 = 70 \qquad \blacksquare \end{aligned}$$

The key idea about a scalar product is: It is a linear combination of the entries in each vector. *Any linear combination of variables or numbers can be expressed as a scalar product*. Consider the linear equation

$$20x_1 + 4x_2 + 4x_3 = 500$$

The left side is a linear combination of variables. If $\mathbf{a} = [20, 4, 4]$ and $\mathbf{x} = [x_1, x_2, x_3]$, the left side can be written as a scalar product

$$20x_1 + 4x_2 + 4x_3 = [20, 4, 4] \cdot [x_1, x_2, x_3] = \mathbf{a} \cdot \mathbf{x}$$

Similarly, any linear equation or system of linear equations can be written in terms of scalar products.

Example 2. **Representing the Refinery System of Equations**

Recall the system of equations for the refinery production problem in Section 1.2.

$$
\begin{aligned}
20x_1 + 4x_2 + 4x_3 &= 500 \\
10x_1 + 14x_2 + 5x_3 &= 850 \\
5x_1 + 5x_2 + 12x_3 &= 1000
\end{aligned}
\tag{1a}
$$

Or making a vector of the quantities on each side of these equations,

$$
\begin{bmatrix}
20x_1 + 4x_2 + 4x_3 \\
10x_1 + 14x_2 + 5x_3 \\
5x_1 + 5x_2 + 12x_3
\end{bmatrix}
=
\begin{bmatrix}
500 \\
850 \\
1000
\end{bmatrix}
\tag{1b}
$$

Let $\mathbf{A}$ be the 3-by-3 matrix of the coefficients on the left side of the equations in (1a) with row vectors $\mathbf{a}_1^R$, $\mathbf{a}_2^R$, $\mathbf{a}_3^R$. Let $\mathbf{b}$ be the right-side vector, and let $\mathbf{x}$ be the vector of unknowns.

$$
\mathbf{A} =
\begin{matrix} \mathbf{a}_1^R \\ \mathbf{a}_2^R \\ \mathbf{a}_3^R \end{matrix}
\begin{bmatrix}
20 & 4 & 4 \\
10 & 14 & 5 \\
5 & 5 & 12
\end{bmatrix},
\qquad
\mathbf{b} =
\begin{bmatrix}
500 \\
850 \\
1000
\end{bmatrix},
\qquad
\mathbf{x} =
\begin{bmatrix}
x_1 \\
x_2 \\
x_3
\end{bmatrix}
\tag{2}
$$

The left sides of the equations in (1b) are a vector of scalar products of $\mathbf{x}$ with the rows of $\mathbf{A}$:

$$
\begin{bmatrix}
20x_1 + 4x_2 + 4x_3 \\
10x_1 + 14x_2 + 5x_3 \\
5x_1 + 5x_2 + 12x_3
\end{bmatrix}
=
\begin{bmatrix}
[20, 4, 4] \cdot \mathbf{x} \\
[10, 14, 5] \cdot \mathbf{x} \\
[5, 5, 12] \cdot \mathbf{x}
\end{bmatrix}
$$

$$
=
\begin{bmatrix}
\mathbf{a}_1^R \cdot \mathbf{x} \\
\mathbf{a}_2^R \cdot \mathbf{x} \\
\mathbf{a}_3^R \cdot \mathbf{x}
\end{bmatrix}
= \mathbf{A}\mathbf{x}
\tag{3}
$$

As noted in (3), we call the result of multiplying each row of **A** times a vector **x** the matrix-vector product **Ax**. Thus in matrix notation, (1) is written simply **Ax** = **b**. ∎

By treating a matrix as a set of row vectors, we can extend our definition of vector-times-vector multiplication to matrix-times-vector multiplication.

Matrix-Vector Multiplication

Let **A** be an m-by-n matrix and **b** be an n-vector. Let $\mathbf{a}_i^R$ be the ith row of **A**. Then the matrix-vector product **Ab** (the multiplication sign is normally omitted) is defined to be the column vector of scalar products $\mathbf{a}_i^R \cdot \mathbf{b}$:

$$\mathbf{Ab} = \begin{bmatrix} \mathbf{a}_1^R \\ \mathbf{a}_2^R \\ \vdots \\ \mathbf{a}_m^R \end{bmatrix} \mathbf{b} = \begin{bmatrix} \mathbf{a}_1^R \cdot \mathbf{b} \\ \mathbf{a}_2^R \cdot \mathbf{b} \\ \vdots \\ \mathbf{a}_m^R \cdot \mathbf{b} \end{bmatrix} \tag{4}$$

For example, if

$$\mathbf{A} = \begin{bmatrix} -1 & 0 & 2 \\ 2 & 1 & 1 \\ 3 & 3 & 3 \end{bmatrix} \quad \text{and} \quad \mathbf{b} = \begin{bmatrix} 1 \\ 2 \\ 1 \end{bmatrix}$$

then

$$\mathbf{Ab} = \begin{bmatrix} -1 & 0 & 2 \\ 2 & 1 & 1 \\ 3 & 3 & 3 \end{bmatrix}\begin{bmatrix} 1 \\ 2 \\ 1 \end{bmatrix} = \begin{bmatrix} -1\times1 + 0\times2 + 2\times1 \\ 2\times1 + 1\times2 + 1\times1 \\ 3\times1 + 3\times2 + 3\times1 \end{bmatrix} = \begin{bmatrix} 1 \\ 5 \\ 12 \end{bmatrix}$$

What if we want to multiply a vector **b** times the columns of **A**? The convention is that *when a vector* **b** *multiplies the rows of* **A**, **b** *is written to the right of* **A** *in the product, as in (4). When* **b** *multiplies the columns of* **A**, *then* **b** *is written to the left of* **A** *as in (5).* The reason for this convention will become evident shortly.

$$\mathbf{bA} = \mathbf{b} \cdot [\mathbf{a}_1^C, \mathbf{a}_2^C, \ldots, \mathbf{a}_n^C] \tag{5}$$
$$= [\mathbf{b} \cdot \mathbf{a}_1^C, \mathbf{b} \cdot \mathbf{a}_2^C, \ldots, \mathbf{b} \cdot \mathbf{a}_n^C]$$

For example, if we reverse the order of **A** and **b** in the previous computation of **Ab**, we have

$$\mathbf{bA} = [1 \quad 2 \quad 1]\begin{bmatrix} -1 & 0 & 2 \\ 2 & 1 & 1 \\ 3 & 3 & 3 \end{bmatrix}$$

$$= [1 \times (-1) + 2 \times 2 + 1 \times 3, \quad 1 \times 0 + 2 \times 1 + 1 \times 3,$$
$$1 \times 2 + 2 \times 1 + 1 \times 3]$$
$$= [6, 5, 7]$$

Remember that the length of **b** *must equal the length of the rows of* **A** *in the product* **Ab**. *Similarly, the length of* **b** *must equal the length of the columns of* **A** *in the product* **bA**.

Example 3. Comparing Computations by Different Computers

In Example 1 we computed how long it would take a Superduper computer to complete a set of jobs. There were three types of jobs and a vector $t = [3, 4, 2]$ of times for Superduper to do each type of job. Suppose that we also have three other brands of computers, Wacko, Whooper, and Ultima, and for each there is a similar vector of times to do the jobs. Let us put all these vectors into a matrix **A**:

$$\mathbf{A} = \begin{matrix} & \text{Type of Job} \\ & \begin{matrix} 1 & 2 & 3 \end{matrix} \\ \begin{matrix} \text{Superduper} \\ \text{Wacko} \\ \text{Whooper} \\ \text{Ultima} \end{matrix} & \begin{bmatrix} 3 & 4 & 2 \\ 5 & 7 & 3 \\ 1 & 2 & 1 \\ 3 & 3 & 3 \end{bmatrix} \end{matrix} \quad \text{Matrix of times}$$

In Example 1 we computed how long it would take a Superduper computer to do 6 type 1, 8 type 2, and 10 type 3 jobs by forming the scalar product of the Superduper time vector $[3, 4, 2]$ and with the number-of-jobs vector $n = [6, 8, 10]$. Now let us find out how long it would take each of the computers to do this set of jobs by multiplying each row of **A** times **n**, that is, by computing **An**.

$$\mathbf{An} = \begin{bmatrix} 3 & 4 & 2 \\ 5 & 7 & 3 \\ 1 & 2 & 1 \\ 3 & 3 & 3 \end{bmatrix} \begin{bmatrix} 6 \\ 8 \\ 10 \end{bmatrix} = \begin{bmatrix} 3 \times 6 + 4 \times 8 + 2 \times 10 \\ 5 \times 6 + 7 \times 8 + 3 \times 10 \\ 1 \times 6 + 2 \times 8 + 1 \times 10 \\ 3 \times 6 + 3 \times 8 + 3 \times 10 \end{bmatrix} = \begin{bmatrix} 70 \\ 116 \\ 32 \\ 72 \end{bmatrix}$$

The final column tells us that the set of jobs takes 70 minutes for Superduper, 116 minutes for Wacko, 32 minutes for Whooper, and 72 minutes for Ultima. ∎

Example 2 (continued). Representing the Refinery Systems of Equations

Let us quickly review our matrix notation for the refinery system of equations.

$$\begin{bmatrix} 20x_1 + 4x_2 + 4x_3 \\ 10x_1 + 14x_2 + 5x_3 \\ 5x_1 + 5x_2 + 12x_3 \end{bmatrix} = \begin{bmatrix} 500 \\ 850 \\ 1000 \end{bmatrix}$$

Let **A** be the 3-by-3 matrix of the coefficients on the left side of the equations in (1a), **b** be the right-side demand vector, and **x** be the vector of unknowns:

$$\mathbf{A} = \begin{bmatrix} 20 & 4 & 4 \\ 10 & 14 & 5 \\ 5 & 5 & 12 \end{bmatrix}, \qquad \mathbf{b} = \begin{bmatrix} 500 \\ 850 \\ 1000 \end{bmatrix}, \qquad \mathbf{x} = \begin{bmatrix} x_1 \\ x_2 \\ x_3 \end{bmatrix}$$

The left sides of the equations are a vector of scalar products of **x** with the rows of **A**. This vector is simply **Ax**:

$$\begin{bmatrix} 20x_1 + 4x_2 + 4x_3 \\ 10x_1 + 14x_2 + 5x_3 \\ 5x_1 + 5x_2 + 12x_3 \end{bmatrix} = \begin{bmatrix} 20 & 4 & 4 \\ 10 & 14 & 5 \\ 5 & 5 & 10 \end{bmatrix} \begin{bmatrix} x_1 \\ x_2 \\ x_3 \end{bmatrix} = \mathbf{Ax} \qquad (6)$$

Note that we can write (6) as a weighted sum of vectors:

$$x_1 \begin{bmatrix} 20 \\ 10 \\ 5 \end{bmatrix} + x_2 \begin{bmatrix} 4 \\ 14 \\ 5 \end{bmatrix} + x_3 \begin{bmatrix} 4 \\ 5 \\ 12 \end{bmatrix} = \begin{bmatrix} 500 \\ 850 \\ 1000 \end{bmatrix} \qquad (7)$$

Or in vector notation, (7) becomes

$$x_1 \mathbf{a}_1^C + x_2 \mathbf{a}_2^C + x_3 \mathbf{a}_3^C = \mathbf{b} \qquad \blacksquare$$

Observe how (7) views **Ax** = **b** in terms of columns, while (6) views **Ax** in terms of rows.

We shall use the notation **Ax** = **b** for a system of equations over and over again herein. For another example we consider the system of equations in the Leontief economic model.

Example 4. **Matrix Representation of Leontief Economic Model**

The supply–demand equations of the Leontief economic model in Section 1.2 can be written

$$\begin{bmatrix} x_1 \\ x_2 \\ x_3 \\ x_4 \end{bmatrix} = \begin{bmatrix} .4x_1 + .2x_2 + .2x_3 + .2x_4 \\ .3x_1 + .3x_2 + .2x_3 + .1x_4 \\ .1x_1 + .1x_2 + \quad\quad + .2x_4 \\ \quad\quad + .1x_2 + .1x_3 \end{bmatrix} + \begin{bmatrix} 100 \\ 50 \\ 100 \\ 0 \end{bmatrix} \qquad (8)$$

Let $\mathbf{x} = [x_1, x_2, x_3, x_4]$ and let $\mathbf{D}$ be the matrix of coefficients on the right-hand sides of (8), and let $\mathbf{c}$ be the rightmost column vector of consumer demands:

$$\begin{bmatrix} x_1 \\ x_2 \\ x_3 \\ x_4 \end{bmatrix} = \begin{array}{c} = \\ = \\ = \\ = \end{array} \begin{bmatrix} .4 & .2 & .2 & .2 \\ .3 & .3 & .2 & .1 \\ .1 & .1 & & .2 \\ & .1 & .1 & \end{bmatrix} \begin{bmatrix} x_1 \\ x_2 \\ x_3 \\ x_4 \end{bmatrix} + \begin{bmatrix} 100 \\ 50 \\ 100 \\ 0 \end{bmatrix} \tag{9a}$$

or

$$\mathbf{x} = \mathbf{Ax} + \mathbf{c} \tag{9b}$$

The system of equations in (8) can also be written as a linear combination of columns:

$$\begin{bmatrix} x_1 \\ x_2 \\ x_3 \\ x_4 \end{bmatrix} = x_1 \begin{bmatrix} .4 \\ .3 \\ .1 \\ 0 \end{bmatrix} + x_2 \begin{bmatrix} .2 \\ .3 \\ .1 \\ .1 \end{bmatrix} + x_3 \begin{bmatrix} .2 \\ .2 \\ 0 \\ .1 \end{bmatrix} + x_4 \begin{bmatrix} .2 \\ .1 \\ .2 \\ 0 \end{bmatrix} + \begin{bmatrix} 100 \\ 50 \\ 100 \\ 0 \end{bmatrix} \tag{10}$$ ∎

Example 5. **Matrix Notation for a Linear Program**

In Example 4 of Section 1.4 we presented the following linear program for maximizing revenue from planting two crops, corn (C) and wheat (W).

$$\begin{aligned} \text{Maximize } & 60C + 40W \\ \text{subject to } & C \geq 0, W \geq 0 \text{ and} \\ \text{Land:} \quad & C + W \leq 200 \\ \text{Labor:} \quad & 4C + W \leq 320 \\ \text{Capital:} \quad & 20C + 10W \leq 2200 \end{aligned} \tag{11}$$

If we let

$$\mathbf{A} = \begin{bmatrix} 1 & 1 \\ 4 & 1 \\ 20 & 10 \end{bmatrix}, \qquad \mathbf{b} = \begin{bmatrix} 200 \\ 320 \\ 2200 \end{bmatrix}$$

and $\mathbf{c} = [60, 40]$, $\mathbf{x} = [C, W]$, the inequality constraints in (11) can be written

$$\begin{bmatrix} 1 & 1 \\ 4 & 1 \\ 20 & 10 \end{bmatrix} \begin{bmatrix} C \\ W \end{bmatrix} \leq \begin{bmatrix} 200 \\ 320 \\ 2200 \end{bmatrix} \qquad \text{or } \mathbf{Ax} \leq \mathbf{b}$$

In matrix notation, (11) can be written

$$\text{Maximize } \mathbf{c} \cdot \mathbf{x}$$
$$\text{subject to } \mathbf{x} \geq \mathbf{0} \text{ and } \mathbf{A}\mathbf{x} \leq \mathbf{b} \tag{12}$$

where $\mathbf{0}$ is a vector of all zeros.   ∎

Example 6. **Matrix Notation for Markov Chains**

In Section 1.3 we introduced the concept of a Markov chain and in Example 2 of Section 1.3 gave the following Markov chain for the random movements of a frog across a highway; the possible locations for the frog were represented as states 1 through 6. The matrix of transition probabilities $\mathbf{A}$ was

$$\mathbf{A} = \begin{bmatrix} .50 & .25 & 0 & 0 & 0 & 0 \\ .50 & .50 & .25 & 0 & 0 & 0 \\ 0 & .25 & .50 & .25 & 0 & 0 \\ 0 & 0 & .25 & .50 & .25 & 0 \\ 0 & 0 & 0 & .25 & .50 & .50 \\ 0 & 0 & 0 & 0 & .25 & .50 \end{bmatrix}$$

We let $\mathbf{p} = [p_1, p_2, \ldots, p_6]$ be the current probability distribution vector (p_i is the probability the frog is currently in the ith state) and $\mathbf{p}'$ be the vector of the probability distribution in the next minute. We developed the following system of linear equations to determine $\mathbf{p}'$ from $\mathbf{A}$ and $\mathbf{p}$.

$$\begin{aligned}
p_1' &= .50p_1 + .25p_2 \\
p_2' &= .50p_1 + .50p_2 + .25p_3 \\
p_3' &= \qquad\quad .25p_2 + .50p_3 + .25p_4 \\
p_4' &= \qquad\qquad\qquad .25p_3 + .50p_4 + .25p_5 \\
p_5' &= \qquad\qquad\qquad\qquad .25p_4 + .50p_5 + .50p_6 \\
p_6' &= \qquad\qquad\qquad\qquad\qquad\quad .25p_5 + .50p_6
\end{aligned} \tag{13}$$

In matrix form, (13) becomes

$$\begin{bmatrix} p_1' \\ p_2' \\ p_3' \\ p_4' \\ p_5' \\ p_6' \end{bmatrix} = \begin{bmatrix} .50 & .25 & 0 & 0 & 0 & 0 \\ .50 & .50 & .25 & 0 & 0 & 0 \\ 0 & .25 & .50 & .25 & 0 & 0 \\ 0 & 0 & .25 & .50 & .25 & 0 \\ 0 & 0 & 0 & .25 & .50 & .50 \\ 0 & 0 & 0 & 0 & .25 & .50 \end{bmatrix} \begin{bmatrix} p_1 \\ p_2 \\ p_3 \\ p_4 \\ p_5 \\ p_6 \end{bmatrix}$$

or

$$\mathbf{p}' = \mathbf{Ap} \tag{14}$$

Note that each individual equation in (13) can be written as

$$p'_i = \mathbf{a}_i^R \cdot \mathbf{p} \tag{14a}$$

Let us recall what (14) represents for a general Markov transition matrix $\mathbf{A}$ and initial distribution $\mathbf{p}$. For simplicity, let $\mathbf{A}$ be 2-by-2.

$$\mathbf{A} = \begin{bmatrix} a_{11} & a_{12} \\ a_{21} & a_{22} \end{bmatrix}, \qquad \mathbf{p} = \begin{bmatrix} p_1 \\ p_2 \end{bmatrix}$$

Then $\mathbf{p}' = \mathbf{Ap}$ becomes

$$\begin{bmatrix} p'_1 \\ p'_2 \end{bmatrix} = \begin{bmatrix} a_{11} & a_{12} \\ a_{21} & a_{22} \end{bmatrix} \begin{bmatrix} p_1 \\ p_2 \end{bmatrix} = \begin{bmatrix} a_{11}p_1 + a_{12}p_2 \\ a_{21}p_1 + a_{22}p_2 \end{bmatrix}$$

Now that we have a concise way of writing Markov chain calculations, we can easily write equations to express the probability distribution vector $\mathbf{p}''$ for the frog 2 minutes from now.

$$\begin{aligned} \mathbf{p}'' = \mathbf{Ap}' &= \mathbf{A}(\mathbf{Ap}) \\ &= \mathbf{A}^2\mathbf{p} \end{aligned} \tag{15}$$

In (15) we have rewritten $\mathbf{AA}$ as $\mathbf{A}^2$, just as one would with single variable. However, we have yet to define what the product of two matrices is. The first line of (15) says that to get $\mathbf{p}''$ we must multiply $\mathbf{p}$ by $\mathbf{A}$ twice. It should be possible to "multiply" $\mathbf{A}$ times $\mathbf{A}$ and then multiply the resulting $\mathbf{A}^2$ times $\mathbf{p}$ to obtain $\mathbf{p}''$. We shall show how to do this matrix multiplication shortly.

With this notation, we can write the probability distribution vector $\mathbf{p}^{(3)}$ for the frog 3 minutes from now as

$$\mathbf{p}^{(3)} = \mathbf{Ap}'' = \mathbf{A}(\mathbf{A}^2\mathbf{p}) = \mathbf{A}^3\mathbf{p} \tag{16}$$

Generalizing this formula, we find that the probability distribution $\mathbf{p}^{(n)}$ for the frog in n minutes is given by

$$\mathbf{p}^{(n)} = \mathbf{A}^n\mathbf{p} \tag{17}$$

Note how concisely we can write the complex calculations for $\mathbf{p}''$, $\mathbf{p}^{(3)}$, and $\mathbf{p}^{(n)}$ using matrix notation. It would be impossible to analyze properties of Markov chains without such notation. ∎

In Example 6 we wrote $\mathbf{A}^2$ and other powers of matrix $\mathbf{A}$ without ever defining what matrix multiplication was. To the question as to how matrix multiplication is defined, our reply is that it should be defined to make matrix multiplication a useful operation. In this instance it should be defined to make the formulas in Example 6 valid.

The next example further motivates matrix multiplication and shows us how to do the computation.

Example 7. A Collection of Computer Computation Times

In Example 2 we computed the time it would take four different computers, Superduper, Wacko, Whooper, and Ultima, to do a set of jobs. We were given a matrix **A** telling how many minutes it took each computer to do each of three types of job.

$$
\begin{array}{c}
 & \text{Type of Job} \\
 & \begin{array}{ccc} 1 & 2 & 3 \end{array} \\
\mathbf{A} = \begin{array}{l} \text{Superduper} \\ \text{Wacko} \\ \text{Whooper} \\ \text{Ultima} \end{array} & \begin{bmatrix} 3 & 4 & 2 \\ 5 & 7 & 3 \\ 1 & 2 & 1 \\ 3 & 3 & 3 \end{bmatrix}
\end{array} \qquad \text{Matrix of times} \qquad (18)
$$

We calculated how long it would take each computer to do 6 type 1, 8 type 2, and 10 type 3 jobs by multiplying **A** times the vector **n** = [6, 8, 10]:

$$
\mathbf{An} = \begin{bmatrix} 3 & 4 & 2 \\ 5 & 7 & 3 \\ 1 & 2 & 1 \\ 3 & 3 & 3 \end{bmatrix} \begin{bmatrix} 6 \\ 8 \\ 10 \end{bmatrix} = \begin{bmatrix} 3\times 6 + 4\times 8 + 2\times 10 \\ 5\times 6 + 7\times 8 + 3\times 10 \\ 1\times 6 + 2\times 8 + 1\times 10 \\ 3\times 6 + 3\times 8 + 3\times 10 \end{bmatrix} = \begin{bmatrix} 70 \\ 116 \\ 32 \\ 72 \end{bmatrix}
$$
(19)

Now let us do this calculation not for one set of jobs, but for three sets of jobs. Set *A* will be the previous set **n** = [6, 8, 10]. Sets *B* and *C* will be [2, 5, 5] and [4, 4, 4]. Let us calculate the times required to do each set on each computer by expanding the vector **n** in (19) into a matrix **N** of three column vectors.

$$
\begin{array}{c}
 & \text{Sets of Jobs} \\
 & \begin{array}{ccc} A & B & C \end{array} \\
\mathbf{N} = \begin{array}{l} \text{Type 1} \\ \text{Type 2} \\ \text{Type 3} \end{array} & \begin{bmatrix} 6 & 2 & 4 \\ 8 & 5 & 4 \\ 10 & 5 & 4 \end{bmatrix}
\end{array} \qquad \text{Matrix of jobs}
$$

The calculation of **An** in (19) required us to multiply each row of **A** times the vector **n**. Now we need to multiply each row of **A** (one for each computer) times each column of **N** (one for each set of jobs):

$$\mathbf{AN} = \begin{bmatrix} 3 & 4 & 2 \\ 5 & 7 & 3 \\ 1 & 2 & 1 \\ 3 & 3 & 3 \end{bmatrix} \begin{bmatrix} 6 & 2 & 4 \\ 8 & 5 & 4 \\ 10 & 5 & 4 \end{bmatrix}$$

$$= \begin{bmatrix} 3\times6 + 4\times8 + 2\times10 & 3\times2 + 4\times5 + 2\times5 & 3\times4 + 4\times4 + 2\times4 \\ 5\times6 + 7\times8 + 3\times10 & 5\times2 + 7\times5 + 3\times5 & 5\times4 + 7\times4 + 3\times4 \\ 1\times6 + 2\times8 + 1\times10 & 1\times2 + 2\times5 + 1\times5 & 1\times4 + 2\times4 + 1\times4 \\ 3\times6 + 3\times8 + 3\times10 & 3\times2 + 3\times5 + 3\times5 & 3\times4 + 3\times4 + 3\times4 \end{bmatrix}$$

Sets of Jobs

$$= \begin{matrix} \text{Superduper} \\ \text{Wacko} \\ \text{Whooper} \\ \text{Ultima} \end{matrix} \begin{bmatrix} A & B & C \\ 70 & 36 & 36 \\ 116 & 60 & 60 \\ 32 & 17 & 16 \\ 72 & 36 & 36 \end{bmatrix} \quad \begin{matrix} \text{Matrix of total} \\ \text{computation times} \end{matrix} \qquad (20)$$

∎

Formalizing the computation process in this example yields a method for extending matrix-vector multiplication to matrix-matrix multiplication.

Matrix Multiplication. Let **A** be an *m*-by-*r* matrix and **B** be an *r*-by-*n* matrix. *The number of columns in* **A** *must equal the number of rows in* **B**. Then the matrix product **AB** is an *m-by-n matrix* obtained by forming the scalar product of each row in **A** with each column in **B**. That is, the (i, j)th entry in **AB** is $\mathbf{a}_i^R \cdot \mathbf{b}_j^C$, where $\mathbf{a}_i^R$ is the *i*th row of **A** and $\mathbf{b}_j^C$ is the *j*th column of **B**.

$$\mathbf{AB} = \begin{bmatrix} \mathbf{a}_1^R \\ \mathbf{a}_2^R \\ \vdots \\ \mathbf{a}_m^R \end{bmatrix} \cdot [\mathbf{b}_1^C, \mathbf{b}_2^C, \ldots, \mathbf{b}_n^C]$$

$$= \begin{bmatrix} \mathbf{a}_1^R \cdot \mathbf{b}_1^C & \mathbf{a}_1^R \cdot \mathbf{b}_2^C & \cdots & \mathbf{a}_1^R \cdot \mathbf{b}_n^C \\ \mathbf{a}_2^R \cdot \mathbf{b}_1^C & \mathbf{a}_2^R \cdot \mathbf{b}_2^C & \cdots & \mathbf{a}_2^R \cdot \mathbf{b}_n^C \\ \vdots & \vdots & \vdots & \vdots \\ \mathbf{a}_m^R \cdot \mathbf{b}_1^C & \mathbf{a}_m^R \cdot \mathbf{b}_2^C & \cdots & \mathbf{a}_m^R \cdot \mathbf{b}_n^C \end{bmatrix} \qquad (A)$$

If

$$A = \begin{bmatrix} 1 & 2 \\ -1 & 0 \end{bmatrix} \quad \text{and} \quad B = \begin{bmatrix} 4 & 5 \\ 6 & 7 \end{bmatrix},$$

$$\text{then } AB = \begin{bmatrix} 1 \cdot 4 + 2 \cdot 6 & 1 \cdot 5 + 2 \cdot 7 \\ -1 \cdot 4 + 0 \cdot 6 & -1 \cdot 5 + 0 \cdot 7 \end{bmatrix}$$

There are several ways to interpret matrix multiplication: first, as the scalar product of each row of A with each column of B, as in (A). Next, we can adopt the point of view of Example 7, where the product AB was an exension of the matrix-vector product Ab to the matrix-vector products of A with each column of B.

$$AB = A[b_1^C, b_2^C, \ldots, b_n^C] = [Ab_1^C, Ab_2^C, \ldots, Ab_n^C] \quad (B)$$

For A, B above, check that the first column of AB is $\begin{bmatrix} 1 & 2 \\ -1 & 0 \end{bmatrix}\begin{bmatrix} 4 \\ 6 \end{bmatrix}$.

Finally, we could also view AB as an extension of the vector-matrix product aB to the vector-matrix products of B with each row of A.

$$AB = \begin{bmatrix} a_1^R \\ a_2^R \\ \vdots \\ a_m^R \end{bmatrix} B = \begin{bmatrix} a_1^R B \\ a_2^R B \\ \vdots \\ a_m^R B \end{bmatrix} \quad (C)$$

For A, B above, check that the first row of AB is $[1 \quad 2]\begin{bmatrix} 4 & 5 \\ 6 & 7 \end{bmatrix}$.

***Equivalent Definitions of Matrix Multiplication* AB**

(A) Entry (i, j) of AB is scalar product $a_i^R \cdot b_j^C$.
(B) Column j of AB is matrix-vector product Ab_j^C.
(C) Row i of AB is vector-matrix product $a_i^R B$.

Remember that for the matrix product AB to make sense, the length of the rows in A (= the number of columns in A) must equal the length of the columns of B (= the number of rows in B). Further, if A is m-by-r and B is r-by-n, then AB is an m-by-n matrix: AB has as many rows as A and as many columns as B.

We shall see shortly that this form of matrix multiplication is exactly what is needed for Markov chain calculations. If this definition of matrix multiplication were given to a reader out of the blue, it would probably seem quite strange and artificial. But from Example 7 and dozens of other examples throughout this book we see that this strange definition is the "natural" definition to give.

Example 8. **Matrix Multiplication Example**

Let $\mathbf{A} = \begin{bmatrix} 4 & 5 \\ 6 & 7 \end{bmatrix}$ and $\mathbf{B} = \begin{bmatrix} 2 & 1 & 2 \\ 3 & 0 & 1 \end{bmatrix}$. Then $\mathbf{AB}$ is

$$\mathbf{AB} = \begin{bmatrix} 4 & 5 \\ 6 & 7 \end{bmatrix} \cdot \begin{bmatrix} 2 & 1 & 2 \\ 3 & 0 & 1 \end{bmatrix}$$

$$= \begin{bmatrix} 4 \times 2 + 5 \times 3 & 4 \times 1 + 5 \times 0 & 4 \times 2 + 5 \times 1 \\ 6 \times 2 + 7 \times 3 & 6 \times 1 + 7 \times 0 & 6 \times 2 + 7 \times 1 \end{bmatrix}$$

$$= \begin{bmatrix} 23 & 4 & 13 \\ 33 & 6 & 19 \end{bmatrix} \qquad \blacksquare$$

Note that the order of the matrices in matrix multiplication makes a big difference. That is, if $\mathbf{A}$ and $\mathbf{B}$ were two square n-by-n matrices, the matrix products $\mathbf{AB}$ and $\mathbf{BA}$ would yield different results (except in unusual cases). In mathematical terms, we say that matrix multiplication is *noncommutative*.

Example 9. **Matrix Multiplication Is Not Commutative**

Let $\mathbf{A} = \begin{bmatrix} 1 & 0 \\ 3 & 1 \end{bmatrix}$ and $\mathbf{B} = \begin{bmatrix} 1 & -1 \\ 0 & 2 \end{bmatrix}$. Then

$$\mathbf{AB} = \begin{bmatrix} 1 \times 1 + 0 \times 0 & 1 \times (-1) + 0 \times 2 \\ 3 \times 1 + 1 \times 0 & 3 \times (-1) + 1 \times 2 \end{bmatrix} = \begin{bmatrix} 1 & -1 \\ 3 & -1 \end{bmatrix}$$

$$\mathbf{BA} = \begin{bmatrix} 1 \times 1 + (-1) \times 3 & 1 \times 0 + (-1) \times 1 \\ 0 \times 1 + \quad 2 \times 3 & 0 \times 0 + \quad 2 \times 1 \end{bmatrix} = \begin{bmatrix} -2 & -1 \\ 6 & 2 \end{bmatrix}$$

Thus $\mathbf{AB} \neq \mathbf{BA}$. $\qquad \blacksquare$

Matrix multiplication is clearly quite tedious. It is easy to make a mistake and multiply the wrong entries together. But with three simple loops, a short computer program for matrix multiplication can be written to do the work for you. (This is a beautiful example of the advantage computers have in speed and accuracy for doing repetitive arithmetic.) Assume that $\mathbf{A}$ is an m-by-r matrix, $\mathbf{B}$ is r-by-n, and so the product $\mathbf{C} = \mathbf{AB}$ is m-by-n.

```
FOR I = 1 TO M
    FOR J = 1 TO N
        FOR K = 1 TO R
            C(I,J) = C(I,J) + A(I,K) * B(K,J)          (21)
        NEXT K
    NEXT J
NEXT I
```

We assume in this program that $C(I,J) = 0$ initially; otherwise, the statement $C(I,J) = 0$ must be inserted just after FOR J = 1 TO N.

Example 10. Powers of Markov Chain Transition Matrices

Let **p** denote the vector of the current probability distribution. In Example 6 we showed that the system of equations to compute the probability distribution vector **p**′ for the next period can be written as **p**′ = **Ap**, and that the probability distribution vector **p**″ after 2 minutes is

$$\mathbf{p}'' = \mathbf{A}\mathbf{p}' = \mathbf{A}(\mathbf{A}\mathbf{p}) \stackrel{?}{=} \mathbf{A}^2\mathbf{p} \tag{22}$$

Similarly, the distribution $\mathbf{p}^{(3)}$ after three periods is

$$\mathbf{p}^{(3)} = \mathbf{A}(\mathbf{A}(\mathbf{A}\mathbf{p})) \stackrel{?}{=} \mathbf{A}^3\mathbf{p}$$

The $\stackrel{?}{=}$ means that the step is yet to be proven. In (22), we want $\mathbf{A}^2\mathbf{p}$ to be the same as $\mathbf{A}(\mathbf{A}\mathbf{p})$—that is, premultiplying **p** by $\mathbf{A}^2$ should be the same as premultiplying **p** twice by **A**; and similarly for $\mathbf{A}^3$.

Let us first compute $\mathbf{A}^2$ and $\mathbf{A}^3$ for the weather Markov chain introduced in Section 1.3.

$$\mathbf{A} = \begin{matrix} \text{Sunny} \\ \text{Cloudy} \end{matrix} \begin{matrix} \text{Sunny} & \text{Cloudy} \\ \begin{bmatrix} \frac{3}{4} & \frac{1}{2} \\ \frac{1}{4} & \frac{1}{2} \end{bmatrix} \end{matrix}$$

Then

$$\mathbf{A}^2 = \begin{bmatrix} \frac{3}{4} & \frac{1}{2} \\ \frac{1}{4} & \frac{1}{2} \end{bmatrix} \cdot \begin{bmatrix} \frac{3}{4} & \frac{1}{2} \\ \frac{1}{4} & \frac{1}{2} \end{bmatrix}$$

$$= \begin{bmatrix} \frac{3}{4}\times\frac{3}{4} + \frac{1}{2}\times\frac{1}{4} & \frac{3}{4}\times\frac{1}{2} + \frac{1}{2}\times\frac{1}{2} \\ \frac{1}{4}\times\frac{3}{4} + \frac{1}{2}\times\frac{1}{4} & \frac{1}{4}\times\frac{1}{2} + \frac{1}{2}\times\frac{1}{2} \end{bmatrix} \tag{23}$$

$$= \begin{bmatrix} \frac{11}{16} & \frac{5}{8} \\ \frac{5}{16} & \frac{3}{8} \end{bmatrix}$$

Next we compute $\mathbf{A}^3$:

$$\mathbf{A}^3 = \mathbf{A} \cdot \mathbf{A}^2 = \begin{bmatrix} \frac{3}{4} & \frac{1}{2} \\ \frac{1}{4} & \frac{1}{2} \end{bmatrix} \cdot \begin{bmatrix} \frac{11}{16} & \frac{5}{8} \\ \frac{5}{16} & \frac{3}{8} \end{bmatrix}$$

$$= \begin{bmatrix} \frac{3}{4} \times \frac{11}{16} + \frac{1}{2} \times \frac{5}{16} & \frac{3}{4} \times \frac{5}{8} + \frac{1}{2} \times \frac{3}{8} \\ \frac{1}{4} \times \frac{11}{16} + \frac{1}{2} \times \frac{5}{16} & \frac{1}{4} \times \frac{5}{8} + \frac{1}{2} \times \frac{3}{8} \end{bmatrix} \tag{24}$$

$$= \begin{bmatrix} \frac{43}{64} & \frac{21}{32} \\ \frac{21}{64} & \frac{11}{32} \end{bmatrix}$$

The entries in $\mathbf{A}^2$ will be transition probabilities for two periods and the entries in $\mathbf{A}^3$ transition probabilities for three periods. For example, the value $\frac{5}{16}$ in entry (2, 1) of $\mathbf{A}^2$ should mean that if we are now in state 1 (Sunny), the chance is $\frac{5}{16}$ that in 2 days we will be in state 2 (Cloudy). The value of $\frac{21}{64}$ in entry (2, 1) of $\mathbf{A}^3$ tells us that if now Sunny, the probability is $\frac{21}{64}$ that in 3 days it will be Cloudy.

The values we obtained in computing $\mathbf{A}^2$ and $\mathbf{A}^3$ look reasonable. In particular, the numbers in each column of $\mathbf{A}^2$ and $\mathbf{A}^3$ sum to 1. ■

We must check that matrix multiplication has given the correct probabilities in $\mathbf{A}^2$ and $\mathbf{A}^3$ in (23) and (24). In general, entry (i, j) in $\mathbf{A}^2$ should be the probability that if we are now in state j, then in two periods we shall be in state i. By the scalar-product definition of matrix multiplication,

$$\mathbf{AA} = \begin{bmatrix} \mathbf{a}_1^R \cdot \mathbf{a}_1^C & \cdots & \mathbf{a}_1^R \cdot \mathbf{a}_j^C & \cdots & \mathbf{a}_1^R \cdot \mathbf{a}_n^C \\ \mathbf{a}_2^R \cdot \mathbf{a}_1^C & \cdots & \mathbf{a}_2^R \cdot \mathbf{a}_j^C & \cdots & \mathbf{a}_2^R \cdot \mathbf{a}_n^C \\ \vdots & \cdots & \vdots & \cdots & \vdots \\ \mathbf{a}_n^R \cdot \mathbf{a}_1^C & \cdots & \mathbf{a}_n^R \cdot \mathbf{a}_j^C & \cdots & \mathbf{a}_n^R \cdot \mathbf{a}_n^C \end{bmatrix} \tag{25}$$

where

$$\mathbf{a}_i^R \cdot \mathbf{a}_j^C = a_{i1}a_{1j} + a_{i2}a_{2j} + \cdots + a_{in}a_{nj} \tag{26}$$

In the case of a 2-by-2 transition matrix, (25) is

$$\begin{bmatrix} a_{11} & a_{12} \\ a_{21} & a_{22} \end{bmatrix} \begin{bmatrix} a_{11} & a_{12} \\ a_{21} & a_{22} \end{bmatrix} = \begin{bmatrix} a_{11}a_{11} + a_{12}a_{21} & a_{11}a_{12} + a_{12}a_{22} \\ a_{21}a_{11} + a_{22}a_{21} & a_{21}a_{12} + a_{22}a_{22} \end{bmatrix}$$

In words, we interpret (26) as follows: The probability of going from state j to state i in two periods is obtained by finding the probability of going from state j to state 1 (in the first period) and then from state 1 to state i (in the second period), plus the probability of going from j to 2 and then from 2 to i, plus from j to 3 and then from 3 to i, and so on. This is exactly the probability of going from state j now to state i in two periods.

This argument extends to show that $\mathbf{A}^3$ is the three-period transition matrix and $\mathbf{A}^k$ is the k-period transition matrix.

Example 10 (continued). **Powers of Markov Transition Matrices**

Let us turn now to the frog Markov chain. Recall that the transition matrix is

$$
\mathbf{A} = \begin{bmatrix}
.50 & .25 & 0 & 0 & 0 & 0 \\
.50 & .50 & .25 & 0 & 0 & 0 \\
0 & .25 & .50 & .25 & 0 & 0 \\
0 & 0 & .25 & .50 & .25 & 0 \\
0 & 0 & 0 & .25 & .50 & .50 \\
0 & 0 & 0 & 0 & .25 & .50
\end{bmatrix} \tag{27}
$$

Using the computer program given above, one can compute $\mathbf{A}^2$ and $\mathbf{A}^3$.

$$
\mathbf{A}^2 = \mathbf{AA} = \begin{bmatrix}
.375 & .25 & .062 & 0 & 0 & 0 \\
.50 & .437 & .25 & .062 & 0 & 0 \\
.125 & .25 & .375 & .25 & .062 & 0 \\
0 & .062 & .25 & .375 & .25 & .125 \\
0 & 0 & .062 & .25 & .437 & .50 \\
0 & 0 & 0 & .062 & .25 & .375
\end{bmatrix} \tag{28}
$$

and

$$
\mathbf{A}^3 = \mathbf{A}^2\mathbf{A} = \begin{bmatrix}
.312 & .234 & .094 & .016 & 0 & 0 \\
.468 & .406 & .250 & .093 & .016 & 0 \\
.187 & .25 & .312 & .234 & .094 & .031 \\
.031 & .094 & .234 & .312 & .25 & .187 \\
0 & .016 & .094 & .250 & .406 & .468 \\
0 & 0 & .016 & .094 & .234 & .312
\end{bmatrix} \tag{29}
$$

Looking at entry (2, 1) in $\mathbf{A}^2$ and $\mathbf{A}^3$, we see that if the frog were now in state 1, then in 2 minutes the frog has probability .5 of being in state 2 and in 3 minutes it has probability .468 of being in state 2. ∎

Optional Example on Linear Filtering (Revisited)

In Example 4 of Section 1.5 we applied linear filtering to a noisy pattern of darkness levels that might have come from the TV eye of a robot. The array of readings, numbers between 0 and 8, represented levels of darkness. The initial array is given below. First we give the matrix **D** of darkness levels and beneath the array of darkness characters.

$$\mathbf{D} = \begin{bmatrix} 1 & 2 & 0 & 3 & 1 & 7 & 8 & 2 & 2 & 4 & 2 & 0 \\ 4 & 1 & 2 & 3 & 7 & 5 & 8 & 6 & 3 & 2 & 4 & 2 \\ 3 & 4 & 1 & 3 & 7 & 5 & 7 & 7 & 3 & 4 & 2 & 2 \\ 7 & 2 & 3 & 4 & 1 & 7 & 5 & 2 & 2 & 4 & 3 & 7 \\ 8 & 4 & 2 & 7 & 5 & 8 & 7 & 6 & 7 & 2 & 3 & 6 \\ 3 & 7 & 8 & 2 & 6 & 8 & 6 & 7 & 3 & 8 & 6 & 2 \\ 4 & 2 & 3 & 3 & 8 & 8 & 4 & 7 & 2 & 4 & 3 & 1 \\ 2 & 6 & 3 & 2 & 6 & 7 & 8 & 7 & 3 & 1 & 7 & 2 \\ 6 & 1 & 2 & 3 & 8 & 2 & 3 & 7 & 1 & 3 & 2 & 5 \\ 1 & 7 & 1 & 4 & 8 & 2 & 3 & 7 & 1 & 1 & 5 & 0 \\ 3 & 7 & 2 & 5 & 7 & 3 & 3 & 7 & 2 & 4 & 4 & 5 \\ 2 & 2 & 8 & 8 & 7 & 3 & 2 & 8 & 7 & 6 & 1 & 4 \end{bmatrix} = \tag{30}$$

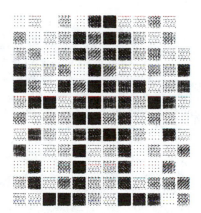

In Section 1.5 we obtained a filtered matrix **D**′ (Data Plot 5) by applying linear filtering to **D**. We computed each entry d'_{ij} in **D**′ individually from the weighted-average formula

$$d'_{ij} = \tfrac{1}{36}\{16d_{ij} + 4(d_{i,j-1} + d_{i,j+1} + d_{i-1,j} + d_{i+1,j}) \\ + 1(d_{i-1,j-1} + d_{i+1,j-1} + d_{i-1,j+1} + d_{i+1,j+1})\} \tag{31}$$

Now we shall give a 12-by-12 filtering matrix **F** that performs the filtering (31) on the whole darkness matrix **D** at once using matrix multiplication.

$$\mathbf{F} = \begin{bmatrix} \frac{5}{6} & \frac{1}{6} & 0 & 0 & 0 & 0 & 0 & 0 & 0 & 0 & 0 & 0 \\ \frac{1}{6} & \frac{2}{3} & \frac{1}{6} & 0 & 0 & 0 & 0 & 0 & 0 & 0 & 0 & 0 \\ 0 & \frac{1}{6} & \frac{2}{3} & \frac{1}{6} & 0 & 0 & 0 & 0 & 0 & 0 & 0 & 0 \\ 0 & 0 & \frac{1}{6} & \frac{2}{3} & \frac{1}{6} & 0 & 0 & 0 & 0 & 0 & 0 & 0 \\ 0 & 0 & 0 & \frac{1}{6} & \frac{2}{3} & \frac{1}{6} & 0 & 0 & 0 & 0 & 0 & 0 \\ 0 & 0 & 0 & 0 & \frac{1}{6} & \frac{2}{3} & \frac{1}{6} & 0 & 0 & 0 & 0 & 0 \\ 0 & 0 & 0 & 0 & 0 & \frac{1}{6} & \frac{2}{3} & \frac{1}{6} & 0 & 0 & 0 & 0 \\ 0 & 0 & 0 & 0 & 0 & 0 & \frac{1}{6} & \frac{2}{3} & \frac{1}{6} & 0 & 0 & 0 \\ 0 & 0 & 0 & 0 & 0 & 0 & 0 & \frac{1}{6} & \frac{2}{3} & \frac{1}{6} & 0 & 0 \\ 0 & 0 & 0 & 0 & 0 & 0 & 0 & 0 & \frac{1}{6} & \frac{2}{3} & \frac{1}{6} & 0 \\ 0 & 0 & 0 & 0 & 0 & 0 & 0 & 0 & 0 & \frac{1}{6} & \frac{2}{3} & \frac{1}{6} \\ 0 & 0 & 0 & 0 & 0 & 0 & 0 & 0 & 0 & 0 & \frac{1}{6} & \frac{5}{6} \end{bmatrix} \tag{32}$$

We claim that the filtered matrix $\mathbf{D}'$ equals the product matrix

$$\mathbf{D}' = \mathbf{FDF} \tag{33}$$

A closer examination of why (33) is true and computation with (33) is left to the Exercises.   ∎

Section 2.2 Exercises

Summary of Exercises

Exercises 1–15 deal with matrix-vector products. Exercises 16–34 deal with matrix multiplication; Exercises 29–34 look at some general classes of matrices. Exercises 35–39 involve writing or using computer programs. Exercises 40 and 41 cover material in the optional final pages of the section.

1. Let $\mathbf{a} = [1, 2, 3]$, $\mathbf{b} = [-1, 3, -1]$, $\mathbf{c} = [2, 5, 8]$. Compute
   (a) $\mathbf{a} \cdot \mathbf{b}$ (b) $\mathbf{b} \cdot \mathbf{c}$ (c) $\mathbf{a} \cdot (\mathbf{b} + \mathbf{c})$ (d) $\mathbf{a} \cdot \mathbf{a}$
   (e) Show that for any $\mathbf{a}$, $\mathbf{a} \cdot \mathbf{a}$ is the sum of squares of entries.

2. In Example 2 of Section 1.5 we smoothed a time series with the transformation $d'_i = (d_{i-1} + d_i + d_{i+1})/3$. If $\mathbf{d}_i$ is the vector $\mathbf{d}_i = [d_{i-1}, d_i, d_{i+1}]$, define a vector $\mathbf{c}$ so that $d'_i = \mathbf{c} \cdot \mathbf{d}_i$.

3. Let $\mathbf{a}$, $\mathbf{b}$, $\mathbf{c}$ be as in Exercise 1. Let

$$\mathbf{A} = \begin{bmatrix} 1 & 2 & 3 & 4 \\ 2 & 4 & 6 & 8 \\ 3 & 5 & 7 & 9 \end{bmatrix}, \quad \mathbf{B} = \begin{bmatrix} 1 & 0 & -1 \\ 2 & -2 & 0 \\ 0 & 1 & -1 \end{bmatrix}, \quad \mathbf{C} = \begin{bmatrix} 5 & 4 & 1 \\ 1 & 0 & 2 \\ 3 & 2 & 1 \\ 0 & 1 & 3 \end{bmatrix}$$

Which of the following matrix calculations are well defined (the sizes match)? If the computation makes sense, perform it.
(a) $\mathbf{aA}$ (b) $\mathbf{bB}$ (c) $\mathbf{cC}$ (d) $\mathbf{Aa}$ (e) $\mathbf{Bb}$ (f) $\mathbf{Cc}$

4. Calculate the following expressions, unless the sizes do not match. The vectors **a**, **b**, **c** are as defined in Exercise 1, and matrices **A**, **B**, **C** are as in Exercise 3.

   (a) $(\mathbf{aB}) \cdot \mathbf{c}$ (b) $(\mathbf{a} + \mathbf{c})\mathbf{A}$ (c) $\mathbf{b}(\mathbf{A} + \mathbf{B})$

5. Suppose that you want to buy 5 cantaloupes, 4 apples, 3 oranges, and 2 pineapples. You comparison shop and find that at store A the costs of these four fruits, respectively, are 30 cents, 10 cents, 10 cents, and 75 cents a piece, while at store B the costs are 25 cents, 15 cents, 8 cents, and 80 cents.

   (a) Express the problem of determining the cost of this set of fruit at each store as a matrix-vector product; write out the matrix and vector.

   (b) Compute the costs of the fruits at the two stores.

6. Suppose that you want to have a party catered and will need 10 hero sandwiches, 6 quarts of fruit punch, 3 quarts of potato salad, and 2 plates of hors d'oeuvres. The following matrix gives the costs of these supplies from three different caterers.

	Caterer A	Caterer B	Caterer C
Hero sandwich	$5	$5	$4
Fruit punch	$1	$1.50	$.75
Potato salad	$.75	$1.00	$1
Hors d'oeuvres	$8	$7	$10

   (a) Express the problem of determining the costs of catering the party by each caterer as a matrix-vector product (be careful whether you place the vector first or second in the product).

   (b) Determine the costs of catering with each caterer.

7. Write the following systems of equations in matrix notation. Define any matrices or vectors you use.

   (a) $3x_1 + 4x_2 = 5$ (b) $2x_1 + x_2 - 2x_3 = 0$ (c) $x_1 = 2x_1 - x_2$
   $\quad\;\; 2x_1 - 5x_2 = 3$ $\qquad x_1 \quad\;\; + 3x_3 = 3 \qquad x_2 = 3x_1 + 2x_2$
   $\qquad\qquad\qquad\qquad\quad 3x_1 - x_2 \quad\;\; = 5 \qquad x_3 = 4x_1 - 3x_2$

8. Write the following linear programs in matrix notation. Define any matrices or vectors that you use.

   (a) Maximize $3x_1 + 10x_2$ (b) Minimize $5x_1 + 5x_2 + 5x_3$

   $\qquad x_1 \geq 0, \quad x_2 \geq 0$ $\qquad\qquad x_1 \geq 0, \quad x_2 \geq 0, \quad x_3 \geq 0$

   $\qquad 2x_1 + x_2 \leq 20$ $\qquad\qquad\quad x_1 + 2x_2 + 3x_3 \geq 20$

   $\qquad\;\; x_1 + 3x_2 \leq 10$ $\qquad\qquad\quad 2x_1 + x_2 + 2x_3 \geq 25$

   $\qquad 4x_1 + 2x_2 \leq 35$ $\qquad\qquad\quad 2x_1 - x_2 + 3x_3 \geq 15$

9. Write the rabbit-fox equations

$$R' = R + .1R - .15F$$
$$F' = F + .2R - .3F$$

in matrix form using $\mathbf{p} = [R, F]$, $\mathbf{p}' = [R', F']$, and $\mathbf{A} = \begin{bmatrix} .1 & -.15 \\ .2 & -.3 \end{bmatrix}$.

10. Consider the system of equations

$$2x_1 + 3x_2 - 2x_3 = 5y_1 + 2y_2 - 3y_3 + 200$$
$$x_1 + 4x_2 + 3x_3 = 6y_1 - 4y_2 + 4y_3 - 120$$
$$5x_1 + 2x_2 - x_3 = 2y_1 \qquad - 2y_3 + 350$$

(a) Write this system of equations in matrix form. Define the vectors and matrices you introduce.
(b) Rewrite in matrix form with all the variables on the left side (and just numbers on the right).

11. Three different types of computers need varying amounts of four different types of integrated circuits. The following matrix $\mathbf{A}$ gives the number of each circuits needed by each computer.

$$\mathbf{A} = \text{Computers} \begin{array}{c} \\ A \\ B \\ C \end{array} \overset{\begin{array}{cccc} \text{Circuits} \\ 1 & 2 & 3 & 4 \end{array}}{\begin{bmatrix} 2 & 3 & 2 & 1 \\ 5 & 1 & 3 & 2 \\ 3 & 2 & 2 & 2 \end{bmatrix}}$$

Let $\mathbf{d} = [10, 20, 30]$ be the computer demand vector (how many of each type of computer is needed). Let $\mathbf{p} = [\$2, \$5, \$1, \$10]$ be the price vector for the circuits (the cost of each type of circuit).

Write an expression in terms of $\mathbf{A}$, $\mathbf{d}$, $\mathbf{p}$ for the total cost of the circuits needed to produce the set of computers in demand; indicate where the matrix-vector product occurs and where the vector product occurs. Compute this total cost.

12. For the frog Markov chain in Example 6 it was noted in Section 1.2 that $p^* = [.1, .2, .2, .2, .2, .1]$ is a stable distribution. In matrix algebra, this means that $\mathbf{p}^* = \mathbf{A}\mathbf{p}^*$, where $\mathbf{A}$ is the frog Markov transition matrix. Verify that $\mathbf{p}^* = \mathbf{A}\mathbf{p}^*$ for this $\mathbf{p}^*$.

13. One can express polynomial multiplication in terms of a matrix-vector product as follows: to multiply the quadratic $2x^2 + 3x + 4$ by $1x^2 - 2x + 5$, we multiply

$$\begin{bmatrix} 2 & 0 & 0 \\ 3 & 2 & 0 \\ 4 & 3 & 2 \\ 0 & 4 & 3 \\ 0 & 0 & 4 \end{bmatrix} \cdot \begin{bmatrix} 1 \\ -2 \\ 5 \end{bmatrix}$$

The resulting vector will give the coefficients in the product. Confirm this. For the polynomial multiplication $(x^3 + 2x^2 + 3x + 4) \times (4x^2 - 3x + 1)$, write the associated matrix-vector product.

14. Let **1** denote a vector of all 1's. Show for a Markov transition matrix **A** that $\mathbf{1} \cdot \mathbf{a}_j^C = 1$ (where $\mathbf{a}_j^C$ is the *j*th column of **A**). Then show that $\mathbf{1A} = \mathbf{1}$.

15. In the problem of smoothing a time series introduced in Example 2 of Section 1.5, we start with a time-series vector **d** of data values. We want to smooth the time series **d** into **d**′ with the transformation on the entries $d_i' = (d_{i-1} + d_i + d_{i+1})/3$. Show that $\mathbf{d}' = \mathbf{Sd}$, where **S** is the matrix

$$\begin{bmatrix} \frac{1}{2} & \frac{1}{2} & 0 & 0 & 0 & \cdots \\ \frac{1}{3} & \frac{1}{3} & \frac{1}{3} & 0 & 0 & \cdots \\ 0 & \frac{1}{3} & \frac{1}{3} & \frac{1}{3} & 0 & \cdots \\ 0 & 0 & \frac{1}{3} & \frac{1}{3} & \frac{1}{3} & \cdots \end{bmatrix}$$

16. Indicate which pairs of the following matrices can be multiplied together and give the size of the resulting product.
    (i) A 3-by-7 matrix **A**
    (ii) A 2-by-3 matrix **B**
    (iii) A 3-by-3 matrix **C**
    (iv) A 2-by-2 matrix **D**
    (v) A 7-by-2 matrix **E**

17. Let $\mathbf{A} = \begin{bmatrix} 1 & 2 \\ 3 & 4 \end{bmatrix}$, $\mathbf{B} = \begin{bmatrix} 1 & -1 & 0 \\ 2 & 10 & -2 \end{bmatrix}$, $\mathbf{C} = \begin{bmatrix} 3 & 1 \\ 2 & 5 \end{bmatrix}$. Compute the following matrix products (if possible).
    (a) **AB**   (b) **AC**   (c) **BC**   (d) **CA**   (e) **(CA)B**

18. Let

$$\mathbf{A} = \begin{bmatrix} 1 & 2 & 3 & 4 \\ 2 & 4 & 6 & 8 \\ 3 & 5 & 7 & 9 \end{bmatrix}, \quad \mathbf{B} = \begin{bmatrix} 1 & 0 & -1 \\ 2 & -2 & 0 \\ 0 & 1 & -1 \end{bmatrix}, \quad \mathbf{C} = \begin{bmatrix} 5 & 4 & 1 \\ 1 & 0 & 2 \\ 3 & 2 & 1 \\ 0 & 1 & 3 \end{bmatrix}$$

Compute these matrix products (if possible).
(a) **AB**   (b) **BA**   (c) **AC**   (d) **CA**   (e) **CB**

19. Compute just one row or column, as requested, in the following matrix products (**A**, **B**, **C** are as in Exercise 18).
(a) Row 1 in **B**^{2} (b) Column 2 in **AC**   (c) Column 3 in **CB**.

20. Show that **AB** = **BA** for the matrices

$$\mathbf{A} = \begin{bmatrix} 1 & 1 \\ 2 & 3 \end{bmatrix} \quad \text{and} \quad \mathbf{B} = \begin{bmatrix} 3 & -1 \\ -2 & 1 \end{bmatrix}$$

(Normally, matrix multiplication does not commute.)

21. For **A**, **B**, **C** in Exercise 18, compute entry (2, 3) in (**BA**)**C**.

22. Suppose that we are given the following matrices involving the costs of fruits at different stores, the amounts of fruit different types of people want, and the numbers of people of different types in different towns.

	Store *A*	Store *B*			Apple	Orange	Pear
Apple	.10	.15		Person *A*	5	10	3
Orange	.15	.20		Person *B*	4	5	5
Pear	.10	.10					

	Person *A*	Person *B*
Town 1	1000	500
Town 2	2000	1000

(a) Compute a matrix that tells how much each person's fruit purchases cost at each store.
(b) Compute a matrix that tells how many of each fruit will be purchased in each town.
(c) Compute a matrix that tells the total cost of everyone's fruit purchases in town 1 and in town 2 when people use store *A* and when they use store *B* (a different number for each town and each store).

23. Express in matrix notation the following operations on these arrays of data: Matrix **A** gives the amount of time each of three jobs requires of I/O (input/output), of Execution time, and System overhead; matrix **B** gives the charges (per unit of time) of different computer activities under two different charging plans; matrix **C** (actually a vector) tells how many jobs of each type there are; and matrix **D** tells the fraction of the time that each time-charging plan (the columns in matrix **B**) is used each day.

	Time		
A	I/O	Execution	System
Job *A*	5	20	10
Job *B*	4	25	8
Job *C*	10	10	5

	Time Charges	
B	Plan I	Plan II
I/O	2	3
Execution	6	5
System	3	4

	Number of Jobs of
C	Each Type
Job *A*	4
Job *B*	5
Job *C*	3

	Fraction
D	of Time
Plan I	.3
Plan II	.7

Compute the following arrays using **A**, **B**, **C**, and **D**.
(a) Total cost of each type of job for each charge plan.
(b) Total amount of I/O, Execution, and System time for all the jobs (all jobs are summarized in matrix **C**).
(c) Total cost of all jobs when run under plan I and under plan II.
(d) Average cost of 1 unit of I/O, of Execution, of System time.
Hint: Use matrix **D**.
(e) Average cost of each type of job (job *A*, job *B*, job *C*).

24. Express in matrix notation the following operations on these arrays of data: Matrix **A** gives the amounts of raw material required to build different products; matrix **B** gives the costs of these raw materials in two different countries; matrix **C** tells how many of the products are needed to build two types of houses; and matrix **D** gives the demand for houses in the two countries.

	Raw Material		
A	Wood	Labor	Steel
Item A	5	20	10
Item B	4	25	8
Item C	10	10	5

	Cost by Country	
B	Spain	Italy
Wood	$2	$3
Labor	$6	$5
Steel	$3	$4

	Items Needed in House		
C	Item *A*	Item *B*	Item *C*
House I	4	8	3
House II	5	5	2

	Demand for Houses	
D	House I	House II
Spain	50,000	200,000
Italy	80,000	500,000

(a) Compute the first row in the matrix product **AB**.
(b) Which matrix product tells how much of the different *items* are needed to meet the demand for houses (types I and II combined) in the different countries? D x C
(c) Which matrix product gives the cost of building each type of house in each country? A B   B C

CAB

(d) Which entry in what matrix product would give the total cost of building all homes in Spain?

> *Note:* If the rows and columns in a matrix must be interchanged in a product, indicate this by using the transpose of the matrix (transposes are not formally introduced until Section 2.4).

25. Express in matrix notation the following operations on these arrays of data: Matrix **A** gives the number of tradesmen needed each day to build different types of small stores; matrix **B** gives the number of days it takes to build each type of store in each state; matrix **C** gives the cost of tradesmen (per day) in New York and Texas; and matrix **D** gives the number of stores of each type needed in two different sorts of shopping centers.

A	Carpenter	Electrician	Bricklayer
Store *A*	5	2	1
Store *B*	4	2	2
Store *C*	3	1	1

B	New York	Texas
Store *A*	20	15
Store *B*	30	25
Store *C*	20	20

C	New York	Texas
Carpenter	$100	$60
Electrician	$80	$50
Bricklayer	$80	$60

D	Shopping Center I	Shopping Center II
Store *A*	10	5
Store *B*	10	10
Store *C*	20	20

(a) Compute the first column in **AC**.

(b) Which matrix product tells how many tradesmen per day are needed to build the stores in each type of shopping center? *Do not compute this product.*

(c) Which entry in what matrix product would give the total cost of building three stores, one of each type, in New York? (Total cost covers all the days of construction.)

> *Note:* If the rows and columns in a matrix must be interchanged in some matrix product, indicate this by using the transpose of the matrix (transposes are defined in Section 2.4).

26. Consider a growth model for the numbers of computers (*C*) and dogs (*D*) from year to year:

$$D' = 3C + D$$
$$C' = 2C + 2D$$

Let $\mathbf{x} = [C, D]$ be the initial vector and let $\mathbf{x}^{(k)}$ denote the vector of computers and dogs after k years. Let $\mathbf{A}$ be the matrix of coefficients in this system. Write $\mathbf{x}^{(k)}$ in terms of $\mathbf{A}$ and $\mathbf{x}$.

27. Consider a variation on the Markov chain for Sunny and Cloudy weather. The new transition matrix $\mathbf{A}$ is

$$\mathbf{A} = \begin{matrix} \\ \text{Sunny} \\ \text{Cloudy} \end{matrix} \begin{matrix} \text{Sunny} & \text{Cloudy} \\ \left[\begin{matrix} \frac{2}{3} & \frac{1}{3} \\ \frac{1}{3} & \frac{2}{3} \end{matrix} \right] \end{matrix}$$

(a) Compute $\mathbf{A}^2$. What probability does entry $(1, 2)$ in $\mathbf{A}^2$ represent?
(b) Compute $\mathbf{A}^3$. What is the probability if sunny today that it is sunny in 3 days?
(c) Compute $\mathbf{A}^4$. What vector do the columns of $\mathbf{A}^k$, for $k = 2, 3, 4$, seem to be approaching?

28. (a) Show that if we multiply any 3-by-3 matrix $\mathbf{A}$ by

$$\mathbf{I} = \begin{bmatrix} 1 & 0 & 0 \\ 0 & 1 & 0 \\ 0 & 0 & 1 \end{bmatrix}$$

then the result $\mathbf{IA}$ (or $\mathbf{AI}$) always equals $\mathbf{A}$.
Hint: Make up a 3-by-3 matrix and multiply it by $\mathbf{I}$.
(b) Show that if we premultiply $\mathbf{A}$ by

$$\mathbf{K} = \begin{bmatrix} 1 & 0 & 0 \\ 0 & 2 & 0 \\ 0 & 0 & 3 \end{bmatrix}$$

the result is $\mathbf{A}$ except that the second row of $\mathbf{A}$ is doubled and the third row of $\mathbf{A}$ is tripled.
(c) Suppose that $\mathbf{K}$ has k_1 in entry $(1, 1)$, k_2 in $(2, 2)$, k_3 in $(3, 3)$, and 0's elsewhere. Describe the effect of premultiplying any 3-by-3 matrix $\mathbf{A}$ by $\mathbf{K}$.

29. A square matrix is called *diagonal* if all its nonzero entries are on the main diagonal.
(a) Find the product $\mathbf{AB}$ of

$$\mathbf{A} = \begin{bmatrix} 2 & 0 & 0 \\ 0 & 1 & 0 \\ 0 & 0 & 3 \end{bmatrix} \quad \text{and} \quad \mathbf{B} = \begin{bmatrix} 3 & 0 & 0 \\ 0 & 2 & 0 \\ 0 & 0 & 5 \end{bmatrix}$$

(b) Suppose that a_{11}, a_{22}, a_{33} are the diagonal entries in the diagonal matrix $\mathbf{A}$ and b_{11}, b_{22}, b_{33} are the diagonal entries in the diagonal matrix $\mathbf{B}$. Then what are the diagonal entries in $\mathbf{AB}$?

30. (a) If we premultiply any 3-by-3 matrix **A** by

$$\mathbf{Q} = \begin{bmatrix} 1 & 0 & 0 \\ 0 & 0 & 1 \\ 0 & 1 & 0 \end{bmatrix}$$

show that the resulting matrix **QA** is just **A** with the second and third rows interchanged.
Hint: To see why this is so, make up a 3-by-3 matrix and premultiply by **Q**.

(b) If we premultiply any matrix **A** by

$$\mathbf{R} = \begin{bmatrix} 0 & 0 & 1 \\ 0 & 2 & 0 \\ 1 & 0 & 0 \end{bmatrix}$$

show that the result **RA** is a matrix with the rows of **A** reversed and the values of all entries in the second row doubled.

(c) If we premultiply any matrix **A** by

$$\mathbf{S} = \begin{bmatrix} 1 & 0 & 0 \\ -2 & 1 & 0 \\ 0 & 0 & 1 \end{bmatrix}$$

show that the result **SA** is **A** except that twice the first row has been subtracted from the second row.

(d) Construct a 3-by-3 matrix which when premultiplying a 3-by-3 matrix has the effect of adding four times the third row to the first row.

(e) Construct a 3-by-3 matrix which when premultiplying a 3-by-3 matrix has the effect of subtracting twice the first row from the second row and also adding three times the first row to the third row.

31. (a) Show that if we postmultiply any 3-by-3 matrix **A** by the matrix **Q** in Exercise 30, the resulting matrix **AQ** is **A** with the second and third *columns* interchanged.

(b) Show that if we postmultiply **A** by **R** in Exercise 30, the resulting matrix **AR** is **A** with the columns reversed and the values of all entries in column two doubled.

(c) Show that if we postmultiply **A** by **S** in Exercise 30, the resulting matrix **AS** is **A** except that twice the second column has been subtracted from the first column.

(d) Construct a 3-by-3 matrix which when postmultiplying a 3-by-3 matrix has the effect of adding two times the third column to the second column.

(e) Construct a 3-by-3 matrix which when postmultiplying a 3-by-3 matrix has the effect of subtracting four times the first column from the second column and also adding twice the first column to the third column.

32. Compute the matrix product

$$
\begin{bmatrix} 1 & 2 & 1 & 0 \\ 0 & 1 & 3 & 1 \\ 0 & 0 & 1 & 2 \\ 0 & 0 & 0 & 1 \end{bmatrix} \cdot \begin{bmatrix} 2 & 0 & 1 & 1 \\ 0 & 1 & 2 & 0 \\ 0 & 0 & 1 & 1 \\ 0 & 0 & 0 & 3 \end{bmatrix}
$$

A matrix is called *upper triangular* if the only nonzero entries are on or above the main diagonal. The previous computation illustrated the fact that the product of two upper triangular matrices is again upper triangular. Give an explanation of why this is true for the product of any two 4-by-4 upper triangular matrices (or more generally for any size matrices).

33. Show that in a 2-by-2 matrix $\begin{bmatrix} a & b \\ c & d \end{bmatrix}$, if one row is a multiple of the other row, then one column is a multiple of the other column.

34. Let **A** and **B** be 2-by-2 matrices. If $\mathbf{C} = \mathbf{AB}$ and the second row of **A** is 3 times the first row of **A** ($\mathbf{a}_2^R = 3\mathbf{a}_1^R$), show that the second row of **C** is 3 times the first row of **C**.

Hint: Compare $c_{11} = \mathbf{a}_1^R \cdot \mathbf{b}_1^C$ with $c_{21} = \mathbf{a}_2^R \cdot \mathbf{b}_1^C$; similarly for c_{21} versus c_{22}.

Exercises Involving Computer Programs
35. Write a program to multiply a matrix times a vector, take the resulting vector, and premultiply it again by the matrix, and so on a specified number of times. (This is the computation needed to follow a Markov chain over many periods.)

36. Write a program to read in two matrices and multiply them together.

37. Write a program to raise a (square) matrix to a specified power.

38. Use the program you wrote in Exercise 35 (or one supplied by the instructor) to compute successive probability distributions for 20 periods for the following Markov chains. Unless otherwise specified, assume that the initial distribution vector is $\mathbf{p} = [1, 0, 0, \ldots]$ (i.e., one starts in state 1).
   (a) The weather Markov chain (in Example 10).
   (b) The frog Markov chain starting with $\mathbf{p} = [0, 0, \frac{1}{2}, \frac{1}{2}, 0, 0]$.
   (c) The rat maze Markov chain in Exercise 8 of Section 1.3.
   (d) The poker Markov chain in Exercise 9 of Section 1.3 with $\mathbf{p} =$

[0, 0, 0, 1, 0, 0, 0]—you start with $3. What are your chances of winning $6 by the end of 20 periods?

(e) The tank battle Markov chain in Exercise 10 of Section 1.3. What are each tank's chances of winning the battle?

39. Use the program you wrote in Exercise 37 to raise the following Markov transition matrices to the twentieth power. In each case explain the pattern of values in the matrix.
   (a) The weather Markov chain (in Example 10).
   (b) The frog Markov chain.
   (c) The rat maze Markov chain in Exercise 8 of Section 1.3.
   (d) The poker Markov chain in Exercise 9 of Section 1.3.
   (e) The tank battle Markov chain in Exercise 10 of Section 1.3

Exercises for Optional Part of Section

40. Compute entry (3, 3) in the matrix product **FDF** in Example 11, and verify that it agrees with the value given in Data Plot 5 of Section 1.5.

41. (Difficult) (a) Verify that the matrix product **FDF** performs the filtering given by formula (31) in Example 11 [e.g., that entry (i, j) equals that formula].
   (b) Build a filtering matrix **G** such that **GDG** performs the filtering transformation

$$d'_{ij} = \tfrac{1}{16}\{4d_{ij} + 2(d_{i,j-1} + d_{i,j+1} + d_{i-1,j} + d_{i+1,j}) + 1(d_{i-1,j-1} + d_{i+1,j-1} + d_{i-1,j+1} + d_{i+1,j+1})\}$$

Section 2.3 0–1 Matrices

Most of the mathematics that is studied in high school involves numbers or geometric objects. However, there are important fields of discrete mathematics that work with sets of nonnumeric objects. One such field is graph theory. These graphs are different from the graphs for plotting functions. Graph theory is used extensively in computer science and systems analysis.

A **graph** $G = (N, E)$ consists of a set N of **nodes** and a collection E of **edges** that are pairs of nodes. There is a natural way to "draw" a graph. We make a point for each node and draw lines linking the pairs of nodes in the edges. For example, the graph G with node set $N = \{a, b, c, d, e\}$ and edge set $E = \{(a, b), (a, c), (a, d), (b, c), (b, d), (d, e), (e, e)\}$ is drawn in Figure 2.2. An edge may link a node with itself, called a *loop* edge, as at node e in Figure 2.2.

Figure 2.2

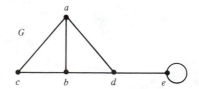

A flowchart for a computer program is a form of graph. The data structures that are used to organize complex sets of data are graphs. Organizational charts, electrical circuits, telephone networks, and road maps are other examples of graphs. Billions of dollars are spent every year analyzing problems that are modeled in terms of graphs.

One class of questions asked about graphs concerns paths. A **path** is a sequence of nodes with edges linking consecutive nodes. We may want to find the shortest path between two nodes, or determine whether or not any path exists between a given pair of nodes. Finding paths through graphs arises when one wants to route a set of telephone calls through a network between prescribed cities without exceeding the capacity of any edge. The question of whether a path exists between two nodes arises over and over again in studying the effect on networks of random disruption, say, due to lightning. For example, in a given 1000-edge network one might want to know the probability that if five randomly chosen edges are destroyed, the network will become disconnected.

The purpose of mentioning all these graph problems is to motivate the importance of having good methods to represent and manipulate graphs in a computer. We frequently use matrices for representing graphs.

Adjacency Matrix A(G) of a Graph G. **A**(*G*) tells which pairs of nodes are adjacent (i.e., which pairs form edges). Entry $a_{ij} = 1$ if there is an edge linking the *i*th and *j*th nodes; otherwise, $a_{ij} = 0$. Note that $a_{ii} = 0$ unless there is a loop at the *i*th node.

The adjacency matrix **A**(*G*) of the graph *G* in Figure 2.2 is

$$\mathbf{A}(G) = \begin{array}{c} \\ a \\ b \\ c \\ d \\ e \end{array} \begin{array}{c} \begin{array}{ccccc} a & b & c & d & e \end{array} \\ \left[\begin{array}{ccccc} 0 & 1 & 1 & 1 & 0 \\ 1 & 0 & 1 & 1 & 0 \\ 1 & 1 & 0 & 0 & 0 \\ 1 & 1 & 0 & 0 & 1 \\ 0 & 0 & 0 & 1 & 1 \end{array} \right] \end{array} \tag{1}$$

Matrix **A**(*G*) is *symmetric*; that is, $a_{ij} = a_{ji}$.

Let us see how a question about graphs can be solved in terms of this matrix.

Example 1. Paths in Graphs

A path is a sequence of nodes such that each consecutive pair of nodes in the path is linked by an edge. The *length* of a path is the number of edges along it. For example, in Figure 2.2, (*a*, *b*, *d*, *e*) is a path of length 3 between *a* and *e*. A single edge is a path of length 1.

We claim that the *i*th and *j*th nodes can be joined by a path of length 2 if and only if entry (i, j) in $A^2(G)$, the square of the adjacency matrix $A(G)$, is positive.

First let us compute $A^2(G)$ for the graph in Figure 2.2. To do this, we must find the scalar product of each row of $A(G)$ with each column of $A(G)$. Since $A(G)$ is symmetric, this is equivalent to finding the scalar product of each row with every other row. (Why?) Consider the scalar product of *a*'s row and *b*'s row in $A(G)$:

$$
\begin{array}{c}
a \\
b \\
\\
\\
\\
\end{array}
\begin{bmatrix}
0 & 1 & 1 & 1 & 0 \\
1 & 0 & 1 & 1 & 0 \\
\cdot & \cdot & \cdot & \cdot & \cdot \\
\cdot & \cdot & \cdot & \cdot & \cdot \\
\cdot & \cdot & \cdot & \cdot & \cdot \\
\end{bmatrix}
\tag{2}
$$

$$
[0, 1, 1, 1, 0] \cdot [1, 0, 1, 1, 0] = 0 \times 1 + 1 \times 0 + 1 \times 1 +
$$
$$
1 \times 1 + 0 \times 0 = 2
$$

The product of two entries in this scalar product will be 1 if and only if the two entries are both 1. Thus the value of the scalar product is simply the number of positions where the two vectors both have a 1. With this observation, it is easy to compute all the scalar products that form the entries of $A^2(G)$.

$$
A^2(G) =
\begin{array}{c}
\\
a \\
b \\
c \\
d \\
e \\
\end{array}
\begin{array}{c}
\begin{array}{ccccc}
a & b & c & d & e
\end{array} \\
\begin{bmatrix}
3 & 2 & 1 & 1 & 1 \\
2 & 3 & 1 & 1 & 1 \\
1 & 1 & 2 & 2 & 0 \\
1 & 1 & 2 & 3 & 1 \\
1 & 1 & 0 & 1 & 2 \\
\end{bmatrix}
\end{array}
\tag{3}
$$

We now interpret the computation of the scalar product (2) in terms of adjacencies in the graph. In (2), when rows *a* and *b* have a 1 in *c*'s column, this means that *a* and *b* are both adjacent to *c*. From (2) we see that *a* and *b* are both adjacent to nodes *c* and *d*. In general, when two nodes n_i and n_j are adjacent to a common node n_k, then (n_i, n_k, n_j) will be a path of length 2 between n_i and n_j. This proves that the (i, j) entry in $A^2(G)$ equals the number of paths of length 2 between the *i*th and *j*th nodes.

This property extends to higher powers of $A(G)$. The entries of $A^3(G)$ tell how many paths of length 3 join different pairs of nodes. For any positive integer *m*, the entries of $A^m(G)$ tell how many paths of length *m* join different pairs of nodes. Illustrative examples and mathematical verification of this property of $A^m(G)$ are left to the exercises.   ∎

A graph *G* is **connected** if every pair of nodes in *G* are joined by a path. Using powers of $A(G)$, we can determine whether or not a graph is connected. If G has n nodes, any path between two nodes in G has length $\leq n - 1$. So G is connected when there exist paths of length $\leq n - 1$ between all pairs of nodes. To determine whether all such paths exist, we compute $\mathbf{A}^2, \mathbf{A}^3, \ldots, \mathbf{A}^{n-1}(G)$ and check for each (i, j) pair, $i \neq j$, whether entry (i, j) is positive for some power of $\mathbf{A}$. For example, the graph G in Figure 2.2 is seen to be connected; all entries but (c, e) and (e, c) are positive in $\mathbf{A}^2(G)$, and these two zero entries become positive in $\mathbf{A}^3(G)$.

Summarizing our discussion of paths and connectedness, we have

---

Graphs and Matrix Multiplication

1. Let $\mathbf{A}(G)$ be the adjacency matrix of graph G. Then the entry (i, j) in $\mathbf{A}^2(G)$ tells how many paths of length 2 join node i with node j, and more generally, entry (i, j) in $\mathbf{A}^m(G)$ tells how many paths of length m join node i with node j.
2. Let G be an n-vertex graph. G is connected if and only if for each (i, j) pair, $i \neq j$, entry (i, j) is positive in some power $\mathbf{A}^k$, $k = 1, 2, \ldots, n - 1$.

If D is a directed graph (in which an edge (a, b) goes only *from a to b*), its adjacency matrix $\mathbf{A}(D)$ has a 1 in entry (i, j) if there is an edge from node j to node i. Then (i, j) entry in $\mathbf{A}^m(D)$ will tell how many directed paths of length m there are in D from node j to node i.

There is one important scalar product which we shall use in Example 2 that merits discussion. If $\mathbf{b}$ is some vector $\mathbf{b} = [b_1, b_2, \ldots, b_n]$ and $\mathbf{1}$ is a vector of n 1's, the sum of the b_i can be written

$$\sum b_i = b_1 + b_2 + \cdots + b_n$$
$$= 1 \times b_1 + 1 \times b_2 + \cdots + 1 \times b_n$$
$$= \mathbf{1} \cdot \mathbf{b}$$

Example 2. Ranking Track Teams

Suppose that there are five track teams, named Ants (A), Birds (B), Cats (C), Dogs (D), and Elephants (E), that compete in nine meets. The results of the competition are modeled by a directed graph G with nine edges, in which there is an edge from A to B if A beat B (see Figure 2.3).

One approach, the one we will use, is to give two points to team A for each team that A beats and to give one point to team A for each case where team A beats a team, say D, that in turns beats another team, say B (this way, team A gets several points when A beats a ''good'' team that has beaten other teams). In the directed graph, node A gets two points for each edge directed out from A and gets one point for each path of length 2 directed out from A. Recall that the number of paths of length 2 from node j to node i is entry (i, j) in $\mathbf{A}^2(G)$.

Figure 2.3

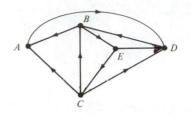

$$\begin{array}{c} \text{From} \\ \begin{array}{ccccc} A & B & C & D & E \end{array} \\ \mathbf{A}(D) = \text{To} \begin{array}{c} A \\ B \\ C \\ D \\ E \end{array} \begin{bmatrix} 0 & 1 & 1 & 0 & 0 \\ 0 & 0 & 1 & 1 & 0 \\ 0 & 0 & 0 & 0 & 1 \\ 1 & 0 & 1 & 0 & 1 \\ 0 & 1 & 0 & 0 & 0 \end{bmatrix} \end{array}$$

The question is: How to rank the five teams?

If **1** denotes a vector of five 1's and **2** denotes a vector of five 2's, then team A's total score will be the scalar product of **2** times A's column in $\mathbf{A}(G)$ plus the scalar product of **1** times A's column in $\mathbf{A}^2(G)$. To get the scores of all the teams at once, we multiply **2** times all columns of $\mathbf{A}(G)$—that is, compute $2\mathbf{A}(G)$— and add it to **1** times all columns of $\mathbf{A}^2(D)$—that is, compute $1\mathbf{A}^2(G)$. Our computations yield (here we have already computed $\mathbf{A}^2(G)$ on a computer):

$$2\mathbf{A}(G) + 1\mathbf{A}^2(G) = [2 \quad 2 \quad 2 \quad 2 \quad 2] \begin{bmatrix} 0 & 1 & 1 & 0 & 0 \\ 0 & 0 & 1 & 1 & 0 \\ 0 & 0 & 0 & 0 & 1 \\ 1 & 0 & 1 & 0 & 1 \\ 0 & 1 & 0 & 0 & 0 \end{bmatrix}$$

$$+ [1 \quad 1 \quad 1 \quad 1 \quad 1] \begin{bmatrix} 0 & 0 & 1 & 1 & 1 \\ 1 & 0 & 1 & 0 & 2 \\ 0 & 1 & 0 & 0 & 0 \\ 0 & 2 & 1 & 0 & 1 \\ 0 & 0 & 1 & 1 & 0 \end{bmatrix}$$

$$= [2, 4, 6, 2, 4] + [1, 3, 4, 2, 4]$$

$$\begin{array}{ccccc} A & B & C & D & E \end{array}$$
$$= [3, 7, 10, 4, 8]$$

Based on this scoring system, the ranking of the teams would be C first, E second, B third, D fourth, A fifth. The reader may want to experiment with other ranking systems. ∎

The fact that matrix multiplication can be used to answer questions about paths in graphs is quite unexpected. But it helps prove the point that a very large number of diverse mathematical problems can be analyzed with matrices and linear models. This example also shows how useful it can be to interpret the meaning of mathematical operations, such as matrix multiplication, in terms of the system being modeled.

Examples 3 and 4 involve mathematical schemes to encode information in a fashion designed to detect, and if possible correct, errors introduced by noise when messages are transmitted. A **binary code** is a scheme for encoding a letter or number as a binary sequence of 0's and 1's and then decoding the binary sequence back into a letter or number. The binary sequence often is transmitted over a communications channel with random noise that may change one of the digits in the binary sequence. That is, when a 1 is sent, a 0 may be received; or vice versa. We assume that the chance of two errors in one binary sequence is small enough that it can be ignored.

The examples about binary codes involve multiplication and addition mod 2. Because computers represent numbers in terms of 0's and 1's (a circuit is open or closed), it is very easy for computers to calculate mod 2. The following tables summarize the rules for addition and multiplication mod 2.

+ mod 2	0	1		× mod 2	0	1
0	0	1		0	0	0
1	1	0		1	0	1

The sum of many 1's is 0 if there are an even number of 1's, and is 1 if there are an odd number of 1's. For example, the following scalar product is calculated in arithmetic mod 2:

$$[1, 1, 0, 1, 1] \cdot [1, 0, 1, 1, 1]$$
$$= 1 \times 1 + 1 \times 0 + 0 \times 1 + 1 \times 1 + 1 \times 1$$
$$= 1 + 0 + 0 + 1 + 1 \quad (\text{mod } 2)$$
$$= 1 \quad (\text{mod } 2)$$

Example 3. **Parity-Bit Code for Error Detection**

Error-detecting binary codes are designed to make it possible to detect an error should one occur during transmission. Then the transmitter would be asked to send the binary sequence again. A standard error-detecting code used to transmit data over telephone lines between computers is a *parity-bit code*.

The basic unit in such communication is usually a *byte*, an 8-bit binary sequence. Let $\mathbf{b} = [b_1, b_2, \ldots, b_8]$ be the byte to be sent. In a parity-bit code, an additional 9th bit p is added to $\mathbf{b}$ to get the sequence to be transmitted, $\mathbf{c} = [b_1, b_2, \ldots, b_8, p]$. This bit p is normally chosen so that the number of 1's in $\mathbf{c}$ is even. Another way to say this is: Pick p so that the sum of the bits in $\mathbf{c}$ equals 0   mod 2:

$$b_1 + b_2 + \cdots + b_8 + p = 0 \quad (\text{mod } 2)$$

or equivalently,

$$\mathbf{1} \cdot \mathbf{c} = 0 \quad \text{mod } 2$$

For example, if the byte to be sent is $\mathbf{b} = [1, 0, 1, 0, 0, 1, 0, 0]$, which has an odd number of 1's, then $p = 1$ and we send $\mathbf{c} = [1, 0, 1, 0, 0, 1, 0, 0, 1]$. Suppose that the message we received was $\mathbf{c}' = [1, 0, 0, 0, 0, 1, 0, 0, 1]$—the third bit was erroneously changed to 0.

Whenever a message $\mathbf{c}'$ is received, we compute $\mathbf{1} \cdot \mathbf{c}'$. If no errors had occurred and $\mathbf{c}' = \mathbf{c}$, then we would find that $\mathbf{1} \cdot \mathbf{c}' = 0$ mod 2. On the other hand, if $\mathbf{1} \cdot \mathbf{c}' = 1$, as in this example, then some digit was altered—and we ask the sender to retransmit.

A simple way to compute the proper value for the parity bit p is to let

$$p = \Sigma\, b_i = \mathbf{1} \cdot \mathbf{b} \quad (\text{mod } 2) \tag{4}$$

That is, let p equal the sum mod 2 of the 1's in $\mathbf{b}$. If $p = \mathbf{1} \cdot \mathbf{b} = 1$ mod 2, $\mathbf{b}$ has an odd number of 1's and making p 1 will give $\mathbf{c}$ an even number of 1's. If $p = \mathbf{1} \cdot \mathbf{b} = 0$ mod 2, then $\mathbf{b}$ already has an even number of 1's and we want p to be 0.   ■

Example 4. Hamming Code for Error Correction

More advanced *error-correcting* codes can actually correct an error and reconstruct the original binary sequence that was sent. The following scheme due to Hamming takes a 4-bit binary sequence and encodes it as a 7-bit sequence. Let $\mathbf{b} = [b_1, b_2, b_3, b_4]$ be the binary message, let p_1, p_2, p_3 be the parity-check bits, and let $\mathbf{c} = [c_1, c_2, \ldots, c_7]$ be the code sequence that will be transmitted. The parity-check bits are chosen to satisfy the following three parity checks:

$$
\begin{array}{llllll}
p_1 & + b_1 + b_2 & + b_4 = 0 & (\text{mod } 2) \\
\quad\ p_2 & + b_1 & + b_3 + b_4 = 0 & (\text{mod } 2) & \quad (5) \\
\quad\quad\ p_3 & + b_2 + b_3 + b_4 = 0 & (\text{mod } 2)
\end{array}
$$

Let us encode these message and parity-check bits in the code sequence $\mathbf{c}$ as follows:

$$c_1 = p_1, \qquad c_2 = p_2, \qquad c_3 = b_1, \qquad c_4 = p_3,$$
$$c_5 = b_2, \qquad c_6 = b_3, \qquad c_7 = b_4 \tag{6}$$

The reason why $c_3 = b_1$, not p_3, will be clear shortly. Now (5) is

$$
\begin{aligned}
c_1 \quad\;\; + c_3 \quad\quad + c_5 \quad\quad + c_7 &= 0 \quad (\text{mod } 2) \\
c_2 + c_3 \quad\quad\quad\; + c_6 + c_7 &= 0 \quad (\text{mod } 2) \\
c_4 + c_5 + c_6 + c_7 &= 0 \quad (\text{mod } 2)
\end{aligned}
\tag{7}
$$

or

$$\mathbf{Mc} = \mathbf{0} \quad (\text{mod } 2)$$

where $\mathbf{0}$ is a vector of all 0's and $\mathbf{M}$ is the matrix of coefficients in (7):

$$
\mathbf{M} = \begin{bmatrix}
1 & 0 & 1 & 0 & 1 & 0 & 1 \\
0 & 1 & 1 & 0 & 0 & 1 & 1 \\
0 & 0 & 0 & 1 & 1 & 1 & 1
\end{bmatrix}
\tag{8}
$$

Each of the c_i's is involved in a different subset of parity checks, so when (exactly) one c_i is altered in transmission, the parity checks allow us to determine which bit was changed.

Since each of the parity bits $c_1 \,(= p_1)$, $c_2 \,(= p_2)$, $c_4 \,(= p_3)$ is in just one of the parity equations in (7), each can be determined as the sum of the other bits in their equation (just as $p = \Sigma\, b_i$ in Example 2). Summarizing how we go from the message vector $\mathbf{b}$ to the code vector $\mathbf{c}$, we obtain:

$$c_1 = b_1 + b_2 + b_4, \qquad c_2 = b_1 + b_3 + b_4, \qquad c_3 = b_1,$$
$$c_4 = b_2 + b_3 + b_4, \qquad c_5 = b_2, \qquad c_6 = b_3, \tag{9}$$
$$c_7 = b_4$$

The following matrix-vector product (mod 2) does the encoding specified by (9).

$$
\mathbf{c} = \mathbf{Qb}, \qquad \text{where } \mathbf{Q} = \begin{bmatrix}
1 & 1 & 0 & 1 \\
1 & 0 & 1 & 1 \\
1 & 0 & 0 & 0 \\
0 & 1 & 1 & 1 \\
0 & 1 & 0 & 0 \\
0 & 0 & 1 & 0 \\
0 & 0 & 0 & 1
\end{bmatrix}
\tag{10}
$$

For example, suppose that $\mathbf{b} = [1, 0, 1, 0]$. Then

$$\mathbf{c} = \mathbf{Qb} = \begin{bmatrix} 1 & 1 & 0 & 1 \\ 1 & 0 & 1 & 1 \\ 1 & 0 & 0 & 0 \\ 0 & 1 & 1 & 1 \\ 0 & 1 & 0 & 0 \\ 0 & 0 & 1 & 0 \\ 0 & 0 & 0 & 1 \end{bmatrix} \begin{bmatrix} 1 \\ 0 \\ 1 \\ 0 \end{bmatrix}$$

$$= \begin{bmatrix} 1\times1 + 1\times0 + 0\times1 + 1\times0 \\ 1\times1 + 0\times0 + 1\times1 + 1\times0 \\ 1\times1 + 0\times0 + 0\times1 + 0\times0 \\ 0\times1 + 1\times0 + 1\times1 + 1\times0 \\ 0\times1 + 1\times0 + 0\times1 + 0\times0 \\ 0\times1 + 0\times0 + 1\times1 + 0\times0 \\ 0\times1 + 0\times0 + 0\times1 + 1\times0 \end{bmatrix} = \begin{bmatrix} 1 \\ 0 \\ 1 \\ 1 \\ 0 \\ 1 \\ 0 \end{bmatrix}$$

$$(11)$$

Suppose that the transmission received was $\mathbf{c}' = [1, 0, 1, 1, 0, 0, 0]$; the sixth bit was changed from 1 to 0. We compute the vector $\mathbf{e}$:

$$\mathbf{e} = \mathbf{Mc}' = \begin{bmatrix} 1 & 0 & 1 & 0 & 1 & 0 & 1 \\ 0 & 1 & 1 & 0 & 0 & 1 & 1 \\ 0 & 0 & 0 & 1 & 1 & 1 & 1 \end{bmatrix} \begin{bmatrix} 1 \\ 0 \\ 1 \\ 1 \\ 0 \\ 0 \\ 0 \end{bmatrix} = \begin{bmatrix} 0 \\ 1 \\ 1 \end{bmatrix} \qquad (12)$$

Note that $\mathbf{Mc}'$ is just the set of left sides in (7) with $\mathbf{c}$ replaced by $\mathbf{c}'$. If no error had occurred and $\mathbf{c}' = \mathbf{c}$, then $\mathbf{e} = \mathbf{Mc} = \mathbf{0}$, as in (7). If $\mathbf{e} \neq \mathbf{0}$, as in (12), an error must have occurred. Depending on which parity equations are now violated, we can figure out which bit in $\mathbf{c}'$ was changed in transmission. We claim that $\mathbf{e}$ ($= \mathbf{Mc}'$) equals the column of $\mathbf{M}$ corresponding to the bit of $\mathbf{c}$ that was changed. This is the case in (12), where $\mathbf{e}$ equals the sixth column of $\mathbf{M}$. The reason is that when the kth bit is altered, exactly those equations involving the kth bit (i.e., those rows of $\mathbf{M}$ with a 1 in the kth column) will now equal 1 (mod 2).

As the reader has probably noticed, for each i, the ith column of $\mathbf{M}$ is simply the binary representation of the number i. Thus the vector $\mathbf{e}$ "spells out" the location of the bit that was changed. To get the correct transmission, we simply change back the bit in the position spelled out by $\mathbf{e}$. In this instance we would change the sixth bit from a 0 back to a 1. ∎

Optional

Another useful matrix associated with a graph is the incidence matrix $\mathbf{M}(G)$.

> ***Incidence Matrix*** $\mathbf{M}(G)$ ***of a Graph*** G. $\mathbf{M}(G)$ has a row for each node
> of G and a column for each edge of G. If the jth edge is incident to
> the ith node (i.e., the ith node is an endpoint of the jth edge), then
> entry $m_{ij} = 1$; entry $m_{ii} = 2$ if there is a loop edge at i; and otherwise,
> $m_{ij} = 0$.

The incidence matrix for the graph in Figure 2.2 is

$$
\mathbf{M}(G) = \begin{array}{c} \\ a \\ b \\ c \\ d \\ e \end{array}
\begin{array}{c}
\begin{array}{ccccccc} e_1 & e_2 & e_3 & e_4 & e_5 & e_6 & e_7 \end{array} \\
\left[\begin{array}{ccccccc}
1 & 1 & 1 & 0 & 0 & 0 & 0 \\
1 & 0 & 0 & 1 & 1 & 0 & 0 \\
0 & 1 & 0 & 1 & 0 & 0 & 0 \\
0 & 0 & 1 & 0 & 1 & 1 & 0 \\
0 & 0 & 0 & 0 & 0 & 1 & 2
\end{array} \right]
\end{array}
\tag{13}
$$

$\mathbf{M}(G)$ will always have exactly two 1's in each column (or one 2), since an
edge has two endpoints.

The next example shows how, using $\mathbf{M}(G)$, one can recast a graph
optimization problem as a linear program. Although the reformulation is not
hard to follow, it required considerable ingenuity to think it up.

Example 5. **Finding a Maximum Independent Set**

An *independent set* I of nodes in a graph is a set of nodes with no
linking edges. For example, $\{a, e\}$ is an independent set of nodes for
the graph in Figure 2.2. Independent sets arise in various settings. For
example, let G be a graph in which each node stands for a letter that
can be transmitted over a noisy communications channel and an edge
joins two nodes if the corresponding letters can be confused when
transmitted (i.e., one letter is sent but another letter is received). An
independent set would represent a set of letters that cannot be confused
with one another.

We shall now recast the property of being an independent set in
terms of a set of linear inequalities. We assume that G has n nodes.
The key step is to represent a set of nodes by a membership vector.
The *membership vector* $\mathbf{x} = [x_1, x_2, \ldots, x_n]$ for a set I is defined to
have $x_i = 1$ if the ith node is in the set I and $x_i = 0$ otherwise.

We claim that $\mathbf{x}$ is the membership vector for an independent set
if and only if the following inequality holds.

$$
\mathbf{x}\mathbf{M}(G) \le \mathbf{1}
\tag{14}
$$

Recall that putting the vector before the matrix means that we are forming the scalar product of **x** with each column of $\mathbf{M}(G)$; the columns of $\mathbf{M}(G)$ correspond to the edges of G. For the graph in Figure 2.2, (14) becomes

$$
\begin{array}{lll}
\text{Edge } e_1: & x_a + x_b & \leq 1 \\
\text{Edge } e_2: & x_a \quad\;\; + x_c & \leq 1 \\
\text{Edge } e_3: & x_a \qquad\quad + x_d & \leq 1 \\
\text{Edge } e_4: & x_b + x_c & \leq 1 \qquad (15) \\
\text{Edge } e_5: & x_b \quad\; + x_d & \leq 1 \\
\text{Edge } e_6: & x_d + \; x_e & \leq 1 \\
\text{Edge } e_7: & 2x_e & \leq 1
\end{array}
$$

Recall that each column of $\mathbf{M}(G)$ has two 1's, which correspond to the pair of nodes that form that edge. Then the left side of the first inequality in (15) says that nodes a and b are the endpoints of edge e_1. The condition that $x_a + x_b \leq 1$ says that not both a and b can be in the independent set I (i.e., not both $x_a = 1$ and $x_b = 1$).

We are ready to pose the problem of finding a *maximum* independent set of G as a linear program. In the graph model for confusing letters sent over a noisy communication channel, a maximum independent set would be the largest possible set of letters that can be sent without one being confused with another.

We must restate the concept of maximizing the size of the independent set in terms of membership vector. What we want is to maximize the number of 1's in the membership vector. Another way to say this is to maximize the sum of the x_i's. But $\Sigma\, x_i = \mathbf{1} \cdot \mathbf{x}$. Combining this fact with (14), we have the linear program

$$
\begin{aligned}
&\text{Maximize } \mathbf{1} \cdot \mathbf{x} \\
&\text{subject to } \mathbf{x}\mathbf{M}(G) \leq 1 \\
&\text{and } x_i = 0 \text{ or } 1
\end{aligned}
$$

Because of the integer constraint, such a linear program is called an **integer program**. There is a large literature about solving integer programs. All optimization problems in graph theory can be posed as integer problems. ∎

Section 2.3 Exercises

Summary of Exercises
Exercises 1–10 deal with adjacency matrices and paths in graphs, Exercises 8–10 being "theoretical." Exercises 11–20 concern coding problems. Exercises 21–23 deal with the optional material at the end of the section.

1. Draw the graphs with the following adjacency matrices.

(a) $\begin{bmatrix} 0 & 0 & 0 & 1 \\ 0 & 0 & 1 & 0 \\ 0 & 1 & 0 & 0 \\ 1 & 0 & 0 & 0 \end{bmatrix}$
(b) $\begin{bmatrix} 0 & 1 & 1 & 1 \\ 1 & 0 & 0 & 1 \\ 1 & 0 & 0 & 1 \\ 1 & 1 & 1 & 0 \end{bmatrix}$
(c) $\begin{bmatrix} 0 & 0 & 1 & 1 & 0 & 0 \\ 0 & 1 & 1 & 0 & 1 & 1 \\ 1 & 1 & 0 & 0 & 0 & 1 \\ 1 & 0 & 0 & 1 & 1 & 1 \\ 0 & 1 & 0 & 1 & 0 & 0 \\ 0 & 1 & 1 & 1 & 0 & 1 \end{bmatrix}$

2. Write the adjacency matrices for the following graphs.

(a) G_1 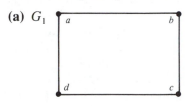   (b) G_2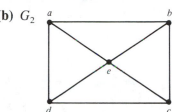

(c) G_3 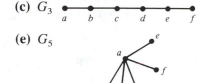   (d) G_4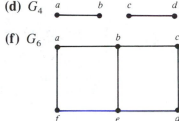

(e) G_5 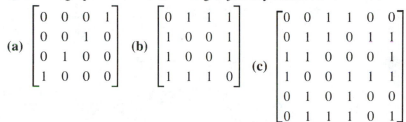   (f) G_6

3. Compute the square of the adjacency matrix for the following graphs from Exercise 2.
   (a) G_1 (b) G_2 (c) G_3 (d) G_4 (e) G_5 (f) G_6
   Use your answer to tell how many paths of length 2 there are in each graph between vertex a and vertex d.

4. Compute the cube of the adjacency matrix for the following graphs from Exercise 2.
   (a) G_1 (b) G_2 (c) G_3 (d) G_4 (e) G_5 (f) G_6

   Note: You may use a computer program to do this computation.
        Use your answer here along with that in Exercise 3 to tell if all vertices in the graph are joined by a path of length ≤ 3.

5. Use your calculation in Exercise 4 to show that G_1 and G_2 are connected.

6. Direct the edges in the graphs G_1, G_2 in Exercise 2 from the earlier node to the later node according to the alphabetical order of the nodes; for example, edge (a, c) goes from a to c.

(a) Write the adjacency matrix for graphs G_1 and G_2 and compute the square of each.

(b) Give an argument to explain why entry (i, j) in the square $\mathbf{A}^2(\mathbf{D})$ of the adjacency matrix $\mathbf{A}(D)$ of a directed graph D tells how many directed paths there are in D from j to i.

7. For the directed versions (see Exercise 6) of graphs G_1, G_2 in Exercise 2, compute the total points for each node according to the method in Example 2.

8. For an undirected graph G, show that the result of the matrix-vector product $\mathbf{A}(G)\mathbf{1}$ (where $\mathbf{1}$ is a vector of 1's) is a vector in which the ith position tells how many nodes are adjacent to node i.

9. (a) Explain in words why, if entry (i, j) in $\mathbf{A}(G)^2$ is the number of paths of length 2 between nodes i and j in graph G, entry (i, j) in $\mathbf{A}(G)^3$ is the number of paths of length 3 between nodes i and j in G.

(b) (Advanced) Extend the argument in part (a) by induction to show that entry (i, j) in $\mathbf{A}(G)^k$ is the number of paths of length k between nodes i and j in G.

10. (a) Suppose that we redefined the adjacency matrix $\mathbf{A}(G)$ so that the diagonal entries (i, i) were all 1. Now what is the interpretation of entry (i, j) being positive in $\mathbf{A}^2(G)$? In $\mathbf{A}^k(G)$?

(b) Compute $\mathbf{A}^2(G)$ for this redefined adjacency matrix for graph G_1 in Exercise 2.

11. If the following bytes (8-bit binary sequences) are being sent in a parity-check code, what should the additional parity bit be (0 or 1)?
(a) 10101010   (b) 11100110   (c) 00000000

12. Suppose that the following 9-bit messages were received in a parity-check code. Which messages are altered during transmission?
(a) 101010101   (b) 101101101   (c) 000000000

13. Explain why if two errors occur during transmission, a parity-check code of the sort in Example 3 will not detect an error.

14. Suppose that in the Hamming code in Example 4, the following 4-bit binary messages $\mathbf{b} = [b_1, b_2, b_3, b_4]$ are to be sent. What will the coded 7-bit message $\mathbf{c} = [c_1, c_2, c_3, c_4, c_5, c_6, c_7]$ be?
(a) $\mathbf{b} = [1, 1, 0, 0]$ (b) $\mathbf{b} = [1, 1, 1, 0]$ (c) $[1, 1, 1, 1]$

15. Suppose that in the Hamming code in Example 4, the following messages $\mathbf{c}'$ are received. In each case compute the error vector $\mathbf{e} = \mathbf{M}\mathbf{c}'$ and from it tell which bit, if any, was changed in transmission.
(a) $[0, 0, 1, 1, 1, 1, 0]$ (b) $[1, 1, 1, 1, 1, 0, 1]$
(c) $[0, 1, 0, 0, 0, 0, 0]$

16. Assume that at most one error occurred in transmission of the Hamming code in Example 4. What message was originally sent if the following message was received?
    (a) [0, 0, 1, 1, 1, 1, 0]   (b) [0, 1, 0, 0, 0, 0, 0, 0]
    (c) [1, 1, 1, 0, 0, 1, 0]

17. Suppose that we let $c_1 = p_1$, $c_2 = p_2$, $c_3 = p_3$, $c_4 = b_1$, $c_5 = b_2$, $c_6 = b_3$, $c_7 = b_4$ instead of the encoding scheme in (6). With this encoding scheme, what is the new **Q** matrix in (10)? If **b** = [1, 1, 0, 0], what would **c** be?

18. (a) Explain why the Hamming code in Example 4 will always detect two errors; that is, if two bits in the code are changed, the error vector **e** cannot be all 0's.
    (b) Give an example to show that the Hamming code in Example 4 cannot correct two errors.
    (c) Give an example to show that the Hamming code cannot always detect three errors.

19. The Hamming code in Example 4 can be extended to a similar code for 15-bit sequences—11 messages bits and 4 parity-check bits. Write out the system of parity-check equations (or equivalently, the matrix for coefficients for these equations) for a 15-bit Hamming code.

20. Another way to encode a binary sequence is by treating the sequence as the coefficients of a polynomial $p(x)$; for example, the sequence [1, 0, 1, 1] yields the polynomial $p(x) = 1 + 0x + 1x^2 + 1x^3$. We encode by multiplying $p(x)$ by some other polynomial $g(x)$ to get the polynomial $p^*(x) = g(x)p(x)$, whose coefficients we transmit. For example, let $g(x) = 1 + x$. Then for $p(x) = 1 + 0x + 1x^2 + 1x^3$, we compute $p^*(x) = g(x)p(x) = (1 + x)(1 + 0x + 1x^2 + 1x^3) = 1 + 1x + 1x^2 + 0x^3 + 1x^4$. (Remember that arithmetic is mod 2.) So we transmit [1, 1, 1, 0, 1]. To decode, we divide the polynomial $p^*(x)$ by $g(x)$. If any error occurred in transmission, there will be a reminder—this tells us that an error occurred.
    (a) Using $g(x) = 1 + x$, perform a polynomial encoding of the sequence [1, 1, 0, 1].
    (b) Suppose that the following messages are received, based on this polynomial encoding with $g(x) = 1 + x$. Which ones have errors?
    (i) [1, 1, 0, 1, 0]   (ii) [1, 0, 1, 0]   (iii) [1, 0, 0, 0]
    (c) (Advanced) Show that the parity of messages transmitted is always even with the polynomial encoding scheme with $g(x) = 1 + x$.

    Hint: By setting $x = 1$ in $p^*(x)$ [$= g(x)p(x)$], one can sum the coefficients (i.e., the message bits).

Exercises for Optional Material

21. Write the incidence matrix **M**(G) for the following graphs in Exercise 2.
    (a) G_1 (b) G_2 (c) G_5

22. Write out the linear program for finding a maximum independent set in the following graphs in Exercise 2.
(a) G_1 (b) G_2 (c) G_5

23. Show that $1 \cdot M(G) = 21 = [2, 2, 2, \ldots, 2]$ for any graph.

Section 2.4 Matrix Algebra

In this section we introduce the algebra of matrices. Algebraic techniques are used in high school to manipulate, simplify, and solve equations, such as rewriting $2x - 2y = 4$ as $y = x - 2$. Similar methods exist for matrices.

In Sections 2.2 and 2.3 we used matrix notation to express systems of linear equations. For example, in Example 2 of Section 2.2 we wrote the refinery equations

$$
\begin{aligned}
20x_1 + 4x_2 + 4x_3 &= 500 \\
10x_1 + 14x_2 + 5x_3 &= 850 \\
5x_1 + 5x_2 + 12x_3 &= 1000
\end{aligned}
\tag{1}
$$

as

$$
\begin{bmatrix} 20 & 4 & 4 \\ 10 & 14 & 5 \\ 5 & 5 & 12 \end{bmatrix} \cdot \begin{bmatrix} x_1 \\ x_2 \\ x_3 \end{bmatrix} = \begin{bmatrix} 20x_1 + 4x_2 + 4x_3 \\ 10x_1 + 14x_2 + 5x_3 \\ 5x_1 + 5x_2 + 12x_3 \end{bmatrix} = \begin{bmatrix} 500 \\ 850 \\ 1000 \end{bmatrix}
$$

or, in matrix notation,

$$
\mathbf{Ax} = \mathbf{b}
\tag{2}
$$

where $\mathbf{A}$ is the 3-by-3 matrix of coefficients, $\mathbf{b}$ is the right-side vector, and $\mathbf{x}$ is the vector of unknowns.

Just as the matrix notation of $\mathbf{Ax} = \mathbf{b}$ gives a concise way to write the system of equations in (1), so matrix algebra provides a concise, powerful way to manipulate and solve matrix equations. As one would expect, the rules of matrix algebra are basically extensions of single-variable high school algebra. We start with some examples that illustrate the power of matrix algebra.

First we need to define the **ones vector 1** and **identity matrix I**. The vector $\mathbf{1}$ is simply a vector of all 1's:

$$
\mathbf{1} = [1, 1, \ldots, 1]
$$

When we write a scalar product such as $\mathbf{1} \cdot \mathbf{b}$, we assume that the ones vector $\mathbf{1}$ has the same length as $\mathbf{b}$ (so that the product makes sense). As noted in Section 2.3, the product $\mathbf{1} \cdot \mathbf{b}$ is simply the sum of the elements in the vector $\mathbf{b} = [b_1, b_2, \ldots, b_n]$.

$$\mathbf{1} \cdot \mathbf{b} = b_1 + b_2 + \cdots + b_n \tag{3}$$

The identity matrix $\mathbf{I}$, always a square matrix, has 1's on the main diagonal and zeros elsewhere.

$$\mathbf{I} = \begin{bmatrix} 1 & 0 & 0 & \cdots & 0 \\ 0 & 1 & 0 & \cdots & 0 \\ 0 & 0 & 1 & \cdots & 0 \\ \vdots & \vdots & \vdots & \vdots\vdots\vdots & \vdots \\ 0 & 0 & 0 & \cdots & 1 \end{bmatrix}$$

If we multiply $\mathbf{I}$ times a vector $\mathbf{b}$, the result is simply $\mathbf{b}$.

$$\mathbf{Ib} = \begin{bmatrix} 1 & 0 & 0 & \cdots & 0 \\ 0 & 1 & 0 & \cdots & 0 \\ 0 & 0 & 1 & \cdots & 0 \\ \vdots & \vdots & \vdots & \vdots\vdots\vdots & \vdots \\ 0 & 0 & 0 & \cdots & 1 \end{bmatrix} \cdot \begin{bmatrix} b_1 \\ b_2 \\ b_3 \\ \vdots \\ b_n \end{bmatrix}$$

$$= \begin{bmatrix} 1 \times b_1 + 0 \times b_2 + 0 \times b_3 + \cdots + 0 \times b_n \\ 0 \times b_1 + 1 \times b_2 + 0 \times b_3 + \cdots + 0 \times b_n \\ 0 \times b_1 + 0 \times b_2 + 1 \times b_3 + \cdots + 0 \times b_n \\ \vdots \qquad \vdots \qquad \vdots \qquad \vdots\vdots\vdots \qquad \vdots \\ 0 \times b_1 + 0 \times b_2 + 0 \times b_3 + \cdots + 1 \times b_n \end{bmatrix} = \begin{bmatrix} b_1 \\ b_2 \\ b_3 \\ \vdots \\ b_n \end{bmatrix} = \mathbf{b}$$

In a similar way, one can verify that $\mathbf{bI} = \mathbf{b}$. So we have

$$\mathbf{Ib} = \mathbf{b} = \mathbf{bI} \tag{4}$$

This is why $\mathbf{I}$ is called the identity matrix. As with the ones vector $\mathbf{1}$, we assume that the size of $\mathbf{I}$ equals the length of $\mathbf{b}$.

Equation (4) extends to matrices. That is,

$$\mathbf{IB} = \mathbf{B} = \mathbf{BI} \tag{5}$$

for any matrix $\mathbf{B}$. Note that if $\mathbf{B}$ is an m-by-r matrix, then the $\mathbf{I}$ on the left side of $\mathbf{B}$ must by m-by-m, while the $\mathbf{I}$ on the right side of $\mathbf{B}$ must be r-by-r.

We can use matrix algebra to verify (5). The columns of the matrix product $\mathbf{IB}$ equal the matrix-vector products of $\mathbf{I}$ with each column $\mathbf{b}_j^C$ of $\mathbf{B}$. For example, the first column in $\mathbf{IB}$ is the matrix-vector product $\mathbf{Ib}_1^C$, which by (4) equals $\mathbf{b}_1^C$. In matrix notation we write

$$\mathbf{IB} = \mathbf{I}[\mathbf{b}_1^C, \mathbf{b}_2^C, \mathbf{b}_3^C, \ldots, \mathbf{b}_n^C] = [\mathbf{Ib}_1^C, \mathbf{Ib}_2^C, \mathit{Ib}_3^C, \ldots, \mathit{Ib}_n^C] \quad (6)$$
$$= [\mathbf{b}_1, \mathbf{b}_2, \mathbf{b}_3, \ldots, \mathbf{b}_n] = \mathbf{B}$$

Example 1. Matrix Algebra in the Leontief Economic Model

In Example 4 of Section 2.2 we used matrix notation to represent the system of equations in the Leontief economic model

	Supply	Industrial Demands				Consumer Demand
		Energy	Constr.	Transp.	Steel	
Energy:	$x_1 =$	$.4x_1 +$	$.2x_2 +$	$.2x_3 +$	$.2x_4 +$	100
Construct.	$x_2 =$	$.3x_1 +$	$.3x_2 +$	$.2x_3 +$	$.1x_4 +$	50
Transport.	$x_3 =$	$.1x_1 +$	$.1x_2 +$		$.2x_4 +$	100
Steel:	$x_4 =$		$.1x_2 +$	$.1x_3$		$+$ 0

$$(7)$$

as

$$\mathbf{x} = \mathbf{Dx} + \mathbf{c} \quad (8)$$

where $\mathbf{D}$ is the matrix of coefficients for interindustry demands, $\mathbf{c} = [100, 50, 100, 0]$ is the vector of consumer demand, and $\mathbf{x}$ is the vector of (unknown) production levels x_i.

The standard way to write a system of linear equations is with the x_i all on the left side and constants on the right. If we bring all the x_i over to the left side, (7) becomes

$$\begin{aligned}
.6x_1 - .2x_2 - .2x_3 - .2x_4 &= 100 \\
-.3x_1 + .7x_2 - .2x_3 - .1x_4 &= 50 \\
-.1x_1 - .1x_2 + x_3 - .2x_4 &= 100 \\
- .1x_2 - .1x_3 + x_4 &= 0
\end{aligned} \quad (9)$$

We can also shift the x_i to the left side in matrix notation. We rewrite (8)

$$\mathbf{x} = \mathbf{Dx} + \mathbf{c} \quad \rightarrow \quad \mathbf{x} - \mathbf{Dx} = \mathbf{c} \quad (10)$$

Recall from (4) that $\mathbf{x} = \mathbf{Ix}$ ($\mathbf{I}$ is the identity matrix). Using this fact, we can rewrite (10)

$$\mathbf{x} - \mathbf{Dx} = \mathbf{c} \quad \rightarrow \quad \mathbf{Ix} - \mathbf{Dx} = \mathbf{c} \quad (11)$$
$$\rightarrow \quad (\mathbf{I} - \mathbf{D})\mathbf{x} = \mathbf{c}$$

Writing $\mathbf{I} - \mathbf{D}$ out, we have

$$\begin{bmatrix} 1 & 0 & 0 & 0 \\ 0 & 1 & 0 & 0 \\ 0 & 0 & 1 & 0 \\ 0 & 0 & 0 & 1 \end{bmatrix} - \begin{bmatrix} .4 & .2 & .2 & .2 \\ .3 & .3 & .2 & .1 \\ .1 & .1 & 0 & .2 \\ 0 & .1 & .1 & 0 \end{bmatrix}$$

$$= \begin{bmatrix} .6 & -.2 & -.2 & -.2 \\ -.3 & .7 & -.2 & -.1 \\ -.1 & -.1 & 1 & -.2 \\ 0 & -.1 & -.1 & 1 \end{bmatrix} \quad (12)$$

The resulting matrix in (12) is the coefficient matrix of (9).

In Section 1.2 we stated Leontief's input constraint that every industry be profitable; that is, making 1 dollar's worth of the *i*th commodity should cost less than 1 dollar. Recall that the input costs of energy are the coefficients in the Energy column of the demand matrix **D**; similarly for other commodities. So Leontief's constraint is that each column sum in **D** should be less than 1.

We now develop a compact matrix inequality that expresses this constraint on **D**. Using the ones vector **1**, we can write Leontief's constraint (where $\mathbf{d}_j^C$ is the *j*th column of **D**):

Input constraint: $\qquad \mathbf{1} \cdot \mathbf{d}_j^C = d_{1j} + d_{2j} + d_{3j} + d_{4j} < 1 \quad (13)$

[This use of **1** in summing was discussed in (3)]. Combining (13) for all columns, we have

$$[\mathbf{1} \cdot \mathbf{d}_1^C, \, \mathbf{1} \cdot \mathbf{d}_2^C, \, \mathbf{1} \cdot \mathbf{d}_3^C, \, \mathbf{1} \cdot \mathbf{d}_4^C] < [1, 1, 1, 1] = \mathbf{1}$$

The vector inequality $<$ means term-by-term inequality. Factoring **1** out in front, as with one-variable expressions, we have

$$\mathbf{1}[\mathbf{d}_1^C, \, \mathbf{d}_2^C, \, \mathbf{d}_3^C, \, \mathbf{d}_4^C] < \mathbf{1}$$

or

$$\mathbf{1D} < \mathbf{1} \qquad\qquad (14)$$

This is the compact mathematical way to say that all column sums are less than 1 in the interindustry matrix **D**. In Section 3.4 we shall use $\mathbf{1D} < \mathbf{1}$ to prove that all Leontief economic models have a solution.

∎

Example 2. **Matrix Algebra in Markov Chains**

We showed in Section 2.2 that the transition equations for a Markov chain can be written in matrix notation as

$$\mathbf{p}' = \mathbf{Ap} \tag{15}$$

or

$$p'_i = \mathbf{a}^R_i \mathbf{p} \tag{16}$$

where **A** is the Markov transition matrix, **p** is the vector of the current probability distribution, and **p**′ is the vector of the next-period probability distribution. For the frog Markov chain, the system of equations **p**′ = **Ap** is

$$
\begin{aligned}
p'_1 &= .50p_1 + .25p_2 \\
p'_2 &= .50p_1 + .50p_2 + .25p_3 \\
p'_3 &= .25p_2 + .50p_3 + .25p_4 \\
p'_4 &= .25p_3 + .50p_4 + .25p_5 \\
p'_5 &= .25p_4 + .50p_5 + .50p_6 \\
p'_6 &= .25p_5 + .50p_6
\end{aligned}
$$

The next-period calculations represented by (15) can be repeated to find the distribution **p**″ after two periods. Using matrix algebra, we have

$$\mathbf{p}'' = \mathbf{A}(\mathbf{Ap}) = \mathbf{A}^2\mathbf{p}$$

and after *n* periods, the distribution vector **p**$^{(n)}$ is

$$\mathbf{p}^{(n)} = \mathbf{A}^n\mathbf{p}$$

The current probabilities p_i, the entries in **p**, must sum to 1. We can express this fact with the ones vector **1** as

$$\mathbf{1} \cdot \mathbf{p} = p_1 + p_2 + \cdots + p_n = 1 \tag{17}$$

The entries in the columns $\mathbf{a}^C_j$ in **A** must also sum to 1.

$$\mathbf{1} \cdot \mathbf{a}^C_j = a_{1j} + a_{2j} + \cdots + a_{nj} = 1 \tag{18}$$

Combining all the columns together into **A**, we see that (18) yields

$$
\begin{aligned}
\mathbf{1A} &= [\mathbf{1} \cdot \mathbf{a}^C_1, \mathbf{1} \cdot \mathbf{a}^C_2, \ldots, \mathbf{1} \cdot \mathbf{a}^C_n] \\
&= [1, 1, \ldots, 1] = \mathbf{1}
\end{aligned} \tag{19}
$$

Matrix algebra allows us first to represent the column sum being 1 concisely and then also allows us to state the fact for all columns at once as **1A** = **1**.

Equations **p**′ = **Ap**, **1** · **p** = 1, and **1A** = **1** can be used to show that in the next-period distribution vector **p**′, the entries p'_i also sum to 1. That is, we want to prove

$$\mathbf{1} \cdot \mathbf{p}' \overset{?}{=} 1: \quad \mathbf{1} \cdot \mathbf{p}' = \mathbf{1} \cdot (\mathbf{Ap}) \qquad \text{since } \mathbf{p}' = \mathbf{Ap}$$
$$= (\mathbf{1A}) \cdot \mathbf{p} = \mathbf{1} \cdot \mathbf{p} \quad \text{since } \mathbf{1A} = \mathbf{1} \qquad (20)$$
$$= 1 \qquad \qquad \text{since } \mathbf{1} \cdot \mathbf{p} = 1$$

This argument can be repeated to show that the entries sum to 1 in $\mathbf{p}''$, the distribution vector in two periods, and more generally in $\mathbf{p}^{(n)}$. ∎

Transpose of a Matrix and Symmetric Matrices

The operation of transposing a matrix has many theoretical and practical uses. In this book, its primary use is in computing pseudoinverses (in Section 5.3).

The **transpose** of a matrix $\mathbf{A}$, written $\mathbf{A}^T$, is the matrix obtained from $\mathbf{A}$ by interchanging rows and columns. Another way to think of it is, flipping $\mathbf{A}$'s entries around the main diagonal. For example, if

$$\mathbf{A} = \begin{bmatrix} 1 & 2 \\ 4 & 5 \\ 7 & 8 \end{bmatrix}, \qquad \text{then } \mathbf{A}^T = \begin{bmatrix} 1 & 4 & 7 \\ 2 & 5 & 8 \end{bmatrix}$$

Transposes have the following properties:

$$\mathbf{A}^T + \mathbf{B}^T = (\mathbf{A} + \mathbf{B})^T \qquad \qquad (21)$$
$$(\mathbf{AB})^T = \mathbf{B}^T\mathbf{A}^T \qquad \text{and} \qquad (\mathbf{Ab})^T = \mathbf{b}^T\mathbf{A}^T \qquad (22)$$
$$(\mathbf{A}^T)^T = \mathbf{A} \qquad \qquad (23)$$

The order of multiplying $\mathbf{A}$ and $\mathbf{B}$ is reversed on the left side of (22) because transposing reverses the roles of rows and columns: If $\mathbf{A}$ is m-by-r and $\mathbf{B}$ is r-by-n, then $\mathbf{A}^T$ is r-by-m and $\mathbf{B}^T$ is n-by-r. We use the notation $\mathbf{b}^T$ in the second part of (22) to emphasize the change of $\mathbf{b}$'s role from a column vector to a row vector.

Proof of $(\mathbf{AB})^T = \mathbf{B}^T\mathbf{A}^T$. We must show for any i, j, that entry (i, j) in $(\mathbf{AB})^T$ equals entry (i, j) in $\mathbf{B}^T\mathbf{A}^T$. Entry (i, j) in $\mathbf{B}^T\mathbf{A}^T$ equals the scalar product of the ith row of $\mathbf{B}^T$ ($=$ ith column of $\mathbf{B}$) times the jth column of $\mathbf{A}^T$ ($=$ jth row of $\mathbf{A}$). So we have

$$\text{entry } (i, j) \text{ of } \mathbf{B}^T\mathbf{A}^T = \mathbf{b}_i^C \cdot \mathbf{a}_j^R$$

and

$$\text{entry } (i, j) \text{ in } (\mathbf{AB})^T = \text{entry } (j, i) \text{ in } \mathbf{AB} = \mathbf{a}_j^R \cdot \mathbf{b}_i^C$$

Since $\mathbf{b}_i^C \cdot \mathbf{a}_j^R = \mathbf{a}_j^R \cdot \mathbf{b}_i^C$, the identity is proved. ∎

One of the most useful properties a matrix may have is symmetry. A matrix $\mathbf{A}$ is **symmetric** if $\mathbf{A} = \mathbf{A}^T$. The adjacency matrix $\mathbf{A}(G)$ of a graph G, introduced in Section 2.3, is a symmetric matrix. Of course, a symmetric matrix must be a square matrix. Symmetric matrices have many nice theoretical and computational properties. If $\mathbf{A}$ is a symmetric matrix, all the information in the matrix is contained on or above the main diagonal.

A familiar example of a symmetric matrix is a mileage chart on a road map. Because of its symmetric structure, only the upper (or lower) triangular portion of this matrix is usually given. Symmetric matrices are very common in physical sciences applications.

There is a useful symmetric matrix associated with any unsymmetric matrix, the matrix $\mathbf{A}^T\mathbf{A}$. Entry (i, j) in $\mathbf{A}^T\mathbf{A}$ will be the scalar product $\mathbf{a}_i^C \cdot \mathbf{a}_j^C$ of the ith and jth columns of $\mathbf{A}$. Since $\mathbf{a}_i^C \cdot \mathbf{a}_j^C = \mathbf{a}_j^C \cdot \mathbf{a}_i^C$, entry (i, j) and entry (j, i) in $\mathbf{A}^T\mathbf{A}$ are the same. So $\mathbf{A}^T\mathbf{A}$ will be symmetric. (One computes scalar products of pairs of *rows* in the related symmetric matrix $\mathbf{A}\mathbf{A}^T$.)

There are many problems where one wants to measure in some informal way how similar various pairs of columns are in a matrix. Scalar products of the columns, as computed in $\mathbf{A}^T\mathbf{A}$, provide one good measure.

Example 3. Scalar Products of Columns as a Similarity Measure

Suppose that five students A, B, C, D, E have been asked to rate six subjects—linguistics, mathematics, necromancy, optometry, philosophy, and quantum mechanics—as subjects they like (rating = 1) or as subjects they do not like (rating = −1). The following rating matrix $\mathbf{R}$ was obtained.

$$
\mathbf{R} = \begin{array}{c@{}c}
 & \begin{array}{ccccc} A & B & C & D & E \end{array} \\
\begin{array}{c} \text{Ling} \\ \text{Math} \\ \text{Necr} \\ \text{Opto} \\ \text{Phil} \\ \text{QM} \end{array} &
\left[\begin{array}{ccccc}
1 & -1 & 1 & -1 & 1 \\
1 & 1 & -1 & -1 & 1 \\
1 & -1 & 1 & -1 & -1 \\
1 & 1 & -1 & -1 & 1 \\
-1 & -1 & -1 & -1 & -1 \\
1 & -1 & 1 & -1 & 1
\end{array} \right]
\end{array}
\tag{24}
$$

To measure the similarity of interests among students, we want to use the scalar product of pairs of columns. Observe that the scalar product will be positive if two students' ratings tend to agree and will be negative if they tend to disagree. To get these scalar products, we simply compute $\mathbf{R}^T\mathbf{R}$. Since $\mathbf{R}^T\mathbf{R}$ is symmetric, we only need to compute the entries on or above the main diagonal.

This computation yields

$$\mathbf{R}^T\mathbf{R} = \begin{array}{c} \\ A \\ B \\ C \\ D \\ E \end{array} \begin{array}{ccccc} A & B & C & D & E \\ \left[\begin{array}{ccccc} 6 & 0 & 2 & -4 & 4 \\ & 6 & -4 & 2 & 2 \\ & & 6 & 0 & 0 \\ & & & 6 & -2 \\ & & & & 6 \end{array}\right] \end{array} \qquad (25)$$

Computing $\mathbf{R}\mathbf{R}^T$ would yield scalar products of pair of rows in $\mathbf{R}$. These products would measure how similar different pairs of subjects are perceived to be; that is, a large positive number would mean that most students give the same rating to the two subjects. ∎

We close our discussion of symmetric matrices with a very special, and simple, type of symmetric matrix. A **diagonal matrix** is a matrix with nonzero entries only on the main diagonal. The identity matrix is a diagonal matrix. If one premultiplies a matrix $\mathbf{A}$ by a diagonal matrix $\mathbf{D}$—$\mathbf{DA}$—the result just multiplies the ith row of $\mathbf{A}$ by the ith diagonal element in $\mathbf{D}$:

$$\begin{bmatrix} 5 & 0 \\ 0 & 8 \end{bmatrix}\begin{bmatrix} 1 & 2 \\ 3 & 4 \end{bmatrix} = \begin{bmatrix} 5\times1 + 0\times3 & 5\times2 + 0\times4 \\ 0\times1 + 8\times3 & 0\times2 + 8\times4 \end{bmatrix} = \begin{bmatrix} 5 & 10 \\ 24 & 32 \end{bmatrix} \qquad (26)$$

Postmultiplying by a diagonal matrix has a similar effect on the columns (see the Exercises).

Rules of Matrix Algebra

In Example 2 the step $\mathbf{1} \cdot (\mathbf{Ap}) = (\mathbf{1A}) \cdot \mathbf{p}$ [going from line 1 to line 2 in (20)] was not justified. Similarly, in Example 1 we wrote $\mathbf{Ix} = \mathbf{Dx} = (\mathbf{I} - \mathbf{D})\mathbf{x}$ without explanation. These are common algebraic manipulations for single-variable equations, the associative and the distributive laws, respectively. We were implicitly assuming that these laws are also valid in matrix algebra.

The rest of this section is devoted to a quick summary of what algebraic manipulations are and are not valid for matrices. In the following we assume that our vectors and matrices have the right sizes (so that operations make sense).

Since all the basic rules are stated in terms of equations, their proofs consist in showing that the (i, j) entry of the matrix on the left side equals the (i, j) entry in the matrix on the right side. One such proof is worked out to show both the technique of proving matrix equality and the power of the notation we have developed. The other proofs are left to the Exercises.

Commutative Law. Matrix addition is commutative, but matrix multiplication is not commutative.

$$\mathbf{A} + \mathbf{B} = \mathbf{B} + \mathbf{A} \qquad \text{but } \mathbf{AB} \neq \mathbf{BA} \qquad (27)$$

Distributive Law

$$A(B + C) = AB + AC \quad \text{and} \quad (B + C)A = BA + CA \quad (28)$$

Proof of $A(B + C) = AB + AC$. We must show that the (i, j) entry of the product matrix $A(B + C)$ equals the (i, j) entry of the sum of products $AB + AC$. This entry in $A(B + C)$ is the scalar product of the ith row of A times the jth column of $B + C$ (which is $\mathbf{b}_j^C + \mathbf{c}_j^C$)—$\mathbf{a}_i^R \cdot (\mathbf{b}_j^C + \mathbf{c}_j^C)$. Now we must write out this symbolic scalar product term by term and use the distributive law for scalars (this gets a bit messy).

$$\mathbf{a}_i^R(\mathbf{b}_j^C + \mathbf{c}_j^C) = \sum_k a_{ik}(b_{kj} + c_{kj}) = \sum_k a_{ik}b_{kj} + \sum_k a_{ik}c_{kj}$$

$$= \mathbf{a}_i^R \mathbf{b}_j^C + \mathbf{a}_i^R \mathbf{c}_j^C \quad (28a)$$

and $\mathbf{a}_i^R \mathbf{b}_j^C + \mathbf{a}_i^R \mathbf{c}_j^C$ is exactly the (i, j) entry in $AB + AC$. ∎

Associative Law

$$(AB)C = A(BC) \quad (29)$$

There is one new property for matrices, scalar factoring. If r is a scalar (a single number), then

Scalar Factoring

$$r(AB) = (rA)B = A(rB) \quad (30)$$

A vector is just a matrix with one row or one column. The rules for matrix multiplication thus apply to matrix-vector multiplication. For completeness, we restate them.

$(AB)\mathbf{c} = A(B\mathbf{c})$	and	$\mathbf{a}(BC) = (\mathbf{a}B)C$	(31)
$A(\mathbf{b} + \mathbf{c}) = A\mathbf{b} + A\mathbf{c}$	and	$(\mathbf{c} + \mathbf{d})A = \mathbf{c}A + \mathbf{d}A$	(32)
$\mathbf{a}(B + C) = \mathbf{a}B + \mathbf{a}C$	and	$(C + D)\mathbf{a} = C\mathbf{a} + D\mathbf{a}$	(33)
$(rA)\mathbf{b} = A(r\mathbf{b}) = r(A\mathbf{b})$	and	$(r\mathbf{b})A = \mathbf{b}(rA) = r(\mathbf{b}A)$	(34)

In this book we have not made a major distinction between a vector $\mathbf{x}$ being a column vector or being a row vector. However, in complex products involving matrices and vectors, it is essential to treat each vector as an n-by-1 or a 1-by-n matrix (whichever is appropriate) and then treat the result of a matrix-vector product, such as $A\mathbf{x}$ (where A is m-by-n), as an m-by-1 matrix. For example, the following equality is false:

$$(A\mathbf{b})C \neq A(\mathbf{b}C) \quad (35)$$

since $A\mathbf{b}$ yields a column vector and C should be premultiplied by a row vector, and similarly for **bC** and **A**. For the same reason

$$(\mathbf{Ab}) \cdot (\mathbf{Cd}) \neq (\mathbf{A(bC)}) \cdot \mathbf{d} \qquad (36)$$

See the Exercises for specific counterexamples of (35) and (36).

On the other hand, the following equality is valid:

$$(\mathbf{aB}) \cdot \mathbf{c} = \mathbf{a} \cdot (\mathbf{Bc}) \qquad (37)$$

because no vector changes roles from a column to a row vector.

Near the beginning of this section, we introduced the vector **1** of all 1's and the identity matrix **I**. We noted

$$\mathbf{1} \cdot \mathbf{a} = a_1 + a_2 + \cdots + a_n \qquad (38)$$

and

$$\mathbf{aI} = \mathbf{a} = \mathbf{Ia} \quad \text{and} \quad \mathbf{IA} = \mathbf{A} = \mathbf{AI} \qquad (39)$$

A related vector and matrix is the 0's vector **0** of all 0's and the **O** matrix of all 0's. It is immediate that

$$\mathbf{O} \cdot \mathbf{a} = 0 \quad \mathbf{Oa} = \mathbf{0} \quad \mathbf{OA} = \mathbf{O} \qquad (40)$$

There is another special vector that will be used in this book. The *i*th **unit vector**, denoted $\mathbf{e}_i$, is the *i*th column in the identity matrix **I**. Vector $\mathbf{e}_i$ has a 1 in the *i*th entry and 0's elsewhere. This vector has the following useful property:

$$\begin{aligned} \mathbf{Ae}_i &= \mathbf{a}_i^C \quad \text{(the *i*th column of **A**)} \\ \mathbf{e}_i\mathbf{A} &= \mathbf{a}_i^R \quad \text{(the *i*th row of **A**)} \end{aligned} \qquad (41)$$

Vector sums and products (scalar products) also behave nicely. In fact, scalar products are commutative as well as distributive and have scalar factoring.

$$\mathbf{a} + \mathbf{b} = \mathbf{b} + \mathbf{a} \qquad (42)$$

$$\mathbf{a} \cdot \mathbf{b} = \mathbf{b} \cdot \mathbf{a} \qquad (43)$$

$$\mathbf{a} \cdot (\mathbf{b} + \mathbf{c}) = \mathbf{a} \cdot \mathbf{b} + \mathbf{a} \cdot \mathbf{c} \qquad (44)$$

$$r(\mathbf{a} \cdot \mathbf{b}) = (r\mathbf{a}) \cdot \mathbf{b} = \mathbf{a} \cdot (r\mathbf{b}) \qquad (45)$$

Note that in the distributive law (44), addition is vector addition on the left and scalar addition on the right.

There is no associative law for scalar products, since the expression $(\mathbf{a} \cdot \mathbf{b}) \cdot \mathbf{c}$ is nonsense: $(\mathbf{a} \cdot \mathbf{b})$ is a scalar, not a vector. There is a related expression that looks reasonable but is not valid.

$$(\mathbf{a} \cdot \mathbf{b})\mathbf{c} \neq \mathbf{a}(\mathbf{b} \cdot \mathbf{c}) \qquad (46)$$

To sum up matters thus far, it is easiest to say what is not true for matrices. The rules that fail are

---

1. $\mathbf{AB} \neq \mathbf{BA}$.
2. Throughout an algebraic manipulation, a vector must always be treated as a 1-by-n matrix (or always as an n-by-1 matrix)—it cannot change from one form to the other.
3. $(\mathbf{a} \cdot \mathbf{b})\mathbf{c} \neq \mathbf{a}(\mathbf{b} \cdot \mathbf{c})$.

Section 2.4 **Exercises**

Summary of Exercises

Exercises 1–16 develop skill with simple matrix algebra manipulations. Exercises 17–23 deal with transposes and symmetric matrices. Exercises 24–34 involve verifying rules of matrix algebra.

1. Evaluate the following products involving the 1's vector **1**, the identity matrix **I**, and the ith unit vector $\mathbf{e}_i = [0, 0, \ldots, 1 \ldots, 0, 0]$. Assume that all have size n.
   (a) $\mathbf{I}^2$ **(b)** $\mathbf{1} \cdot \mathbf{1}$ **(c)** $\mathbf{I1}$ **(d)** $\mathbf{Ie}_i$ **(e)** $\mathbf{1} \cdot \mathbf{e}_i$ **(f)** $\mathbf{e}_i \cdot \mathbf{e}_i$
   (g) $\mathbf{e}_i \cdot \mathbf{e}_j \ (i \neq j)$ **(h)** $\mathbf{1I1}$ **(i)** $\mathbf{e}_i \mathbf{Ie}_j \ (i \neq j)$

2. **(a)** Write the following system of equations in matrix form.

$$3x_1 + 5x_2 + 7x_3 = 8$$
$$2x_1 - x_2 + x_3 = 4$$
$$x_1 + 6x_2 - 2x_3 = 6$$

   (b) Rewrite the matrix equation in part (a) to reflect the operation of bringing the right side over to the left side (so that the right sides are now 0's).

3. **(a)** Write the following system of equations in matrix form.

$$2x_1 - 3x_2 = x_1$$
$$5x_1 + 4x_2 = x_2$$

   (b) Rewrite the matrix equation in part (a) to reflect the operation of bringing the right-side variables over to the left side. Your new equation should be of the form $\mathbf{Qx} = \mathbf{0}$, where $\mathbf{Q}$ is a matrix expression involving $\mathbf{I}$, the identity matrix.

   Hint: See Example 1.

4. Repeat Exercise 3, parts (a) and (b) for the following system of equations.

$$3x_1 + 2x_2 = 2x_1$$
$$4x_1 - 3x_2 = 2x_2$$

5. Consider the system of equations

$$2x_1 + 3x_2 - 2x_3 = 5y_1 + 2y_2 - 3y_3 + 200$$
$$x_1 + 4x_2 + 3x_3 = 6y_1 - 4y_2 + 4y_3 - 120$$
$$5x_1 + 2x_2 - x_3 = 2y_1 \qquad - 2y_3 + 350$$

(a) Write this system of equations in matrix form. (Define the vectors and matrices you introduce.)
(b) Rewrite in matrix form with all the variables on the left side and just numbers on the right.
(c) Rewrite in matrix form so that x_1 is the only term on the left in the first equation, x_2 is the only term on the left in the second equation, and x_3 is the only term on the left in the third equation (similar to the form of the Leontief economic model).

6. (a) Consider the following system of equations for the growth of rabbits and foxes from year to year:

$$R' = 1.5R - .2F + 100$$
$$F' = .3R + .9F + 50$$

Write this system in matrix form, where $\mathbf{p} = [R, F]$ and $\mathbf{p}' = [R', F']$.
(b) Write a matrix equation for $\mathbf{p}''$, the vector of rabbits and foxes after 2 years.
(c) Write a matrix equation for $\mathbf{p}^{(3)}$, the vector of rabbits and foxes after 3 years.
(d) Using summation notation (Σ), write a matrix equation for $\mathbf{p}^{(n)}$, the vector of rabbits and foxes after n years.

7. (a) Write the rabbit–fox equations

$$R' = R + .1R - .15F$$
$$F' = F + .2R - .3F$$

in matrix form using $\mathbf{p} = [R, F]$, $\mathbf{p}' = [R', F']$, and $\mathbf{A} = \begin{bmatrix} .1 & -.15 \\ .2 & -.3 \end{bmatrix}$.

(b) Rewrite the equation from part (a) in the form $\mathbf{p}' = \mathbf{Q}\mathbf{p}$, where $\mathbf{Q}$ is some matrix expression.
Hint: Use the identity matrix $\mathbf{I}$.
(c) Let $\mathbf{p}^{(20)}$ be the vector of population sizes 20 periods later. Express $\mathbf{p}^{(20)}$ in terms of $\mathbf{p}$, $\mathbf{A}$, and $\mathbf{I}$.

8. Show that if $\mathbf{x}^0$ is a solution to $\mathbf{Ax} = \mathbf{b}$, then $r\mathbf{x}^0$ is a solution to $\mathbf{Ax} = r\mathbf{b}$.

9. Show that if $\mathbf{x}^0$ is a solution to $\mathbf{Ax} = \mathbf{b}$ and if $\mathbf{x}^*$ is a solution to $\mathbf{Ax} = \mathbf{0}$, then $\mathbf{x}^0 + \mathbf{x}^*$ is also a solution to $\mathbf{Ax} = \mathbf{b}$.

10. Let $\mathbf{A}$ and $\mathbf{B}$ be 2-by-2 matrices and $\mathbf{x}$, $\mathbf{y}$, $\mathbf{z}$ be 2-vectors such that $\mathbf{Ax} = \mathbf{By} = [1, 1]$, $\mathbf{Ay} = [1, 0]$, $\mathbf{Bx} = [0, 1]$. Determine $\mathbf{z}$ when
    (a) $\mathbf{z} = \mathbf{A}(2\mathbf{x} - \mathbf{y})$ (b) $\mathbf{z} = (\mathbf{A} - \mathbf{B})\mathbf{x}$
    (c) $\mathbf{z} = (\mathbf{A} + \mathbf{B})\mathbf{x} - 2(\mathbf{A} + \mathbf{B})\mathbf{y}$ (d) $\mathbf{z} = (3\mathbf{A} + \mathbf{B})(\mathbf{x} + \mathbf{y})$
    (e) $\mathbf{z} = [(\mathbf{A} + \mathbf{B})\mathbf{y}] \cdot [(\mathbf{A} + 3\mathbf{B})(\mathbf{x} - \mathbf{y})]\mathbf{1}$

11. Given a linear model of the form $\mathbf{x}' = \mathbf{Ax} + \mathbf{b}$, let us expand the n-by-n matrix $\mathbf{A}$ into an $(n + 1)$-by-$(n + 1)$ matrix $\mathbf{A}^*$ by including $\mathbf{b}$, row vector $\mathbf{0}$ of 0's and a 1 in entry $(n + 1, n + 1)$ so that $\mathbf{A}^*$ has

    the form $\mathbf{A}^* = \begin{bmatrix} \mathbf{A} & \mathbf{b} \\ \mathbf{0} & 1 \end{bmatrix}$.

    We should also add to $\mathbf{x}$ an $(n + 1)$st entry equal to 1; call the new vector $\mathbf{x}^*$, and now our linear model has the form $\mathbf{x}^* = \mathbf{A}^*\mathbf{x}^*$. Give the new $\mathbf{A}^*$ for the following linear models.
    (a) $x_1' = 3x_1 + 2x_2 + 10$ (b) $x_1' = x_1 + 2x_2 + 5x_3 + 20$
    $ x_2' = 4x_1 - 5x_2 + 8$ $ x_2' = 2x_1 - x_2 - 2x_3 - 10$
    $ x_3' = 3x_1 + 4x_2 + 6x_3 + 30$

    (c) Leontief model in Example 1.

12. Show that if the second row of $\mathbf{A}$ is all 0's, the second row in the product $\mathbf{AB}$ (if defined) is all 0's.

13. Show that if $\mathbf{A}$ is the transition matrix of a Markov chain with five states, $\mathbf{1A1} = 5$.

    Hint: Write $\mathbf{1A1} = (\mathbf{1A})\mathbf{1}$ and use the result in Example 2.

14. (a) Extend the reasoning in Example 2 to show that $\mathbf{1} \cdot \mathbf{p}'' = 1$, $\mathbf{p}''$ is the distribution after two periods.
       Hint: Write $\mathbf{p}''$ as $\mathbf{A}(\mathbf{Ap})$ and use the steps in equations (20) twice.
    (b) Prove by induction that the sum of the probabilities in $\mathbf{p}^{(n)}$, the distribution after n periods, equals 1.

15. (a) State the fact that for a Markov transition matrix $\mathbf{A}$, the column sums in $\mathbf{A}^2$ equal 1 with a matrix equation involving $\mathbf{1}$.
       Hint: See equation (19).
    (b) Use equation (19) to prove the equation you wrote in part (a).
    (c) Prove that the column sums in $\mathbf{A}^3$ equal 1 using matrix algebra [follow the reasoning in parts (a) and (b)].
    (d) Use induction to prove column sums in $\mathbf{A}^n$ equal 1.

16. Let **A** and **B** be *n*-by-*n* matrices. If **C** = **AB** and the second row of **A** is five times the first row of **A**($\mathbf{a}_2^R = 5\mathbf{a}_1^R$), show that the second row of **C** is five times the first row of **C**.

    Hint: Compare $c_{11} = \mathbf{a}_1^R \cdot \mathbf{b}_1^C$ with $c_{21} = \mathbf{a}_2^R \cdot \mathbf{b}_1^C$, similarly for c_{21} versus c_{22}, and so on.

17. Verify $(\mathbf{Ab})^T = \mathbf{b}^T\mathbf{A}^T$ for $\mathbf{b} = [b_1, b_2]$ and $\mathbf{A} = \begin{bmatrix} a_{11} & a_{12} \\ a_{21} & a_{22} \end{bmatrix}$.

18. Why are the entries all equal to 6 on the main diagonal of $\mathbf{R}^T\mathbf{R}$ in equation (25)?

19. Compute $\mathbf{RR}^T$ in Example 3 to find a measure of how much different students share common views of their subjects.

20. The faculties in the four divisions of the College of Arts and Sciences at Wayward University (Natural Science/Mathematics, Biological Science, Arts & Humanities, Social Science) have taken stands for or against the following five issues:

	NS/M	Bio	A&H	SS
(a) Wayward needs to change its name	No	Yes	Yes	Yes
(b) Wayward has a friendly campus	Yes	Yes	No	No
(c) CompSci 112 is too hard	No	No	Yes	Yes
(d) The Alfred E. Neuman dorm is ugly	No	Yes	Yes	No
(e) Wayward athletes should be better	No	No	No	Yes

    Compute a matrix of similarities *between the divisions* (remember that, by symmetry, you only have to compute the entries on or above the main diagonal).

21. Let **A**(*G*) denote the adjacency matrix of the graph in Figure 2.2 and let **M**(*G*) denote the incidence matrix of that graph [see equation (13) of Section 2.3]. Show that entry (*i*, *j*) of $\mathbf{M}(G)\mathbf{M}(G)^T$ equals entry (*i*, *j*) of **A**(*G*), for $i \neq j$. Explain in words why this result is always true.

22. (a) Compute **AD** for the matrices **A** and **D** given in the discussion of diagonal matrices.
    (b) Show that if **D** is a diagonal matrix, **AD** has the effect of multiplying the *i*th column of **A** by the *i*th diagonal entry of **D**.

23. Show that if **A** is symmetric, then $\mathbf{A}^2$ is symmetric.

24. Prove that $(\mathbf{B} + \mathbf{C})\mathbf{A} = \mathbf{BA} + \mathbf{CA}$ by mimicking the argument in the text used to show $\mathbf{A}(\mathbf{B} + \mathbf{C}) = \mathbf{AB} + \mathbf{AC}$.

25. (a) Verify $(\mathbf{AB})\mathbf{c} = \mathbf{A}(\mathbf{Bc})$ for $\mathbf{A}$ and $\mathbf{B}$ arbitrary 2-by-2 matrices and $\mathbf{c} = [c_1, c_2]$.
    (b) Extend the result in part (a) to show $(\mathbf{AB})\mathbf{C} = \mathbf{A}(\mathbf{BC})$ using the reasoning in equation (6).

26. Show that the identity $\mathbf{A}(\mathbf{b} \cdot \mathbf{c}) \overset{?}{=} (\mathbf{Ab}) \cdot \mathbf{c}$ makes no sense by making up a 3-by-2 matrix $\mathbf{A}$, a 2-vector $\mathbf{b}$, and a 2-vector $\mathbf{c}$. Compute the value of $\mathbf{A}(\mathbf{b} \cdot \mathbf{c})$ and then try to compute $(\mathbf{Ab}) \cdot \mathbf{c}$.

27. Show that the identity $\mathbf{A}(\mathbf{bC}) \overset{?}{=} (\mathbf{Ab})\mathbf{C}$ makes no sense by making up matrices $\mathbf{A}$, $\mathbf{C}$ and vector $\mathbf{b}$ with $\mathbf{A}$ 3-by-2 so that the matrix expression $\mathbf{A}(\mathbf{bC})$ makes sense (the sizes fit together properly), and then show that the sizes are wrong for $(\mathbf{Ab})\mathbf{C}$.

28. Verify that $\mathbf{bI} = \mathbf{b}$ for any $\mathbf{b} = [b_1, b_2, \ldots, b_n]$.

29. Verify that $\mathbf{IB} = \mathbf{B}$ for a 3-by-3 matrix

$$\mathbf{B} = \begin{bmatrix} b_{11} & b_{12} & b_{13} \\ b_{21} & b_{22} & b_{23} \\ b_{31} & b_{32} & b_{33} \end{bmatrix}$$

by performing the matrix multiplication $\mathbf{IB}$.

30. (a) For a given matrix $\mathbf{B}$, let $\mathbf{b}_i^R$ denote the ith row of $\mathbf{B}$. Show that $\mathbf{1} \cdot \mathbf{b}_i^R$ equals the sum of entries in the ith row of $\mathbf{B}$. Show that $\mathbf{B1}$ yields a vector whose ith position is the sum of the entries in the ith row.
    (b) Show that $\mathbf{1B}$ yields a vector whose jth position is the sum of the entries in the jth column of $\mathbf{B}$.
    (c) Show that $\mathbf{1B1}$ equals the sum of all the entries in $\mathbf{B}$.

31. Give an example involving three 2-vectors to show that $(\mathbf{a} \cdot \mathbf{b})\mathbf{c} \neq \mathbf{a}(\mathbf{b} \cdot \mathbf{c})$.

32. Let $\mathbf{e}_i$ denote the vector with a 1 in the ith position and 0's elsewhere. What is the value of $\mathbf{1Ae}_i$?

33. (a) Why is the following identity false: $(\mathbf{AB})^2 = \mathbf{A}^2\mathbf{B}^2$? What is $(\mathbf{AB})^2$ actually equal to?
    (b) Can you find two nonzero matrices $\mathbf{A}$, $\mathbf{B}$ for which $(\mathbf{AB})^2 = \mathbf{A}^2\mathbf{B}^2$?

34. Why is the following identity false: $(\mathbf{A} + \mathbf{B})^2 = \mathbf{A}^2 + 2\mathbf{AB} + \mathbf{B}^2$? What is $(\mathbf{A} + \mathbf{B})^2$ actually equal to?

2-27-92

Section 2.5 Scalar Measures of a Matrix: Norms and Eigenvalues

The goal of this section is to express with a single number the magnifying effect a matrix $\mathbf{A}$ has when $\mathbf{A}$ multiplies a vector, as in $\mathbf{p}' = \mathbf{A}\mathbf{p}$. For example, will $\mathbf{p}'$ be about twice the size of $\mathbf{p}$? We want to capture in one number the "essence" of multiplication by $\mathbf{A}$. The first problem is how to measure the size of a vector.

In matrix algebra, the word **norm** is the name used for the size of a quantity. For scalars, the standard norm is the absolute value $|a|$. The most common norm for an n-vector $\mathbf{a} = [a_1, a_2, \ldots, a_n]$ is the euclidean distance $|\mathbf{a}|$ of the point $[a_1, a_2, \ldots, a_n]$ from the origin. This distance is the square root of the sum of the squares of the a_i's.

$$|\mathbf{a}| = \sqrt{a_1^2 + a_2^2 + \cdots + a_n^2} \tag{1}$$

For example, if $\mathbf{a} = [-1, 2, 2]$, then $|\mathbf{a}| = \sqrt{1^2 + 2^2 + 2^2} = \sqrt{9} = 3$. A set of vectors with easily computed norms are the unit vectors $\mathbf{e}_i$, which have a 1 in the ith position and 0 elsewhere. Clearly, formula (1) gives $|\mathbf{e}_i| = 1$.

Because it uses the euclidean distance, the norm in (1) is called the **euclidean norm**. Although the euclidean norm has a nice geometrical interpretation, this norm is often tedious to compute. Since there are other vector norms that are easier to compute and more natural for our work, the euclidean norm has limited value in linear models considered in this book. Two natural ways to measure the size of a vector are the sum of its entries and its largest entry. Since norms need to be nonnegative, we use absolute values in defining the sum and largest-entry norms.

$$\textbf{Sum norm:}\quad |\mathbf{a}|_s = \Sigma\, |a_i| \qquad \textbf{Maximum norm:}\quad |\mathbf{a}|_{mx} = \max_i \{|a_i|\}$$

For example, $|[-1, 2, 2]|_s = 1 + 2 + 2 = 5$ and $|[-1, 2, 2]|_{mx} = \max\{1, 2, 2\} = 2$. Any probability vector will have sum norm $= 1$—this is what we would expect for such a vector. The unit vectors $\mathbf{e}_i$ have a value of 1 for both these norms, just as they did for the euclidean norm. We now write the euclidean norm as $|\mathbf{a}|_e$ to avoid confusion.

The norm $\|\mathbf{A}\|$ of a matrix $\mathbf{A}$ is a bound on the magnifying effect $\mathbf{A}$ has when it multiplies some vector. We define $\|\mathbf{A}\|$ to be the (smallest) bound so that

$$|\mathbf{A}\mathbf{x}| \leq \|\mathbf{A}\| \cdot |\mathbf{x}| \tag{2}$$

Thus

$$\|\mathbf{A}\| = \max_{\text{all } \mathbf{x}} \left(\frac{|\mathbf{Ax}|}{|\mathbf{x}|} \right) \tag{3}$$

We assume that $\mathbf{x} \neq \mathbf{0}$ in (3).

There is an immediate extension of (2) to powers of $\mathbf{A}$.

$$|\mathbf{A}^k \mathbf{x}| \leq \|\mathbf{A}\|^k \cdot |\mathbf{x}| \tag{4}$$

Since the magnifying effect depends on how the size of the vector is measured, for each of our three vector norms we get a corresponding matrix norm: a euclidean matrix norm $\|\mathbf{A}\|_e$, a sum matrix norm $\|\mathbf{A}\|_s$, and a max matrix norm $\|\mathbf{A}\|_{mx}$. Each of these three matrix norms has its own special properties.

As with the euclidean norm for vectors, the euclidean norm for matrices is the most commonly used matrix norm in linear algebra and has the best theoretical properties. However, the euclidean norm of a matrix is very difficult to calculate, while the sum and max norms are easy to determine.

Theorem 1. The sum matrix norm $\|\mathbf{A}\|_s$ equals the largest column sum of $\mathbf{A}$ (in absolute value), and the max matrix norm $\|\mathbf{A}\|_{mx}$ equals the largest row sum of $\mathbf{A}$. That is,

$$\text{(i) } \|\mathbf{A}\|_s = \max_j \left(|\mathbf{a}_j^C|_s \right) \qquad \text{(ii) } \|\mathbf{A}\|_{mx} = \max_i \left(|\mathbf{a}_i^R|_s \right)$$

The proof of Theorem 1, part (i) is given below [part (ii) is left to the Exercises]. First let us illustrate these formulas.

Example 1. Sum and Max Norms of a Matrix

Use Theorem 1 to determine the sum and max norms of $\mathbf{A}$.

$$\mathbf{A} = \begin{bmatrix} 1 & 2 & 3 \\ 4 & 5 & 6 \\ 7 & 8 & 9 \end{bmatrix}$$

The last column has the largest column sum, so the sum matrix norm of $\mathbf{A}$ is $\|\mathbf{A}\|_s = 3 + 6 + 9 = 18$. The last row has the largest row sum, so the max matrix norm of $\mathbf{A}$ is $\|\mathbf{A}\|_{mx} = 7 + 8 + 9 = 24$.

Let us see how these norms bound the magnifying effect of multiplying a vector $\mathbf{x}$ by $\mathbf{A}$. Since $|\mathbf{Ax}| \leq \|\mathbf{A}\| \cdot |\mathbf{x}|$, using the sum and max norms, we have

$$|\mathbf{Ax}|_s \le 18|\mathbf{x}|_s \qquad |\mathbf{Ax}|_{mx} \le 24|\mathbf{x}|_{mx} \tag{5}$$

We can attain the sum norm bound in (5) using $\mathbf{e}_3 = [0, 0, 1]$, with $|\mathbf{e}_3|_s = 1$. The bound for $\mathbf{Ae}_3$ is

$$|\mathbf{Ae}_3|_s \le \|\mathbf{A}\|_s \cdot |\mathbf{e}_3|_s = 18 \times 1 = 18$$

and $|\mathbf{Ae}_3|_s$ equals this bound:

$$\mathbf{Ae}_3 = \begin{bmatrix} 1 & 2 & 3 \\ 4 & 5 & 6 \\ 7 & 8 & 9 \end{bmatrix}\begin{bmatrix} 0 \\ 0 \\ 1 \end{bmatrix} = \begin{bmatrix} 3 \\ 6 \\ 9 \end{bmatrix} \qquad \text{so} \tag{6}$$

$$|\mathbf{Ae}_3|_s = 3 + 6 + 9 = 18$$

The reader should check that we attain the max norm bound of 24 with the 1's vector $\mathbf{x} = \mathbf{1} = [1, 1, 1]$. ∎

The use of a unit vector and the 1's vector to achieve the norm bounds in (5) always works for positive matrices.

Theorem 2. Let **A** be a matrix with nonnegative entries.
    (i) Sum norm: If the jth column of **A** has the largest sum, unit vector $\mathbf{e}_j$ achieves the sum norm bound: $|\mathbf{Ae}_j|_s = \|\mathbf{A}\|_s|\mathbf{e}_j|_s$.
    (ii) Max norm: The 1's vector **1** achieves the max norm bound: $|\mathbf{A1}|_{mx} = \|\mathbf{A}\|_{mx}|\mathbf{1}|_{mx}$.

Proof of Theorems 1 and 2, part (i): First we note that in the definition (3) of the sum norm $\|\mathbf{A}\| = \max(|\mathbf{Ax}|_s/|\mathbf{x}|_s)$, it is sufficient to consider only vectors **x** with $|\mathbf{x}|_s = 1$. For if $|\mathbf{x}|_s = k$, then $\mathbf{y} = (1/k)\mathbf{x}$ has sum norm 1 and $|\mathbf{Ay}|_s/|\mathbf{y}|_s = |\mathbf{Ax}|_s/|\mathbf{x}|_s$.
    For concreteness, we work with the matrix **A** in Example 1. We want to find an **x**, $|\mathbf{x}|_s = 1$, that maximizes $|\mathbf{Ax}|_s \,(= |\mathbf{Ax}|_s/|\mathbf{x}|_s)$. Let us write the matrix-vector product **Ax** in the following form:

$$\mathbf{Ax} = \begin{bmatrix} 1 & 2 & 3 \\ 4 & 5 & 6 \\ 7 & 8 & 9 \end{bmatrix}\begin{bmatrix} x_1 \\ x_2 \\ x_3 \end{bmatrix} = x_1\begin{bmatrix} 1 \\ 4 \\ 7 \end{bmatrix} + x_2\begin{bmatrix} 2 \\ 5 \\ 8 \end{bmatrix} + x_3\begin{bmatrix} 3 \\ 6 \\ 9 \end{bmatrix} \tag{7}$$

With $|x_1| + |x_2| + |x_3| = 1$, we must pick the x_i's to make the linear combination of column vectors in (7) as large as possible. Clearly, (7) is maximized when the x_i associated with the largest column—column 3—is 1 (and the other x_j's are 0). Thus the maximizing **x** is $[0, 0, 1]$ and the sum norm of (7) with this **x** is $|\mathbf{a}_3^C|_s$, the sum of the third column's entries. This reasoning is valid for any matrix. ∎

A simple alteration of the 1's vector is required to achieve the max norm when **A** has negative entries (see Exercise 25).

Example 2. Norm Bound on Growth in Rabbit–Fox Population Model

Example 3 of Section 1.3 started with the rabbit–fox growth model

$$R' = R + .2R - .3F$$
$$F' = F - .1F + .1R$$
(8a)

or

$$\begin{bmatrix} R' \\ F' \end{bmatrix} = \begin{bmatrix} 1.2 & -.3 \\ .1 & .9 \end{bmatrix} \begin{bmatrix} R \\ F \end{bmatrix}$$
(8b)

Let $\mathbf{p} = [R, F]$, $\mathbf{p}' = [R', F']$ and $\mathbf{A}$ be the matrix of coefficients in (8b). The sum norm (largest absolute-value column sum) and max norm (largest absolute-value row sum) of $\mathbf{A}$ are

$$\|\mathbf{A}\|_s = 1.3 \qquad \|\mathbf{A}\|_{mx} = 1.5$$
(9)

Using (2), we have $|\mathbf{p}'| = |\mathbf{Ap}| \leq \|\mathbf{A}\| \cdot |\mathbf{p}|$, or by (9),

$$|\mathbf{p}'|_s \leq 1.3|\mathbf{p}|_s \qquad \text{and} \qquad |\mathbf{p}'|_{mx} \leq 1.5|\mathbf{p}|_{mx}$$
(10)

When we started with $R = 100$, $F = 100$, we found that the populations declined to extinction—we did not get close to the norm bounds. When we started with $R = 100$, $F = 50$, the populations grew initially. Let us get a bound from (10) on this growth. For $\mathbf{p} = [100, 50]$, we have $|\mathbf{p}|_s = 150$. So (10) yields the sum norm bound

$$|\mathbf{p}'|_s \leq 1.3|\mathbf{p}|_s = 1.3 \times 150 = 195$$

From (8a) we compute $R' = 105$, $F' = 55$, so $|\mathbf{p}'|_s = 160$. Thus for $\mathbf{p} = [100, 50]$, the sum norm bound is a decent estimate.

Bound (4) can be used for the population $\mathbf{p}^{(k)}$ after k periods:

$$|\mathbf{p}^{(k)}|_s = |\mathbf{A}^k\mathbf{p}|_s \leq \|\mathbf{A}\|_s^k|\mathbf{p}|_s = (1.3)^k \times 150 \qquad \blacksquare$$

Example 3. Sum Norm of a Markov Transition Matrix

Since the sum norm equals the largest column sum and the entries in every column $\mathbf{a}_j^C$ sum to 1 in a Markov transition matrix, it follows that such a matrix $\mathbf{A}$ has sum norm $\|\mathbf{A}\|_s = 1$. Because powers of a transition matrix have column sums of 1, all powers also have a sum norm of 1.

Any probability vector $\mathbf{p}$ achieves the sum norm bound:

$$|\mathbf{Ap}|_s = \|\mathbf{A}\|_s|\mathbf{p}|_s \quad \text{or} \quad |\mathbf{Ap}|_s = |\mathbf{p}|_s \quad (\text{since } \|\mathbf{A}\|_s = 1)$$

since the vectors $\mathbf{p}$ and $\mathbf{Ap}$ ($= \mathbf{p}'$) both have sum norm 1. ■

Example 4. **Sum Norm of Demand Matrix in Leontief Economic Model**

The Leontief economic model, given in Example 2 of Section 1.2, has an input constraint that the (nonnegative) entries in each column $\mathbf{d}_j^C$ of the demand matrix $\mathbf{D}$ sum to < 1. This meant that it cost less than 1 dollar (of inputs) to produce 1 dollar worth of the jth product. It follows immediately that $\|\mathbf{D}\|_s < 1$.

In Section 3.4 we will see how $\|\mathbf{D}\|_s < 1$ guarantees that the Leontief model always has a solution. ■

The following properties are true for any matrix norm:

$$|r|\|\mathbf{x}\| = |r\mathbf{x}| \quad \text{and} \quad \|r\mathbf{A}\| = |r| \cdot \|\mathbf{A}\|$$
$$\|\mathbf{AB}\| \le \|\mathbf{A}\| \cdot \|\mathbf{B}\| \tag{11}$$
$$\|\mathbf{A}^k\| \le (\|\mathbf{A}\|)^k \tag{12}$$
$$\|\mathbf{A} + \mathbf{B}\| \le \|\mathbf{A}\| + \|\mathbf{B}\|$$

One of the most important uses of norms is to determine error bounds.

Example 5. **Use of Matrix Norm in Error Bounds**

Consider the following growth model for the numbers of computers C and dogs D in successive years:

$$C' = 3C + D \tag{13}$$
$$D' = 2C + 2D$$

The sum norm $\|\mathbf{A}\|_s$ of the coefficient matrix $\mathbf{A}$ is 5 ($=$ the sum of coefficients in the first column). If $\mathbf{c} = [C, D]$ is the initial numbers of computers and dogs and $\mathbf{c}' = [C', D']$, then $\mathbf{c}' = \mathbf{Ac}$.

Suppose that there is an error in determining $\mathbf{c}$ and we mistakenly use the initial vector $\mathbf{b}$, where $\mathbf{b} = \mathbf{c} + \mathbf{e}$—here $\mathbf{e}$ is the vector of errors. Then the error 1 year later is

$$\mathbf{b}' - \mathbf{c}' = \mathbf{Ab} - \mathbf{Ac} \tag{14}$$
$$= \mathbf{A}(\mathbf{b} - \mathbf{c}) = \mathbf{Ae}$$

Taking (sum) norm bounds, we have the *error bound*

$$|\mathbf{b}' - \mathbf{c}'|_s \le \|\mathbf{A}\|_s|\mathbf{e}|_s = 5|\mathbf{e}|_s \tag{15}$$

If we know that the sum of the errors is no more than 2, then our error after 1 year is at most $5|\mathbf{e}|_s = 5 \cdot 2 = 10$.

Following the model for n years, we let $\mathbf{c}^{(n)}$ denote the numbers of computers and dogs after n years. Then

$$\mathbf{c}^{(n)} = \mathbf{A}^n \mathbf{c}$$

and after n years, the error is

$$
\begin{aligned}
\mathbf{b}^{(n)} - \mathbf{c}^{(n)} &= \mathbf{A}^n \mathbf{b} - \mathbf{A}^n \mathbf{c} \\
&= \mathbf{A}^n(\mathbf{b} - \mathbf{c}) \qquad\qquad (16) \\
&= \mathbf{A}^n \mathbf{e}
\end{aligned}
$$

Taking norms, we have the error bound

$$\left|\mathbf{b}^{(n)} - \mathbf{c}^{(n)}\right|_s \leq \|\mathbf{A}^n\|_s |\mathbf{e}|_s \leq \|\mathbf{A}\|_s^n |\mathbf{e}|_s = 5^n |\mathbf{e}|_s \qquad (17)$$

For example, suppose that $\mathbf{c} = [20, 10]$ but $\mathbf{b} = [22, 11]$, so $\mathbf{e} = [2, 1]$ with $|\mathbf{e}|_s = 3$. Then computing $\mathbf{c}'$ and $\mathbf{b}'$, we find that

$$\mathbf{c}' = \mathbf{Ac} = \begin{bmatrix} 3 & 1 \\ 2 & 2 \end{bmatrix} \begin{bmatrix} 20 \\ 10 \end{bmatrix} = \begin{bmatrix} 70 \\ 60 \end{bmatrix}$$

and

$$\mathbf{b}' = \mathbf{Ab} = \begin{bmatrix} 3 & 1 \\ 2 & 2 \end{bmatrix} \begin{bmatrix} 22 \\ 11 \end{bmatrix} = \begin{bmatrix} 77 \\ 66 \end{bmatrix}$$

and $\mathbf{b}' - \mathbf{c}' = [7, 6]$, with $|\mathbf{b}' - \mathbf{c}'|_s = 13$. The sum norm bound on this error is, from (15),

$$|\mathbf{b}' - \mathbf{c}'|_s \leq \|\mathbf{A}\|_s |\mathbf{e}|_s = 5 \cdot 3 = 15$$

This bound of 15 compares well with the observed error of 13. If we had iterated n times, the error between $\mathbf{b}^{(n)} = \mathbf{A}^n \mathbf{b}$ and $\mathbf{c}^{(n)} = \mathbf{A}^n \mathbf{c}$ would be bounded, using (17), by $\|\mathbf{A}\|_s^n |\mathbf{e}|_s = 5^n \cdot 3$.

Repeating the analysis above using the max norm yields

$$|\mathbf{b}' - \mathbf{c}'|_{mx} \leq \|\mathbf{A}\|_{mx} |\mathbf{e}|_{mx} = 4 \cdot 2 = 8$$

This max bound of 8 compares well with the observed max error of 7. ∎

The norm $\|\mathbf{A}\|$ provides a single-number bound for the magnifying effect of multiplying a vector $\mathbf{x}$ by a matrix $\mathbf{A}$. When $\mathbf{A}$ is a square matrix, sometimes multiplying by $\mathbf{A}$ has exactly the same effect as multiplying by a single number λ; that is, for some vector $\mathbf{u}$, $\mathbf{Au}$ equals $\lambda\mathbf{u}$. When $\mathbf{Au} = \lambda\mathbf{u}$ (and $\mathbf{u} \neq \mathbf{0}$), the vector $\mathbf{u}$ is called an **eigenvector** of $\mathbf{A}$ (*eigen* is the German word for "proper") and the scalar λ is called an **eigenvalue** of $\mathbf{A}$.

Example 6. Eigenvalues in a Growth Model

Consider again the growth model (13) for computers (C) and dogs (D) from year to year.

$$
\begin{aligned}
C' &= 3C + D \\
D' &= 2C + 2D
\end{aligned}
\tag{18}
$$

If initially we had $C = 1$, $D = 1$, then we compute $C' = 4$, $D' = 4$. Letting $C = 4$, $D = 4$, we obtain $C' = 16$, $D' = 16$. Whenever $[C, D] = [a, a]$, then $[C', D'] = [4a, 4a]$. So 4 is an eigenvalue of (18) and any vector of the form $[a, a]$ is an eigenvector.

Observe that

$$
\mathbf{A}^2[a, a] = \mathbf{A}(\mathbf{A}[a, a]) = \mathbf{A}([4a, 4a]) = [16a, 16a]
$$

and in general,

$$
\mathbf{A}^k[a, a] = 4^k[a, a]
$$

Note that if initially we had the (nonsense) vector $[C, D] = [1, -2]$, then $[C', D'] = [1, -2]$. So 1 is also an eigenvalue of (18) with eigenvector $[1, -2]$ (or any multiple of $[1, -2]$). ∎

Example 7. Stable Probability Vector for Weather Markov Chain

In Example 1 of Section 1.3 we introduced the following Markov chain for sunny and cloudy weather:

$$
\begin{array}{c}
 & \text{Today} \\
 & \begin{array}{cc} \text{Sunny} & \text{Cloudy} \end{array} \\
\text{Tomorrow} \begin{array}{c} \text{Sunny} \\ \text{Cloudy} \end{array} & \begin{bmatrix} \frac{3}{4} & \frac{1}{2} \\ \frac{1}{4} & \frac{1}{2} \end{bmatrix}
\end{array}
$$

We claim that $\mathbf{p}^* = [\frac{2}{3}, \frac{1}{3}]$ is a stable probability distribution for this transition matrix $\mathbf{A}$. That is,

$$
\mathbf{Ap}^* = \begin{bmatrix} \frac{3}{4} & \frac{1}{2} \\ \frac{1}{4} & \frac{1}{2} \end{bmatrix} \begin{bmatrix} \frac{2}{3} \\ \frac{1}{3} \end{bmatrix} = \begin{bmatrix} \frac{3}{4} \times \frac{2}{3} + \frac{1}{2} \times \frac{1}{3} \\ \frac{1}{4} \times \frac{2}{3} + \frac{1}{2} \times \frac{1}{3} \end{bmatrix} = \begin{bmatrix} \frac{2}{3} \\ \frac{1}{3} \end{bmatrix} = \mathbf{p}^*
$$

So $\mathbf{Ap}^* = 1\mathbf{p}^*$, and $\mathbf{p}^*$ is an eigenvector of $\mathbf{A}$ with eigenvalue 1. ∎

In the frog Markov chain (Example 2 of Section 1.3) we found by experimental computation that the probability vector **p*** = (.1, .2, .2, .2, .2, .1) had the property that **Ap*** = **p*** for the frog transition matrix **A**. Again this **p*** was an eigenvector with eigenvalue 1.

This property of matrix multiplication acting like scalar multiplication for certain vectors happens for all matrices. It is the key to understanding the behavior of many linear models. An n-by-n matrix usually has n eigenvalues, each with an infinite collection of eigenvectors.

Observe that if **u** is an eigenvector of **A** with **Au** = λ**u**, then *any multiple r***u** *of* **u** *is also an eigenvector*, since **A**(r**u**) = r(**Au**) = r(λ**u**) = λ(r**u**).

The following example shows how eigenvectors provide a simplifying way to understand matrix-vector computations.

Example 8. **Eigenvectors as a Coordinate System**

The computer–dog growth model from Example 6 has the form

$$\mathbf{x}' = \mathbf{Ax}, \qquad \text{where } \mathbf{A} = \begin{bmatrix} 3 & 1 \\ 2 & 2 \end{bmatrix} \qquad \begin{array}{l} \mathbf{x} = [C, D] \\ \mathbf{x}' = [C', D'] \end{array}$$

Earlier we saw that the two eigenvalues and associated eigenvectors of **A** are $\lambda_1 = 4$ with **u** = [1, 1] and $\lambda_2 = 1$ with **v** = [1, −2].

Suppose that we want to determine the effects of this growth model over 20 periods with the starting vector **x** = [1, 7]. Let us express **x** as a linear combination of **u** and **v**. By a method to be explained shortly, we find that

$$\mathbf{x} = 3\mathbf{u} - 2\mathbf{v} \qquad (\text{i.e., } [1, 7] = 3[1, 1] - 2[1, -2])$$

With matrix algebra, we can write

$$\begin{aligned} \mathbf{Ax} = \mathbf{A}(3\mathbf{u} - 2\mathbf{v}) &= 3\mathbf{Au} - 2\mathbf{Av} \\ &= 3(4\mathbf{u}) - 2(1\mathbf{v}) \text{ (since } \mathbf{u}, \mathbf{v} \text{ are eigenvectors)} \\ &= 12\mathbf{u} - 2\mathbf{v} \end{aligned} \qquad (19)$$

For 20 periods, we have

$$\begin{aligned} \mathbf{A}^{20}\mathbf{x} = \mathbf{A}^{20}(3\mathbf{u} - 2\mathbf{v}) &= 3\mathbf{A}^{20}\mathbf{u} - 2\mathbf{A}^{20}\mathbf{v} \\ &= 3(4^{20}\mathbf{u}) - 2(1^{20}\mathbf{v}) \\ &= 3 \cdot 4^{20}[1, 1] - 2[1, -2] \\ &= [3 \cdot 4^{20}, 3 \cdot 4^{20}] - [2, -4] \end{aligned} \qquad (20)$$

Note how the eigenvector with the larger eigenvalue swamps the other eigenvector. The relative effect of the other eigenvector is so small that it can be neglected. So after n periods we have

$$\mathbf{A}^n\mathbf{x} \simeq \mathbf{A}^n(3\mathbf{u}) = 3\cdot4^n\mathbf{u} = [3\cdot4^n, 3\cdot4^n] \tag{21}$$

This is a lot easier than multiplying $\mathbf{A}^n\mathbf{x}$ out directly for various n. ■

Let us generalize the result in Example 8. Suppose that $\mathbf{u}$ and $\mathbf{v}$ are eigenvectors of a 2-by-2 matrix $\mathbf{A}$ with eigenvalues λ_1 and λ_2, respectively, with $\lambda_1 > \lambda_2 > 0$. Suppose we can express the vector $\mathbf{x}$ as a linear combination of $\mathbf{u}$ and $\mathbf{v}$ (e.g., $\mathbf{x} = a\mathbf{u} + b\mathbf{v}$). Then using the laws of matrix algebra, the matrix-vector products $\mathbf{A}\mathbf{x}$ and $\mathbf{A}^2\mathbf{x}$ can be calculated as

$$\mathbf{A}\mathbf{x} = \mathbf{A}(a\mathbf{u} + b\mathbf{v}) = a\mathbf{A}\mathbf{u} + b\mathbf{A}\mathbf{v} = a\lambda_1\mathbf{u} + b\lambda_2\mathbf{v} \tag{22}$$
$$\mathbf{A}^2\mathbf{x} = \mathbf{A}^2(a\mathbf{u} + b\mathbf{v}) = a\mathbf{A}^2\mathbf{u} + b\mathbf{A}^2\mathbf{v} = a\lambda_1^2\mathbf{u} + b\lambda_2^2\mathbf{v}$$

and more generally,

$$\mathbf{A}^n\mathbf{x} = a\lambda_1^n\mathbf{u} + b\lambda_2^n\mathbf{v} \tag{23}$$

As noted in Example 8, for large n, λ_1^n will be much larger than λ_2^n, since $\lambda_1 > \lambda_2$, so we have

$$\mathbf{A}^n\mathbf{x} \simeq a\lambda_1^n\mathbf{u} \tag{24}$$

This is clearly a very simple way to follow growth models over many periods.

Theorem 3. Let λ^* be the largest eigenvalue of $\mathbf{A}$ (strictly larger *in absolute value* than other λ's) and $\mathbf{u}^*$ a corresponding eigenvector. Then for any vector $\mathbf{x}$, the expression $\mathbf{A}^n\mathbf{x}$ approaches a multiple of $\mathbf{u}^*$ as n becomes large.

There is an implicit message about $\mathbf{A}$ in writing a vector $\mathbf{x}$ as a linear combination of eigenvectors and in saying that $\mathbf{A}^n$ approaches a multiple of $\mathbf{u}^*$. The latter statement says that somehow, $\mathbf{A}^n$ must be closely related to $\mathbf{u}^*$. In Section 5.5 we show when $\mathbf{A}$ is symmetric that $\mathbf{A}$ can be decomposed into a set of "simple" matrices generated by the different eigenvectors, and that $\mathbf{A}^n$ approaches the "simple" matrix generated by $\mathbf{u}^*$. In Sections 3.1, 3.4, and 5.5 we learn different ways to find eigenvalues and eigenvectors of a square matrix.

The one missing step for us at this point is how to determine a and b so that $\mathbf{x} = a\mathbf{u} + b\mathbf{v}$. Recall from Chapter 1 that anytime two variables must be determined, the calculations are bound to involve two linear equations in these two unknowns.

If $\mathbf{x} = [x_1, x_2]$, $\mathbf{u} = [u_1, u_2]$, and $\mathbf{v} = [v_1, v_2]$, the statement $\mathbf{x} = a\mathbf{u} + b\mathbf{v}$ is actually a system of equations

$$\begin{bmatrix} x_1 \\ x_2 \end{bmatrix} = a\begin{bmatrix} u_1 \\ u_2 \end{bmatrix} + b\begin{bmatrix} v_1 \\ v_2 \end{bmatrix} \tag{25}$$

or

$$x_1 = u_1a + v_1b$$
$$x_2 = u_2a + v_2b$$

In the computer–dog growth model with $\mathbf{x} = [1, 7]$, $\mathbf{u} = [1, 1]$, $\mathbf{v} = [1, -2]$, this system of equations becomes

$$1 = 1a + 1b$$
$$7 = 1a - 2b$$

which can be solved by elimination, yielding $a = 3$, $b = -2$.

In closing, we observe that the norm of a matrix (any norm) provides an upper bound on the size of eigenvalues. The norm is a bound on the magnifying effect of a matrix $\mathbf{A}$: $|\mathbf{Ax}| \leq \|\mathbf{A}\| \cdot |\mathbf{x}|$. If $\mathbf{Au} = \lambda\mathbf{u}$, so $|\mathbf{Au}| = |\lambda||\mathbf{u}|$, it follows that $|\lambda| \leq \|\mathbf{A}\|$. Check that this was true in Example 6. Typically, the largest eigenvalue (in absolute value) $|\lambda|$ is very close to $\|\mathbf{A}\|$ (in any norm). One can show (see Exercise 30) that the sum and max norms of a 2-by-2 symmetric matrix $\mathbf{A}$ of the form $\begin{bmatrix} a & b \\ b & a \end{bmatrix}$ equal the largest (absolute) eigenvalue.

Example 9. Norm and Eigenvalues of a Symmetric 2-by-2 Matrix

We claim that $\mathbf{x}_1 = [1, 1]$ and $\mathbf{x}_2 = [-1, 1]$ are eigenvectors of the matrix

$$\mathbf{A} = \begin{bmatrix} 3 & 2 \\ 2 & 3 \end{bmatrix}$$

Computation shows that $\mathbf{Ax}_1 = [5, 5] = 5\mathbf{x}_1$ and $\mathbf{Ax}_2 = [-1, 1] = \mathbf{x}_2$. Thus 5 and 1 are the eigenvalues associated with $\mathbf{x}_1$ and $\mathbf{x}_2$.

As asserted above, the larger eigenvalue of such a symmetric 2-by-2 matrix equals its norm (for all three matrix norms). The larger eigenvalue is 5, so $\|\mathbf{A}\| = 5$. Checking, we see that 5 is the sum of each column and row, so by Theorem 1, 5 is the sum and max norm of $\mathbf{A}$.   ■

Section 2.5 Exercises

Summary of Exercises

Exercises 1–25 involve the norms of vectors and matrices, with Exercises 13–25 being of a more "theoretical" nature. Exercises 26–31 discuss the determination of eigenvalues and eigenvectors and their use in computing $\mathbf{A}^k\mathbf{x}$.

1. Show that the euclidean norm of $\mathbf{a}$ equals $\sqrt{\mathbf{a} \cdot \mathbf{a}}$.

2. Give the euclidean norm, sum norm, and max norm of the following vectors.
   (a) [1, 1, 1]   **(b)** [3, 0, 0]   **(c)** [−1, 1, 4]   **(d)** [−1, 4, 3]
   (e) [4, 4, 4, 4]

3. The distance between two vectors **a**, **b** is defined to be the norm of their distance $|\mathbf{a} - \mathbf{b}|$.
   (a) What is the distance between the following vectors [2, 5, 7] and [3, −1, 4] using the euclidean norm, sum norm, and max norm?
   (b) Explain in words what the distance between two vectors in the sum norm measures.
   (c) Repeat part (b) for the max norm.

4. Give the sum and max norms of the following matrices.

   (a) $\begin{bmatrix} 1 & 4 \\ 5 & 3 \end{bmatrix}$ **(b)** $\begin{bmatrix} 0 & 3 \\ -5 & 3 \end{bmatrix}$ **(c)** $\begin{bmatrix} 1 & 2 & 2 \\ 6 & 1 & 3 \\ 5 & 1 & 2 \end{bmatrix}$

   (d) $\begin{bmatrix} -5 & 4 & 6 \\ 8 & 0 & 2 \\ -6 & 7 & 7 \end{bmatrix}$

5. **(a)** For each of the matrices in Exercise 4, give the vector **x*** such that
   $|\mathbf{Ax*}|_s = \|\mathbf{A}\|_s \cdot |\mathbf{x*}|_s$.
   (b) For each of the matrices in Exercise 4, give the vector **x*** such that
   $|\mathbf{Ax*}|_{mx} = \|\mathbf{A}\|_{mx}|\mathbf{x*}|_{mx}$.

6. **(a)** What is the sum norm of the following matrix?

   $$\mathbf{A} = \begin{bmatrix} 2 & 4 & -5 \\ -3 & 3 & 3 \\ 4 & 1 & -1 \end{bmatrix}$$

   (b) If **v** is a vector with sum norm $= 3$, give an upper bound on the sum norm of **Av**.
   (c) Give a vector with sum norm $= 3$ for which the bound in part (b) is achieved.
   (d) If **w** is a vector with sum norm $= 5$, give an upper bound on the sum norm of $\mathbf{A}^2\mathbf{w}$.

7. **(a)** What is the max norm of the following matrix?

   $$\mathbf{A} = \begin{bmatrix} 1 & 3 & 2 \\ 2 & 1 & 3 \\ 1 & 1 & 1 \end{bmatrix}$$

(b) If **v** is a vector with max norm = 4, give an upper bound on the max norm of **Av**.

(c) Give a vector with max norm = 4 for which this bound in part (b) is achieved.

(d) If **w** is a vector with max norm = 6, give an upper bound on the max norm of $\mathbf{A}^3\mathbf{w}$.

8. In the rabbit–fox model in Example 2, give a bound on the size of $\mathbf{p}^{(5)} = \mathbf{A}^5\mathbf{p}$ in the sum and max norms when $\mathbf{p} = [100, 100]$.

9. (a) In the following rabbit–fox models,

(i) $R' = R + .1R - .15F$ (ii) $R' = R + .2R - .5F$

$\quad\; F' = F + .1R - .15F$ $F' = F + .1R - .2F$

Determine the sum and max norms of the coefficient matrix **A**.

(b) If the current vector of population sizes is $\mathbf{p} = [20, 20]$, determine bounds (in sum and max norms) for the size of $\mathbf{p}' = \mathbf{Ap}$. Compute $\mathbf{p}'$ and see how close it is to the norm bounds.

(c) Compute a sum norm bound on the size of the population vector after three periods, $\mathbf{p}^{(3)} = \mathbf{A}^3\mathbf{p}$.

10. (a) In the following model for the growth of rabbits, foxes, and humans,

$$R' = R + .3R - .1F - .2H$$
$$F' = F + .4R - .2F - .3H$$
$$H' = H + .1R + .1F + .1H$$

determine the sum and max norms of the coefficient matrix **A**.

(b) If the current vector of population sizes is $\mathbf{p} = [10, 10, 10]$, determine bounds (in sum and max norms) for the size of $\mathbf{p}' = \mathbf{Ap}$. Compute $\mathbf{p}'$ and see how close it is to the norm bounds.

(c) Give a sum norm bound on the size of population vector after four periods, $\mathbf{p}^{(4)}$.

11. In Example 5, suppose that we assume $\mathbf{c} = [15, 5]$ when the correct value is actually $[14, 7]$. What is the maximum size that the error could be after 3 years (using the sum norm)?

12. In the rabbit–fox model in Example 2, suppose the initial vector of $\mathbf{p} = [100, 100]$ actually should have been $[95, 103]$. How large an error is possible in $\mathbf{p}'$, in $\mathbf{p}^{(3)}$ (using the sum norm)?

13. Whereas Example 5 discussed the absolute size of errors, it is often more interesting to consider the relative size of errors. The relative error in **b** is $|\mathbf{b} - \mathbf{c}|/|\mathbf{c}|$ if **b** is used when **c** really should be used. (Use the sum norm.)

(a) If $R' = R + F$, $F' = 3R - 4F$ and $\mathbf{p} = [R, F]$ was set equal to $[3, 1]$ when it really should have been $[2, 2]$, what is the relative error in **p** and what is the relative error in $\mathbf{p}'$?

(b) If $R' = 2R - 3F$, $F' = -5R + 7F$ and $\mathbf{p} = [R, F]$ was set equal to $[5, 5]$ when it really should have been $[6, 4]$, what is the relative error in $\mathbf{p}$ and what is the relative error in $\mathbf{p}'$?

14. **(a)** Describe those vectors for which the euclidean norm and max norm are equal. Explain the reason for your answer.
    (b) Describe those vectors for which the sum norm and max norm are equal. Explain the reason for your answer.
    (c) Describe those vectors for which the euclidean norm, sum norm, and max norm are all equal. Why must these be the only vectors with this property?

15. We prove that $|\mathbf{a}|_{mx} \leq |\mathbf{a}|_e \leq |\mathbf{a}|_s$.
    (a) Show that the max norm of a vector $\mathbf{a}$ is always less than or equal to the euclidean norm of $\mathbf{a}$.
    (b) Show that the euclidean norm of a vector $\mathbf{a}$ is always less than or equal to the sum norm of $\mathbf{a}$.

16. Let $\mathbf{a}$ be an n-vector.
    (a) Show that $|\mathbf{a}|_s \leq n|\mathbf{a}|_{mx}$.   **(b)** Show that $|\mathbf{a}|_s \leq \sqrt{n}|\mathbf{a}|_e$.

17. Show that $|\mathbf{a} \cdot \mathbf{b}| \leq |\mathbf{a}|_s \cdot |\mathbf{b}|_{mx}$.

18. **(a)** Show that $|\mathbf{a} + \mathbf{b}|_s \leq |\mathbf{a}|_s + |\mathbf{b}|_s$.
    (b) Repeat part (a) for the max norm.

19. Explain why the sum norm and max norm of a symmetric matrix are the same (symmetric means $a_{ij} = a_{ji}$).

20. Let $\mathbf{A}^T$ be the transpose of $\mathbf{A}$ ($\mathbf{A}^T$ is obtained from $\mathbf{A}$ by interchanging rows and columns). Show that $\|\mathbf{A}^T\|_{mx} = \|\mathbf{A}\|_s$ and $\|\mathbf{A}^T\|_s = \|\mathbf{A}\|_{mx}$.

21. **(a)** Show that $\|\mathbf{A} + \mathbf{B}\|_s \leq \|\mathbf{A}\|_s + \|\mathbf{B}\|_s$.
    (b) Repeat part (a) for the max norm.

22. **(a)** Use the fact $|\mathbf{Ax}|_s \leq \|\mathbf{A}\|_s|\mathbf{x}|_s$ to show that $\|\mathbf{AB}\|_s \leq \|\mathbf{A}\|_s\|\mathbf{B}\|_s$.
    (b) Use part (a) and Exercise 20 to show that $\|\mathbf{AB}\|_{mx} \leq \|\mathbf{A}\|_{mx}\|\mathbf{B}\|_{mx}$.

    Necessary fact: $(\mathbf{AB})^T = \mathbf{B}^T\mathbf{A}^T$.

23. If $\mathbf{A}$ is a matrix that is all 0's except one entry that has value a, show that $\|\mathbf{A}\|_s = \|\mathbf{A}\|_{mx} = |a|$.

24. **(a)** Give the adjacency matrix $\mathbf{A}(G)$ for this graph G.

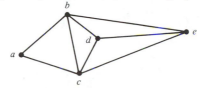

(b) What is the sum norm of $A(G)$?

(c) Explain in words an interpretation that can be given to the sum norm of the adjacency matrix of a graph.

25. (a) Show that for the matrix $A = \begin{bmatrix} 2 & 5 \\ 1 & 3 \end{bmatrix}$, a 2-vector $v = \lfloor a, b \rfloor$ with

$0 \le a \le 1, 0 \le b \le 1$ will maximize the max norm of Av by setting $a = b = 1$.

(b) Generalize your argument to show that for any 2-by-2 matrix A with nonnegative entries and any 2-vector v with max norm $|v|_{mx} = 1$, the value of $|Av|_{mx}$ is maximized when $v = [1, 1]$ and $|Av|_{mx} =$ maximum row sum (in absolute value) of A.

(c) Explain how to modify v if A has negative entries.
Hint: Possibly change one (or both) of the 1's in v to -1's.

(d) Generalize parts (b) and (c) to n-by-n matrices.

26. (a) The matrix $\begin{bmatrix} 1 & 1 \\ 0 & 2 \end{bmatrix}$ has eigenvectors $u_1 = [1, 1]$ and $u_2 = [1, 0]$. What are the corresponding eigenvalues for these eigenvectors?

(b) The matrix $\begin{bmatrix} 1 & -2 \\ -2 & 1 \end{bmatrix}$ has eigenvectors $u_1 = [1, 1]$ and $u_2 = [1, -1]$. What are the corresponding eigenvalues for these eigenvectors?

(c) The matrix $\begin{bmatrix} 1 & 6 \\ -2 & -6 \end{bmatrix}$ has eigenvectors $u_1 = [-2, 1]$ and $u_2 = [-3, 2]$. What are the corresponding eigenvalues for these eigenvectors?

(d) The matrix

$$\begin{bmatrix} -4 & 4 & 4 \\ -1 & 1 & 2 \\ -3 & 2 & 4 \end{bmatrix}$$

has eigenvectors $u_1 = [2, 0, 1]$, $u_2 = [6, 4, 5]$, and $u_3 = [4, 3, 2]$. What are the corresponding eigenvalues?

27. Verify for each Markov transition matrix A that the given vector is a stable probability vector.

(a) $A = \begin{bmatrix} \frac{1}{3} & \frac{2}{3} \\ \frac{2}{3} & \frac{1}{3} \end{bmatrix}$, $p = [\frac{1}{2}, \frac{1}{2}]$

(b) $A = \begin{bmatrix} \frac{1}{4} & \frac{1}{2} \\ \frac{3}{4} & \frac{1}{2} \end{bmatrix}$, $p = [\frac{2}{5}, \frac{3}{5}]$

(c) $\mathbf{A} = \begin{bmatrix} \frac{1}{3} & \frac{1}{2} & 0 \\ \frac{2}{3} & 0 & \frac{2}{3} \\ 0 & \frac{1}{2} & \frac{1}{3} \end{bmatrix}$, $\mathbf{p} = [\frac{3}{11}, \frac{4}{11}, \frac{3}{11}]$

28. The matrix $\begin{bmatrix} 2 & 5 \\ 6 & 1 \end{bmatrix}$ has eigenvalue $\lambda_1 = 7$ with eigenvector $\mathbf{u}_1 = [1, 1]$ and $\lambda_2 = -4$ with $\mathbf{u}_2 = [-5, 6]$.

(a) We want to compute $\mathbf{A}^3\mathbf{v}$, where $\mathbf{v} = [-2, 9]$. Writing $\mathbf{v}$ as $\mathbf{v} = 3\mathbf{u}_1 + \mathbf{u}_2$, compute $\mathbf{A}^3\mathbf{v}$ indirectly as in Example 8.

(b) Give an approximate formula for $\mathbf{A}^n\mathbf{v}$.

(c) Use the method discussed following Example 8 to determine a and b so that the vector $\mathbf{v} = [2, 13]$ can be written as $\mathbf{v} = a\mathbf{u}_1 + b\mathbf{u}_2$, and use this representation of $\mathbf{v}$ to compute $\mathbf{A}^3\mathbf{v}$.

29. The matrix $\mathbf{A} = \begin{bmatrix} 1 & 1 \\ 0 & 2 \end{bmatrix}$ has eigenvectors $\mathbf{u}_1 = [1, 1]$ and $\mathbf{u}_2 = [1, 0]$.

(a) We want to compute $\mathbf{A}^4\mathbf{v}$, where $\mathbf{v} = [3, 1]$. Writing $\mathbf{v}$ as $\mathbf{v} = \mathbf{u}_1 + 2\mathbf{u}_2$, compute $\mathbf{A}^4\mathbf{v}$ indirectly as in Example 8.

(b) Give an approximate formula for $\mathbf{A}^n\mathbf{v}$.

(c) Use the method discussed following Example 8 to determine a and b so that the vector $\mathbf{v} = [6, 9]$ can be written as $\mathbf{v} = a\mathbf{u}_1 + b\mathbf{u}_2$, and use this representation of $\mathbf{v}$ to compute $\mathbf{A}^5\mathbf{v}$.

30. Show that if $\mathbf{A}$ is a 2-by-2 symmetric matrix of the form $\begin{bmatrix} a & b \\ b & a \end{bmatrix}$, then the eigenvalues of $\mathbf{A}$ are $a + b$ and $a - b$. Verify that $[1, 1]$ and $[1, -1]$ are the associated eigenvectors.

Note: $a + b = \|\mathbf{A}\|_s = \|\mathbf{A}\|_{mx}$.

31. Show that if λ is an eigenvalue of a matrix $\mathbf{A}$, then λ^2 is an eigenvalue of $\mathbf{A}^2$.

Section 2.6 **Efficient Matrix Computation**

Computational Complexity and Error Analysis

In this section we discuss computational details of matrix multiplication. A lot of arithmetic must be done when two matrices are multiplied, and despite the great speed of modern computers, theoretical shortcuts are still needed when large matrices must be multiplied repeatedly. It is also important to know the relative complexity of different basic matrix operations. For example, which is faster, squaring an n-by-n matrix $\mathbf{A}$ or solving the system

$\mathbf{Ax} = \mathbf{b}$ of n equations in n unknowns? (Would you have guessed that normally solving $\mathbf{Ax} = \mathbf{b}$ is faster?)

Before looking for shortcuts, let us first determine the computational complexity of matrix multiplication, that is, count how many entry-by-entry multiplications and additions are required to multiply an m-by-r matrix $\mathbf{A}$ times an r-by-n matrix $\mathbf{B}$. Each entry in the product $\mathbf{AB}$ is obtained by forming a scalar product $\mathbf{a}_i^R \cdot \mathbf{b}_j^C$ of a row $\mathbf{a}_i^R$ of $\mathbf{A}$ with a column $\mathbf{b}_j^C$ of $\mathbf{B}$. Since $\mathbf{a}_i^R$ and $\mathbf{b}_j^C$ are both r-vectors, their product requires r multiplications and $r - 1$ additions. There are mn entries to be computed in $\mathbf{AB}$, each requiring r multiplications and $r - 1$ additions. Thus in total we have

Theorem 1. The matrix product $\mathbf{AB}$ of the m-by-r matrix $\mathbf{A}$ and the r-by-n matrix $\mathbf{B}$ requires mnr multiplications and $mn(r - 1)$ additions.

Corollary A. (i) The matrix product of two n-by-n matrices requires n^3 multiplications and approximately n^3 additions.

(ii) The matrix-vector product $\mathbf{Ax}$ of an m-by-n matrix $\mathbf{A}$ times an n-vector $\mathbf{x}$ requires nm multiplications and $m(n - 1)$ additions.

Proof of (ii). Treat $\mathbf{x}$ as an n-by-1 matrix. ∎

Corollary B. If the sizes (numbers of rows and columns) of two matrices are doubled, the number of multiplications in the matrix product increases by a factor of 8.

Proof. If $\mathbf{A}'$ is $2m$-by-$2r$ and $\mathbf{B}'$ is $2r$-by-$2n$, then by Theorem 1, the number of multiplications is $(2m)(2n)(2r) = 8mnr$. ∎

The reader should verify Corollary B by squaring a 4-by-4 matrix and then squaring an 8-by-8 matrix on a computer. The second calculation should take eight times as long as the first.

A matrix is called **sparse** if most of its entries are 0. Although the percentage of 0-entries to qualify as sparse is not defined precisely, most people use the figure of 80%. Large matrices in practical problems often have over 99% 0-entries. The point is that if a matrix is sparse, substantial savings in computation should be possible by forming only nonzero products.

We shall consider two approaches for reducing the computation time in sparse matrix multiplication. The first approach is symbolic, using matrix algebra, and it also works on nonsparse matrices with special patterns. The second approach involves data structures to represent sparse matrices efficiently.

First let us say a few words about the numerical stability of matrix multiplication. We want to know how much small errors, both errors in estimating the values of matrix entries and roundoff errors in computation, can influence the result of a single matrix multiplication or a sequence of multiplications, as in computing powers of a matrix.

There is nothing inherently bad about a single matrix multiplication except for the magnification of errors inherent in subtraction (if some terms in a scalar product are positive and some negative). Subtraction can be very lethal if some small numbers in the data have only one or two significant digits. For example, consider the scalar product

$$[245, -149] \cdot [.2, .3] = 245 \times .2 + (-149) \times .3$$
$$= 49.0 - 44.7 = 4.3$$

The result of 4.3 is meaningless for the following reason. Since both the terms .2 and .3 have only one-digit accuracy, $245 \times .2 = 49.0$ and $149 \times .3 = 44.7$ are only accurate to one significant digit. That is, the 9.0 in 49.0 and the 4.7 in 44.7 are essentially random numbers. Hence the subtraction result, $49.0 - 44.7 = 4.3$, has no meaning. An indication that subtraction-induced error could have occurred is if the magnitude of an entry in the matrix product is less than the magnitude of entries in the input matrices.

A long series of matrix multiplications may result in small errors building up into large errors, just as large errors can occur in repeated scalar multiplication. For example, if we need to multiply 1.15 times itself 10 times, the correct answer is $1.15^{10} = 4.045558$. But if we round 1.15 to two significant digits as 1.1 or 1.2, we get answers of $1.1^{10} \simeq 2.6$ and $1.2^{10} \simeq 6.2$.

Partitioning of Matrices

Any vector **a** can be partitioned into two or more subvectors, such as

$$\mathbf{a} = [\mathbf{a}_1, \mathbf{a}_2, \mathbf{a}_3] \tag{1}$$

For example, if $\mathbf{a} = [1, 2, 3, 0, 0, 0, 1, 2, 3]$ and if $\mathbf{a}' = [1, 2, 3]$ and $\mathbf{0}_3$ is the three-entry zero vector, we can write **a** as $[\mathbf{a}', \mathbf{0}_3, \mathbf{a}']$.

A matrix **A** can be partitioned into submatrices, such as

$$\mathbf{A} = [\mathbf{A}_1 \quad \mathbf{A}_2] \tag{2}$$

or

$$\mathbf{A} = \begin{bmatrix} \mathbf{A}_{11} & \mathbf{A}_{12} \\ \mathbf{A}_{21} & \mathbf{A}_{22} \end{bmatrix} \tag{3}$$

or

$$\mathbf{A} = \begin{bmatrix} & & \mathbf{A}_1 \\ \mathbf{A}^* & | & \\ - & - & \mathbf{A}_2 \\ \mathbf{A}_1' & \mathbf{A}_2' & \mathbf{A}_3 \end{bmatrix} \tag{4}$$

For example, we might partition

$$
\begin{bmatrix}
1 & 2 & 3 & 4 & | & 1 & 1 \\
2 & 3 & 4 & 5 & 1 & 1 \\
3 & 4 & 5 & 6 & | & 0 & 0 \\
4 & 5 & 6 & 7 & 0 & 0 \\
1 & 1 & 0 & 0 & | & 1 & 1 \\
1 & 1 & 0 & 0 & 1 & 1
\end{bmatrix}
=
\begin{bmatrix}
 & & & & | & \mathbf{A}_1 \\
 & \mathbf{A}^* & & & & \\
 & & & & | & \mathbf{A}_0 \\
 & & & & & \\
\mathbf{A}_1 & | & \mathbf{A}_0 & | & \mathbf{A}_1
\end{bmatrix}
$$

where $\mathbf{A}_1$ is a 2-by-2 matrix of 1's and $\mathbf{A}_0$ is a 2-by-2 matrix of 0's.

The partition of a matrix $\mathbf{A}$ will typically correspond to different components of the underlying model. A partition of $\mathbf{A}$ in form of (3) would arise in a Markov chain transition matrix if the states divided in some natural way into two groups, group S_1 and group S_2. The partition in (2) arises naturally if the columns of $\mathbf{A}$ represent two different types of variables. For example, the Leontief supply–demand equations (see Example 3 of Section 2.2) were written in matrix notation as

$$
\mathbf{x} = \mathbf{D}\mathbf{x} + \mathbf{c} \quad \text{or} \quad
\begin{bmatrix} x_1 \\ x_2 \\ x_3 \\ x_4 \end{bmatrix}
=
\begin{bmatrix}
.4 & .2 & .2 & .2 \\
.3 & .3 & .2 & .1 \\
.1 & .1 & 0 & .2 \\
0 & .1 & .1 & 0
\end{bmatrix}
\begin{bmatrix} x_1 \\ x_2 \\ x_3 \\ x_4 \end{bmatrix}
+
\begin{bmatrix} 100 \\ 50 \\ 100 \\ 0 \end{bmatrix}
\tag{5}
$$

The right-hand-side numbers in these equations could be combined into one expanded matrix by appending $\mathbf{c}$ as the last column:

$$
\mathbf{D}' = [\mathbf{D} \quad \mathbf{c}]
\tag{6}
$$

Correspondingly, we expand $\mathbf{x}$ by adding an additional entry with value 1, $\mathbf{x}' = [\mathbf{x}, 1]$. Now (5) becomes

$$
\mathbf{x} = \mathbf{D}'\mathbf{x}' \quad \text{or} \quad
\begin{bmatrix} x_1 \\ x_2 \\ x_3 \\ x_4 \end{bmatrix}
=
\begin{bmatrix}
.4 & .2 & .2 & .2 & 100 \\
.3 & .3 & .2 & .1 & 50 \\
.1 & .1 & 0 & .2 & 100 \\
0 & .1 & .1 & 0 & 0
\end{bmatrix}
\begin{bmatrix} x_1 \\ x_2 \\ x_3 \\ x_4 \\ 1 \end{bmatrix}
\tag{7}
$$

Example 1. **Partitioning the Adjacency Matrix of a Graph**

Figure 2.4

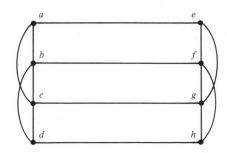

The graph G in Figure 2.4 has the following adjacency matrix:

$$\mathbf{A}(G) = \begin{array}{c} \\ a \\ b \\ c \\ d \\ e \\ f \\ g \\ h \end{array} \begin{array}{c} \begin{array}{cccccccc} a & b & c & d & e & f & g & h \end{array} \\ \left[\begin{array}{cccccccc} 0 & 1 & 1 & 0 & 1 & 0 & 0 & 0 \\ 1 & 0 & 1 & 1 & 0 & 1 & 0 & 0 \\ 1 & 1 & 0 & 1 & 0 & 0 & 1 & 0 \\ 0 & 1 & 1 & 0 & 0 & 0 & 0 & 1 \\ 1 & 0 & 0 & 0 & 0 & 1 & 1 & 0 \\ 0 & 1 & 0 & 0 & 1 & 0 & 1 & 1 \\ 0 & 0 & 1 & 0 & 1 & 1 & 0 & 1 \\ 0 & 0 & 0 & 1 & 0 & 1 & 1 & 0 \end{array} \right] \end{array} \tag{8}$$

$\mathbf{A}(G)$ has the nice partitioned form

$$\mathbf{A}(G) = \begin{bmatrix} \mathbf{A}' & \mathbf{I} \\ \mathbf{I} & \mathbf{A}' \end{bmatrix}, \qquad \text{where } \mathbf{A}' = \begin{bmatrix} 0 & 1 & 1 & 0 \\ 1 & 0 & 1 & 1 \\ 1 & 1 & 0 & 1 \\ 0 & 1 & 1 & 0 \end{bmatrix} \tag{9}$$

and $\mathbf{I}$ is the 4-by-4 identity matrix. ∎

Partitioning is very useful in matrix multiplication because we can initially treat the submatrices like scalar entries. For example, if

$$\mathbf{A} = \begin{bmatrix} \mathbf{A}_{11} & \mathbf{A}_{12} \\ \mathbf{A}_{21} & \mathbf{A}_{22} \end{bmatrix} \qquad \text{and} \qquad \mathbf{B} = \begin{bmatrix} \mathbf{B}_{11} & \mathbf{B}_{12} \\ \mathbf{B}_{21} & \mathbf{B}_{22} \end{bmatrix} \tag{10}$$

then

$$\mathbf{AB} = \begin{bmatrix} \mathbf{A}_{11}\mathbf{B}_{11} + \mathbf{A}_{12}\mathbf{B}_{21} & \mathbf{A}_{11}\mathbf{B}_{12} + \mathbf{A}_{12}\mathbf{B}_{22} \\ \mathbf{A}_{21}\mathbf{B}_{11} + \mathbf{A}_{22}\mathbf{B}_{21} & \mathbf{A}_{21}\mathbf{B}_{12} + \mathbf{A}_{22}\mathbf{B}_{22} \end{bmatrix} \tag{11}$$

Verification of (11) is left as an exercise. Of course, (11) requires that the number of columns in the $\mathbf{A}$ submatrices equal the number of rows in the appropriate $\mathbf{B}$ submatrices. The situation with partitioning is similar to all the matrix algebra rules represented in Section 2.4. Unless it is expressly prohibited, anything you would like to be true about partitioning is probably true.

If some of the submatrices of $\mathbf{A}$ and $\mathbf{B}$ have nice forms (e.g., $\mathbf{O}$ or $\mathbf{I}$), the amount of work needed to compute the matrix product $\mathbf{AB}$ is greatly reduced.

Example 1 (continued). **Powers of a Partitioned
Adjacency Matrix**

Let us use (11) to compute the square of the partitioned adjacency matrix $\mathbf{A}(G)$ in (9).

$$A^2(G) = \begin{bmatrix} A'A' + II & A'I + IA' \\ A'I + IA' & A'A' + II \end{bmatrix}$$

$$= \begin{bmatrix} A'^2 + I & 2A' \\ 2A' & A'^2 + I \end{bmatrix} \tag{12}$$

Computing A'^2 can be done by inspection faster than entering the numbers in a computer program. It just involves counting how many 1-entries each pair of rows in A' have in common (this was explained in Example 1 of Section 2.3). So

$$A'^2 = \begin{bmatrix} 2 & 1 & 1 & 2 \\ 1 & 3 & 2 & 1 \\ 1 & 2 & 3 & 1 \\ 2 & 1 & 1 & 2 \end{bmatrix} \quad \text{and} \quad A'^2 + I = \begin{bmatrix} 3 & 1 & 1 & 2 \\ 1 & 4 & 2 & 1 \\ 1 & 2 & 4 & 1 \\ 2 & 1 & 1 & 3 \end{bmatrix} \tag{13}$$

Inserting $A'^2 + I$ and $2A'$ into the partition product in (12), we obtain

$$A^2(G) = \begin{array}{c} \\ a \\ b \\ c \\ d \\ e \\ f \\ g \\ h \end{array} \begin{array}{c} \begin{array}{cccccccc} a & b & c & d & e & f & g & h \end{array} \\ \left[\begin{array}{cccc|cccc} 3 & 1 & 1 & 2 & 0 & 2 & 2 & 0 \\ 1 & 4 & 2 & 1 & 2 & 0 & 2 & 2 \\ 1 & 2 & 4 & 1 & 2 & 2 & 0 & 2 \\ 2 & 1 & 1 & 3 & 0 & 2 & 2 & 0 \\ \hline 0 & 2 & 2 & 0 & 3 & 1 & 1 & 2 \\ 2 & 0 & 2 & 2 & 1 & 4 & 2 & 1 \\ 2 & 2 & 0 & 2 & 1 & 2 & 4 & 1 \\ 0 & 2 & 2 & 0 & 2 & 1 & 1 & 3 \end{array} \right] \end{array} \tag{14}$$

Using partitioning, the only matrix product we had to calculate was A'^2, a 4-by-4 problem. By Corollary B, this is an eightfold savings over computing A^3 directly.

Suppose that we want to compute higher powers of $A(G)$. The result will not be as nice, since the four 4-by-4 submatrices of $A^2(G)$ are not as simple as those of $A(G)$. Still the problem is easier than multiplying without partitioning. The computation of $A^3(G)$ using (11) is left as an exercise. ∎

The pattern of nonzero entries in $A(G)$ is typical of patterns found in a Markov chain transition matrix, a Leontief supply–demand matrix, or a constraint matrix in a linear program, where there are interrelated clusters of states or industries, such as node sets $\{a, b, c, d\}$ and $\{e, f, g, h\}$ in G, with a small number of links between states in different clusters. See the Exercises for specific examples.

Data Structures for Sparse Matrices (Optional)

Various schemes can be used to store a sparse matrix (with most entries equal to 0). The objectives are

1. To minimize the space needed to store the matrix by storing only the nonzero entries.
2. To speed matrix multiplication (and other matrix calculations) by enabling us to compute only those terms in the product that will be nonzero.

There are two general categories of sparse matrices. The first type of sparse matrix is a matrix in which the nonzero entries form a particular pattern. An example of such a pattern is the transition matrix A of the frog Markov chain.

$$A = \begin{bmatrix} .50 & .25 & 0 & 0 & 0 & 0 \\ .50 & .50 & .25 & 0 & 0 & 0 \\ 0 & .25 & .50 & .25 & 0 & 0 \\ 0 & 0 & .25 & .50 & .25 & 0 \\ 0 & 0 & 0 & .25 & .50 & .50 \\ 0 & 0 & 0 & 0 & .25 & .50 \end{bmatrix} \tag{15}$$

All the nonzero entries in A are grouped around the main diagonal. A matrix with such a pattern is called a **band matrix**. The **bandwidth** of a band matrix is the smallest number w such that if a_{ij} is a nonzero entry, then $|i - j| < w$. The bandwidth of the matrix A in (15) is 2.

In the ith row of a band matrix with bandwidth w, the nonzero entries occur in positions $i - w + 1$ through $i + w - 1$. For obvious reasons, band matrices such as A with $w = 2$ are called *tridiagonal* matrices. Band matrices arise in many different settings.

The matrix in Example 11 of Section 2.2 for filtering digital pictures was the following 12-by-12 tridiagonal matrix:

$$F = \begin{bmatrix} \frac{5}{6} & \frac{1}{6} & 0 & 0 & 0 & 0 & 0 & 0 & 0 & 0 & 0 & 0 \\ \frac{1}{6} & \frac{2}{3} & \frac{1}{6} & 0 & 0 & 0 & 0 & 0 & 0 & 0 & 0 & 0 \\ 0 & \frac{1}{6} & \frac{2}{3} & \frac{1}{6} & 0 & 0 & 0 & 0 & 0 & 0 & 0 & 0 \\ 0 & 0 & \frac{1}{6} & \frac{2}{3} & \frac{1}{6} & 0 & 0 & 0 & 0 & 0 & 0 & 0 \\ 0 & 0 & 0 & \frac{1}{6} & \frac{2}{3} & \frac{1}{6} & 0 & 0 & 0 & 0 & 0 & 0 \\ 0 & 0 & 0 & 0 & \frac{1}{6} & \frac{2}{3} & \frac{1}{6} & 0 & 0 & 0 & 0 & 0 \\ 0 & 0 & 0 & 0 & 0 & \frac{1}{6} & \frac{2}{3} & \frac{1}{6} & 0 & 0 & 0 & 0 \\ 0 & 0 & 0 & 0 & 0 & 0 & \frac{1}{6} & \frac{2}{3} & \frac{1}{6} & 0 & 0 & 0 \\ 0 & 0 & 0 & 0 & 0 & 0 & 0 & \frac{1}{6} & \frac{2}{3} & \frac{1}{6} & 0 & 0 \\ 0 & 0 & 0 & 0 & 0 & 0 & 0 & 0 & \frac{1}{6} & \frac{2}{3} & \frac{1}{6} & 0 \\ 0 & 0 & 0 & 0 & 0 & 0 & 0 & 0 & 0 & \frac{1}{6} & \frac{2}{3} & \frac{1}{6} \\ 0 & 0 & 0 & 0 & 0 & 0 & 0 & 0 & 0 & 0 & \frac{1}{6} & \frac{5}{6} \end{bmatrix}$$

In Section 4.7 a tridiagonal 100-by-100 matrix arises when we approximate a differential equation with a finite-difference system.

The natural way to store a band matrix $\mathbf{A}$ is to store just the subvectors $\mathbf{a}_i^*$ of nonzero entries in each row, so that the ith row of $\mathbf{A}$ is of the form $\mathbf{a}_i^R = [\mathbf{0}, \mathbf{a}_i^*, \mathbf{0}]$. For example, for the frog Markov chain $\mathbf{a}_3^* = \mathbf{a}_4^* = [.25, .5, .25]$. We also must store the subvectors of nonzero entries in each column.

When we are multiplying two band matrices $\mathbf{A}$ and $\mathbf{B}$ together (with bandwidths w and w', respectively), there will be many cases where the bands of $\mathbf{A}$ and $\mathbf{B}$ do not overlap, so that the vector product will be 0. This will happen if $i + w - 1 \geq j - w'$ or if $j + w' - 1 \geq i - w$. [Check this for $\mathbf{A} = \mathbf{B} = \mathbf{F}$, the matrix in (15).] When the bands do overlap, the lower bound for k in the summation will be $\max(1, i - w + 1, j - w' + 1)$; finding the upper bound is left as an exercise.

Note that in calculating powers of the transition matrix $\mathbf{A}$ in (15), we could compute $\mathbf{A}^2$ as the product of two tridiagonal matrices. However, the result $\mathbf{A}^2$ is not tridiagonal, so to compute $\mathbf{A}^3 = \mathbf{A}\mathbf{A}^2$ and higher powers we would be multiplying a tridiagonal matrix $\mathbf{A}$ times a regular matrix.

The second type of sparse matrix data structure is for a matrix whose nonzero entries are randomly located.

Example 2. **Squaring a Sparse Matrix**

Consider the example of the 16-by-16 0–1 matrix $\mathbf{M}$ in (16) that has only 32 1's among its $16 \times 16 = 256$ entries (about 12%). $\mathbf{M}$ partitions into four identical 8-by-8 matrices, which we call $\mathbf{N}$.

$$\mathbf{M} = \begin{bmatrix} \mathbf{N} & | & \mathbf{N} \\ \overline{} & \overline{} & \overline{} \\ \mathbf{N} & | & \mathbf{N} \end{bmatrix} = \begin{array}{c|cccccccc|cccccccc} & 1 & 2 & 3 & 4 & 5 & 6 & 7 & 8 & 9 & A & B & C & D & E & F & G \\ 1 & 0 & 0 & 0 & 0 & 0 & 1 & 0 & 0 & 0 & 0 & 0 & 0 & 0 & 1 & 0 & 0 \\ 2 & 0 & 0 & 1 & 0 & 0 & 0 & 0 & 1 & 0 & 0 & 1 & 0 & 0 & 0 & 0 & 1 \\ 3 & 0 & 0 & 0 & 0 & 0 & 0 & 0 & 0 & 0 & 0 & 0 & 0 & 0 & 0 & 0 & 0 \\ 4 & 0 & 0 & 0 & 0 & 0 & 0 & 0 & 1 & 0 & 0 & 0 & 0 & 0 & 0 & 0 & 1 \\ 5 & 1 & 0 & 0 & 0 & 0 & 0 & 0 & 0 & 1 & 0 & 0 & 0 & 0 & 0 & 0 & 0 \\ 6 & 1 & 0 & 0 & 1 & 0 & 0 & 0 & 0 & 1 & 0 & 0 & 1 & 0 & 0 & 0 & 0 \\ 7 & 0 & 0 & 0 & 0 & 1 & 0 & 0 & 0 & 0 & 0 & 0 & 0 & 1 & 0 & 0 & 0 \\ 8 & 0 & 0 & 0 & 0 & 0 & 0 & 0 & 0 & 0 & 0 & 0 & 0 & 0 & 0 & 0 & 0 \\ 9 & 0 & 0 & 0 & 0 & 0 & 1 & 0 & 0 & 0 & 0 & 0 & 0 & 0 & 1 & 0 & 0 \\ A & 0 & 0 & 1 & 0 & 0 & 0 & 0 & 1 & 0 & 0 & 1 & 0 & 0 & 0 & 0 & 1 \\ B & 0 & 0 & 0 & 0 & 0 & 0 & 0 & 0 & 0 & 0 & 0 & 0 & 0 & 0 & 0 & 0 \\ C & 0 & 0 & 0 & 0 & 0 & 0 & 0 & 1 & 0 & 0 & 0 & 0 & 0 & 0 & 0 & 1 \\ D & 1 & 0 & 0 & 0 & 0 & 0 & 0 & 0 & 1 & 0 & 0 & 0 & 0 & 0 & 0 & 0 \\ E & 1 & 0 & 0 & 1 & 0 & 0 & 0 & 0 & 1 & 0 & 0 & 1 & 0 & 0 & 0 & 0 \\ F & 0 & 0 & 0 & 0 & 1 & 0 & 0 & 0 & 0 & 0 & 0 & 0 & 1 & 0 & 0 & 0 \\ G & 0 & 0 & 0 & 0 & 0 & 0 & 0 & 0 & 0 & 0 & 0 & 0 & 0 & 0 & 0 & 0 \end{array} \tag{16}$$

To limit row and column names to one symbol, we use the notation $A = 10, B = 11, C = 12, D = 13, E = 14, F = 15, G = 16$ in (16).

For each row and each column of **N**, we make lists R_i and C_j of the nonzero positions. For example, the first few row lists for **N** would be $R_1 = (6)$, $R_2 = (3, 8)$, $R_3 = ( )$.

Suppose that we want to compute the square of **M**. In terms of **N**,

$$\mathbf{M}^2 = \begin{bmatrix} \mathbf{N} & | & \mathbf{N} \\ \hline \mathbf{N} & | & \mathbf{N} \end{bmatrix} \begin{bmatrix} \mathbf{N} & | & \mathbf{N} \\ \hline \mathbf{N} & | & \mathbf{N} \end{bmatrix}$$

$$= \begin{bmatrix} \mathbf{N}^2 + \mathbf{N}^2 & | & \mathbf{N}^2 + \mathbf{N}^2 \\ \hline \mathbf{N}^2 + \mathbf{N}^2 & | & \mathbf{N}^2 + \mathbf{N}^2 \end{bmatrix} = \begin{bmatrix} \mathbf{N}' & | & \mathbf{N}' \\ \hline \mathbf{N}' & | & \mathbf{N}' \end{bmatrix} \quad (17)$$

where $\mathbf{N}' = 2\mathbf{N}^2$. So to square **M** we only have to compute the square of **N**.

Recall that when we multiply **N** times **N**, entry (i, j) in $\mathbf{N}^2$ is simply the number of positions where the row $\mathbf{n}_i^R$ and column $\mathbf{n}_j^C$ of **N** are both 1 (see Example 1 of Section 2.3). To determine how many 1's $\mathbf{n}_i^R$ and $\mathbf{n}_j^C$ have in common, we simply look at our lists of 1-entries for $\mathbf{n}_i^R$ and $\mathbf{n}_j^C$ and see how many positions these two lists have in common. Notice that no actual multiplication ever occurs in finding the matrix product of two sparse 0–1 matrices. Using the lists of 1-entries to perform the matrix multiplication typically requires only one-tenth the time of normal matrix multiplication, when multiplying two 15-by-15 0–1 matrices with around 10% 1-entries (savings are greater for larger or sparser matrices).

Using lists of 1-entries, we compute

$$\mathbf{N}^2 = \begin{matrix} 0 \\ 0 \\ 0 \\ 0 \\ 1 \\ 1 \\ 0 \\ 0 \end{matrix} \begin{bmatrix} 0 & 0 & 0 & 0 & 1 & 0 & 0 \\ 0 & 1 & 0 & 0 & 0 & 0 & 1 \\ 0 & 0 & 0 & 0 & 0 & 0 & 0 \\ 0 & 0 & 0 & 0 & 0 & 0 & 1 \\ 0 & 0 & 0 & 0 & 0 & 0 & 0 \\ 0 & 0 & 1 & 0 & 0 & 0 & 0 \\ 0 & 0 & 0 & 1 & 0 & 0 & 0 \\ 0 & 0 & 0 & 0 & 0 & 0 & 0 \end{bmatrix} \begin{bmatrix} 0 & 0 & 0 & 0 & 0 & 1 & 0 & 0 \\ 0 & 0 & 1 & 0 & 0 & 0 & 0 & 1 \\ 0 & 0 & 0 & 0 & 0 & 0 & 0 & 0 \\ 0 & 0 & 0 & 0 & 0 & 0 & 0 & 1 \\ 1 & 0 & 0 & 0 & 0 & 0 & 0 & 0 \\ 1 & 0 & 0 & 1 & 0 & 0 & 0 & 0 \\ 0 & 0 & 0 & 0 & 1 & 0 & 0 & 0 \\ 0 & 0 & 0 & 0 & 0 & 0 & 0 & 0 \end{bmatrix}$$

$$
= \begin{array}{c} 1 \\ 0 \\ 0 \\ 0 \\ 0 \\ 0 \\ 1 \\ 0 \end{array}
\begin{bmatrix}
0 & 0 & 1 & 0 & 0 & 0 & 0 \\
0 & 0 & 0 & 0 & 0 & 0 & 0 \\
0 & 0 & 0 & 0 & 0 & 0 & 0 \\
0 & 0 & 0 & 0 & 0 & 0 & 0 \\
0 & 0 & 0 & 0 & 1 & 0 & 0 \\
0 & 0 & 0 & 0 & 1 & 0 & 1 \\
0 & 0 & 0 & 0 & 0 & 0 & 0 \\
0 & 0 & 0 & 0 & 0 & 0 & 0
\end{bmatrix}
\tag{18}
$$

The entries in $\mathbf{M}^2$ are obtained from $\mathbf{N}^2$ as shown in (17). ∎

If the nonzero entries in $\mathbf{M}$ were not always 1, then for each position in a row or column list we must record the number together with its position. For example, if the first three rows of $\mathbf{M}$ were

$$
\begin{array}{c}
\\ 1 \\ 2 \\ 3
\end{array}
\begin{array}{cccccccccccccccc}
1 & 2 & 3 & 4 & 5 & 6 & 7 & 8 & 9 & A & B & C & D & E & F & G \\
\end{array}
$$

$$
\begin{array}{c}
1 \\ 2 \\ 3
\end{array}
\begin{bmatrix}
0 & 0 & 0 & 0 & 0 & 4 & 0 & 0 & 0 & 0 & 0 & 0 & 0 & 0 & 0 & 0 \\
0 & 0 & 2 & 0 & 0 & 0 & 0 & 7 & 0 & 0 & 0 & 0 & -2 & 0 & 0 & 0 \\
0 & 0 & 0 & 0 & 0 & 0 & 0 & 0 & 0 & 1 & 0 & 0 & 9 & 0 & 0 & 0
\end{bmatrix}
$$

then the first three row lists would each be a pair of lists: $R_1 = (6)$, $R_1' = (4)$; $R_2 = (3, 8, D)$, $R_2' = (2, 7, -2)$; $R_3 = (A, D)$, $R_3 = (1, 9)$, where R_i is the list of nonzero positions in row i and R_i' is the list of values of these nonzero positions.

When we multiply two sparse matrices $\mathbf{M}$ and $\mathbf{N}$ that are not 0–1 matrices, we start with the procedure above of comparing lists for $\mathbf{m}_i^R$ and $\mathbf{n}_j^C$ to find out in which positions $\mathbf{m}_j^R$ and $\mathbf{n}_j^C$ are both nonzero. But now we multiply the two numbers when two nonzero positions match and add up all such products to obtain entry (i, j) in $\mathbf{MN}$.

The main work in multiplying two sparse matrices stored in this fashion is comparing the lists R_i, and C_j for matching positions (there is lots of testing for matches and few cases of an actual match when we multiply). If list R_j has s positions (of nonzero entries) and C_j has t positions, the two lists can be checked for all possible matches with $s + t - 1$ comparisons (this basic fact about data structures is left as an exercise). If $\mathbf{A}$ is an m-by-r matrix with probability p_1 of an entry being nonzero, then on average, row $\mathbf{a}_i^R$ has $p_1 r$ nonzero entries. Similarly, if $\mathbf{B}$ is an r-by-n matrix with probability p_2 of an entry being nonzero, then, on average, column $\mathbf{b}_j^C$ has $p_2 r$ nonzero entries (r is the size of $\mathbf{a}_i^R$ and $\mathbf{b}_j^C$). So on average, the scalar product $\mathbf{a}_i^R \cdot \mathbf{b}_j^C$ requires $p_1 r + p_2 r - 1$ comparisons.

We have to perform mn such scalar products to obtain all entries in the product $\mathbf{AB}$, so in total this is $mn(p_1 r + p_2 r - 1) = mnr(p_1 + p_2 - 1/r)$ comparisons. Recall that mnr operations (entry-by-entry multiplications) are required for normal matrix multiplication. For example, if p_1 and $p_2 = .1$ and $r = 15$, then $(p_1 + p_2 - 1/r) \approx .13$, so the computational effort will be around one-eighth as much using these data structures.

The methods presented here make it possible to construct a mathematical model with a 100,000-by-20,000 matrix to, say, describe the distribution and use of all forms of energy by all sectors in the U.S. economy. Often, no more than 1 in 1000 of the entries in such a matrix is nonzero; further, the matrix can be partitioned extensively. The combination of partitioning and data structures can reduce the time of sparse matrix operations by a factor of 1,000,000.

Section 2.6 Exercises

Summary of Exercises
Exercises 1 and 2 involve the speed of matrix multiplication. Exercises 3–15 deal with partitioned matrices. Exercises 16–21 involve band matrices, and Exercises 22 and 23 involve sparse matrices.

1. How many multiplications are required to perform the following matrix operations?
   (a) Square a 10-by-10 matrix.
   (b) Square a 100-by-100 matrix.
   (c) Multiply a 20-by-5 matrix times a 5-by-20 matrix.
   (d) Multiply a 5-by-20 matrix times a 20-by-5 matrix.
   (e) Cube a 10-by-10 matrix.
   (f) Multiply a 10-by-10 matrix times itself 10 times.

2. (a) Suppose that you had to compute the product **ABC**, where **A** is 8-by-4, **B** is 4-by-8, and **C** is 8-by-5. You can either multiply **A** times **B** and then multiply the product **AB** times **C**, or you can multiply **B** times **C** and then multiply **A** times the product **BC**. How many operations are involved each way in computing **ABC**? Which way is faster?
   (b) Repeat part (a) for the product **ABC**, where **A** is 10-by-8, **B** is 8-by-4, and **C** is 4-by-6.

3. Partition the following matrices into appropriate submatrices.

$$
\text{(a)}\ \left[\begin{array}{cccc|cccc}
2 & 2 & 2 & 2 & 1 & 1 & 1 & 1 \\
2 & 2 & 2 & 2 & 1 & 1 & 1 & 1 \\
1 & 1 & 1 & 1 & 2 & 2 & 2 & 2 \\
1 & 1 & 1 & 1 & 2 & 2 & 2 & 2 \\
\hline
2 & 2 & 2 & 2 & 1 & 1 & 1 & 1 \\
2 & 2 & 2 & 2 & 1 & 1 & 1 & 1 \\
1 & 1 & 1 & 1 & 2 & 2 & 2 & 2 \\
1 & 1 & 1 & 1 & 2 & 2 & 2 & 2
\end{array}\right]
\qquad
\text{(b)}\ \left[\begin{array}{cccc|cccc}
1 & 0 & 1 & 0 & 0 & 1 & 0 & 1 \\
0 & 1 & 0 & 1 & 1 & 0 & 1 & 0 \\
1 & 0 & 1 & 0 & 0 & 1 & 0 & 1 \\
0 & 1 & 0 & 1 & 1 & 0 & 1 & 0 \\
\hline
2 & 0 & 2 & 0 & 0 & 1 & 0 & 1 \\
0 & 2 & 0 & 2 & 1 & 0 & 1 & 0 \\
2 & 0 & 2 & 0 & 0 & 1 & 0 & 1 \\
0 & 2 & 0 & 2 & 1 & 0 & 1 & 0
\end{array}\right]
$$

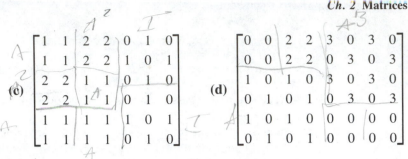

$$\text{(c)} \begin{bmatrix} 1 & 1 & 2 & 2 & 0 & 1 & 0 \\ 1 & 1 & 2 & 2 & 1 & 0 & 1 \\ 2 & 2 & 1 & 1 & 0 & 1 & 0 \\ 2 & 2 & 1 & 1 & 0 & 1 & 0 \\ 1 & 1 & 1 & 1 & 1 & 0 & 1 \\ 1 & 1 & 1 & 1 & 0 & 1 & 0 \end{bmatrix} \qquad \text{(d)} \begin{bmatrix} 0 & 0 & 2 & 2 & 3 & 0 & 3 & 0 \\ 0 & 0 & 2 & 2 & 0 & 3 & 0 & 3 \\ 1 & 0 & 1 & 0 & 3 & 0 & 3 & 0 \\ 0 & 1 & 0 & 1 & 0 & 3 & 0 & 3 \\ 1 & 0 & 1 & 0 & 0 & 0 & 0 & 0 \\ 0 & 1 & 0 & 1 & 0 & 0 & 0 & 0 \end{bmatrix}$$

4. Write the following systems of equations in matrix notation as $\mathbf{x} = \mathbf{Dx}^*$; define $\mathbf{D}$, $\mathbf{x}$, and $\mathbf{x}^*$.

   (a) $x_1 = 3x_1 + 4x_2 + 100$
   $ x_2 = 2x_1 - 3x_2 + 200$

   (b) $x_1 = x_1 + .3x_1 - .4x_2 + 100$
   $ x_2 = x_2 - .2x_1 + .3x_2 + 200$

   (c) $p_1 = 2p_1 - p_2 + 100$
   $ p_2 = 5p_1 + 3p_2 + 50$
   $ 1 = p_1 + p_2.$

5. In equation (7), alter D' by adding another row so that the equation has the form $\mathbf{x}' = \mathbf{D}''\mathbf{x}'$.

6. Write the adjacency matrices of the following graphs and define a partitioned form of the matrices.

   (a) G_1 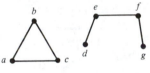   (b) G_2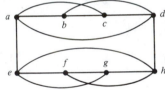

   (c) G_3 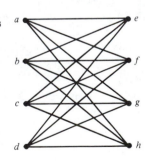   (d) G_4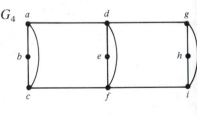

7. Consider the following Markov chain model.

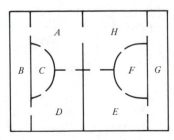

In the maze, a person in any given room has equal chances of leaving by any door out of the room (but never remains in a room). Write the Markov transition matrix for this maze. Write the matrix in partitioned form.

8. In a Leontief economic model, we might consider three commodities A, B, C in two different countries. Suppose that there is the following interindustry demand matrix **D** for the three commodities in each country.

$$
\mathbf{D} = \begin{array}{c} \\ A \\ B \\ C \end{array}
\begin{array}{c} \begin{array}{ccc} A & B & C \end{array} \\
\begin{bmatrix} .2 & .3 & .2 \\ .1 & 0 & .3 \\ .2 & .2 & .1 \end{bmatrix} \end{array}
$$

In addition, to produce a dollar's worth of each commodity in the second country requires as input .1 dollar of commodity A from country 1. Similarly, each first country commodity requires .1 dollar of the second country's commodity A. The consumer demand in the first country for the three commodities is [50, 100, 50], and the consumer demand in the second country is [100, 200, 100].

Write out the system of equations for this Leontief model. Also write the right side of these equations as $\mathbf{D'x'}$; define $\mathbf{D'}$.

9. Determine the square of the matrix in Exercise 3, part (b).

10. Determine $\mathbf{AA}^T$ for the matrix in Exercise 3, part (a) ($\mathbf{A}^T$ denotes the transpose of $\mathbf{A}$ — with rows and columns interchanged).

11. **(a)** Partition the matrix

$$
\mathbf{A} = \begin{bmatrix}
1 & 1 & 0 & 0 & 0 & 0 & 0 & 0 & 0 \\
1 & 0 & 1 & 0 & 0 & 0 & 0 & 0 & 0 \\
0 & 1 & 1 & 0 & 0 & 0 & 0 & 0 & 0 \\
0 & 0 & 0 & 2 & 2 & 0 & 0 & 0 & 0 \\
0 & 0 & 0 & 2 & 0 & 2 & 0 & 0 & 0 \\
0 & 0 & 0 & 0 & 2 & 2 & 0 & 0 & 0 \\
0 & 0 & 0 & 0 & 0 & 0 & 3 & 3 & 0 \\
0 & 0 & 0 & 0 & 0 & 0 & 3 & 0 & 3 \\
0 & 0 & 0 & 0 & 0 & 0 & 0 & 3 & 3
\end{bmatrix}
$$

in terms of the matrix

$$
\mathbf{B} = \begin{bmatrix} 1 & 1 & 0 \\ 1 & 0 & 1 \\ 0 & 1 & 1 \end{bmatrix}
$$

and the zero matrix **0**.

(b) Write $\mathbf{A}^2$ and $\mathbf{A}^3$ in partitioned form in terms of $\mathbf{B}$ and $\mathbf{0}$.

(c) Write out $\mathbf{A}^2$ entry by entry. How many multiplication operations are required to write out $\mathbf{A}^2$ using the advantages of the partitioned form? How many multiplication operations if $\mathbf{A}^2$ was done normally?

(d) Compute $\mathbf{A}^2\mathbf{v}$ where $\mathbf{v} = [1, -1, 1, 0, 1, 0, 1, -1, 1]$.

12. Compute the square of each adjacency matrix in Exercise 6 using the partitioned form of the matrix.

13. Suppose that the adjacency matrix $\mathbf{A}(G)$ of a graph G has the partitioned form

$$\mathbf{A}(G) = \begin{bmatrix} \mathbf{J}_3 & \mathbf{O} \\ \mathbf{O} & \mathbf{J}_3 \end{bmatrix}$$

where $\mathbf{J}_3$ is a 3-by-3 matrix with each entry 1.

(a) Draw G.

(b) Write out all the entries in $\mathbf{A}^3(G)$.

14. Determine the partitioned form of $\mathbf{A}^3(G)$ in Example 1 [in terms of $\mathbf{A}'$ and $\mathbf{I}$, just as $\mathbf{A}^2(G)$ is expressed in (12)].

15. (a) Let

$$\mathbf{A} = \begin{bmatrix} 1 & 2 & 0 \\ -1 & 3 & 1 \\ 0 & 1 & 2 \end{bmatrix} \quad \text{and} \quad \mathbf{B} = \begin{bmatrix} 1 & 2 \\ 3 & 4 \\ 1 & 0 \end{bmatrix}$$

Partition $\mathbf{A}$ into three 3-by-1 submatrices and $\mathbf{B}$ into three 1-by-2 submatrices, and use this partition to compute the product $\mathbf{AB}$. Compute the matrix product $\mathbf{AB}$ the normal way and compare the arithmetic in the two methods.

(b) Extend part (a) to show that in any matrix product $\mathbf{AB}$, the m-by-r matrix $\mathbf{A}$ can be partitioned into r submatrices each consisting of one of $\mathbf{A}$'s columns and the r-by-n matrix $\mathbf{B}$ partitioned into r submatrices each consisting of one of $\mathbf{B}$'s rows. Explain in words the effect of this partitioned product [generalize the comparison you made in part (a)].

16. Compute the square of the frogger transition matrix in (15).

17. Compute the square of the filtering matrix **F**.

18. Compute the square of the following band matrices.

(a)
$$\begin{bmatrix} 2 & 1 & 0 & 0 & 0 & 0 \\ 0 & 2 & 1 & 0 & 0 & 0 \\ 0 & 0 & 2 & 1 & 0 & 0 \\ 0 & 0 & 0 & 2 & 1 & 0 \\ 0 & 0 & 0 & 0 & 2 & 1 \\ 0 & 0 & 0 & 0 & 0 & 2 \end{bmatrix}$$

(b)
$$\begin{bmatrix} 2 & 0 & 1 & 0 & 0 & 0 & 0 \\ 0 & 2 & 0 & 1 & 0 & 0 & 0 \\ 1 & 0 & 2 & 0 & 1 & 0 & 0 \\ 0 & 1 & 0 & 2 & 0 & 1 & 0 \\ 0 & 0 & 1 & 0 & 2 & 0 & 1 \\ 0 & 0 & 0 & 1 & 0 & 2 & 0 \\ 0 & 0 & 0 & 0 & 1 & 0 & 2 \end{bmatrix}$$

19. Suppose that in the frog Markov chain there were 12 lanes in the superhighway, not 4. In addition, there are still the left and right sides of the road.
    (a) Describe the new Markov transition matrix by telling the bandwidth and the values of the entries on, and just off, the main diagonal. The entries in the first and fourteenth columns should also be described.
    (b) Describe the square of this Markov transition matrix in the same terms as given in part (a).

20. Suppose that we define a sequence of values $a_1, a_2, a_3, \ldots, a_8$ such that the weighted running average $(a_{i-1} + 2a_i + a_{i+1})/4 = 5i$, for $i = 2, 3, \ldots, 7$ and for $i = 1$, we use $(a_1 + a_2)/2 = 5$ and similarly for $i = 8$, $(a_7 + a_8)/2 = 35$.
    (a) Write out this system of equations for the unknown a_i.
    (b) Given that a_8 should be 35, solve the system in part (a).

21. What is the bandwidth of the square of a k-bandwidth matrix?

22. (a) Compute the squares of the following sparse matrices.

(i)
$$\begin{bmatrix} 1 & 0 & 0 & 0 & 1 & 0 \\ 0 & 0 & 0 & 1 & 0 & 0 \\ 0 & 1 & 0 & 0 & 0 & 0 \\ 1 & 0 & 0 & 0 & 0 & 1 \\ 1 & 0 & 0 & 0 & 0 & 0 \\ 0 & 0 & 1 & 0 & 0 & 0 \end{bmatrix}$$

(ii)
$$\begin{bmatrix} 0 & 1 & 0 & 0 & 0 & 0 \\ 0 & 0 & 1 & 0 & 0 & 0 \\ 0 & 0 & 0 & 1 & 0 & 0 \\ 0 & 0 & 0 & 0 & 1 & 0 \\ 0 & 0 & 0 & 0 & 0 & 1 \\ 1 & 0 & 0 & 0 & 0 & 0 \end{bmatrix}$$

    (b) Compute the cubes of the matrices in part (a).

23. (a) Give a partitioned form of this 12-by-12 sparse matrix.

$$\begin{bmatrix} 2 & 0 & 2 & 0 & 0 & 0 & 0 & 0 & 0 & 0 & 0 & 1 \\ 0 & 2 & 0 & 2 & 0 & 0 & 0 & 0 & 0 & 0 & 1 & 0 \\ 0 & 0 & 2 & 0 & 2 & 0 & 0 & 0 & 0 & 1 & 0 & 0 \\ 0 & 0 & 0 & 2 & 0 & 2 & 0 & 0 & 1 & 0 & 0 & 0 \\ 2 & 0 & 0 & 0 & 2 & 0 & 0 & 1 & 0 & 0 & 0 & 0 \\ 0 & 2 & 0 & 0 & 0 & 2 & 1 & 0 & 0 & 0 & 0 & 0 \\ 0 & 0 & 0 & 0 & 0 & 1 & 2 & 0 & 2 & 0 & 0 & 0 \\ 0 & 0 & 0 & 0 & 1 & 0 & 0 & 2 & 0 & 2 & 0 & 0 \\ 0 & 0 & 0 & 1 & 0 & 0 & 0 & 0 & 2 & 0 & 2 & 0 \\ 0 & 0 & 1 & 0 & 0 & 0 & 0 & 0 & 2 & 0 & 2 \\ 0 & 1 & 0 & 0 & 0 & 0 & 2 & 0 & 0 & 0 & 2 & 0 \\ 1 & 0 & 0 & 0 & 0 & 0 & 0 & 2 & 0 & 0 & 0 & 2 \end{bmatrix}$$

(b) Compute the square of this matrix using the partitioned form and sparse matrix multiplication.

(c) Give a formula for the nth power of the upper right 6-by-6 submatrix.

3

Solving Systems of Linear Equations

Solving Systems of Equations with Determinants

In this chapter we discuss the central mathematical problem of linear models, solving a system of linear equations. The models introduced in previous chapters will be used to motivate and illustrate the computational techniques and mathematical concepts. Much of the theory about solutions to systems of linear equations will be delayed until Chapter 5. First, in Chapter 4, we use the methods from this chapter to solve systems of linear equations arising in various applications.

In this first section we consider an algebraic approach involving determinants for solving a system of linear equations. Determinants produce a useful formula for solving two equations in two unknowns. Although the method does not yield efficient computational schemes for larger systems, it does yield important information about when such systems have solutions and about eigenvalues of the coefficient matrix. A more general method for solving linear equations is discussed in Section 3.2.

Recall that the quadratic equation $ax^2 + bx + c = 0$ has the solutions $x = (1/2a)(-b \pm \sqrt{b^2 - 4ac})$. We seek similar formulas for solving a system of linear equations. Consider the following system of two equations in two unknowns:

$$ax + by = e$$
$$cx + dy = f$$
(1)

157

Let us solve (1) for x and y. Multiplying the first equation by d and the second by b and then subtracting, we obtain

$$
\begin{array}{r}
adx + bdy = de \\
-(bcx + bdy = bf) \\
\hline
(ad - bc)x = de - bf
\end{array}
$$

Solving for x, we have

$$x = \frac{de - bf}{ad - bc} \tag{2}$$

Substituting (2) in the first equation of (1) and simplifying, we obtain

$$y = \frac{af - ce}{ad - bc} \tag{3}$$

Formulas (2) and (3) give us immediate solutions to any system of two equations in two unknowns. For example, the system

$$
\begin{aligned}
2x - 3y &= 4 \\
x + 2y &= 9
\end{aligned} \tag{4}
$$

is solved using (2) and (3) (with $a = 2$, $b = -3$, $c = 1$, $d = 2$, $e = 4$, $f = 9$).

$$x = \frac{de - bf}{ad - bc} = \frac{2 \times 4 - (-3) \times 9}{2 \times 2 - (-3) \times 1} = \frac{35}{7} = 5$$

$$y = \frac{af - ce}{ad - bc} = \frac{2 \times 9 - 1 \times 4}{2 \times 2 - (-3) \times 1} = \frac{14}{7} = 2$$

By using various techniques, it is possible to extend these formulas to obtain expressions for the solutions to three equations in three unknowns and more generally to n equations in n knowns. However, these expressions become huge, and evaluating them takes far longer than Gaussian elimination (which is presented in Section 3.2).

Formulas (2) and (3) have important theoretical uses. The critical part of the formulas is their denominators, which are the same: $ad - bc$. This denominator is called the **determinant** of the system. It turns out that the expressions for the solution of three equations in three unknowns also have common denominators. This result holds in general for the solution of any system of n equations in n unknowns. *The determinant of the n-by-n matrix* **A** *is defined to be the denominator in the algebraic expressions for the solution of the system* **Ax** = **b**. Formulas such as (2) and (3) give a unique solution to the system of equations, *provided that the determinant is nonzero.*

The determinant is like the expression $b^2 - 4ac$ under the square-root sign in the quadratic formula; recall that $b^2 - 4ac$ is called the discriminant.

The discriminant must be nonnegative for the quadratic equation to have real solutions. Here the determinant must be nonzero for a unique solution.

Rewriting system (1) with matrix subscripts, we have

$$a_{11}x_1 + a_{12}x_2 = b_1$$
$$a_{21}x_1 + a_{22}x_2 = b_2$$

or in matrix notation,

$$\mathbf{Ax} = \mathbf{b}$$

where

$$\mathbf{A} = \begin{bmatrix} a_{11} & a_{12} \\ a_{21} & a_{22} \end{bmatrix}, \qquad \mathbf{b} = \begin{bmatrix} b_1 \\ b_2 \end{bmatrix}, \qquad \mathbf{x} = \begin{bmatrix} x_1 \\ x_2 \end{bmatrix}$$

Now (2) and (3) become

$$x_1 = \frac{a_{22}b_1 - a_{12}b_2}{a_{11}a_{22} - a_{12}a_{21}}, \qquad x_2 = \frac{a_{11}b_2 - a_{21}b_1}{a_{11}a_{22} - a_{12}a_{21}} \qquad (5)$$

The determinant $\det(\mathbf{A})$ of the 2-by-2 matrix $\mathbf{A}$ is written as

$$\det(\mathbf{A}) = \begin{vmatrix} a_{11} & a_{12} \\ a_{21} & a_{22} \end{vmatrix} = a_{11}a_{22} - a_{12}a_{21} \qquad (6)$$

In the 2-by-2 case, $\det(\mathbf{A})$ is simply the product of the two main diagonal entries minus the product of the two off-diagonal entries. Since every square matrix can be interpreted as the coefficient matrix of a system of linear equations, every square matrix has a determinant.

We can write the numerators in the expression for x_1 and x_2 as determinants of the matrices obtained by replacing the first and second columns, respectively, of $\mathbf{A}$ by the vector $\mathbf{b}$. That is, let

$$\mathbf{A}_1 = [\mathbf{b} \quad \mathbf{a}_2^C] \qquad \text{and} \qquad \mathbf{A}_2 = [\mathbf{a}_1^C \quad \mathbf{b}] \qquad (7)$$

Then

$$\det(\mathbf{A}_1) = \begin{vmatrix} b_1 & a_{12} \\ b_2 & a_{22} \end{vmatrix} = a_{22}b_1 - a_{12}b_2$$

$$\det(\mathbf{A}_2) = \begin{vmatrix} a_{11} & b_1 \\ a_{21} & b_2 \end{vmatrix} = a_{11}b_2 - a_{21}b_1 \qquad (8)$$

The expressions in (8) are exactly the numerators in (5). So using $\det(A_1)$ and $\det(A_2)$, our formulas for x_1 and x_2 are

$$x_1 = \frac{\det(A_1)}{\det(A)} \quad \text{and} \quad x_2 = \frac{\det(A_2)}{\det(A)}$$

The numerators in systems of n equations in n unknowns turn out to have the same form as for two equations. That is, if we define A_i to be the matrix obtained from A by replacing the ith column a_i^C by the right-hand-side vector b,

$$A_i = [a_1^C, a_2^C, \ldots, b, a_{i+1}^C, \ldots, a_n^C] \tag{9}$$

then the solution to $Ax = b$ is

Cramer's Rule

$$x_i = \frac{\det(A_i)}{\det(A)}, \quad i = 1, 2, \ldots, n \tag{10}$$

Applying Cramer's rule to the system of equations in (4),

$$2x - 3y = 4$$
$$x + 2y = 9$$

we obtain

$$x_1 = \frac{\begin{vmatrix} 4 & -3 \\ 9 & 2 \end{vmatrix}}{\begin{vmatrix} 2 & -3 \\ 1 & 2 \end{vmatrix}} = \frac{4 \times 2 - (-3) \times 9}{2 \times 2 - (-3) \times 1} = \frac{35}{7} = 5,$$

$$x_2 = \frac{\begin{vmatrix} 2 & 4 \\ 1 & 9 \end{vmatrix}}{\begin{vmatrix} 2 & -3 \\ 1 & 2 \end{vmatrix}} = \frac{2 \times 9 - 4 \times 1}{7} = \frac{14}{7} = 2$$

the same solution as that we obtained earlier.

As long as the denominator does not vanish, the determinant formula in (10) provides a unique solution to the equation.

Theorem 1. Let A be an n-by-n matrix and let b be an arbitrary n-vector. If $\det(A) \neq 0$, the system of equations $Ax = b$ has a unique solution given by Cramer's rule.

Theorem 1 says nothing about what happens if $\det(\mathbf{A}) = 0$. If $\det(\mathbf{A}) = 0$ but one or more $\det(\mathbf{A}_i) \neq 0$, then no solution is possible. However, if all $\det(\mathbf{A}_i) = 0$ as well as $\det(\mathbf{A}) = 0$, then the formulas in (10) become $0/0$—undefined—and solutions may be possible. The following examples from Chapter 1 illustrate this situation.

Example 1. **Canoe with Sail Revisited**

In Example 4 of Section 1.1 we modified the standard high school algebra problem about the speed of a canoe and the speed of the stream by placing a sail at the front of the canoe and giving equations for the canoe's speed C and the wind's speed W when going upwind (into the wind) and downwind.

$$\text{Upwind:} \qquad C + kW = U \qquad (11)$$
$$\text{Downwind:} \qquad C + W = D$$

In Section 1.1 we solved these equations by elimination. Now we can solve (11) by using Cramer's rule, where $\mathbf{b} = (U, D)$ and

$$\mathbf{A} = \begin{bmatrix} 1 & k \\ 1 & 1 \end{bmatrix}$$

We calculate

$$\det(\mathbf{A}) = \begin{vmatrix} 1 & k \\ 1 & 1 \end{vmatrix} = 1 \cdot 1 - k \cdot 1 = 1 - k,$$

$$\det(\mathbf{A}_1) = \begin{vmatrix} U & k \\ D & 1 \end{vmatrix} = U \cdot 1 - k \cdot D = U - kD$$

$$\det(\mathbf{A}_2) = \begin{vmatrix} 1 & U \\ 1 & D \end{vmatrix} = 1 \times D - U \times 1 = D - U$$

and so, by Cramer's rule,

$$C = \frac{U - kD}{1 - k} \qquad \text{and} \qquad W = \frac{D - U}{1 - k} \qquad (12)$$

In Section 1.1 we tried the value of $k = 1$ and found that the equations (11) had no solution (for $U \neq D$); the formulas in (12) have zero denominators. That is, for $k = 1$, $\det(\mathbf{A}) = 1 - 1 = 0$. When $k = 1$, *the two rows of* $\mathbf{A}$ *are equal* and the two equations in (11) represent parallel lines that never intersect.

We also tried letting $k = .75$. For values of $U = 5, D = 7$ the system becomes

$$\text{Upwind:} \qquad C + .75W = 5$$
$$\text{Downwind:} \qquad C + W = 7$$

We obtained the solution $C = -1$, $W = 8$, but a negative value makes no sense in the real world. The problem is that although $\det(\mathbf{A}) = 1 - .75 \neq 0$, $\det(\mathbf{A})$ is still close to 0. Informally, the system almost has no solution, so the answer is unreliable. In any 2-by-2 system, when $\det(\mathbf{A})$ is much smaller than $\det(\mathbf{A}_1)$ and $\det(\mathbf{A}_2)$, the answer should be treated with suspicion.   ∎

The link between equal rows and $\det(\mathbf{A}) = 0$ mentioned in Example 1 is true for all square matrices. By a symmetry argument, the result also holds for columns.

Proposition 1. If one row (column) of an n-by-n matrix $\mathbf{A}$ equals, or is a multiple of, another row (column), then $\det(\mathbf{A}) = 0$.

Example 2. **Stable Rabbit–Fox Populations Revisited**

In Example 3 of Section 1.3 we considered a linear growth model for rabbit and fox populations:

$$
\begin{aligned}
R' &= R + bR - eF \\
F' &= F - dF + e'R
\end{aligned} \tag{13}
$$

Here R, F are current population sizes and R', F' are the sizes 1 month later. We set $R' = R$ and $F' = F$ to solve for stable (unchanging) population sizes, and obtained

$$
\begin{aligned}
bR - eF &= 0 \\
e'R - dF &= 0
\end{aligned} \tag{14}
$$

Obviously, $R = F = 0$ is a solution to (14), but we want "non-trivial" (nonzero) solutions. We found them in Chapter 1 by ad hoc means. Now we can use determinants. Let

$$
\mathbf{A} = \begin{bmatrix} b & -e \\ e' & -d \end{bmatrix} \quad \text{so} \quad \det(\mathbf{A}) = \begin{vmatrix} b & -e \\ e' & -d \end{vmatrix} = b(-d) - (-e)e'
$$
$$
= -bd + ee'
$$

Then by Theorem 1, if $\det(\mathbf{A}) \neq 0$, (14) has a unique solution. Clearly, $R = F = 0$ is such a solution. For other solutions we must have $\det(\mathbf{A}) = 0$. Cramer's rule would now give $R = \det(\mathbf{A}_1)/\det(\mathbf{A}) = 0/0$ (undetermined), and similarly for F [note $\det(\mathbf{A}_1) = \det(\mathbf{A}_2) = 0$ since the right side in (14) is 0]. To have $\det(\mathbf{A}) = 0$, we require that

$$
\det(\mathbf{A}) = -bd + ee' = 0, \quad \text{or} \quad \frac{e}{b} = \frac{d}{e'}
$$

The equality of these ratios is just an algebraic way of saying that the first row in (14) must be a multiple of the second row. Such a relation between the rows when $\det(\mathbf{A}) = 0$ was predicted by Proposition 1. When $e/b = d/e'$, one can check that any R, F pair with $R = (e/b)F = (d/e')F$ is a solution to (14).

In Section 1.3 we considered the system

$$.1R - .15F = 0 \qquad (15)$$
$$.1R - .15F = 0$$

In (15), solutions are of the form $R = (.15/.1)F = \frac{3}{2}F$. ∎

We now consider determinants of a 3-by-3 matrix and more generally an n-by-n matrix. Remember that the determinant of a square matrix $\mathbf{A}$ is defined to be the algebraic expression that appears in the denominator when we algebraically solve the matrix equation $\mathbf{Ax} = \mathbf{b}$.

One can show that the determinant of a 3-by-3 matrix $\mathbf{A}$ is calculated by multiplying the numbers lying on the 6 "diagonals" in the augmented 3-by-5 array shown below. The products marked by solid lines have plus signs and the products marked by dashed lines have minus signs.

$$\det(\mathbf{A}) = \begin{array}{ccccc} a_{11} & a_{12} & a_{13} & a_{11} & a_{12} \\ a_{21} & a_{22} & a_{23} & a_{21} & a_{22} \\ a_{31} & a_{32} & a_{33} & a_{31} & a_{32} \end{array} = \begin{array}{l} a_{11}a_{22}a_{33} + a_{12}a_{23}a_{31} + a_{13}a_{21}a_{32} \\ \\ -a_{13}a_{22}a_{31} - a_{12}a_{21}a_{33} - a_{11}a_{23}a_{32} \end{array}$$

$$(16)$$

Warning: This process does not apply when $n > 3$.

Example 3. Solving the Refinery Equations by Cramer's Rule

In Section 1.2 we discussed a system of equations for controlling the production of three refineries:

$$20x_1 + 4x_2 + 4x_3 = 500$$
$$10x_1 + 14x_2 + 5x_3 = 850$$
$$5x_1 + 5x_2 + 12x_3 = 1000$$

$$\det(\mathbf{A}) = \begin{vmatrix} 20 & 4 & 4 \\ 10 & 14 & 5 \\ 5 & 5 & 12 \end{vmatrix} = \begin{array}{l} 20 \times 14 \times 12 + 4 \times 5 \times 5 \\ \quad + 4 \times 10 \times 5 - 4 \times 14 \times 5 \\ \quad - 4 \times 10 \times 12 - 20 \times 5 \times 5 \end{array}$$

$$= 3360 + 100 + 200 - 280 - 480 - 500$$

$$= 2400$$

To determine x_1 using Cramer's rule, we need $\det(\mathbf{A}_1)$ (recall that $\mathbf{A}_1$ is obtained from $\mathbf{A}$ by replacing the first column of $\mathbf{A}$ by the numbers on the right side of the equations).

$$\det(\mathbf{A}_1) = \begin{vmatrix} 500 & 4 & 4 \\ 850 & 14 & 5 \\ 1000 & 5 & 12 \end{vmatrix} = \begin{aligned} & 500 \times 14 \times 12 + 4 \times 5 \times 1000 \\ & + 4 \times 850 \times 5 - 4 \times 14 \times 1000 \\ & - 4 \times 850 \times 12 - 500 \times 5 \times 5 \end{aligned}$$

$$= 84{,}000 + 20{,}000 + 17{,}000$$
$$- 56{,}000 - 40{,}800 - 12{,}500$$
$$= 11{,}700$$

We compute x_1:

$$x_1 = \frac{\det(\mathbf{A}_1)}{\det(\mathbf{A})} = \frac{11{,}700}{2{,}400} = 4\frac{7}{8}$$

It is left as an exercise for the reader to determine x_2 and x_3. ∎

Even for 3-by-3 determinants, the calculations are messy. It gets so complicated beyond 3-by-3 that one has to resort to a general form of description of a determinant (we are lucky that such a description even exists).

Computational Definition. The *determinant of an n-by-n matrix* $\mathbf{A}$ is formed by adding or subtracting all possible products of n entries involving one entry from each row and each column (there is a technical rule of signs for determining whether the product gets a plus or minus sign).

The reader should check that our formulas for 2-by-2 and 3-by-3 determinants involved all products of this sort. A counting argument shows that there are $n!$ [$= n(n - 1)(n - 2) \cdots 3 \times 2 \times 1$] such products in an n-by-n determinant. For example, a 10-by-10 determinant has $10! = 3{,}628{,}800$ products. For this reason, one never solves a large system of equations using determinants.

There is one special class of matrices that arises frequently in theory and applications for which the determinant is very easy to compute. A square matrix is **upper triangular** if all entries below the main diagonal are zero, such as

$$\mathbf{A} = \begin{bmatrix} 2 & 4 & 1 \\ 0 & 1 & 7 \\ 0 & 0 & 2 \end{bmatrix} \tag{17}$$

A **lower triangular** matrix has 0's above the main diagonal.

Proposition 2. If $\mathbf{A}$ is an upper or a lower triangular matrix, $\det(\mathbf{A}) = a_{11}a_{22} \cdots a_{nn}$, the product of entries on the main diagonal.

Proof. Except for the product of main-diagonal entries, any other product of n entries, each in a different row and column, will have to contain an 0 entry below (or above) the main diagonal, so all such other products are 0. ∎

By Proposition 2, $\det(\mathbf{A}) = 2 \times 1 \times 2 = 4$ for the matrix $\mathbf{A}$ in (17). A special upper (and lower) triangular matrix is the identity matrix $\mathbf{I}$ (with 1's on the main diagonal and 0's elsewhere). Then by Proposition 2, $\det(\mathbf{I}) = 1$.

In Section 3.2 we shall learn how to transform any square matrix $\mathbf{A}$ into an upper triangular matrix $\mathbf{U}$ in a manner that does not change the value of the determinant. Using Proposition 2, we will then be able to compute $\det(\mathbf{A})$ simply by taking the product of the main-diagonal entries of $\mathbf{U}$.

There is one additional nice property of determinants that we will need later in Section 3.2.

Proposition 3. The determinant of a matrix product is the product of the determinants:

$$\det(\mathbf{AB}) = \det(\mathbf{A}) \det(\mathbf{B})$$

It was noted at the start of this section that one of the chief reasons for studying determinants was their role in finding eigenvalues of a matrix. Recall from Section 2.5 that the defining equation for an eigenvalue λ and its eigenvector $\mathbf{u}$ is

$$\mathbf{Au} = \lambda\mathbf{u} \qquad \text{or} \qquad \mathbf{Au} - \lambda\mathbf{u} = 0$$

or

$$(\mathbf{A} - \lambda\mathbf{I})\mathbf{u} = 0 \tag{18}$$

Given the eigenvalue λ, we can determine $\mathbf{u}$ by solving the system of equations in (18). More important, we can use (18) to determine the eigenvalues of $\mathbf{A}$. To do this, we recall from Theorem 1 that if $\det(\mathbf{A} - \lambda\mathbf{I}) \neq 0$, then (18) has only one solution, namely $\mathbf{u} = \mathbf{0}$. Since an eigenvector cannot be the zero vector $\mathbf{0}$, to get eigenvalues we need to choose λ so that $\det(\mathbf{A} - \lambda\mathbf{I}) = 0$.

Theorem 2. The values λ that make the $\det(\mathbf{A} - \lambda\mathbf{I}) = 0$ are eigenvalues of $\mathbf{A}$. The associated eigenvector(s) for λ are the nonzero solutions to $(\mathbf{A} - \lambda\mathbf{I})\mathbf{x} = \mathbf{0}$.

To prove Theorem 2 requires vector space theory developed in Chapter 5. For any matrix $\mathbf{A}$, $\det(\mathbf{A} - \lambda\mathbf{I})$ will be a polynomial in λ, called the **characteristic polynomial** of $\mathbf{A}$. The zeros of the characteristic polynomial of $\mathbf{A}$ are the eigenvalues of $\mathbf{A}$. Remember that the eigenvector associated with an eigenvalue λ is actually a family of eigenvectors: If $\mathbf{Au} = \lambda\mathbf{u}$, then for any r, $\mathbf{A}(r\mathbf{u}) = \lambda r\mathbf{u}$.

Example 4. Determining Eigenvalues and Eigenvectors

Consider the system of computer–dog growth equations from Section 2.5.

$$C' = 3C + D \quad \text{or} \quad \mathbf{c}' = \mathbf{Ac}, \quad \text{where } \mathbf{A} = \begin{bmatrix} 3 & 1 \\ 2 & 2 \end{bmatrix}$$
$$D' = 2C + 2D$$

In Section 2.5 the eigenvalues and eigenvectors were given without any explanation of how they were found. Let us calculate them now. By Theorem 2 the eigenvalues are the zeros of the characteristic polynomial $\det(\mathbf{A} - \lambda \mathbf{I})$:

$$\det(\mathbf{A} - \lambda \mathbf{I}) = \begin{vmatrix} 3 - \lambda & 1 \\ 2 & 2 - \lambda \end{vmatrix} = (3 - \lambda)(2 - \lambda) - 1 \cdot 2$$
$$= (6 - 5\lambda + \lambda^2) - 2$$
$$= 4 - 5\lambda + \lambda^2$$
$$= (4 - \lambda)(1 - \lambda) \qquad (19)$$

So the zeros of $\det(\mathbf{A} - \lambda \mathbf{I}) = (4 - \lambda)(1 - \lambda)$ are 4 and 1.

To find an eigenvector $\mathbf{u}$ for the eigenvalue 4, we must solve the system $\mathbf{Au} = 4\mathbf{u}$ or, by matrix algebra, $(\mathbf{A} - 4\mathbf{I})\mathbf{u} = \mathbf{0}$, where

$$\mathbf{A} - 4\mathbf{I} = \begin{bmatrix} 3 - 4 & 1 \\ 2 & 2 - 4 \end{bmatrix} = \begin{bmatrix} -1 & 1 \\ 2 & -2 \end{bmatrix}$$

We find that

$$-u_1 + u_2 = 0 \rightarrow u_1 = u_2$$
$$2u_1 - 2u_2 = 0$$

The second equation here is just -2 times the first equation (so it is superfluous). Then $\mathbf{u}$ is an eigenvector if $u_1 = u_2$, or equivalently if $\mathbf{u}$ is a multiple of $[1, 1]$.

It is left as an exercise for the reader to verify that $\mathbf{v} = [1, -2]$ is an eigenvector for $\lambda = 1$ by showing that this $\mathbf{v}$ is a solution to $\mathbf{Av} = \mathbf{v}$ or $(\mathbf{A} - \mathbf{I})\mathbf{v} = \mathbf{0}$. ∎

Example 5. Eigenvalues and Eigenvectors for Rabbit–Fox Population Model

Consider the rabbit–fox growth model

$$R' = R + .1R - .15F,$$
$$F' = F + .1R - .15F$$

or

$$\begin{bmatrix} R' \\ F' \end{bmatrix} = \begin{bmatrix} 1 + .1 & -.15 \\ .1 & 1 - .15 \end{bmatrix} \begin{bmatrix} R \\ F \end{bmatrix} \tag{20}$$

which we studied in Section 1.3. Let us find both eigenvalues and associated eigenvectors **u**, **v**, so that we can write a starting population vector **p** in terms of **u** and **v**, $\mathbf{p} = a\mathbf{u} + b\mathbf{v}$ and use these eigenvectors to compute $\mathbf{p}^{(k)}$, the population vector for k periods.

We first compute $\det(\mathbf{A} - \lambda\mathbf{I})$, the characteristic polynomial of **A**.

$$\det(\mathbf{A} - \lambda\mathbf{I}) = \begin{vmatrix} 1.1 - \lambda & -.15 \\ .1 & .85 - \lambda \end{vmatrix} = (1.1 - \lambda)(.85 - \lambda)$$

$$- .1(-.15)$$

$$= \lambda^2 - 1.95\lambda + .95 \tag{21}$$

By factoring or the quadratic formula, we find the zeros to be $\lambda = 1$ and $\lambda = .95$.

To find an eigenvector **u** associated with $\lambda = 1$, we solve

$$(\mathbf{A} - \mathbf{I})\mathbf{u} = \mathbf{0}: \quad \begin{aligned} .1u_1 - .15u_2 = 0 \\ .1u_1 - .15u_2 = 0 \end{aligned} \rightarrow u_1 = \tfrac{3}{2}u_2 \quad \mathbf{u} = [3, 2] \tag{22}$$

So nonzero multiples of $\mathbf{u} = [3, 2]$ are eigenvectors for $\lambda = 1$—that is, stable population vectors.

For completeness, we solve for the eigenvector of $\lambda = .95$:

$$(\mathbf{A} - .95\mathbf{I})\mathbf{v}: \quad \begin{aligned} .15v_1 - .15v_2 = 0 \\ .1v_1 - .1v_2 = 0 \end{aligned} \rightarrow v_1 = v_2 \quad \mathbf{v} = [1, 1] \tag{23}$$

So nonzero multiples of $\mathbf{v} = [1, 1]$ are eigenvectors of $\lambda = .95$.

With the eigenvalues and eigenvectors, we can now explain the behavior of this model that we observed in Section 1.3. In doing so, we illustrate the basic role of eigenvalues and eigenvectors in describing the long-term behavior of dynamic systems.

In Section 1.3 we started with the $[R, F]$ pair $= [50, 40]$ and followed our model [equations (20)] for many periods:

$$\begin{array}{llllll}
0 \text{ months:} & 50 & \text{rabbits,} & 40 & \text{foxes} \\
1 \text{ month:} & 49 & \text{rabbits,} & 39 & \text{foxes} \\
2 \text{ months:} & 48 & \text{rabbits,} & 38 & \text{foxes} \\
3 \text{ months:} & 47 & \text{rabbits,} & 37 & \text{foxes} \\
& \vdots & & \vdots & & \vdots \\
10 \text{ months:} & 42 & \text{rabbits,} & 32 & \text{foxes} & \tag{24} \\
& \vdots & & \vdots & & \vdots
\end{array}$$

$$20 \text{ months:} \qquad 37 \quad \text{rabbits,} \quad 27 \quad \text{foxes}$$

$$\vdots \qquad\qquad \vdots \qquad\qquad \vdots$$

$$50 \text{ months:} \qquad 31.5 \text{ rabbits,} \quad 21.5 \text{ foxes}$$

$$\vdots \qquad\qquad \vdots \qquad\qquad \vdots$$

$$100 \text{ months:} \qquad 30.1 \text{ rabbits,} \quad 20.1 \text{ foxes}$$

Let us express the starting vector $\mathbf{p} = [50, 40]$ in terms of the eigenvectors $\mathbf{u} = [3, 2]$ and $\mathbf{v} = [1, 1]$: $\mathbf{p} = a\mathbf{u} + b\mathbf{v}$:

$$\begin{bmatrix} 50 \\ 40 \end{bmatrix} = a \begin{bmatrix} 3 \\ 2 \end{bmatrix} + b \begin{bmatrix} 1 \\ 1 \end{bmatrix} \qquad \text{or} \qquad \begin{array}{c} 3a + b = 50 \\ 2a + b = 40 \end{array} \qquad (25)$$

By Cramer's rule, we find that $a = 10$ and $b = 20$. Thus

$$\mathbf{p} = 10\mathbf{u} + 20\mathbf{v} \qquad (26)$$

Then using (26) to compute the population sizes in (24) gives

$$\begin{aligned} \mathbf{p}' = \mathbf{A}\mathbf{p} = 10\mathbf{A}\mathbf{u} + 20\mathbf{A}\mathbf{v} &= 10(1\mathbf{u}) + 20(.95\mathbf{v}) \\ &= 10[3, 2] + 19[1, 1] = [49, 39] \end{aligned}$$

and

$$\begin{aligned} \mathbf{p}^{(k)} = \mathbf{A}^k\mathbf{p} = 10\mathbf{A}^k\mathbf{u} + 20\mathbf{A}^k\mathbf{v} &= 10(1^k\mathbf{u}) + 20(.95^k\mathbf{v}) \\ &= 10[3, 2] + 20 \times .95^k[1, 1] \\ &= [30, 20] + .95^k[20, 20] \qquad (27) \end{aligned}$$

The second term $.95^k[20, 20]$ in the last line of (26) slowly goes to 0, leaving the stable population vector [30, 20]. With (27), the behavior in table (24) is completely explained!

If we generalize the calculation in (27) and the starting vector $\mathbf{p}$ has the eigenvector representation $\mathbf{p} = a\mathbf{u} + b\mathbf{v}$, then

$$\begin{aligned} \mathbf{p}^{(k)} = \mathbf{A}^k\mathbf{p} = a\mathbf{A}^k\mathbf{u} + b\mathbf{A}^k\mathbf{v} &= a\mathbf{u} + b \times .95^k\mathbf{v} \qquad (28) \\ &= [3a, 2a] + .95^k[b, b] \end{aligned}$$

So the long-term stable population is $[3a, 2a]$. The critical number is a. To find a for the general starting vector $\mathbf{p} = [R, F]$, we substitute $[R, F]$ for [50, 40] in (25) and apply Cramer's rule.

$$a = \frac{\det(\mathbf{A}_1)}{\det(\mathbf{A})} = \frac{\begin{vmatrix} R & 1 \\ F & 1 \end{vmatrix}}{\begin{vmatrix} 3 & 1 \\ 2 & 1 \end{vmatrix}} = \frac{R - F}{3 - 2} = R - F$$

∎

Section 3.1 **Exercises**

Summary of Exercises

Exercises 1–21 involve properties of determinants and their use in solving systems of equations, with Exercises 13–21 involving theory. Exercises 22–28 involve using determinants to find eigenvalues. Exercises 29–34 present theoretical properties of the characteristic polynomial, including the Cayley–Hamilton theorem. Exercises 35 and 36 introduce the euclidean norm of a matrix.

1. Compute the determinant of the following matrices.

    (a) $\begin{bmatrix} 1 & 3 \\ 5 & -2 \end{bmatrix}$
    (b) $\begin{bmatrix} 2 & 4 \\ -3 & -6 \end{bmatrix}$
    (c) $\begin{bmatrix} 3 & 2 \\ 1 & 0 \end{bmatrix}$

2. Find the (unique) solution to the following systems of equations, if possible, using Cramer's Rule.

    (a) $x + y = 34$
     $2x - y = 30$

    (b) $2x - 3y = 5$
     $-4x + 6y = 10$

    (c) $3x + y = 7$
     $2x - 2y = 7$

3. Consider the two-refinery production of diesel oil and gasoline. The second refinery has not been built, but when it is built it will produce twice as much gas as diesel oil from each barrel of crude oil. We have

$$\text{Diesel oil:} \quad 10x_1 + ax_2 = D$$
$$\text{Gasoline:} \quad 5x_1 + 2ax_2 = G$$

where a is to be determined, D is demand for diesel oil, and G is demand for gasoline (and x_i is number of barrels of crude oil processed by refinery i, $i = 1, 2$). Solve this system of equations to determine x_1 and x_2 in terms of a, D, G using Cramer's rule.

4. Consider the two-refinery production of diesel oil and gasoline. The second refinery has not been built but when it is built it will produce 15 gallons of gasoline and k gallons of diesel oil from each barrel of crude oil. We have

$$\text{Diesel oil:} \quad 10x_1 + kx_2 = D$$
$$\text{Gasoline:} \quad 5x_1 + 15x_2 = G$$

where k is to be determined, D is demand for diesel oil, and G is demand for gasoline (and x_i is number of barrels of crude oil processed by refinery i, $i = 1, 2$). Solve this system of equations to determine x_1 and x_2 in terms of k, D, G using Cramer's rule. What value of k yields a nonunique solution? In practical terms, what does this nonuniqueness mean?

5. Which of the following systems of equations have nonzero solutions? If the solution is not unique, give the set of all possible solutions.

$$\textbf{(a)}\ \begin{matrix} 3x + 4y = 0 \\ 6x + 2y = 0 \end{matrix} \qquad \textbf{(b)}\ \begin{matrix} 4x - y = 0 \\ 4x - y = 0 \end{matrix} \qquad \textbf{(c)}\ \begin{matrix} 2x - 6y = 0 \\ -x + 3y = 0 \end{matrix}$$

6. When the right-hand side is nonzero and the determinant is 0, there may be no solution to the system of equations. Which of the following systems of equations have no solution?

$$\textbf{(a)}\ \begin{matrix} 3x + 2y = 2 \\ 6x + 4y = 2 \end{matrix} \qquad \textbf{(b)}\ \begin{matrix} 2x - 3y = 2 \\ 2x - 3y = 2 \end{matrix} \qquad \textbf{(c)}\ \begin{matrix} 2x - 6y = 4 \\ -x + 3y = -2 \end{matrix}$$

7. Compute the determinant of the following matrices.

$$\textbf{(a)}\ \begin{bmatrix} 1 & 1 & 1 \\ 1 & 2 & 1 \\ 3 & 1 & 1 \end{bmatrix} \qquad \textbf{(b)}\ \begin{bmatrix} 2 & 0 & -1 \\ 0 & 0 & 3 \\ 2 & 0 & -1 \end{bmatrix} \qquad \textbf{(c)}\ \begin{bmatrix} 0 & 1 & 0 \\ 1 & 2 & 1 \\ 2 & 5 & 2 \end{bmatrix}$$

$$\textbf{(d)}\ \begin{bmatrix} 1 & 0 & 1 \\ 0 & 2 & 2 \\ 1 & 2 & 3 \end{bmatrix} \qquad \textbf{(e)}\ \begin{bmatrix} 2 & 1 & 0 \\ 0 & 0 & 2 \\ 2 & 2 & 2 \end{bmatrix} \qquad \textbf{(f)}\ \begin{bmatrix} \frac{1}{6} & \frac{1}{7} & \frac{1}{8} \\ \frac{1}{7} & \frac{1}{8} & \frac{1}{9} \\ \frac{1}{8} & \frac{1}{9} & \frac{1}{10} \end{bmatrix}$$

8. In which matrices in Exercise 7 is one row or column a multiple of another (so that by Proposition 1, the determinant will be 0)?

9. Use Cramer's rule to solve for x_2 and x_3 in Example 3.

10. Solve the following systems of equations using Cramer's rule.

$$\textbf{(a)}\ \begin{matrix} 2x - & y + 2z = 4 \\ x & + 3z = 6 \\ & 2y - z = 1 \end{matrix} \qquad \textbf{(b)}\ \begin{matrix} x + & y + z = 3 \\ 2x + 3y + z = 9 \\ -x - & y - z = -4 \end{matrix}$$

$$\textbf{(c)}\ \begin{matrix} -x + 3y - z = 4 \\ 2x - y = 6 \\ x + z = 3 \end{matrix}$$

11. Consider the following system of equations for the growth of rabbits (*R*), foxes (*F*), and humans (*H*).

$$R' = R + .3R - .1F - .2H$$
$$F' = F + .4R - .2F - .3H$$
$$H' = H + .1R + .1F + .1H$$

We want to see if stable population sizes are possible (when $R' = R$, $F' = F$, $H' = H$). Set up the stable population system of equations [similar to (15) in Example 2] and compute the determinant to see if a nonzero solution is possible (*do not try to find a stable solution*).

12. Repeat Exercise 11 with the system

$$R' = R + .3R - .2F - .2H$$
$$F' = F + .4R - .2F - .4H$$
$$H' = H + .2R + .1F + .1H$$

13. If you double the first row in the system

$$ax + by = e$$
$$cx + dy = f$$

show using Cramer's rule that the solution does not change.

14. If you double the first column in the system

$$ax + by = e$$
$$cx + dy = f$$

show using Cramer's rule that the value of x is half as large and the value of y is unchanged.

15. (a) If you interchange the rows of a 2-by-2 matrix $\mathbf{A}$, show that the determinant of the new matrix is -1 times the $\det(\mathbf{A})$.
   Hint: Use (6). The same is true for interchanging columns.
   (b) If you interchange the first two rows of a 3-by-3 matrix $\mathbf{A}$, show that the determinant of the new matrix is -1 times the $\det(\mathbf{A})$.
   Hint: Use (16).

16. From the computational definition for a determinant, deduce that for any square matrix $\mathbf{A}$, its transpose $\mathbf{A}^T$ (obtained by interchanging rows and columns) has the same determinant as $\mathbf{A}$. (Thus $\mathbf{A}^T\mathbf{x} = \mathbf{b}$ has a unique solution if and only if $\mathbf{A}\mathbf{x} = \mathbf{b}$ does.)

17. From the computational definition for a determinant, deduce that any square matrix $\mathbf{A}$ with a row (or column) of all 0's has $\det(\mathbf{A}) = 0$.

18. (a) From the computational definition for a determinant, deduce that if $\mathbf{B}$ is a square 3-by-3 matrix obtained from $\mathbf{A}$ by doubling every entry in the second row of matrix $\mathbf{A}$, then $\det(\mathbf{B}) = 2 \cdot \det(\mathbf{A})$.
   (b) More generally, if every entry in a row (or column) of an n-by-n matrix $\mathbf{A}$ is multiplied by a constant k, the determinant of the resulting matrix equals $k \cdot \det(\mathbf{A})$.

19. Compute the determinant of the following matrices.

(a) $\begin{bmatrix} 2 & 3 & 7 & 8 \\ 0 & 3 & 9 & 1 \\ 0 & 0 & 1 & 5 \\ 0 & 0 & 0 & 4 \end{bmatrix}$
(b) $\begin{bmatrix} 0 & 0 & 0 & 0 \\ 2 & 0 & 0 & 0 \\ 4 & 5 & 0 & 0 \\ 3 & 4 & 5 & 0 \end{bmatrix}$
(c) $\begin{bmatrix} 1 & 2 & 0 & 0 & 0 \\ 0 & 0 & 3 & 0 & 0 \\ 0 & 0 & 0 & 1 & 0 \\ 0 & 0 & 0 & 0 & 2 \\ 2 & 0 & 0 & 0 & 2 \end{bmatrix}$

20. Let **A** and **B** be arbitrary 2-by-2 matrices. Using (6), show that $\det(\mathbf{AB}) = \det(\mathbf{A})\det(\mathbf{B})$.

21. In the following figure, the area of the triangle ABC can be expressed as

$$\text{area } ABC = \text{area } ABB'A' + \text{area } BCC'B' \tag{*}$$
$$- \text{area } ACC'A'$$

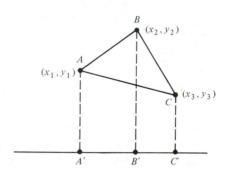

Using (*) and the fact that the area of a trapezoid is one-half of the distance between the parallel sides times the sum of the lengths of the parallel sides, show that

$$\text{area } ABC = \frac{1}{2}\begin{vmatrix} x_1 & y_1 & 1 \\ x_2 & y_2 & 1 \\ x_3 & y_3 & 1 \end{vmatrix}$$

22. Determine an eigenvector associated with $\lambda = 1$ in Example 4.

23. (a) Compute the eigenvalues of each of the following matrices.

(i) $\begin{bmatrix} 4 & 0 \\ 2 & 2 \end{bmatrix}$ (ii) $\begin{bmatrix} 1 & 2 \\ 3 & 4 \end{bmatrix}$ (iii) $\begin{bmatrix} 2 & 1 \\ 2 & 3 \end{bmatrix}$

(iv) $\begin{bmatrix} 4 & -1 \\ 1 & 2 \end{bmatrix}$ (v) $\begin{bmatrix} 0 & 2 & 1 \\ 3 & -1 & -3 \\ -2 & 2 & 3 \end{bmatrix}$

(b) Determine an eigenvector associated with the largest eigenvalue, using the method in Example 4, for the matrices in part (a).

24. (a) For the following rabbit–fox models, determine both eigenvalues.
(i) $R' = R + .1R - .3F$ (ii) $R' = R + .3R - .2F$
 $F' = F + .2R - .6F$ $F' = F + .15R - .1F$

(b) Determine an eigenvector **u** associated with the largest eigenvalue in each system in part (a).
(c) Determine the other eigenvector **v** (associated with the smaller eigenvalue) for each system in part (a).

(d) If the initial population is $\mathbf{x} = [10, 10]$, express $\mathbf{x}^{(k)}$ as a linear combination of $\mathbf{u}$ and $\mathbf{v}$, as in equation (27), for each system in part (a). Use this expression to describe in words the behavior of this model over time.

25. The following system of equations was the first rabbit–fox model analyzed in Section 1.3.

$$R' = R + .2R - .3F$$

$$F' = F + .1R - .1F$$

Determine the dominant eigenvalue and an associated eigenvector.

26. **(a)** For the following rabbit–fox models, determine both eigenvalues.
    (i) $R' = R + .1R + .1F$ (ii) $R' = R + 2R - 3F$
          $F' = F + .2R + .1F$ $F' = F + 1.5R - 4.5F$

    (b) Determine an eigenvector $\mathbf{u}$ associated with the largest eigenvalue in each system in part (a).
    (c) Determine the other eigenvector $\mathbf{v}$ (associated with the smaller eigenvalue) for each system in part (a).
    (d) If the initial population is $\mathbf{x} = [10, 10]$ express $\mathbf{x}^{(k)}$ as a linear combination of $\mathbf{u}$ and $\mathbf{v}$ as in equation (27), for each system in part (a). Use this expression to describe in words the behavior of this model over time.

27. The following growth model for elephants (E) and mice (M) predicts population changes from decade to decade.

$$E' = 3E + M$$

$$M' = 2E + 4M$$

    (a) Determine the eigenvalues and associated eigenvectors for this system.
    (b) Suppose initially that we have $\mathbf{p} = [E, M] = [5, 5]$. Write $\mathbf{p}$ as a linear combination of the eigenvectors.
    (c) Use the information in part (b) to determine an approximate value for the population sizes in eight decades.

28. The following growth model for computer science teachers (T) and programmers (P) predicts population changes from decade to decade.

$$T' = T - P$$

$$P' = 2T + 4P$$

    (a) Determine the eigenvalues and associated eigenvectors for this system.
    (b) Suppose initially that we have $\mathbf{p} = [T, P] = [10, 100]$. Write $\mathbf{p}$ as a linear combination of the eigenvectors.

(c) Use the information in part (b) to determine an approximate value for the population sizes in 12 decades.

29. Verify that the constant term in a characteristic polynomial is det($\mathbf{A}$).

30. Let $\mathbf{A}$ be a 2-by-2 matrix with all positive entries and det($\mathbf{A}$) $\neq 0$. Show that the eigenvalues of $\mathbf{A}$ must be positive real numbers.

31. Show that the product of the eigenvalues of a 2-by-2 matrix $\mathbf{A}$ equals det($\mathbf{A}$).

    Hint: The product of the eigenvalues is the constant term in the characteristic polynomial det($\mathbf{A} - \lambda\mathbf{I}$). Note that this result is true for matrices of any size.

32. Show that the sum of the eigenvalues of a 2-by-2 matrix $\mathbf{A}$ equals the sum of the main-diagonal entries of $\mathbf{A}$.

    Hint: These quantities are both the coefficient of λ in the characteristic polynomial det($\mathbf{A} - \lambda\mathbf{I}$). Note that this result is true for matrices of any size.

33. **(a)** For a matrix $\mathbf{A} = \begin{bmatrix} a & q \\ 0 & b \end{bmatrix}$, show that eigenvalues are a and b.
    (b) Generalize the result in part (a) to show that in any upper triangular matrix, the eigenvalues are just the entries on the main diagonal.

    Hint: Recall that the determinant of such a matrix is simply the product of the main-diagonal entries.

34. This exercise illustrates a famous result in linear algebra known as the Cayley–Hamilton theorem, which says that a square matrix $\mathbf{A}$ satisfies its characteristic equation, det($\mathbf{A} - \lambda\mathbf{I}$) = 0.

    (a) Let $\mathbf{A} = \begin{bmatrix} 2 & 1 \\ 1 & 2 \end{bmatrix}$. So det($\mathbf{A} - \lambda\mathbf{I}$) = $(2 - \lambda)(2 - \lambda) - 1 \cdot 1 = \lambda^2 - 4\lambda + 3$. The characteristic equation of $\mathbf{A}$ is then $\lambda^2 - 4\lambda + 3 = 0$. Verify that $\mathbf{A}$ satisfies its characteristic equation by setting $\lambda = \mathbf{A}$ and showing that $\mathbf{A}^2 - 4\mathbf{A} + 3\mathbf{I} = \mathbf{O}$.
    (b) The characteristic equation can be factored to $(\lambda - 3)(\lambda - 1) = 0$. Check that $(\mathbf{A} - 3\mathbf{I})(\mathbf{A} - \mathbf{I}) = \mathbf{O}$.
    (c) Following the same steps as in part (a), check that the matrix $\mathbf{A} = \begin{bmatrix} 3 & 1 \\ 2 & 2 \end{bmatrix}$ for the computer–dog model satisfies its characteristic equation.

35. The euclidean norm $\|\mathbf{A}\|_e$ of $\mathbf{A}$ satisfies $|\mathbf{Ax}|_e \leq \|\mathbf{A}\|_e \cdot |\mathbf{x}|_e$, where $|\ |_e$ denotes the euclidean distance norm of a vector. If $\mathbf{A}$ is a symmetric matrix, it can be proved that $\|\mathbf{A}\|_e$ equals the largest eigenvalue of $\mathbf{A}$ (in absolute value). Compute the euclidean norm of the following matrices

and compare this value with the sum norm and max norm of these matrices.

(a) $\begin{bmatrix} 2 & -1 \\ -1 & 2 \end{bmatrix}$ (b) $\begin{bmatrix} -5 & 2 \\ 2 & -5 \end{bmatrix}$ (c) $\begin{bmatrix} 0 & 3 \\ 3 & 0 \end{bmatrix}$

36. The euclidean norm (see Exercise 35) of a nonsymmetric matrix $\mathbf{A}$ is equal to the square root of the largest eigenvalue (in absolute value) of the symmetric matrix $\mathbf{A}^T\mathbf{A}$ (where $\mathbf{A}^T$ is the transpose of $\mathbf{A}$). Compute the euclidean norm of the following matrices and compare this value with the sum norm and max norm of these matrices.

(a) $\begin{bmatrix} 1 & 2 \\ 2 & 0 \end{bmatrix}$ (b) $\begin{bmatrix} 3 & 0 \\ 1 & 1 \end{bmatrix}$ (c) $\begin{bmatrix} -1 & 1 \\ 2 & 4 \end{bmatrix}$ (d) $\begin{bmatrix} 0 & 4 \\ 3 & 0 \end{bmatrix}$

Section 3.2 Solving Systems of Equations by Elimination

In this section we develop the general procedure of elimination for solving any system of n equations in n unknowns—to find the unique solution, if one exists, or to show that no unique solution exists. Elimination was devised by Karl Friedrich Gauss around 1820 to solve systems of linear equations that arose while solving a regression model (such as the one introduced in Section 1.4) to estimate locations in survey mapping. The method of elimination was used in the beginning of Section 3.1 to find a general solution to a system of two equations in two unknowns.

The solution by elimination involves two stages. The first is to transform the given system (as far as possible) into an upper triangular system such as

$$
\begin{aligned}
2x_1 + x_2 - x_3 &= 1 \\
x_2 + 4x_3 &= 5 \\
x_3 &= 2
\end{aligned}
$$

The second stage is to use back substitution to obtain values for the unknowns.

$$x_3 = 2 \rightarrow x_2 + 4(2) = 5 \quad \text{or} \quad x_2 = -3; \quad \text{and}$$

$$x_2 = -3, x_3 = 2 \rightarrow 2x_1 + (2) - (-3) = 1 \quad \text{or} \quad x_1 = 3$$

The solution vector is thus $\mathbf{x} = [3, -3, 2]$.

The elimination transformations in the first stage are based on the following two simple properties of equations:

1. If we multiply both sides of an equation by a constant, this does not affect the possible solutions to the equation.

2. If we add two equations together (add the left sides together and add the right sides together), any solution to *both* of the original equations is also a solution to the combined equation.

Combining these two properties repeatedly, we construct a new set of easily solved equations whose solution will be a solution to the original system of equations.

Principle of Gaussian Elimination. Subtract multiples of the ith equation to eliminate the ith variable from the remaining equations, for $i = 1, 2, \ldots, n - 1$.

The best way to show how Gaussian elimination works is with some examples. Then we state the procedure in algebraic terms.

Example 1. Gaussian Elimination Example

We start with a very simple system of two equations in two unknowns.

$$\begin{array}{ll} \text{(a)} & x + y = 4 \\ \text{(b)} & 2x - y = -1 \end{array}$$

To eliminate the $2x$ term from (b), we subtract 2 times (a) from (b), and obtain the following new second equation:

$$\begin{array}{ll} \text{(b)} & 2x - y = -1 \\ -2\text{(a)} & -\ (2x + 2y = 8) \\ \hline \text{(b}') = \text{(b)} - 2\text{(a)} & 0 - 3y = -9 \end{array}$$

Our new system of equations is

$$\begin{array}{ll} \text{(a)} & x + y = 4 \\ \text{(b}') & -3y = -9 \end{array}$$

By properties 1 and 2, any solution to (a) and (b) is also a solution to (a) and (b′). Further, we can reverse the step creating (b′). That is, (b′) = (b) − 2(a) implies that (b) = (b′) + 2(a). Thus (b) is formed from (b′) and a multiple of (a), so any solution to (a) and (b′) is a solution to (a) and (b).

But (b′) is trivial to solve, and gives

$$y = 3$$

Substituting $y = 3$ in (a), we have

$$x + 3 = 4 \ \rightarrow \ x = 4 - 3 = 1$$

The reader should check that $x = 1$, $y = 3$ is a solution to (a) and (b). ∎

Example 2. **Gaussian Elimination for Refinery Problem**

Recall the refinery problem introduced in Section 1.2 with three refineries whose production levels had to be chosen to meet the demands for heating oil, diesel oil, and gasoline.

$$
\begin{array}{lll}
\text{Heating oil:} & \text{(a)} & 20x_1 + 4x_2 + 4x_3 = 500 \\
\text{Diesel oil:} & \text{(b)} & 10x_1 + 14x_2 + 5x_3 = 850 \qquad (1) \\
\text{Gasoline:} & \text{(c)} & 5x_1 + 5x_2 + 12x_3 = 1000
\end{array}
$$

Use multiples of equation (a) to eliminate x_1 from (b) and (c). First, subtract $\frac{1}{2}$ times (a) from (b) to eliminate the $10x_1$ term from (b) and obtain a new second equation (b′).

$$
\begin{array}{ll}
\text{(b)} & 10x_1 + 14x_2 + 5x_3 = 850 \\
-\frac{1}{2}\text{(a)} & -\,(10x_1 + 2x_2 + 2x_3 = 250) \\
\hline
\text{(b′)} = \text{(b)} - \frac{1}{2}\text{(a)} & 0 + 12x_2 + 3x_3 = 600
\end{array}
$$

In a similar fashion we subtract $\frac{1}{4}$ times (a) from (c) to eliminate the $5x_1$ term from (c) and obtain a new equation (c′):

$$
\begin{array}{ll}
\text{(c)} & 5x_1 + 5x_2 + 12x_3 = 1000 \\
-\frac{1}{4}\text{(a)} & -\,(5x_1 + x_2 + x_3 = 125) \\
\hline
\text{(c′)} = \text{(c)} - \frac{1}{4}\text{(a)} & 4x_2 + 11x_3 = 875
\end{array}
$$

Our new system of equations is now

$$
\begin{array}{lll}
\text{(a)} & 20x_1 + 4x_2 + 4x_3 = 500 \\
\text{(b′)} & 12x_2 + 3x_3 = 600 \qquad (2) \\
\text{(c′)} & 4x_2 + 11x_3 = 875
\end{array}
$$

Next we use equation (b′) to eliminate the $4x_2$ term from (c′) and obtain a new third equation (c″).

$$
\begin{array}{ll}
\text{(c′)} & 4x_2 + 11x_3 = 875 \\
-\frac{1}{3}\text{(b′)} & -\,(4x_2 + x_3 = 200) \\
\hline
\text{(c″)} = \text{(c′)} - \frac{1}{3}\text{(b′)} & 10x_3 = 675
\end{array}
$$

Our new system of equations is

$$
\begin{array}{lll}
\text{(a)} & 20x_1 + 4x_2 + 4x_3 = 500 \\
\text{(b′)} & 12x_2 + 3x_3 = 600 \qquad (3) \\
\text{(c″)} & 10x_3 = 675
\end{array}
$$

By properties 1 and 2, any solution to the original system (1) is a solution to the new system (3). Furthermore, by reversing the steps in going from (1) to (3) [so that (1) is formed from linear combinations of the equations in (3)], we also have that any solution to (3) is a solution to (1).

Now (3) is in upper triangular form and we can solve using back substitution. From (c″) we have

$$x_3 = \tfrac{675}{10} = 67\tfrac{1}{2}$$

and giving this value for x_3 in (b′), we have

$$12x_2 + 3(67\tfrac{1}{2}) = 600$$

or

$$12x_2 = 600 - 202\tfrac{1}{2} \rightarrow x_2 = 33\tfrac{1}{8}$$

and substituting these values for x_3 and x_2 in (a), we have

$$20x_1 + 4(33\tfrac{1}{8}) + 4(67\tfrac{1}{2}) = 500$$

or

$$x = \frac{500 - 402\tfrac{1}{2}}{20} = 4\tfrac{7}{8}$$

So the vector of production levels of the three respective refineries is $(4\tfrac{7}{8}, 33\tfrac{1}{8}, 67\tfrac{1}{2})$. Recall that in Section 1.2, by trial and error we had obtained the estimated solution vector (5, 33, 68)—a pretty good guess. ∎

Example 3. **System of Equations Without Unique Solution**

Suppose that we change the third equation in Example 2 so that our system is now

$$
\begin{array}{lrcl}
\text{(a)} & 20x_1 + 4x_2 + 4x_3 &=& 500 \\
\text{(b)} & 10x_1 + 14x_2 + 5x_3 &=& 850 \\
\text{(c)} & 10x_1 - 10x_2 - x_3 &=& -350
\end{array}
\quad (4)
$$

After eliminating x_1 from (b) and (c) as above, we have

$$
\begin{array}{lrcl}
\text{(a)} & 20x_1 + 4x_2 + 4x_3 &=& 500 \\
\text{(b′)} & 12x_2 + 3x_3 &=& 600 \\
\text{(c′)} & -12x_2 - 3x_3 &=& -600
\end{array}
\quad (5)
$$

Next we add (b′) to (c′) to eliminate the $-12x_2$ term, but this eliminates all of (c′).

$$(c'') = (c') + (b') \qquad 0 = 0$$

That is, equation (c′) is just minus (b′). We have only two equations in three unknowns. This system has an infinite number of solutions, since we can pick any value for x_3 and then knowing x_3 we can determine x_2 from (b′) and then x_1 from (a).

Let us reconsider (4) with the third equation replaced by

$$(c) \qquad 10x_1 - 10x_2 - x_3 = 300$$

then (c′) would have been

$$(c') \qquad -12x_2 - 3x_3 = 50$$

Now when we use (b′) to eliminate the $-12x_2$ term in (c), we get

$$(c'') = (c') - (b') \qquad 0 = 650$$

That is, (b′) and (c′) are inconsistent equations, and this new system has no solution.

The reader should check that the coefficient matrix in (4) has determinant 0. The reason is that the first row minus the second row equals the third row. ∎

Suppose that we have a system of n equations in n unknowns

$$
\begin{aligned}
a_{11}x_1 + a_{12}x_2 + \cdots + a_{1n}x_n &= b_1 \\
a_{21}x_1 + a_{22}x_2 + \cdots + a_{2n}x_n &= b_2 \\
\vdots \qquad \vdots \qquad \vdots \qquad \vdots \qquad \vdots \\
a_{i1}x_1 + a_{i2}x_2 + \cdots + a_{in}x_n &= b_i \\
\vdots \qquad \vdots \qquad \vdots \qquad \vdots \qquad \vdots \\
a_{n1}x_1 + a_{n2}x_2 + \cdots + a_{nn}x_n &= b_n
\end{aligned}
\tag{6}
$$

Since the first equation begins $a_{11}x_1 + \cdots$ and the ith equation begins $a_{i1}x_1 + \cdots$, then multiplying the first equation by a_{i1}/a_{11} will yield a new equation that begins $a_{i1}x_1 + \cdots$, that is,

$$\frac{a_{i1}}{a_{11}}(a_{11}x_1 + a_{12}x_2 + \cdots + a_{1n}x_n = b_1)$$

equals

$$a_{i1}x_1 + \frac{a_{i1}}{a_{11}} a_{12}x_2 + \cdots + \frac{a_{i1}}{a_{11}} a_{1n}x_n = \frac{a_{i1}}{a_{11}} b_1 \qquad (7)$$

Subtracting (7) from the ith equation in (6) yields

$$\left(a_{i2} - \frac{a_{i1}}{a_{11}} a_{12}\right)x_2 + \cdots + \left(a_{in} - \frac{a_{i1}}{a_{11}} a_{1n}\right)x_n = b_i - \frac{a_{i1}}{a_{11}} b_1 \quad (8)$$

By performing the steps in (7) and (8) for $i = 2, 3, \ldots, n$, we can eliminate the x_1 term from every equation except the first, so that now, the second through nth equations will form a system of $n - 1$ equations in $n - 1$ unknowns. We repeat the elimination process with this $n-1$-by-$n-1$ system, eliminating the x_2 term from the third through nth equations. We continue this method of eliminating variables until we finally have one equation in x_n—which is trivial to solve.

Once x_n is known, we can work backwards to determine the value of x_{n-1}, then of x_{n-2}, and so on, as in the previous examples. We are assuming here that when it is time to eliminate x_j from equations $j + 1$ through n, the coefficient of x_j in the current jth equation is nonzero; otherwise, we cannot use the jth equation to eliminate x_j from other equations. We discuss the case where this coefficient is zero shortly.

Since Gaussian elimination involves only coefficients, the variables are just excess baggage. Thus, after stating a problem in equation form, we can perform the elimination algorithm on the coefficient matrix augmented with the right-side vector.

Let us try out Gaussian elimination in this format on a familiar larger system, of four equations in four unknowns.

Example 4. **Solving Leontief Supply–Demand Equations**

Use Gaussian elimination to solve the supply–demand equations introduced in Section 1.2 for a sample Leontief economic model.

Supply	Industrial Demands				Consumer Demand
	Energy	Constr.	Transp.	Steel	
Energy: $x_1 =$	$.4x_1$ +	$.2x_2$ +	$.2x_3$ +	$.2x_4$ +	100
Construct.: $x_2 =$	$.3x_1$ +	$.3x_2$ +	$.2x_3$ +	$.1x_4$ +	50
Transport.: $x_3 =$	$.1x_1$ +	$.1x_2$ +		$.2x_4$ +	100
Steel: $x_4 =$		$.1x_2$ +	$.1x_3$		

$$(9)$$

Bringing the x_i's over to the left side of the equations, we have the system

$$
\begin{array}{lrl}
\text{(a)} & .6x_1 - .2x_2 - .2x_3 - .2x_4 &= 100 \\
\text{(b)} & -.3x_1 + .7x_2 - .2x_3 - .1x_4 &= 50 \\
\text{(c)} & -.1x_1 - .1x_2 + x_3 - .2x_4 &= 100 \\
\text{(d)} & -.1x_2 - .1x_3 + x_4 &= 0
\end{array}
\tag{10}
$$

Changing notation to the augmented coefficient matrix yields

$$
\begin{array}{l}
\text{(a)} \\
\text{(b)} \\
\text{(c)} \\
\text{(d)}
\end{array}
\left[
\begin{array}{rrrr|r}
.6 & -.2 & -.2 & -.2 & 100 \\
-.3 & .7 & -.2 & -.1 & 50 \\
-.1 & -.1 & 1 & -.2 & 100 \\
0 & -.1 & -.1 & 1 & 0
\end{array}
\right]
\tag{10$'$}
$$

First we use multiples of equation (a) to eliminate the x_1 term from equations (b), (c), and (d), that is, to make entries (2, 1), (3, 1), and (4, 1) zero.

$$
\begin{array}{l}
\text{(a)} \\
\text{(b$'$)} = \text{(b)} + \tfrac{1}{2}\text{(a)} \\
\text{(c$'$)} = \text{(c)} + \tfrac{1}{6}\text{(a)} \\
\text{(d)}
\end{array}
\left[
\begin{array}{rrrr|r}
.6 & -.2 & -.2 & -.2 & 100 \\
0 & .6 & -.3 & -.2 & 100 \\
0 & -.133 & .967 & -.233 & 116.67 \\
0 & -.1 & -.1 & 1 & 0
\end{array}
\right]
\tag{11}
$$

Next we make entries (3, 2) and (4, 2) zero.

$$
\begin{array}{l}
\text{(a)} \\
\text{(b$'$)} \\
\text{(c$''$)} = \text{(c$'$)} + \tfrac{2}{9}\text{(b$'$)} \\
\text{(d$''$)} = \text{(d$'$)} + \tfrac{1}{6}\text{(b$'$)}
\end{array}
\left[
\begin{array}{rrrr|r}
.6 & -.2 & -.2 & -.2 & 100 \\
0 & .6 & -.3 & -.2 & 100 \\
0 & 0 & .9 & -.278 & 138.86 \\
0 & 0 & -.15 & .967 & 16.67
\end{array}
\right]
\tag{12}
$$

Finally, we make entry (4, 3) zero.

$$
\begin{array}{l}
\text{(a)} \\
\text{(b$'$)} \\
\text{(c$''$)} \\
\text{(d$'''$)} = \text{(d$''$)} + \tfrac{1}{6}\text{(c$''$)}
\end{array}
\left[
\begin{array}{rrrr|r}
.6 & -.2 & -.2 & -.2 & 100 \\
0 & .6 & -.3 & -.2 & 100 \\
0 & 0 & .9 & -.278 & 138.86 \\
0 & 0 & 0 & .920 & 39.81
\end{array}
\right]
\tag{13}
$$

System (13) is an upper triangular system that is equivalent to (10), in the sense that both systems have the same solutions. In equation form, (13) is

$$
\begin{array}{lrl}
\text{(a$''$)} & .6x_1 - .2x_2 - .2x_3 - .2x_4 &= 100 \\
\text{(b$'''$)} & .6x_2 - .3x_3 - .2x_4 &= 100 \\
\text{(c$'''$)} & .9x_3 - .278x_4 &= 138.86 \\
\text{(d$'''$)} & .920x_4 &= 39.81
\end{array}
\tag{13$'$}
$$

Using back substitution, we obtain

$$x_1 = 325.3, \qquad x_2 = 264.9, \qquad x_3 = 167.7, \qquad x_4 = 43.3$$

As expected, these numbers are close to the estimated answer we obtained by iterated trial and error in Section 1.2.   ■

When we have another right-side b^* for which the system $Ax = b$ must be solved, it is natural to save some of the information used from the solution of $Ax = b$, since the operations performed in elimination depend only on the left-side coefficients, not on the right-side vector. For example, for a new set of consumer demands in the Leontief model above, all that would change in (13) would be the numbers on the right side.

There are two natural sets of information to save. First is the final reduced set of the coefficients [e.g., the coefficients in (13)]; there is no need to compute these numbers again. Second is the collection of multipliers used to subtract the jth equation from the ith equation ($i > j$), since we must perform these subtractions on the new right numbers.

Let U denote the upper triangular matrix of coefficients in the final reduced system, and let L be the matrix of multipliers l_{ij} telling how many times equation j is subtracted from equation i. For reasons to be explained shortly, we set $l_{ii} = 1$.

Example 5. L and U Matrices for Re-solving Refinery Equations

Consider the system of equations from Example 2:

$$
\begin{array}{lrcrcrcl}
\text{(a)} & 20x_1 & + & 4x_2 & + & 4x_3 & = & 500 \\
\text{(b)} & 10x_1 & + & 14x_2 & + & 5x_3 & = & 850 \\
\text{(c)} & 5x_1 & + & 5x_2 & + & 12x_3 & = & 1000
\end{array}
\tag{14}
$$

whose final reduced system was

$$
\begin{array}{lrcrcrcl}
\text{(a)} & 20x_1 & + & 4x_2 & + & 4x_3 & = & 500 \\
\text{(b')} & & & 12x_2 & + & 3x_3 & = & 600 \\
\text{(c'')} & & & & & 10x_3 & = & 675
\end{array}
\tag{15}
$$

The reduced system matrix U is

$$
U = \begin{bmatrix} 20 & 4 & 4 \\ 0 & 12 & 3 \\ 0 & 0 & 10 \end{bmatrix}
\tag{16}
$$

Recall that in Example 2 we eliminated x_1 from equations (b) and (c) in (16) by subtracting $\frac{1}{2}$ times (a) from (b) and $\frac{1}{4}$ times (a) from (c). Thus $l_{21} = \frac{1}{2}$ and $l_{31} = \frac{1}{4}$. Next we eliminated x_2 from the last equation by subtracting $\frac{1}{3}$ times (b′) from (c′). Thus $l_{32} = \frac{1}{3}$. Putting 1's on the main diagonal, we have

$$\mathbf{L} = \begin{bmatrix} 1 & 0 & 0 \\ \frac{1}{2} & 1 & 0 \\ \frac{1}{4} & \frac{1}{3} & 1 \end{bmatrix} \tag{17}$$

To solve (14) for another right-side vector **b***, simply perform the elimination steps on **b*** specified by **L** to get the final right-side vector **b**** and then solve the reduced system **Ux** = **b**** by back substitution.

For example, suppose that **b*** = [400, 500, 600]. Then repeating the elimination steps (using **L**) on the new right sides, we have

$$\begin{array}{ll} \text{(a)} & = 400 \\ \text{(b)} & = 500 \\ \text{(c)} & = 600 \\[6pt] \text{(a)} & = 400 \\ \text{(b′)} = \text{(b)} - \tfrac{1}{2}\text{(a)} & = 300 \\ \text{(c′)} = \text{(c)} - \tfrac{1}{4}\text{(a)} & = 500 \\[6pt] \text{(a)} & = 400 \\ \text{(b′)} & = 300 \\ \text{(c″)} = \text{(c′)} - \tfrac{1}{3}\text{(b′)} & = 400 \end{array}$$

The new reduced system is

$$\begin{array}{lrcl} \text{(a)} & 20x_1 + 4x_2 + 4x_3 & = & 400 \\ \text{(b′)} & 12x_2 + 3x_3 & = & 300 \\ \text{(c″)} & 10x_3 & = & 400 \end{array} \tag{18}$$

Using back substitution, we find

$$x_3 = \tfrac{400}{10} = 40$$

Then

$$x_2 = \frac{300 - 3(40)}{12} = 15$$

and finally

$$x_1 = \frac{400 - 4(15) - 4(40)}{20} = \frac{180}{20} = 9$$

So the new solution is [9, 15, 40]. ∎

The reader can check that for the Leontief system in Example 4, the matrices **U** and **L** are

$$\mathbf{U} = \begin{bmatrix} .6 & -.2 & -.2 & -.2 \\ 0 & .6 & -.3 & -.2 \\ 0 & 0 & .9 & -.278 \\ 0 & 0 & 0 & .920 \end{bmatrix}, \quad \mathbf{L} = \begin{bmatrix} 1 & 0 & 0 & 0 \\ -\frac{1}{2} & 1 & 0 & 0 \\ -\frac{1}{6} & -\frac{2}{9} & 1 & 0 \\ 0 & -\frac{1}{6} & -\frac{1}{6} & 1 \end{bmatrix} \quad (19)$$

Note that ignoring the 1's on **L**'s main diagonal, the data in **L** and **U** can be stored together in one square matrix.

Now we state a remarkable theorem.

Theorem 1. Given any n-by-n **A**, let the matrices **L**, **U** be as defined above. Then

$$\mathbf{A} = \mathbf{LU} \qquad (20)$$

Note that we are assuming that **A**'s rows are arranged so that no 0's occur on the main diagonal during elimination. Theorem 1 is proved at the end of Section 5.2.

Let us check (20) for **L** and **U** in Example 5. We want to multiply:

$$\mathbf{LU} = \begin{bmatrix} 1 & 0 & 0 \\ \frac{1}{2} & 1 & 0 \\ \frac{1}{4} & \frac{1}{3} & 1 \end{bmatrix} \begin{bmatrix} 20 & 4 & 4 \\ 0 & 12 & 3 \\ 0 & 0 & 10 \end{bmatrix} \qquad (21)$$

Let us compute **LU** by the definition of matrix multiplication, which says that the ith row in **LU** equals $\mathbf{l}_i^R \mathbf{U}$ (where $\mathbf{l}_i^R$ is the ith row of **L**). So the first row of the product **LU** in (21) is

$$\mathbf{l}_1^R \mathbf{U} = \begin{bmatrix} 1 & 0 & 0 \end{bmatrix} \begin{bmatrix} 20 & 4 & 4 \\ 0 & 12 & 3 \\ 0 & 0 & 10 \end{bmatrix} = \begin{bmatrix} 20 & 4 & 4 \end{bmatrix} \qquad (22a)$$

The second row is

$$\mathbf{l}_2^R \mathbf{U} = \begin{bmatrix} \frac{1}{2} & 1 & 0 \end{bmatrix} \begin{bmatrix} 20 & 4 & 4 \\ 0 & 12 & 3 \\ 0 & 0 & 10 \end{bmatrix} = \frac{1}{2}\begin{bmatrix} 20 & 4 & 4 \end{bmatrix} + 1\begin{bmatrix} 0 & 12 & 3 \end{bmatrix} \qquad (22b)$$
$$= \begin{bmatrix} 10 & 14 & 5 \end{bmatrix}$$

The third row is

$$\mathbf{l}_1^R \mathbf{U} = \begin{bmatrix} \frac{1}{4} & \frac{1}{3} & 1 \end{bmatrix} \begin{bmatrix} 20 & 4 & 4 \\ 0 & 12 & 3 \\ 0 & 0 & 10 \end{bmatrix} = \frac{1}{4}\begin{bmatrix} 20 & 4 & 4 \end{bmatrix} + \frac{1}{3}\begin{bmatrix} 0 & 12 & 3 \end{bmatrix}$$
$$+ 1\begin{bmatrix} 0 & 0 & 10 \end{bmatrix}$$
$$= \begin{bmatrix} 5 & 5 & 12 \end{bmatrix} \qquad (22c)$$

Putting the three rows of **LU** computed in (22a), (22b), and (22c) together, we have **A**.

The proof of this theorem is a generalization of the computation done in (22a), (22b), and (22c). In the elimination process, we are forming new equations as linear combinations of the original equations. Conversely, the original equations are linear combinations of the final reduced equations. The latter property is exactly what the computations in (22a), (22b), and (22c) illustrate. For example, (22b) shows that a_2^R, the second row of **A**, is the following linear combination of **U**'s rows: $a_2^R = \frac{1}{2}u_1^R + u_2^R$.

The **LU** decomposition of a square matrix has many important uses. It also yields a simple formula for the determinant of a square matrix and also allows us to prove that elimination always finds a solution to **Ax** = **b** if one exists.

Theorem 2. For any square matrix **A**,

$$\det(\mathbf{A}) = u_{11} \cdot u_{22} \cdots u_{nn} \tag{23}$$

That is, $\det(\mathbf{A})$ equals the product of main diagonal entries in **U**, where **U** is the reduced-system matrix in the decomposition **A** = **LU**.

Proof. (i) Since **A** = **LU**, then $\det(\mathbf{A}) = \det(\mathbf{L}) \cdot \det(\mathbf{U})$, by the product rule for determinants (Proposition 3 of Section 3.1).

(ii) $\det(\mathbf{L}) = 1$, and $\det(\mathbf{U}) =$ product of **U**'s main diagonal entries, since the determinant of a lower or upper triangular matrix (like **L** or **U**) is just the product of the main diagonal entries (Proposition 2 of Section 3.1).

Combining parts (i) and (ii), we have formula (23). ∎

Implicit in this theorem is the fact that if one (or more) of the main diagonal entries in **U** is 0, then $\det(\mathbf{A}) = 0$ and **Ax** = **b** does not have a solution or the solution is nonunique. (Remember that we are assuming that the rows of **A** are arranged to avoid 0's on the main diagonal during elimination, unless a whole row of 0's occurs.) When this happens, the elimination process fails, as happened in Example 3. Conversely, if $\det(\mathbf{A}) \neq 0$, the elimination cannot fail. Thus we have proven

Theorem 3. For any n-by-n matrix **A** and any n-vector **b**, Gaussian elimination finds the unique solution to **Ax** = **b** if such a unique solution exists.

We close this section by presenting a variation on Gaussian elimination that is a little slower but eliminates the need to do back substitution. This method is known as **Gauss–Jordan elimination,** but in this book we shall call it the method of **elimination by pivoting.** Pivoting yields a convenient way to calculate the inverse of a matrix (in Section 3.3) and arises in the solution of linear programs.

Elimination by pivoting uses the equation i to *eliminate x_i from all other equations* before, as well as after, equation i (Gaussian elimination only eliminates x_i in equations after equation i). It also divides equation i by a_{ii}, so that the coefficient of x_i in the new equation i is 1.

We use the term **pivot on entry** a_{ij} (the coefficient of x_j in equation i) to denote the process of using equation i to eliminate x_j from all other equations (and make 1 be the new coefficient of x_j in equation i).

Example 6. **Elimination by Pivoting**

Let us rework Example 2 using elimination by pivoting.

$$
\begin{array}{lrcrcrcl}
\text{(a)} & 20x_1 & + & 4x_2 & + & 4x_3 & = & 500 \\
\text{(b)} & 10x_1 & + & 14x_2 & + & 5x_3 & = & 850 \\
\text{(c)} & 5x_1 & + & 5x_2 & + & 12x_3 & = & 1000
\end{array}
\tag{24}
$$

We begin by expressing (24) in terms of an augmented coefficient matrix.

$$
\begin{array}{l}
\text{(a)} \\
\text{(b)} \\
\text{(c)}
\end{array}
\left[
\begin{array}{ccc|c}
20 & 4 & 4 & 500 \\
10 & 14 & 5 & 850 \\
5 & 5 & 12 & 1000
\end{array}
\right]
\tag{24'}
$$

Now we make entry $(1, 1)$ one and the rest of the first column zeros.

$$
\begin{array}{l}
\text{(a')} = \text{(a)}/20 \\
\text{(b')} = \text{(b)} - 10\text{(a')} \\
\text{(c')} = \text{(c)} - 5\text{(a')}
\end{array}
\left[
\begin{array}{ccc|c}
1 & \frac{1}{5} & \frac{1}{5} & 25 \\
0 & 12 & 3 & 600 \\
0 & 4 & 11 & 875
\end{array}
\right]
\tag{25}
$$

Next we make entry $(2, 2)$ one and the rest of the second column zeros.

$$
\begin{array}{l}
\text{(a'')} = \text{(a')} - \frac{1}{5}\text{(b'')} \\
\text{(b'')} = \text{(b')}/12 \\
\text{(c'')} = \text{(c')} - 4\text{(b'')}
\end{array}
\left[
\begin{array}{ccc|c}
1 & 0 & \frac{3}{20} & 15 \\
0 & 1 & \frac{1}{4} & 50 \\
0 & 0 & 10 & 675
\end{array}
\right]
\tag{26}
$$

Finally, we make entry $(3, 3)$ one and the rest of the third column zeros.

$$
\begin{array}{l}
\text{(a'')} = \text{(a')} - \frac{3}{20}\text{(c''')} \\
\text{(b''')} = \text{(b'')} - \frac{1}{4}\text{(c''')} \\
\text{(c''')} = \text{(c'')}/10
\end{array}
\left[
\begin{array}{ccc|c}
1 & 0 & 0 & 4\frac{7}{8} \\
0 & 1 & 0 & 33\frac{1}{8} \\
0 & 0 & 1 & 67\frac{1}{2}
\end{array}
\right]
\tag{27}
$$

Note that the upper triangular system of equations corresponding to (27) yields a solution directly, without back substitution.

$$
\begin{array}{rcl}
x_1 & = & 4\frac{7}{8} \\
x_2 & = & 33\frac{1}{8} \\
x_3 & = & 67\frac{1}{2}
\end{array}
\tag{27'}
$$

■

Actually, in elimination, one can use any equation to eliminate x_i from the other equations. We illustrate the idea with elimination by pivoting, but it also applies to Gaussian elimination.

Example 7. **Solution with Off-Diagonal Pivoting**

Let us repeat Example 6 but with the numbers in equations (b) and (c) changed:

$$
\begin{array}{l}
\text{(a)} \\
\text{(b)} \\
\text{(c)}
\end{array}
\qquad
\left[\begin{array}{ccc|c}
20 & 4 & 4 & 500 \\
10 & 2 & 5 & 850 \\
5 & 5 & 9 & 525
\end{array}\right]
\qquad (28)
$$

We want to make entry $(1, 1)$ one and the rest of the first column zeros.

$$
\begin{array}{l}
\text{(a')} = \text{(a)}/20 \\
\text{(b')} = \text{(b)} - 10\text{(a')} \\
\text{(c')} = \text{(c)} - 5\text{(a')}
\end{array}
\qquad
\left[\begin{array}{ccc|c}
1 & \frac{1}{5} & \frac{1}{5} & 25 \\
0 & 0 & 3 & 600 \\
0 & 4 & 8 & 400
\end{array}\right]
\qquad (29)
$$

Since entry $(2, 2)$ is zero, we cannot pivot on it. Furthermore, we cannot pivot on entry $(1, 2)$ since we have already pivoted on an entry in the first row. Thus we must pivot on entry $(3, 2)$ and make it one while the rest of the second column becomes zeros.

$$
\begin{array}{l}
\text{(a'')} = \text{(a')} - \frac{1}{5}\text{(c'')} \\
\text{(b'')} = \text{(b')} \\
\text{(c'')} = \text{(c')}/4
\end{array}
\qquad
\left[\begin{array}{ccc|c}
1 & 0 & -\frac{1}{5} & 5 \\
0 & 0 & 3 & 600 \\
0 & 1 & 2 & 100
\end{array}\right]
\qquad (30)
$$

Finally, we pivot on entry $(2, 3)$.

$$
\begin{array}{l}
\text{(a''')} = \text{(a'')} + \frac{1}{5}\text{(b'')} \\
\text{(b''')} = \text{(b'')}/3 \\
\text{(c''')} = \text{(c'')} - 2\text{(b'')}
\end{array}
\qquad
\left[\begin{array}{ccc|c}
1 & 0 & 0 & 45 \\
0 & 0 & 1 & 200 \\
0 & 1 & 0 & -300
\end{array}\right]
\qquad (31)
$$

We read off the solution, $x_1 = 45$, $x_2 = -300$, $x_3 = 200$. ∎

We close with two important comments about elimination. The first is *how to handle the problem of entry (i, i) being 0 when we want to use it to eliminate x_i from the following equations*—this occurred in (29). The solution is to pivot on another nonzero entry, say entry (h, i), in the ith column; in (29), we pivoted on entry $(3, 2)$. An equivalent step is to interchange the ith equation with the hth equation; after the interchange, entry (i, i) is nonzero. In (29) we would interchange equations (b') and (c'). Such an interchange works for Gaussian elimination as well as elimination by pivoting.

Second, note that *one cannot pivot twice in the same row*. For example, in system (30), if we pivoted on entry (3, 3), then when we used the third equation to eliminate x_3 terms from other equations, we would be reintroducing x_2 terms into the other equations (see Exercise 21).

The **LU** decomposition exemplifies a very important aspect of computer science. In the **LU** decomposition, we transfer much of the work in solving the system $\mathbf{Ax} = \mathbf{b}$ into a "data structure" problem. We "store" **A** in the decomposed form of a lower and an upper triangular matrix, **L** and **U**. Computer science examines how one processes complex sets of information (in Europe, the subject is often called *informatics*). How data are organized (or preprocessed) into data structures is often more important in information processing than the subsequent computations.

The **LU** decomposition is also a matrix algebra example of the computer science insight—that computer programs can be viewed as a special form of data. That is, programs are stored as a string of 0's and 1's just like other data (before programs were stored as data, computers had to be rewired for each new set of computations). The matrix **L** contains the elimination multipliers, part of the Gaussian elimination "program," which are used to reduce any right-side vector **b** to **b*** after which back substitution solves $\mathbf{Ux} = \mathbf{b}^*$. Premultiplying the reduced-form matrix **U** by **L** to obtain **A** is another instance where **L** acts like a program—the *i*th row of **L** times **U** reconstructs the *i*th row of **A** as a linear combination of **U**'s rows:

$$\mathbf{a}_i^R = \mathbf{l}_i^R \mathbf{U} = l_{i1}\mathbf{u}_1^R + l_{i2}\mathbf{u}_2^R + \cdots + l_{ni}\mathbf{u}_n^R$$

Matrix multiplication makes it possible to use matrices as both data and programs, just like a computer. By pre- and postmultiplying data matrix by the proper "program" matrices, one can do almost anything. At the core of such computations is having the right data representation or data structure, be it a matrix decomposition or some other transformed form of the matrix. The key stage in virtually all modern numerical algorithms involving matrices is the preprocessing, to get the right representation of the data.

Section 3.2 Exercises

Summary of Exercises
Exercises 1–16 involve Gaussian elimination computations. Exercises 17–20 involve word problems. Exercises 21–25 are theoretical.

1. Solve the following systems of equations by Gaussian elimination.
   (a) $x + y = 5$
       $x - 2y = 4$
   (b) $2x - 3y = 4$
       $3x + 2y = 5$
   (c) $3x - y = 0$
       $-2x + y = 2$

2. In each of the following sets of three equations, show that the third equation equals the second equation minus some multiple of the first equation: (c) = (b) − r(a) for some r.

$$\text{(i) (a)} \quad x + 2y = 4$$
$$\text{(b)} \quad 3x + y = 9$$
$$\text{(c)} \quad x - 3y = 1$$

$$\text{(ii) (a)} \quad x - y + z = 2$$
$$\text{(b)} \quad x + y - z = 3$$
$$\text{(c)} \quad -2x + 4y - 4z = -3$$

$$\text{(iii) (a)} \quad 2x + y - 2z = -5$$
$$\text{(b)} \quad 3x - y + z = 8$$
$$\text{(c)} \quad 6x + .5y - 2z = .5$$

3. Solve the following systems of equations using Gaussian elimination.

$$\text{(a)} \quad 2x_1 - 3x_2 + 2x_3 = 0$$
$$x_1 - x_2 + x_3 = 7$$
$$-x_1 + 5x_2 + 4x_3 = 4$$

$$\text{(b)} \quad -x_1 - x_2 + x_3 = 2$$
$$2x_1 + 2x_2 - 4x_3 = -4$$
$$x_1 - 2x_2 + 3x_3 = 5$$

$$\text{(c)} \quad -x_1 - 3x_2 + 2x_3 = -2$$
$$2x_1 + x_2 + 3x_3 = \tfrac{9}{2}$$
$$5x_1 + 4x_2 + 6x_3 + 12$$

$$\text{(d)} \quad 2x_1 + 4x_2 - 2x_3 = 4$$
$$x_1 - 2x_2 - 4x_3 = -1$$
$$-2x_1 - x_2 - 3x_3 = -4$$

$$\text{(e)} \quad x_1 + x_2 + 4x_3 = 4$$
$$2x_1 + x_2 + 3x_3 = 5$$
$$5x_1 + 2x_2 + 5x_3 = 11$$

$$\text{(f)} \quad 2x_1 - 3x_2 - x_3 = 2$$
$$3x_1 - 5x_2 - 2x_3 = -1$$
$$9x_1 + 6x_2 + 4x_3 = 1$$

4. Solve the problems in Exercise 3 using elimination by pivoting (Gauss–Jordan elimination).

5. (a) Write the **LU** decomposition for each coefficient matrix **A** in Exercise 3.
   (b) Multiply **L** times **U** to show that the product is **A**, for each coefficient matrix **A** in Exercise 3.

6. Find the determinant of each matrix in Exercise 3 using Theorem 2.

7. Re-solve each system in Exercise 3 with the new right-hand-side vector [10, 5, 10] using the numbers in the **L** and **U** matrices you found in Exercise 5.

8. For the right-side vector **b** = [1, 2, 3], solve the system of equations **Ax** = **b**, where instead of **A**, you are given the **LU** decomposition of **A**.

$$\text{(a)} \quad \mathbf{L} = \begin{bmatrix} 1 & 0 & 0 \\ 1 & 1 & 0 \\ 2 & 3 & 1 \end{bmatrix}, \quad \mathbf{U} = \begin{bmatrix} 2 & 1 & 1 \\ 0 & 3 & 2 \\ 0 & 0 & -2 \end{bmatrix}$$

$$\text{(b)} \quad \mathbf{L} = \begin{bmatrix} 1 & 0 & 0 \\ -2 & 1 & 0 \\ 4 & -1 & 1 \end{bmatrix}, \quad \mathbf{U} = \begin{bmatrix} 1 & -2 & 2 \\ 0 & 5 & 2 \\ 0 & 0 & 2 \end{bmatrix}$$

9. Solve the following systems of equations using Gaussian elimination and give the **LU** decomposition of the coefficient matrix.

(a) $x_1 + 3x_2 + 2x_3 - x_4 = 7$
$x_1 + x_2 + x_3 + x_4 = 3$
$2x_1 - 2x_2 + x_3 - x_4 = -5$
$x_1 - 3x_2 - x_3 + 2x_4 = -4$

(b) $3x_1 + 2x_2 + x_3 = 3$
$x_1 + x_2 - x_4 = 2$
$2x_1 + x_2 - x_3 + x_4 = -3$
$x_1 + x_2 + x_3 + x_4 = 0$

(c) $x_1 + x_2 - x_3 - x_4 = 4$
$2x_1 + x_4 = 8$
$3x_1 - 2x_4 = 3$
$4x_1 - 2x_2 + x_3 + 3x_4 = 15$

10. Exercise 9, part (c) can be simplified by first solving the second and third equations for x_1 and x_4, and afterward solving for x_3 and x_4. Solve Exercise 9, part (c) this way.

11. Given the **LU** decomposition of an n-by-n matrix **A**, how many multiplications are required to compute **A** as the matrix product **LU** (allowing for known 0's in **L** and **U**)?

12. Determine whether each of the following systems of equations has a unique solution, multiple solutions, or is inconsistent.

(a) $2x - 3y = 6$
$-6x + 9y = 12$

(b) $x_1 + 2x_2 + 3x_3 = 10$
$2x_1 - x_2 + 4x_3 = 20$
$5x_2 + 2x_3 = 0$

(c) $x_1 - x_2 + x_3 = 5$
$x_1 + 3x_2 + 6x_3 = 9$
$-x_1 + 5x_2 + 4x_3 = 10$

(d) $x_1 + x_2 + 2x_3 = 0$
$2x_1 + x_2 - 3x_3 = 0$
$-x_1 + 2x_2 + x_3 = 0$

(e) $x_1 + x_2 + 2x_3 = 3$
$-x_1 - 2x_2 + x_3 = 8$
$x_1 - x_2 + 8x_3 = 25$

(f) $x_1 + 2x_2 + 3x_3 = 5$
$3x_1 - x_2 - 2x_3 = -3$
$-5x_1 + 4x_2 + 10x_3 = 14$

13. Use Gaussian elimination to solve the following variations on the refinery problem in Example 2. Sometimes the variation will have no solution, sometimes multiple solutions (express such an infinite family of solutions in terms of x_3), and sometimes the solution will involve negative numbers (a real-world impossibility).

(a) $20x_1 + 4x_2 + 4x_3 = 500$
$8x_1 + 3x_2 + 5x_3 = 850$
$4x_1 + 5x_2 + 11x_3 = 2050$

(b) $6x_1 + 5x_2 + 6x_3 = 500$
$10x_1 + 10x_2 = 850$
$2x_1 + 12x_3 = 1000$

(c) $6x_1 + 2x_2 + 2x_3 = 500$ (d) $8x_1 + 4x_2 + 3x_3 = 500$
  $3x_1 + 6x_2 + 3x_3 = 300$ $4x_1 + 8x_2 + 5x_3 = 500$
  $3x_1 + 2x_2 + 6x_3 = 1000$ $12x_2 + 6x_3 = 500$

14. Solve each system of equations in Exercise 3 with elimination by pivoting in which off-diagonal pivots are used—to be exact, pivot on entry (2, 1), then on (3, 2), and finally on (1, 3).

15. Solve the following systems of equations by Gaussian elimination. When you come to a zero entry on the main diagonal, interchange equations as appropriate.

(a) $x_1 + 2x_2 + 3x_3 = 6$ (b) $x_1 - 3x_2 + x_3 = 4$
  $2x_1 + 4x_2 + 5x_3 = 12$ $-2x_1 + 6x_2 - 2x_3 = 7$
  $2x_1 + 5x_2 - 3x_3 = 10$ $x_1 + x_2 - x_3 = 3$

(c) $x_1 - x_2 + x_3 + x_4 = 6$ (d) $x_2 + x_3 = 0$
  $2x_1 - x_3 - x_4 = 5$ $x_3 + x_4 = 1$
  $2x_1 - 2x_2 + x_4 = 4$ $x_1 - x_4 = 2$
  $x_2 + x_3 - x_4 = 3$ $x_1 + x_2 = 3$

16. The following systems of equations are large, but their special tridiagonal form makes them easy to solve. Solve them.

(a) $x_1 - x_2 = 2$
  $-x_1 + 2x_2 - x_3 = 0$
  $- x_2 + 2x_3 - x_4 = 0$
  $- x_3 + 2x_4 - x_5 = 0$
  $- x_4 + x_5 = 2$

(b) $2x_1 + x_2 = 1$
  $x_1 + 2x_2 + x_3 = 1$
  $x_2 + 2x_3 + x_4 = 1$
  $x_3 + 2x_4 + x_5 = 1$
  $x_4 + 2x_5 = 1$

17. The staff dietician at the California Institute of Trigonometry has to make up a meal with 600 calories, 20 grams of protein, and 200 milligrams of vitamin C. There are three food types to choose from: rubbery jello, dried fish sticks, and mystery meat. They have the following nutritional content per ounce.

	Jello	Fish Sticks	Mystery Meat
Calories	10	50	200
Protein	1	3	.2
Vitamin C	30	10	0

Set up and solve a system of equations to determine how much of each food should be used.

18. A furniture manufacturer makes tables, chairs, and sofas. In one month, the company has available 300 units of wood, 350 units of labor, and 225 units of upholstery. The manufacturer wants a production schedule for the month that uses all of these resources. The different products require the following amounts of the resources.

	Table	Chair	Sofa
Wood	4	1	3
Labor	3	2	5
Upholstery	2	0	4

Set up and solve a system of equations to determine how much of each product should be manufactured.

19. A company has a budget of $280,000 for computing equipment. Three types of equipment are available: microcomputers at $2000 a piece, terminals at $500 a piece, and word processors at $5000 a piece. There should be five times as many terminals as microcomputers and two times as many microcomputers as word processors. Set this problem up as a system of 3 linear equations and solve to determine how many machines of each type should there be.

20. An investment analyst is trying to find out how much business a secretive TV manufacturer has. The company makes three brands of TV: Brand A, Brand B, and Brand C. The analyst learns that the manufacturer has ordered from suppliers 450,000 type-1 circuit boards, 300,000 type-2 circuit boards and 350,000 type-3 circuit boards. Brand A uses 2 type-1 boards, 1 type-2 board, and 2 type-3 boards. Brand B uses 3 type-1 boards, 2 type-2 boards, and 1 type-3 board. Brand C uses 1 board of each type. How many TV's of each brand are being manufactured?

21. This exercise shows why each pivot (in elimination by pivoting) must be in a different row.
    (a) In Example 7, make the third pivot on entry (3, 3) instead of on entry (3, 2). Can you still read off the solution?
    (b) In Exercise 3, part (b), make the following sequence of pivots, entry (3, 1), entry (2, 2), then entry (2, 3). Does this provide a solution to the system of equations?

22. For what values of k does the following refinery-type system of equations have a unique solution with all x_i nonnegative?

$$6x_1 + 5x_2 + 3x_3 = 500$$
$$4x_1 + x_2 + 7x_3 = 600$$
$$5x_1 + kx_2 + 5x_3 = 1000$$

23. For an arbitary 2-by-2 system of equations

$$ax + by = r$$
$$cx + dy = s$$

(a) Determine the **LU** decomposition of the coefficient matrix **A**.
(b) Verify that **L** times **U** equals **A**.

24. Use Theorem 2 to show that in $\mathbf{Ax} = \mathbf{b}$, if one row of **A** is all 0's, then $\det(\mathbf{A}) = 0$.

25. Consider the following 3-by-3 matrix whose entries are functions. Find the **LU** decomposition of this matrix and find its determinant.

$$\begin{bmatrix} 3x & 6x^2 & e^x \\ x^2 & x^3 & xe^x \\ 6 & 3x & e^x/x \end{bmatrix}$$

Computer Projects
26. Write a computer program to perform Gaussian elimination on a system of n equations in n unknowns (watch out for 0's on the main diagonal).

27. Write a computer program to perform elimination by pivoting on a system of n equations in n unknowns (watch out for 0 pivots).

Section 3.3 The Inverse of a Matrix

In this section we study a general method for solving a system of equations $\mathbf{Ax} = \mathbf{b}$ for any **b**, instead of for one particular **b** as in Sections 3.1 and 3.2.

Any matrix **A** has an additive inverse, the negative $-\mathbf{A}$ (obtained by changing the sign of all entries), such that

$$\mathbf{A} + (-\mathbf{A}) = \mathbf{O}$$

A multiplicative inverse $\mathbf{A}^{-1}$ has the property

$$\mathbf{AA}^{-1} = \mathbf{I} \quad \text{and} \quad \mathbf{A}^{-1}\mathbf{A} = \mathbf{I}$$

where **I** is the identity matrix. Inverses allow us to "solve" a system of equations symbolically the way one solves the scalar equation $ax = b$ by dividing both sides by a, obtaining $x = a^{-1}b$.

Theorem 1. If **A** has an inverse $\mathbf{A}^{-1}$, then the system of equations $\mathbf{Ax} = \mathbf{b}$ has the solution $\mathbf{x} = \mathbf{A}^{-1}\mathbf{b}$.

Proof. As in the one-variable case, we divide both sides of $\mathbf{Ax} = \mathbf{b}$ by **A**, that is, multiply both sides by $\mathbf{A}^{-1}$:

$$\mathbf{A}^{-1}(\mathbf{Ax}) = \mathbf{A}^{-1}\mathbf{b} \tag{1}$$

Using matrix algebra and the fact that $\mathbf{A}^{-1}\mathbf{A} = \mathbf{I}$, we can rewrite the left side of (1), $\mathbf{A}^{-1}(\mathbf{Ax})$, as **x**. The details of this rewriting are

$$\mathbf{A}^{-1}(\mathbf{Ax}) = (\mathbf{A}^{-1}\mathbf{A})\mathbf{x} = \mathbf{Ix} = \mathbf{x} \tag{2}$$

Combining (1) and (2), we have the desired result: $\mathbf{x} = \mathbf{A}^{-1}\mathbf{b}$. ∎

A matrix **A** is **invertible** if it has an inverse. In many books the term *nonsingular* is used instead of invertible; a *singular* matrix has no inverse. Some matrices are invertible and some are not. Much of the theory of linear algebra centers around conditions that will make a matrix invertible. We note that if a matrix **A** has an inverse $\mathbf{A}^{-1}$, *the inverse is unique* (see Exercise 17).

In Section 3.1 we saw that finding a (unique) solution to a system of linear equations was dependent on whether the associated coefficient matrix **A** had a nonzero determinant. Now we have another sufficient condition, the existence of $\mathbf{A}^{-1}$.

We will show shortly how to calculate inverses, when they exist. First, let us verify that certain matrices do and do not have inverses.

Example 1. Matrices With and Without Inverses

(i) Matrix $\mathbf{A} = \begin{bmatrix} 3 & 1 \\ 4 & 2 \end{bmatrix}$ has the inverse $\mathbf{A}^{-1} = \begin{bmatrix} 1 & -\frac{1}{2} \\ -2 & \frac{3}{2} \end{bmatrix}$.

(For the present, do not worry how this inverse was found.) Multiplying **A** times $\mathbf{A}^{-1}$, we have

$$\begin{bmatrix} 3 & 1 \\ 4 & 2 \end{bmatrix}\begin{bmatrix} 1 & -\frac{1}{2} \\ -2 & \frac{3}{2} \end{bmatrix} = \begin{bmatrix} 3\times 1 + 1\times -2 & 3\times -\frac{1}{2} + 1\times\frac{3}{2} \\ 4\times 1 + 2\times -2 & 4\times -\frac{1}{2} + 2\times\frac{3}{2} \end{bmatrix}$$

$$= \begin{bmatrix} 1 & 0 \\ 0 & 1 \end{bmatrix}$$

The reader can verify that if $\mathbf{A}^{-1}$ precedes **A**, again $\mathbf{A}^{-1}\mathbf{A} = \mathbf{I}$.

(ii) We claim that the matrix $\mathbf{B} = \begin{bmatrix} 1 & 4 \\ 2 & 8 \end{bmatrix}$ has no inverse.

The key to our claim is the observation that the second row is twice the first row.

$$\mathbf{b}_2^R = 2\mathbf{b}_1^R \tag{3}$$

where $\mathbf{b}_i^R$ denotes the ith row of $\mathbf{B}$.

Suppose that $\mathbf{C}$ were the inverse of $\mathbf{B}$, so that $\mathbf{BC} = \mathbf{I} = \begin{bmatrix} 1 & 0 \\ 0 & 1 \end{bmatrix}$. If $\mathbf{c}_1^C$ and $\mathbf{c}_2^C$ are the two columns of $\mathbf{C}$, the matrix product $\mathbf{BC}$ is the following collection of scalar products:

$$\mathbf{BC} = \begin{bmatrix} \mathbf{b}_1^R \cdot \mathbf{c}_1^C & \mathbf{b}_1^R \cdot \mathbf{c}_2^C \\ \mathbf{b}_2^R \cdot \mathbf{c}_1^C & \mathbf{b}_2^R \cdot \mathbf{c}_2^C \end{bmatrix} = \begin{bmatrix} 1 & 0 \\ 0 & 1 \end{bmatrix} \tag{4}$$

From (3), $\mathbf{b}_2^R \cdot \mathbf{c}_1^C = 2\mathbf{b}_1^R \cdot \mathbf{c}_1^C$. So the second row of $\mathbf{BC}$, $[\mathbf{b}_2^R \cdot \mathbf{c}_1^C, \mathbf{b}_2^R \cdot \mathbf{c}_2^C]$, must be twice the first row, $[\mathbf{b}_1^R \cdot \mathbf{c}_1^C \quad \mathbf{b}_1^R \cdot \mathbf{c}_2^C]$, but the second row of $\mathbf{I}$ is not twice its first row. This contradiction shows that no inverse can exist. ∎

Example 1, part (ii) shows that if one row of $\mathbf{A}$ is a multiple of another row, no inverse can exist. This result complements Proposition 1 of Section 3.1, which says that if one row is a multiple of another, $\det(\mathbf{A}) = 0$ (so no unique solution to $\mathbf{Ax} = \mathbf{b}$ exists).

Remember that the inverse of a matrix, when it exists, is unique. The following example shows how to compute the inverse of a matrix.

Example 2. **Computing Inverse of a 2-by-2 Matrix**

Consider the 2-by-2 matrix $\mathbf{A}$ and its (unknown) inverse $\mathbf{X}$:

$$\mathbf{A} = \begin{bmatrix} 3 & 1 \\ 4 & 2 \end{bmatrix}, \qquad \mathbf{X} = \begin{bmatrix} x_{11} & x_{12} \\ x_{21} & x_{22} \end{bmatrix}$$

We require that $\mathbf{AX} = \mathbf{I}$:

$$\mathbf{AX} = \begin{bmatrix} 3 & 1 \\ 4 & 2 \end{bmatrix}\begin{bmatrix} x_{11} & x_{12} \\ x_{21} & x_{22} \end{bmatrix} = \begin{bmatrix} 1 & 0 \\ 0 & 1 \end{bmatrix} = \mathbf{I} \tag{5}$$

We determine $\mathbf{X} \ (= \mathbf{A}^{-1})$ a column at a time. First we set $\mathbf{Ax}_1^C$, the first column in product $\mathbf{AX}$ of (5), equal to the first column of $\mathbf{I}$:

$$\mathbf{Ax}_1^C = \mathbf{e}_1: \qquad \begin{bmatrix} 3 & 1 \\ 4 & 2 \end{bmatrix}\begin{bmatrix} x_{11} \\ x_{21} \end{bmatrix} = \begin{bmatrix} 1 \\ 0 \end{bmatrix}$$

or

$$3x_{11} + x_{21} = 1 \qquad (6a)$$
$$4x_{11} + 2x_{21} = 0$$

Similarly, the second column yields the system

$$3x_{12} + x_{22} = 0 \qquad (6b)$$
$$4x_{12} + 2x_{22} = 1$$

Using elimination by pivoting on the augmented coefficient matrix for (6a), we obtain

$$\begin{bmatrix} 3 & 1 & | & 1 \\ 4 & 2 & | & 0 \end{bmatrix} \rightarrow \begin{bmatrix} 1 & \frac{1}{3} & | & \frac{1}{3} \\ 0 & \frac{2}{3} & | & -\frac{4}{3} \end{bmatrix} \rightarrow \begin{bmatrix} 1 & 0 & | & 1 \\ 0 & 1 & | & -2 \end{bmatrix} \qquad (7a)$$

so $x_{11} = 1$, $x_{21} = -2$. For (6b) we obtain

$$\begin{bmatrix} 3 & 1 & | & 0 \\ 4 & 2 & | & 1 \end{bmatrix} \rightarrow \begin{bmatrix} 1 & \frac{1}{3} & | & 0 \\ 0 & \frac{2}{3} & | & 1 \end{bmatrix} \rightarrow \begin{bmatrix} 1 & 0 & | & -\frac{1}{2} \\ 0 & 1 & | & \frac{3}{2} \end{bmatrix} \qquad (7b)$$

so $x_{12} = -\frac{1}{2}$, $x_{22} = \frac{3}{2}$.

Substituting these values for x_{ij} back into $\mathbf{X}$ ($=\mathbf{A}^{-1}$), we have

$$\mathbf{A}^{-1} = \begin{bmatrix} 1 & -\frac{1}{2} \\ -2 & \frac{3}{2} \end{bmatrix}$$

∎

Although elimination is the preferred way to solve a system of equations, Cramer's rule yields an easy-to-remember formula for solving the equations for the inverse of a 2-by-2 matrix. For a general 2-by-2 matrix $\mathbf{A}$, the system of equations $\mathbf{A}\mathbf{x}_1^C = \mathbf{e}_1$ [like (6a)] has the solution by Cramer's rule:

$$x_{11} = \frac{\begin{vmatrix} 1 & a_{12} \\ 0 & a_{22} \end{vmatrix}}{\det(\mathbf{A})} = \frac{a_{22}}{\det(\mathbf{A})} \qquad x_{21} = \frac{\begin{vmatrix} a_{11} & 1 \\ a_{21} & 0 \end{vmatrix}}{\det(\mathbf{A})} = \frac{-a_{21}}{\det(\mathbf{A})} \qquad (8)$$

The simple form of the numerator comes from having a right-side vector of $[1, 0]$. The same simplification occurs in solving $\mathbf{A}\mathbf{x}_2^C = \mathbf{e}_2$ by Cramer's rule (we have the same system of equations except that the right-side vector is now $[0, 1]$). The reader should check that with $[0, 1]$, (8) now yields the solution: $x_{12} = -a_{12}/\det(\mathbf{A})$, $x_{22} = a_{11}/\det(\mathbf{A})$.

These single-number numerators lead to the following general formula for the inverse of a 2-by-2 matrix:

Formula for Inverse of a 2-by-2 Matrix

If $\mathbf{A} = \begin{bmatrix} a_{11} & a_{12} \\ a_{21} & a_{22} \end{bmatrix}$, then $\mathbf{A}^{-1} = \dfrac{1}{\det(\mathbf{A})} \begin{bmatrix} a_{22} & -a_{12} \\ -a_{21} & a_{11} \end{bmatrix}$ (9)

In words, a 2-by-2 inverse of $\mathbf{A}$ is obtained as follows: *Divide all entries of $\mathbf{A}$ by the determinant, then interchange the two diagonal entries and change the sign of the two off-diagonal entries.*

The method in Example 2 for finding the inverse of $\mathbf{A}$ a column at a time can be applied to any matrix.

Theorem 2. Let $\mathbf{A}$ be an n-by-n matrix and $\mathbf{e}_j$ be the jth unit n-vector, $\mathbf{e}_j = [0, 0, \ldots, 1, \ldots, 0]$. If the n-vector $\mathbf{x}_j$ is the solution to the matrix equation

$$\mathbf{A}\mathbf{x}_j = \mathbf{e}_j \qquad (10)$$

for $i = 1, 2, \ldots, n$, then the n-by-n matrix $\mathbf{X}$ with column vectors $\mathbf{x}_j$ is the inverse of $\mathbf{A}$.

$$\mathbf{A}^{-1} = \mathbf{X} = [\mathbf{x}_1, \mathbf{x}_2, \ldots, \mathbf{x}_n] \qquad (11)$$

Note: *If the systems $\mathbf{A}\mathbf{x}_j = \mathbf{e}_j$ do not have solutions, $\mathbf{A}$ does not have an inverse.*

Recall that when we solve a system of equations by elimination, the right sides play a passive role. That is, *using a different right side $\mathbf{b}$ does not change any of the calculations involving the coefficients. If affects only the final values that appear on the right side.* Thus, when we performed elimination by pivoting on the coefficient matrix in (7a) and (7b) of Example 2, we could have simultaneously applied the elimination steps to an augmented coefficient matrix $[\mathbf{A} \quad \mathbf{I}]$ that contained both right-side vectors. The computations would be

$$\begin{bmatrix} 3 & 1 & | & 1 & 0 \\ 4 & 2 & | & 0 & 1 \end{bmatrix} \rightarrow \begin{bmatrix} 1 & \frac{1}{3} & | & \frac{1}{3} & 0 \\ 0 & \frac{2}{3} & | & -\frac{4}{3} & 1 \end{bmatrix} \rightarrow \begin{bmatrix} 1 & 0 & | & 1 & -\frac{1}{2} \\ 0 & 1 & | & -2 & \frac{3}{2} \end{bmatrix} \qquad (12)$$

So starting with $[\mathbf{A} \quad \mathbf{I}]$, elimination by pivoting yields $[\mathbf{I} \quad \mathbf{A}^{-1}]$.

Computation of the Inverse of an n-by-n Matrix A. Pivot on entries $(1, 1), (2, 2), \ldots, (n, n)$ in the augmented matrix $[\mathbf{A} \quad \mathbf{I}]$. The resulting array will be $[\mathbf{I} \quad \mathbf{A}^{-1}]$.

In Section 3.2 we learned that the **LU** decomposition can be used to solve **Ax** = **b** for several different **b**'s. The **LU** decomposition and pivoting on the augmented matrix [**A**   **I**] are equally fast ways to find the inverse. In hand computation, the augmented matrix method is easier.

Example 3. Inverse of a 3-by-3 Matrix

Let

$$\mathbf{A} = \begin{bmatrix} 1 & 0 & 2 \\ 2 & 4 & 2 \\ 1 & 2 & 6 \end{bmatrix}.$$

We compute the inverse using pivoting on the augmented matrix [**A**   **I**].

$$
\begin{array}{ll}
\begin{array}{l}(a) \\ [\mathbf{A} \quad \mathbf{I}] = (b) \\ (c)\end{array} &
\left[\begin{array}{ccc|ccc} 1 & 0 & 2 & 1 & 0 & 0 \\ 2 & 4 & 2 & 0 & 1 & 0 \\ 1 & 2 & 6 & 0 & 0 & 1 \end{array}\right]
\end{array}
$$

$$
\begin{array}{ll}
\begin{array}{l}(a') = (a) \\ (b') = (b) - 2(a) \\ (c') = (c) - (a)\end{array} &
\left[\begin{array}{ccc|ccc} 1 & 0 & 2 & 1 & 0 & 0 \\ 0 & 4 & -2 & -2 & 1 & 0 \\ 0 & 2 & 4 & -1 & 0 & 1 \end{array}\right]
\end{array} \tag{13}
$$

$$
\begin{array}{ll}
\begin{array}{l}(a'') = (a') \\ (b'') = (b')/4 \\ (c'') = (c') - 2(b'')\end{array} &
\left[\begin{array}{ccc|ccc} 1 & 0 & 2 & 1 & 0 & 0 \\ 0 & 1 & -\frac{1}{2} & -\frac{1}{2} & \frac{1}{4} & 0 \\ 0 & 0 & 5 & 0 & -\frac{1}{2} & 1 \end{array}\right]
\end{array}
$$

$$
\begin{array}{ll}
\begin{array}{l}(a''') = (a'') - 2(c''') \\ (b''') = (b'') + (c''')/2 \\ (c''') = (c'')/5\end{array} &
\left[\begin{array}{ccc|ccc} 1 & 0 & 0 & 1 & \frac{1}{5} & -\frac{2}{5} \\ 0 & 1 & 0 & -\frac{1}{2} & \frac{1}{5} & \frac{1}{10} \\ 0 & 0 & 1 & 0 & -\frac{1}{10} & \frac{1}{5} \end{array}\right]
\end{array}
$$

Thus

$$\mathbf{A}^{-1} = \begin{bmatrix} 1 & \frac{1}{5} & -\frac{2}{5} \\ -\frac{1}{2} & \frac{1}{5} & \frac{1}{10} \\ 0 & -\frac{1}{10} & \frac{1}{5} \end{bmatrix} \qquad \blacksquare$$

If we have to solve a system of equations **Ax** = **b** for many different right-hand sides, it is useful to know **A**$^{-1}$. For each new **b'**, we find the solution of **Ax** = **b'** as **x** = **A**$^{-1}$**b'**. The inverse also lets us determine how a small change Δ**b** in **b** will affect our solution **x**.

Example 4. **Use of Inverse in Multiple Right-Hand Sides**

In Example 5 of Section 3.2 we solved the refinery system of equations by pivoting along the diagonal. Let us use the same sequence of pivots with the augmented matrix [**A**   **I**] to compute the inverse.

$$
\begin{array}{l}
\text{(a)} \\
\text{(b)} \\
\text{(c)}
\end{array}
\left[
\begin{array}{ccc|ccc}
20 & 4 & 4 & 1 & 0 & 0 \\
10 & 14 & 5 & 0 & 1 & 0 \\
5 & 5 & 12 & 0 & 0 & 1
\end{array}
\right]
$$

$$
\begin{array}{l}
\text{(a')} = \text{(a)}/20 \\
\text{(b')} = \text{(b)} - 10\text{(a')} \\
\text{(c')} = \text{(c)} - 5\text{(a')}
\end{array}
\left[
\begin{array}{ccc|ccc}
1 & \frac{1}{5} & \frac{1}{5} & \frac{1}{20} & 0 & 0 \\
0 & 12 & 3 & -\frac{1}{2} & 1 & 0 \\
0 & 4 & 11 & -\frac{1}{4} & 0 & 1
\end{array}
\right] \tag{14}
$$

$$
\begin{array}{l}
\text{(a'')} = \text{(a')} - \text{(b'')}/5 \\
\text{(b'')} = \text{(b')}/12 \\
\text{(c'')} = \text{(c')} - 4\,\text{(b'')}
\end{array}
\left[
\begin{array}{ccc|ccc}
1 & 0 & \frac{3}{20} & \frac{7}{120} & -\frac{1}{60} & 0 \\
0 & 1 & \frac{1}{4} & -\frac{1}{24} & \frac{1}{12} & 0 \\
0 & 0 & 10 & -\frac{1}{12} & -\frac{1}{3} & 1
\end{array}
\right]
$$

$$
\begin{array}{l}
\text{(a''')} = \text{(a'')} - \frac{3}{20}\text{(c''')} \\
\text{(b''')} = \text{(b'')} - \frac{1}{4}\text{(c''')} \\
\text{(c''')} = \text{(c'')}/10
\end{array}
\left[
\begin{array}{ccc|ccc}
1 & 0 & 0 & \frac{143}{2400} & -\frac{7}{600} & -\frac{3}{200} \\
0 & 1 & 0 & -\frac{19}{480} & \frac{11}{120} & -\frac{1}{40} \\
0 & 0 & 1 & -\frac{1}{120} & -\frac{1}{30} & \frac{1}{10}
\end{array}
\right]
$$

The inverse is, in decimals,

$$
\mathbf{A}^{-1} =
\begin{bmatrix}
.05958 & -.01166 & -.015 \\
-.03958 & .09167 & -.025 \\
-.00833 & -.03333 & .1
\end{bmatrix} \tag{15}
$$

If we were given a right-hand-side vector for the refinery system, say the vector $\mathbf{b}' = [300, 200, 100]$, then the solution can be obtained by computing $\mathbf{x} = \mathbf{A}^{-1}\mathbf{b}'$.

$$
\mathbf{x} = \mathbf{A}^{-1}\mathbf{b}' =
\begin{bmatrix}
.05958 & -.01166 & -.015 \\
-.03958 & .09167 & -.025 \\
-.00833 & -.03333 & .1
\end{bmatrix}
\begin{bmatrix}
300 \\
200 \\
100
\end{bmatrix}
$$

$$
=
\begin{bmatrix}
.05958 \times 300 - .01166 \times 200 - .015 \times 100 \\
-.03958 \times 300 + .09167 \times 200 - .025 \times 100 \\
-.00833 \times 300 - .03333 \times 200 + .1 \times 100
\end{bmatrix}
$$

$$
=
\begin{bmatrix}
17.87 - 2.33 - 1.5 \\
-11.87 + 18.33 - 2.5 \\
-2.50 - 6.67 + 10
\end{bmatrix}
=
\begin{bmatrix}
14.0 \\
4.0 \\
.8
\end{bmatrix} \tag{16}
$$

Observe that if $\Delta \mathbf{b} = [1, 0, 0]$, the solution $\Delta \mathbf{x} = \mathbf{A}^{-1}\Delta \mathbf{b} = (\mathbf{A}^{-1})_1^C$, where $(\mathbf{A}^{-1})_1^C$ denotes the first column of $\mathbf{A}^{-1}$. If we wanted to increase production by one unit of the first product—change $\mathbf{b}$ to $\mathbf{b} + \Delta \mathbf{b}$—the solution changes from $\mathbf{x}$ to $\mathbf{A}^{-1}\mathbf{b} + \mathbf{A}^{-1}\Delta \mathbf{b} = \mathbf{x} + \Delta \mathbf{x}$. Thus the change equals $\Delta \mathbf{x}$, which is $(\mathbf{A}^{-1})_1^C$. Similarly, the second and third columns tell how the solution will change if we need 1 more unit of the second or third product. In sum, the columns of $\mathbf{A}^{-1}$ show us how the solution $\mathbf{x}$ changes when the right-side vector $\mathbf{b}$ changes.

For example, to find the solution $\mathbf{x}^*$ when we change from $\mathbf{b} = [300, 200, 100]$ to $\mathbf{b}' = [300, 300, 100]$, we take the solution $\mathbf{x} = \mathbf{A}^{-1}\mathbf{b} = [14.0, 4.0, .8]$ computed in (16) for $\mathbf{b}$ and change it by $\mathbf{A}^{-1}\Delta \mathbf{b}$, where $\Delta \mathbf{b} = \mathbf{b}' - \mathbf{b} = [0, 100, 0]$. So

$$\mathbf{x}^* = \mathbf{A}^{-1}\begin{bmatrix} 300 \\ 300 \\ 100 \end{bmatrix} = \mathbf{A}^{-1}\begin{bmatrix} 300 \\ 200 \\ 100 \end{bmatrix} + \mathbf{A}^{-1}\begin{bmatrix} 0 \\ 100 \\ 0 \end{bmatrix}$$

$$= \begin{bmatrix} 14.0 \\ 4.0 \\ .8 \end{bmatrix} + \begin{bmatrix} -1.2 \\ 9.2 \\ -3.3 \end{bmatrix} = \begin{bmatrix} 12.8 \\ 13.2 \\ -2.5 \end{bmatrix} \qquad (17)$$

■

The next two examples interpret the role of the inverse in two familiar linear models.

Example 5. **Decoding Alphabetic Messages**

In Example 1 of Section 1.5 we introduced a scheme for encoding a pair of letters L_1, L_2 as a coded pair C_1, C_2. Recall that each letter is treated as a number between 1 and 26 (e.g., BABY is the numeric sequence 2, 1, 2, 25) and arithmetic is done mod 26. We considered the following instance of this scheme:

$$\begin{aligned} C_1 &\equiv 9L_1 + 17L_2 \quad (\text{mod } 26) \\ C_2 &\equiv 7L_1 + 2L_2 \quad (\text{mod } 26) \end{aligned} \qquad (18)$$

If $L_1 = $ E ($= 5$) and $L_2 = $ C ($= 3$), this pair of letters would be encoded as the following pair C_1, C_2:

$$\begin{aligned} C_1 &\equiv 9 \times 5 + 17 \times 3 = 96 \equiv 18 \quad (\text{mod } 26) = \text{R} \\ C_2 &\equiv 7 \times 5 + 2 \times 3 = 41 \equiv 15 \quad (\text{mod } 26) = \text{O} \end{aligned}$$

In matrix form, with $\mathbf{c} = (C_1, C_2)$, $\mathbf{l} = (L_1, L_2)$, and $\mathbf{E} = \begin{bmatrix} 9 & 17 \\ 7 & 2 \end{bmatrix}$, (18) becomes

$$\mathbf{c} = \mathbf{El} \quad (\text{mod } 26)$$

The person who receives the coded pair $\mathbf{c}$ will decode $\mathbf{c}$ back into the two original message pair **l** by using the inverse of **E**:

$$\mathbf{l} = \mathbf{E}^{-1}\mathbf{c}$$

To use the formula (9) for a 2-by-2 inverse, we first compute

$$\det(\mathbf{E}) = 9 \times 2 - 7 \times 17 = -101 \equiv 3 \quad (\text{mod } 26)$$

since $-101 = -4 \cdot 26 + 3$.

Observe that $3 \times 9 = 27 \equiv 1$ (mod 26), and therefore $1/\det(\mathbf{E}) = \frac{1}{3} \equiv 9$ (mod 26) (note that division mod 26 is not always well defined, but the numbers in this case were chosen so that division would work). By (9), we have

$$\mathbf{E}^{-1} = 9 \begin{bmatrix} 2 & -17 \\ -7 & 9 \end{bmatrix} = \begin{bmatrix} 9 \times 2 & -9 \times 17 \\ -9 \times 7 & 9 \times 9 \end{bmatrix}$$

$$= \begin{bmatrix} 18 & 3 \\ 15 & 3 \end{bmatrix} \quad (\text{mod } 26)$$

So the decoding equations are

$$L_1 \equiv 18C_1 + 3C_2 \quad (\text{mod } 26) \tag{19}$$
$$L_2 \equiv 15C_1 + 3C_2 \quad (\text{mod } 26)$$

For example, the pair R, O $(= 18, 15)$ is decoded using (19) as

$$L_1 \equiv 18 \times 18 + 3 \times 15 = 324 + 45 = 369$$
$$\equiv 5 \quad (\text{mod } 26) = E$$
$$L_2 \equiv 15 \times 18 + 3 \times 15 = 270 + 45 = 315$$
$$\equiv 3 \quad (\text{mod } 26) = C$$

So R, O decode back to the original pair E, C, as required. ∎

Example 6. **Reversing a Markov Chain**

The transition matrix **A** in a Markov chain is used to compute the probability distribution **p**′ in the next period from the current probability distribution **p** according to the matrix equation

$$\mathbf{p}' = \mathbf{A}\mathbf{p} \tag{20}$$

Suppose that we want to run the Markov chain backwards—earlier in time—so that the relation in (20) becomes reversed and **p**′ is used to determine **p**. Then the new transition matrix should just be $\mathbf{A}^{-1}$, since solving (20) for **p** yields

$$\mathbf{p} = \mathbf{A}^{-1}\mathbf{p}'$$

We now try to invert the Markov chain for the frog-in-highway model introduced in Section 1.3. The transition matrix augmented with the identity matrix is

$$
\left[
\begin{array}{cccccc|cccccc}
.50 & .25 & 0 & 0 & 0 & 0 & 1 & 0 & 0 & 0 & 0 & 0 \\
.50 & .50 & .25 & 0 & 0 & 0 & 0 & 1 & 0 & 0 & 0 & 0 \\
0 & .25 & .50 & .25 & 0 & 0 & 0 & 0 & 1 & 0 & 0 & 0 \\
0 & 0 & .25 & .50 & .25 & 0 & 0 & 0 & 0 & 1 & 0 & 0 \\
0 & 0 & 0 & .25 & .50 & .50 & 0 & 0 & 0 & 0 & 1 & 0 \\
0 & 0 & 0 & 0 & .25 & .50 & 0 & 0 & 0 & 0 & 0 & 1
\end{array}
\right] \tag{21}
$$

As usual we pivot down the main diagonal on entry (1, 1), then on (2, 2), then (3, 3), then (4, 4), then (5, 5), and finally (6, 6). After the first five pivots, we have

$$
\left[
\begin{array}{cccccc|cccccc}
1 & 0 & 0 & 0 & 0 & -1 & 10 & -8 & 6 & -4 & 2 & 0 \\
0 & 1 & 0 & 0 & 0 & -2 & -16 & 16 & -12 & 8 & -4 & 0 \\
0 & 0 & 1 & 0 & 0 & -2 & 12 & -12 & 12 & -8 & 4 & 0 \\
0 & 0 & 0 & 1 & 0 & -2 & -8 & 8 & -8 & 8 & -4 & 0 \\
0 & 0 & 0 & 0 & 1 & -2 & 4 & -4 & 4 & -4 & 4 & 0 \\
0 & 0 & 0 & 0 & 0 & 0 & -1 & 1 & -1 & 1 & -1 & 1
\end{array}
\right] \tag{22}
$$

The elimination process fails because entry (6, 6) is 0. Recall that a similar difficulty arose when we were performing elimination in Example 2 of Section 3.2. The failure of the elimination process means that the transition matrix **A** is not invertible. Some Markov transition matrices are invertible, but their inverse will have negative entries and not make sense as a Markov chain—see the Exercises.

A Markov chain cannot run backward. To see why, let **p**' be a unit vector, say, $\mathbf{e}_6 = [0, 0, 0, 0, 0, 1]$. A moment's thought shows that there is no vector **p** such that $\mathbf{Ap} = \mathbf{e}_6$. That is, there is no distribution for the frog in the previous period that would force the frog, with certainty, to be in state 6 now. ■

There is a simple algebraic way to explain the computation of the inverse in Examples 3 to 6. We write the original system of equations **Ax** = **b** as

$$
\mathbf{Ax} = \mathbf{Ib} \tag{23}
$$

In Example 4 with the refinery problem, (23) is

$$
\begin{bmatrix}
20 & 4 & 4 \\
10 & 14 & 5 \\
5 & 5 & 12
\end{bmatrix}
\begin{bmatrix}
x_1 \\
x_2 \\
x_3
\end{bmatrix}
=
\begin{bmatrix}
1 & 0 & 0 \\
0 & 1 & 0 \\
0 & 0 & 1
\end{bmatrix}
\begin{bmatrix}
b_1 \\
b_2 \\
b_3
\end{bmatrix} \tag{24}
$$

Then we perform elimination by pivoting to convert **A** to **I** and get

$$\mathbf{Ix} = \mathbf{A}^{-1}\mathbf{b} \tag{25}$$

In Example 4 this is

$$\begin{bmatrix} 1 & 0 & 0 \\ 0 & 1 & 0 \\ 0 & 0 & 1 \end{bmatrix} \begin{bmatrix} x_1 \\ x_2 \\ x_3 \end{bmatrix} = \begin{bmatrix} .05958 & -.01166 & -.015 \\ -.03958 & .09167 & -.025 \\ -.00833 & -.03333 & .1 \end{bmatrix} \begin{bmatrix} b_1 \\ b_2 \\ b_3 \end{bmatrix} \tag{26}$$

Similarly trying to reverse the Markov chain, we wanted to convert $\mathbf{Ap} = \mathbf{Ip'}$ into $\mathbf{Ip} = \mathbf{A}^{-1}\mathbf{p'}$.

We next turn to some theory about inverses.

Theorem 3. Properties of the Inverse

(i) If **A** and **B** are invertible matrices, **AB** is invertible and

$$(\mathbf{AB})^{-1} = \mathbf{B}^{-1}\mathbf{A}^{-1}$$

(ii) If **A** is an invertible matrix, $\mathbf{A}^{-1}$ is invertible and

$$(\mathbf{A}^{-1})^{-1} = \mathbf{A}$$

(iii) If **A** is an invertible matrix, so is its transpose $\mathbf{A}^T$ and

$$(\mathbf{A}^T)^{-1} = (\mathbf{A}^{-1})^T$$

Proof of (i). The reasoning given here is typical of proofs involving inverses. Since the inverse of a matrix is unique—this fact is critical—we only need to check that **AB** times $\mathbf{B}^{-1}\mathbf{A}^{-1}$ is **I**.

$$(\mathbf{AB})(\mathbf{B}^{-1}\mathbf{A}^{-1}) = \mathbf{A}(\mathbf{BB}^{-1})\mathbf{A}^{-1} = \mathbf{AIA}^{-1} = \mathbf{AA}^{-1} = \mathbf{I} \quad \blacksquare$$

Next we show the links between inverses, determinants, and solutions of systems of linear equations. First note the following relation between the determinants of **A** and of $\mathbf{A}^{-1}$ (assuming that $\mathbf{A}^{-1}$ exists). The identities

$$\det(\mathbf{AB}) = \det(\mathbf{A}) \cdot \det(\mathbf{B}) \qquad \text{and} \qquad \mathbf{AA}^{-1} = \mathbf{I}$$

together imply

$$\det(\mathbf{A}) \cdot \det(\mathbf{A}^{-1}) = \det(\mathbf{AA}^{-1}) = \det(\mathbf{I}) = 1$$

Thus

$$\det(\mathbf{A}^{-1}) = \frac{1}{\det(\mathbf{A})} \qquad \text{and} \qquad \det(\mathbf{A}) = \frac{1}{\det(\mathbf{A}^{-1})} \tag{27}$$

Theorem 4. Fundamental Theorem for Solving Ax = b. The follow-ing four statements are equivalent for any *n*-by-*n* matrix **A**.

   (i)   For all **b**, the system of equations **Ax** = **b** always has a unique solution.
   (ii) The system of equations **Ax** = **Ib** can be converted, using elimi-nation by pivoting, to the system **Ix** = $\mathbf{A}^{-1}\mathbf{b}$.
   (iii) **A** has an inverse.
   (iv) $\det(\mathbf{A}) \neq 0$.

Proof
   (i) $\rightarrow$ (ii):   The elimination by pivoting in (ii) is equivalent to si-multaneously solving **Ax** = $\mathbf{e}_j$ for $j = 1, 2, \ldots, n$ and **Ax** = e_j can be solved by (i).
   (ii) $\rightarrow$ (iii): Obvious.
   (iii) $\rightarrow$ (iv): If $\mathbf{A}^{-1}$ exists, then formula (27) says that $\det(\mathbf{A}) = 1/\det(\mathbf{A}^{-1}) \neq 0$.
   (iv) $\rightarrow$ (i):   As noted in Theorem 1 of Section 3.1, when $\det(\mathbf{A}) \neq 0$, Cramer's rule gives a unique solution to **Ax** = **b**. ∎

We have the following useful corollary.

Corollary
   (i)   If for some **b**, the system of equations **Ax** = **b** has two solutions, then **A** is not invertible.
   (ii) Conversely, if **A** is not invertible, then for all **b**, **Ax** = **b** has either no solution or else multiple solutions.

We conclude this section by incorporating inverses into the eigenvalue-based analysis of growth models that was developed in Sections 2.5 and 3.1.

***Example 7.* Computer–Dog Growth
              Model Revisited**

In Section 2.5 we introduced the computer–dog growth model

$$\mathbf{x}' = \mathbf{Ax}, \qquad \text{where } \mathbf{A} = \begin{bmatrix} 3 & 1 \\ 2 & 2 \end{bmatrix} \qquad \begin{array}{l} \mathbf{x} = [C, D] \\ \mathbf{x}' = [C', D'] \end{array}$$

The two eigenvalues and associated eigenvectors of **A** were $\lambda_1 = 4$ with **u** = [1, 1] and $\lambda_2 = 1$ with **v** = [1, −2].

Our starting vector was **x** = [1, 7]. We expressed **x** as a linear combination, **x** = a**u** + b**v** of **u** and **v**. For this **x**, it is

$$\mathbf{x} = 3\mathbf{u} - 2\mathbf{v} \qquad (\text{i.e., } [1, 7] = 3[1, 1] - 2[1, -2]) \quad (28)$$

Then

$$\mathbf{Ax} = \mathbf{A}(3\mathbf{u} - 2\mathbf{v}) = 3\mathbf{Au} - 2\mathbf{Av}$$
$$= 3(4\mathbf{u}) - 2(1\mathbf{v}) \quad \text{(since } \mathbf{u}, \mathbf{v} \text{ are eigenvectors)}$$
$$= 12\mathbf{u} - 2\mathbf{v} \tag{29}$$

After k periods, we have

$$\mathbf{A}^k\mathbf{x} = \mathbf{A}^k(3\mathbf{u} - 2\mathbf{v}) = 3\mathbf{A}^k\mathbf{u} - 2\mathbf{A}^k\mathbf{v}$$
$$= 3(4^k\mathbf{u}) - 2(1^k\mathbf{v}) \tag{30}$$
$$= 3 \cdot 4^k[1, 1] - [2, -4] \qquad \blacksquare$$

We now describe in matrix notation the three basic steps in the eigenvalue-based analysis of growth models, as illustrated in Example 7.

Step 1. *Express* $\mathbf{x}$ *as a linear combination of eigenvectors,* as in (28). This step, which involves solving a system of equations, can be expressed in terms of an inverse. If $\mathbf{x} = [x_1, x_2]$, $\mathbf{u} = [u_1, u_2]$, and $\mathbf{v} = [v_1, v_2]$, the statement $\mathbf{x} = a\mathbf{u} + b\mathbf{v}$ is equivalent to

$$\begin{bmatrix} x_1 \\ x_2 \end{bmatrix} = a \begin{bmatrix} u_1 \\ u_2 \end{bmatrix} + b \begin{bmatrix} v_1 \\ v_2 \end{bmatrix}$$

or

$$\mathbf{x} = \mathbf{Uc}, \quad \text{where } \mathbf{U} = \begin{bmatrix} u_1 & v_1 \\ u_2 & v_2 \end{bmatrix}, \quad \mathbf{c} = \begin{bmatrix} a \\ b \end{bmatrix}$$

By using inverses, the solution to $\mathbf{x} = \mathbf{Uc}$ is

$$\mathbf{c} = \mathbf{U}^{-1}\mathbf{x} \tag{31}$$

Step 2. *Given* $\mathbf{c} = [a, b]$, *multiply* a *by* λ_1 *and* b *by* λ_2. We can write this step in matrix notation as follows:

$$\begin{bmatrix} a\lambda_1 \\ b\lambda_2 \end{bmatrix} = \begin{bmatrix} \lambda_1 & 0 \\ 0 & \lambda_2 \end{bmatrix} \begin{bmatrix} a \\ b \end{bmatrix} \quad \text{or} \quad \mathbf{c}' = \mathbf{D}_\lambda \mathbf{c} \tag{32}$$

where $\mathbf{D}_\lambda$ is the diagonal matrix of eigenvalues.

Step 3. *Express* $\mathbf{Ax}$ *as a linear combination of eigenvectors.*

$$\mathbf{Ax} = a\lambda_1 \begin{bmatrix} u_1 \\ u_2 \end{bmatrix} + b\lambda_2 \begin{bmatrix} v_1 \\ v_2 \end{bmatrix} = \mathbf{U} \begin{bmatrix} a\lambda_1 \\ b\lambda_2 \end{bmatrix} = \mathbf{Uc}' \tag{33}$$

where $\mathbf{U}$ is the matrix with eigenvectors as columns (see above).

Combining the three steps (31), (32), and (33), we have

$$\mathbf{Ax} = \mathbf{Uc'} = \mathbf{U}(\mathbf{D}_\lambda \mathbf{c}) = \mathbf{U}\mathbf{D}_\lambda \mathbf{U}^{-1}\mathbf{x} \tag{34}$$

This equation is true for any $\mathbf{x}$, and hence we have

$$\mathbf{A} = \mathbf{U}\mathbf{D}_\lambda \mathbf{U}^{-1} \tag{35}$$

Furthermore, as in (30), powers of $\mathbf{A}$ have a similar form (verification is left as an exercise):

$$\mathbf{A}^k = \mathbf{U}\mathbf{D}_\lambda^k \mathbf{U}^{-1} \tag{36}$$

For the computer–dog matrix, (35) becomes [we compute the inverse $\mathbf{U}^{-1}$ using the determinant-based formula (9)].

$$\mathbf{A} = \begin{bmatrix} 3 & 1 \\ 2 & 2 \end{bmatrix} = \begin{bmatrix} 1 & 1 \\ 1 & -2 \end{bmatrix} \begin{bmatrix} 4 & 0 \\ 0 & 1 \end{bmatrix} \begin{bmatrix} \frac{2}{3} & \frac{1}{3} \\ \frac{1}{3} & -\frac{1}{3} \end{bmatrix} \tag{37}$$

and (36) becomes

$$\mathbf{A}^k = \begin{bmatrix} 1 & 1 \\ 1 & -2 \end{bmatrix} \begin{bmatrix} 4^k & 0 \\ 0 & 1^k \end{bmatrix} \begin{bmatrix} \frac{2}{3} & \frac{1}{3} \\ \frac{1}{3} & -\frac{1}{3} \end{bmatrix} \tag{38}$$

Equations (35) and (36) formalize in single matrix equations our eigenvector-based computations for a growth model. They also give us a simple way to compute powers of any matrix $\mathbf{A}$—provided that we know the eigenvalues and eigenvectors of $\mathbf{A}$.

Theorem 5. If $\mathbf{A}$ is an n-by-n matrix with n distinct eigenvectors $\mathbf{u}_1, \mathbf{u}_2, \ldots, \mathbf{u}_n$ and associated eigenvalues $|\lambda_1| \geq |\lambda_2| \geq \cdots \geq |\lambda_n|$, then

$$\mathbf{A} = \mathbf{U}\mathbf{D}_\lambda \mathbf{U}^{-1} \qquad \text{and} \qquad \mathbf{A}^k = \mathbf{U}\mathbf{D}_\lambda^k \mathbf{U}^{-1} \tag{39}$$

where $\mathbf{U}$ is an n-by-n matrix whose jth column is $\mathbf{u}_j$.

Stating (39) in words,

multiplication by $\mathbf{A}$

is equivalent to:

(i) converting to eigenvector coordinates—multiplying by $\mathbf{U}^{-1}$ does this; then

(ii) multiplying those coordinates by the eigenvalues—multiplying by $\mathbf{D}_\lambda$ does this; and finally

(iii) converting back to standard coordinates—multiplying by $\mathbf{U}$ does this.

The formula $\mathbf{A} = \mathbf{U}\mathbf{D}_\lambda\mathbf{U}^{-1}$ for computing $\mathbf{Ax}$ can be visualized with the following diagram:

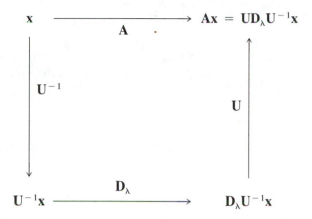

This eigenvalue decomposition of a matrix, often called diagonalization of $\mathbf{A}$, is extremely important. Beside simplifying the computation of powers of a matrix, it can also be applied to other functions of a matrix $\mathbf{A}$. In differential equations (Section 4.3), the expression $e^{\mathbf{A}t}$ arises frequently and can be evaluated with the help of (39). This decomposition will be discussed further in Section 5.5, where we also present methods to find all the eigenvalues of a matrix.

We note that like the **LU** decomposition, the eigenvalue decomposition is an example of using proper representation of the matrix and an example of a matrix product being a "program" (see the end of Section 3.2). If we want to solve systems such as $\mathbf{Ax} = \mathbf{b}$, the **LU** decomposition is the right way to "store" $\mathbf{A}$. If we want to raise $\mathbf{A}$ to various powers, the eigenvalue decomposition is the appropriate way to "store" $\mathbf{A}$. Computing $\mathbf{A}^k\mathbf{x}$ as $(\mathbf{U}\mathbf{D}_\lambda^k\mathbf{U}^{-1})\mathbf{x}$ is a case where the matrix product $\mathbf{U}\mathbf{D}_\lambda^k\mathbf{U}^{-1}$ is a "program" telling us how to compute $\mathbf{A}^k\mathbf{x}$ through a sequence of three distinct steps.

We now apply Theorem 3 to our rabbit–fox model to rework the computations done at the end of Section 3.1.

Example 8. Rabbit–Fox Growth Model Revisited

The growth model we have used repeatedly is

$$R' = R + .1R - .15F \qquad \text{or} \qquad \mathbf{p}' = \mathbf{Ap}$$

$$F' = F + .1R - .15F \qquad \text{where } \mathbf{A} = \begin{bmatrix} 1.1 & -.15 \\ .1 & .85 \end{bmatrix}$$

$$(40)$$

In Section 3.1 we found the eigenvalues and eigenvectors to be

$$\lambda_1 = 1, \quad \mathbf{u}_1 = [3, \quad 2] \qquad \lambda_2 = .95, \quad \mathbf{u}_2 = [1, \quad 1]$$

So

$$\mathbf{U} = \begin{bmatrix} 3 & 1 \\ 2 & 1 \end{bmatrix}_s, \quad \mathbf{D}_\lambda = \begin{bmatrix} 1 & 0 \\ 0 & .95 \end{bmatrix}, \quad \mathbf{U}^{-1} = \begin{bmatrix} 1 & -1 \\ -2 & 3 \end{bmatrix} \tag{41}$$

where again the 2-by-2 inverse formula (9) is used for $\mathbf{U}^{-1}$.

Suppose that our starting vector is $\mathbf{p} = [50, 40]$ and we want to determine the population vector after 20 periods. Then we must compute

$$\mathbf{p}^{(20)} = \mathbf{A}^{20}\mathbf{p} = (\mathbf{U}\mathbf{D}_\lambda^{20}\mathbf{U}^{-1})\mathbf{p}, \quad \text{where } \mathbf{D}_\lambda^{20} = \begin{bmatrix} 1 & 0 \\ 0 & .3585 \end{bmatrix} \tag{42}$$

We evaluate (42) from right to left. That is, first compute

$$\mathbf{c} = \mathbf{U}^{-1}\mathbf{p} = \begin{bmatrix} 1 & -1 \\ -2 & 3 \end{bmatrix} \begin{bmatrix} 50 \\ 40 \end{bmatrix} = \begin{bmatrix} 10 \\ 20 \end{bmatrix} \tag{43}$$

Recall from Example 7 that this vector $\mathbf{c}$ is the set of weights a, b on the eigenvectors. Next we multiply this vector of weights times $\mathbf{D}_\lambda^{20}$:

$$\mathbf{c}' = \mathbf{D}_\lambda^{20}\mathbf{c} = \begin{bmatrix} 1 & 0 \\ 0 & .3585 \end{bmatrix} \begin{bmatrix} 10 \\ 20 \end{bmatrix} = \begin{bmatrix} 10 \\ 7.17 \end{bmatrix} \tag{44}$$

Finally, we use $\mathbf{c}'$ to form a linear combination of the eigenvectors:

$$\mathbf{p}^{(20)} = \mathbf{A}^{20}\mathbf{p} = \mathbf{U}\mathbf{c}' = \begin{bmatrix} 3 & 1 \\ 2 & 1 \end{bmatrix} \begin{bmatrix} 10 \\ 7.17 \end{bmatrix}$$

$$= 10 \begin{bmatrix} 3 \\ 2 \end{bmatrix} + 7.17 \begin{bmatrix} 1 \\ 1 \end{bmatrix} \approx \begin{bmatrix} 37 \\ 7 \end{bmatrix} \quad \blacksquare$$

Section 3.3 Exercises

Summary of Exercises
Exercises 1–17 involve computation of inverses and interpretation of entries in an inverse; Exercise 4 gives an important geometric picture of inverses. Exercises 18–27 examine properties of inverses. Exercises 28 and 29 deal with the existence of solutions and applications of Theorem 4. Exercises 30–37 deal with eigenvalues and Theorem 5.

1. Verify for the matrix $\mathbf{A}$ in Example 1, part (i) that $\mathbf{A}^{-1}\mathbf{A} = \mathbf{I}$.

2. Write the system of equations that entries in the inverse of the following matrices must satisfy. Then find inverses (as in Example 2) or show that none can exist (following the reasoning in Example 1).

(a) $\begin{bmatrix} 1 & 1 \\ 1 & 0 \end{bmatrix}$ (b) $\begin{bmatrix} 0 & 1 \\ 1 & 0 \end{bmatrix}$ (c) $\begin{bmatrix} 2 & 1 \\ 7 & 4 \end{bmatrix}$ (d) $\begin{bmatrix} -1 & 3 \\ 2 & -6 \end{bmatrix}$

(e) $\begin{bmatrix} 1 & 2 & 1 \\ 2 & 4 & 2 \\ 2 & 5 & 1 \end{bmatrix}$ (f) $\begin{bmatrix} 1 & 1 & 0 \\ 0 & 1 & 1 \\ 1 & 2 & 1 \end{bmatrix}$ (g) $\begin{bmatrix} 3 & 2 & 1 \\ 1 & 1 & 1 \\ 7 & 6 & 5 \end{bmatrix}$

3. **(a)** Write out the system of equations that the *first column* of the inverse of **A** must satisfy, where

$$A = \begin{bmatrix} 1 & 0 & 2 \\ 0 & 1 & 3 \\ 1 & 0 & 4 \end{bmatrix}$$

(b) Determine the first column of A^{-1}; use part (a) and Cramer's rule.

4. This exercise gives a "picture" of how when two columns of **A** are almost the same, the inverse of **A** almost does not exist. For the following matrices **A**, solve the system $A \begin{bmatrix} x_1 \\ x_2 \end{bmatrix} = \begin{bmatrix} 1 \\ 0 \end{bmatrix}$. Then plot $x_1 a_1^C$ and $x_2 a_2^C$ in a two-dimensional coordinate system and show geometrically how the sum of vectors $x_1 a_1^C$ and $x_2 a_2^C$ is $\begin{bmatrix} 1 \\ 0 \end{bmatrix}$ (here a_1^C, a_2^C denote the two columns of **A**).

(a) $\begin{bmatrix} 2 & 3 \\ 1 & 2 \end{bmatrix}$ (b) $\begin{bmatrix} 2 & 3 \\ 2 & 2 \end{bmatrix}$ (c) $\begin{bmatrix} 8 & 10 \\ 7 & 7 \end{bmatrix}$ (d) $\begin{bmatrix} 8 & 9 \\ 7 & 7 \end{bmatrix}$

5. **(a)** Find the inverse of the transition matrix **A** for the weather Markov chain (introduced in Example 1 of Section 1.3), where

$$A = \begin{bmatrix} \frac{3}{4} & \frac{1}{2} \\ \frac{1}{4} & \frac{1}{2} \end{bmatrix}.$$

(b) If **p** is today's weather probability distribution and p^0 is yesterday's distribution, show that $p^0 = A^{-1}p$.

(c) Find yesterday's weather probability distribution if today's weather probability distribution is
(i) $p = [\frac{1}{2}, \frac{1}{2}]$ (ii) $[\frac{2}{3}, \frac{1}{3}]$ (iii) $[0, 1]$

(d) Use a computer program to determine the weather probability distribution 20 days ago for the current distributions in part (c).

6. **(a)** Find the system of equations for decoding the following encoding schemes.

(i) $C_1 \equiv 3L_1 + 5L_2$ (ii) $C_1 \equiv 11L_1 + 6L_2$
$C_2 \equiv 5L_1 + 8L_2$ $C_2 \equiv 8L_1 + 5L_2$

(iii) $C_1 \equiv 2L_1 + 3L_2$
$C_2 \equiv 7L_1 + 5L_2$

Hint: The inverse of 7 is -11; the inverse of -11 is 7.

(b) Decode the coded pair EF in each of these schemes.

7. Use elimination by pivoting to find the inverse of the following matrices.

(a) $\begin{bmatrix} 2 & -3 & 2 \\ 1 & -1 & 1 \\ -1 & 5 & 4 \end{bmatrix}$ (b) $\begin{bmatrix} -1 & -1 & 1 \\ 2 & 2 & -4 \\ 1 & -2 & 3 \end{bmatrix}$

(c) $\begin{bmatrix} -1 & -3 & 2 \\ 2 & 1 & 3 \\ 5 & 4 & 6 \end{bmatrix}$ (d) $\begin{bmatrix} 2 & 4 & -2 \\ 1 & -2 & -4 \\ -2 & -1 & -3 \end{bmatrix}$

(e) $\begin{bmatrix} 1 & 1 & 4 \\ 2 & 1 & 3 \\ 5 & 2 & 5 \end{bmatrix}$ (f) $\begin{bmatrix} 2 & -3 & -1 \\ 3 & -5 & -2 \\ 9 & 6 & 4 \end{bmatrix}$

8. For each matrix A in Exercise 7, solve $Ax = b$, where $b = [10, 10, 10]$.

9. For each matrix A in Exercise 7, how much will the solution of $Ax = b$ change if b is changed
   (a) From the vector $[b_1, b_2, b_3]$ to the vector $[b_1, b_2 + 1, b_3]$?
   (b) From the vector $[b_1, b_2, b_3]$ to the vector $[b_1, b_2, b_3 - 2]$?
   (c) From the vector $[b_1, b_2, b_3]$ to the vector $[b_1, b_2 + 1, b_3 - 1]$?

10. Use elimination by pivoting to find the inverse of the following matrices.

(a) $\begin{bmatrix} 1 & 1 & -1 & -1 \\ 2 & 0 & 0 & 1 \\ 3 & 0 & 0 & -2 \\ 4 & -2 & 1 & 3 \end{bmatrix}$ (b) $\begin{bmatrix} 3 & 2 & 1 & 0 \\ 1 & 1 & 0 & -1 \\ 2 & 1 & -1 & 1 \\ 1 & 1 & 1 & 1 \end{bmatrix}$

(c) $\begin{bmatrix} 1 & 3 & 2 & -1 \\ 1 & 1 & 1 & 1 \\ 2 & -2 & 1 & -1 \\ 1 & -3 & -1 & 2 \end{bmatrix}$

11. **(a)** Find the inverse of the tridiagonal matrix

$$\begin{bmatrix} 1 & -1 & 0 & 0 & 0 \\ -1 & 2 & -1 & 0 & 0 \\ 0 & -1 & 2 & -1 & 0 \\ 0 & 0 & -1 & 2 & -1 \\ 0 & 0 & 0 & -1 & 2 \end{bmatrix}$$

Note that the inverse is not tridiagonal or in any way sparse.

(b) Change entry (1, 1) from a 1 to a 2 and repeat part (a). Does this small change affect the inverse substantially?

12. Reverse the following Markov chains. Then find the "distribution" in the previous period if the current distribution is [.5 0 .5]. Is this distribution really a probability distribution?

(a) $\begin{bmatrix} .5 & 0 & 0 \\ .5 & 1 & .5 \\ 0 & 0 & .5 \end{bmatrix}$
(b) $\begin{bmatrix} .5 & .25 & 0 \\ .5 & .5 & .5 \\ 0 & .25 & .5 \end{bmatrix}$
(c) $\begin{bmatrix} .4 & .3 & .3 \\ .3 & .4 & .3 \\ .3 & .3 & .4 \end{bmatrix}$

13. Try to find the inverse of the frogger Markov chain when there are five, not six states (three lanes of highway).

14. (a) Describe those n-by-n Markov transition matrices $A*$ (for each n) that have an inverse $A*^{-1}$ such that if $\mathbf{p}$ is a probability distribution, then $\mathbf{p}^\circ = A*^{-1}\mathbf{p}$ is always a probability distribution.
(b) Give an informal argument why no other such reversible Markov chain can exist (the reasoning is similar to that used in Example 6).

15. (Continuation of Exercise 17 in Section 3.2) The staff dietician at the California Institute of Trigonometry has to make up a meal with 600 calories, 20 grams of protein, and 200 milligrams of vitamin C. There are three food types to choose from: rubbery jello, dried fish sticks, and mystery meat. They have the following nutritional content per ounce.

	Jello	Fish Sticks	Mystery Meat
Calories	10	50	200
Protein	1	3	.2
Vitamin C	30	10	0

(a) Find the inverse of this data matrix and use it to compute the amount of jello, fish sticks, and mystery meat required.
(b) If the protein requirement is increased by 4, how will this change the number of units of jello in the meal?
(c) If the vitamin C requirement is decreased by k milligrams, how much will this change the number of fish sticks in a meal?

16. (Continuation of Exercise 18 in Section 3.2) A furniture manufacturer makes tables, chairs, and sofas. In one month, the company has available 300 units of wood, 350 units of labor, and 225 units of upholstery. The manufacturer wants a production schedule for the month that uses all of these resources. The different products require the following amounts of the resources.

	Table	Chair	Sofa
Wood	4	1	3
Labor	3	2	5
Upholstery	2	0	4

(a) Find the inverse of this data matrix and use it to determine how much of each product should be manufactured.

(b) If the amount of wood is increased by 30 units, how will this change the number of sofas produced?

(c) If the amount of labor is decreased by k, how much will this change your answer in part (a)?

17. (Continuation of Exercise 20 of Section 3.2) An investment analyst is trying to find out how much business a secretive TV manufacturer has. The company makes three brands of TV set: brand A, brand B, and brand C. The analyst learns that the manufacturer has ordered from suppliers 450,000 type 1 circuit boards, 300,000 type 2 circuit boards, and 350,000 type 3 circuit boards. Brand A uses 2 type-1 boards, 1 type-2 board, and 2 type-3 boards. Brand B uses 3 type-1 boards, 2 type-2 boards, and 1 type-3 board. Brand C uses 1 board of each type.

(a) Set up this problem as a system $\mathbf{Ax} = \mathbf{b}$. Find the inverse of $\mathbf{A}$ and use it to determine how many TV sets of each brand are being manufactured.

(b) If the number of type 2 boards used is increased by 100,000, how will this change your answer in part (a)?

(c) If the number of type 1 boards is decreased by $10,000k$, how much will this change your answer in part (a)?

18. Why must a matrix be square if it has an inverse?

19. Verify that for any invertible matrix $\mathbf{A}$, the inverse of the inverse $\mathbf{A}^{-1}$ is $\mathbf{A}$.

20. Verify that the inverse of $\mathbf{A}^T$ is $(\mathbf{A}^{-1})^T$.

    Hint: Use the multiplication rule for tranposes, $(\mathbf{CD})^T = \mathbf{D}^T\mathbf{C}^T$.

21. Show that the inverse of a matrix is unique.

    Hint: If $\mathbf{B}$ and $\mathbf{C}$ are inverses of the matrix $\mathbf{A}$, compute $\mathbf{BAC}$ two different ways as $(\mathbf{BA})\mathbf{C}$ and as $\mathbf{B}(\mathbf{AC})$.

22. Show, by the reasoning in Example 1, that if a matrix has a row (or column) that is all 0's, then the matrix cannot have an inverse.

23. (a) Following the reasoning in Example 1, show that

$$\begin{bmatrix} 1 & 2 & 3 \\ 1 & 4 & 5 \\ 1 & 6 & 7 \end{bmatrix}$$

cannot have an inverse because the third column is the sum of the other two columns.

(b) Generalize the argument in part (a) to show that if one row (column) is a linear combination of two others, $\mathbf{a}_i^R = c\mathbf{a}_k^R + d\mathbf{a}_h^R$, then the matrix cannot have an inverse.

24. In Theorem 1, show that the solution $\mathbf{x} = \mathbf{A}^{-1}\mathbf{b}$ is unique, that is, there cannot exist a different vector $\mathbf{x}'$ with $\mathbf{Ax}' = \mathbf{b}$.

Hint: Multiply both sides of $\mathbf{Ax}' = \mathbf{b}$ by $\mathbf{A}^{-1}$.

25. Find the inverse of a diagonal matrix

$$\begin{bmatrix} a_{11} & 0 & 0 \\ 0 & a_{22} & 0 \\ 0 & 0 & a_{33} \end{bmatrix}$$

Hint: The inverse is also diagonal.

26. (a) Use the following fact: The inverse of an upper triangular matrix (if the inverse exists) is itself upper triangular, to determine what the main diagonal entries must be in the inverse of the upper triangular matrix

$$\begin{bmatrix} 2 & 3 & 4 \\ 0 & 4 & 2 \\ 0 & 0 & 5 \end{bmatrix}$$

(Do not use elimination by pivoting.)

(b) Use the main-diagonal entries in the inverse from part (a) and the fact that the inverse is upper triangular. Determine the other entries in the inverse.

(c) Consider how the computations to find the inverse in elimination by pivoting would go to show that the inverse of an upper triangular matrix must be upper triangular.

27. (a) Determine the inverse of the matrix

$$\begin{bmatrix} 1 & 0 & 0 & 0 \\ 0 & 1 & 0 & 0 \\ a & 0 & 1 & 0 \\ 0 & 0 & 0 & 1 \end{bmatrix}$$

Hint: The inverse has a simple form; try trial-and-error guesswork.
   (b) Generalize your result in part (a) to give the inverse of an *n*-by-*n* matrix with 1's on the main diagonal and 0's elsewhere except one position, entry (i, j), $i \neq j$, whose value is *a*.

28. Which of the following conditions guarantees that the system of equations $\mathbf{Ax} = \mathbf{b}$ has a unique solution; which guarantees that the system does not have a solution or that it is not unique, or guarantees nothing? Explain the reason for your answer. Assume that $\mathbf{A}$ is a square matrix.
   (a) $\mathbf{A}$ has an inverse.
   (b) $\det(\mathbf{A}) = 0$.
   (c) $\mathbf{Ax} = \mathbf{b}'$ has a unique solution for some other $\mathbf{b}'$.
   (d) $\mathbf{Ax} = \mathbf{b}'$ has two solutions for some other $\mathbf{b}'$.
   (e) $\mathbf{b}$ equals a column of $\mathbf{A}$.
   (f) The first row of $\mathbf{A}$ is twice the last row of $\mathbf{A}$.

29. Which of the following conditions guarantees that a matrix $\mathbf{A}$ has an inverse; which guarantees that it does not have an inverse? Explain the reason for your answer briefly.
   (a) The determinant of the matrix equals 17.
   (b) $\mathbf{A}$ has twice as many rows as columns.
   (c) $\mathbf{A}$ is a 4-by-4 Markov chain matrix.
   (d) The first row of $\mathbf{A}$ is twice the last row.
   (e) The system of equations $\mathbf{Ax} = \mathbf{b}$ can be solved for any $\mathbf{b}$.

30. The matrix $\mathbf{B} = \begin{bmatrix} \frac{2}{3} & -\frac{1}{3} \\ -\frac{1}{3} & \frac{2}{3} \end{bmatrix}$ is the inverse of $\mathbf{A} = \begin{bmatrix} 2 & 1 \\ 1 & 2 \end{bmatrix}$.
   (a) Verify that $\mathbf{u}_1 = [1, 1]$ and $\mathbf{u}_2 = [1, -1]$ are eigenvectors of both $\mathbf{A}$ and $\mathbf{B}$.
   (b) Determine the eigenvalues of $\mathbf{A}$ and $\mathbf{B}$. How are the eigenvalues of $\mathbf{A}$ and $\mathbf{B}$ related?

31. Show that the eigenvectors of $\mathbf{A}^{-1}$ must be the same as the eigenvectors of $\mathbf{A}$.

   Hint: Use the fact that $\mathbf{A}^{-1}(\mathbf{Au}) = \mathbf{u}$.

32. Assuming that $\mathbf{A}$ and $\mathbf{A}^{-1}$ have the same eigenvectors, show that the eigenvalues of $\mathbf{A}^{-1}$ must be the reciprocals of the eigenvalues of $\mathbf{A}$ (i.e., $1/\lambda$).

   Hint: Use the fact that $\mathbf{A}^{-1}(\mathbf{Au}) = \mathbf{u}$.

33. Compute the representation $\mathbf{UD}_\lambda\mathbf{U}^{-1}$ of Theorem 5 for the following matrices whose eigenvalues and largest eigenvector you were asked to determine in Exercise 23 of Section 3.1.

   (a) $\begin{bmatrix} 4 & 0 \\ 2 & 2 \end{bmatrix}$ **(b)** $\begin{bmatrix} 1 & 2 \\ 3 & 4 \end{bmatrix}$ **(c)** $\begin{bmatrix} 2 & 1 \\ 2 & 3 \end{bmatrix}$ **(d)** $\begin{bmatrix} 4 & -1 \\ 1 & 2 \end{bmatrix}$

34. For a starting vector of $\mathbf{p} = [10, 10]$, compute $p^{(10)} = \mathbf{A}^{10}\mathbf{p}$ for each matrix $\mathbf{A}$ in Exercise 33 (use your representation of $\mathbf{A}$ found in Exercise 33).

35. **(a)** Given that $\mathbf{A} = \mathbf{UD}_\lambda\mathbf{U}^{-1}$, prove that $\mathbf{A}^2 = \mathbf{UD}_\lambda^2\mathbf{U}^{-1}$.
   (b) Use induction to prove $\mathbf{A}^k = \mathbf{UD}_\lambda^k\mathbf{U}^{-1}$.

36. **(a)** Obtain a formula for $\mathbf{A}^{-1}$ similar to $\mathbf{A} = \mathbf{UD}_\lambda\mathbf{U}^{-1}$.
      Hint: Only the matrix $\mathbf{D}_\lambda$ will be different.
   (b) Verify your formula in part (a) for $\mathbf{A} = \begin{bmatrix} 3 & 1 \\ 2 & 2 \end{bmatrix}$.

37. Show that $\mathbf{A}$ is not invertible if 0 is an eigenvalue.

Section 3.4 Solving Matrix Problems by Iteration

In this section we show how simple iteration methods can be used first to determine eigenvalues and eigenvectors, and then to solve systems of linear equations. We want to use an iterative method to find the largest eigenvalue (in absolute value) of a matrix $\mathbf{A}$ and an associated eigenvector. The largest eigenvalue is the largest root (in absolute value) of the characteristic polynomial $\det(\mathbf{A}-2\mathbf{I})$. However, it is difficult to find the roots of polynomials beyond quadratics. Iterative methods are easier to use and yield both the largest eigenvalue and an associated eigenvector.

In Section 2.5 we saw how the largest eigenvalue and its associated eigenvector dominate the long-term behavior of a linear growth model. We briefly review the results we obtained for the computer–dog growth model.

Example 1. Review of Role of Largest Eigenvalues in a Growth Model

The growth model for computers (C) and dogs (D) from year to year was

$$\begin{aligned} C' &= 3C + D \\ D' &= 2C + 2D \end{aligned} \tag{1}$$

or

$$\mathbf{x}' = \mathbf{Ax}, \qquad \text{where } \mathbf{A} = \begin{bmatrix} 3 & 1 \\ 2 & 2 \end{bmatrix}$$

We saw in Section 2.5 that 4 is the larger eigenvalue with eigenvector **u** = [1, 1] (or any multiple of [1, 1]), and 1 is the other eigenvalue with eigenvector **v** = [1, −2].

We can write any vector **x** as a linear combination of **u** and **v**. (see Section 2.5 for details on how to do this). If the initial vector is **x** = [1, 7], we find that **x** = 3**u** − 2**v** (in other words, [1, 7] = 3[1, 1] − 2[1, −2]). If we want to iterate this model for 20 years, we can use **u** and **v** to compute $\mathbf{A}^{20}\mathbf{x}$ as follows:

$$\mathbf{A}^{20}\mathbf{x} = \mathbf{A}^{20}(3\mathbf{u} - 2\mathbf{v}) = 3\mathbf{A}^{20}\mathbf{u} - 2\mathbf{A}^{20}\mathbf{v}$$
$$= 3(4^{20}\mathbf{u}) - 2(1^{20}\mathbf{v}) \tag{2}$$
$$= [3 \cdot 4^{20}, 3 \cdot 4^{20}] - [2, -4]$$

The term $[3 \cdot 4^{20}, 3 \cdot 4^{20}]$ in (2) swamps $[2, -4]$. So in general, after n periods, we have

$$\mathbf{A}^{n}\mathbf{x} \simeq \mathbf{A}^{n}(3\mathbf{u}) = 3 \cdot 4^{n}\mathbf{u}$$

or

$$\mathbf{A}^{n}\mathbf{x} \simeq [3 \cdot 4^{n}, 3 \cdot 4^{n}] \qquad\qquad \blacksquare$$

Since $\mathbf{A}^{n}\mathbf{x} \simeq [3 \cdot 4^{n}, 3 \cdot 4^{n}]$, $\mathbf{A}^{n}\mathbf{x}$ is approximately a multiple of the eigenvector [1, 1]. This means that we can reverse the previous reasoning and find an eigenvector associated with the larger eigenvalue simply by iterating a growth model for many periods.

Example 2. Determining Largest Eigenvalue and Its Eigenvector by Iteration

In the computer–dog model of Example 1, let us iterate starting with $\mathbf{x}^{(0)} = [1, 7]$. We keep track of the growth rate from one period to the next (the sum norm is used).

$$\mathbf{x}^{(0)} = [1, 7]$$

$$\mathbf{x}^{(1)} = [10, 16] \qquad \frac{|\mathbf{x}^{(1)}|}{|\mathbf{x}^{(0)}|} = 3.25$$

$$\mathbf{x}^{(2)} = [46, 52] \qquad \frac{|\mathbf{x}^{(2)}|}{|\mathbf{x}^{(1)}|} = 3.77$$

$$\mathbf{x}^{(3)} = [190, 196] \qquad \frac{|\mathbf{x}^{(3)}|}{|\mathbf{x}^{(2)}|} = 3.94$$

$$\mathbf{x}^{(4)} = [766, 772] \qquad \frac{|\mathbf{x}^{(4)}|}{|\mathbf{x}^{(3)}|} = 3.98$$

$$\mathbf{x}^{(5)} = [3070, 3076] \qquad \frac{|\mathbf{x}^{(5)}|}{|\mathbf{x}^{(4)}|} = 3.996$$

$$\mathbf{x}^{(6)} = [12{,}286, 12{,}292] \qquad \frac{|\mathbf{x}^{(6)}|}{|\mathbf{x}^{(5)}|} = 3.999$$

$$\mathbf{x}^{(7)} = [49{,}150, 49{,}156] \qquad \frac{|\mathbf{x}^{(7)}|}{|\mathbf{x}^{(6)}|} = 3.9998$$

From the ratios of the norms of the successive iterates, it is clear that the growth rate is converging to 4. The two components of the successive iterates are approximately equal; that is, they are multiples of [1, 1]. So we conclude that the largest eigenvalue is 4 and its eigenvectors are multiples of [1, 1]. ∎

When the sizes of the iterates $\mathbf{x}^{(k)}$ get large, one can scale them back by dividing their entries by the largest entry (so that their max norm is 1). For example, $\mathbf{x}^{(6)} = [12{,}286, 12{,}292]$ would be scaled by dividing by the larger entry, 12,292, to obtain $\mathbf{x}^{*(6)} = [.9995, 1]$.

It is common practice to call the largest eigenvalue (in absolute value) the **dominant** eigenvalue because of the way it dominates the behavior of a growth model. Summarizing our method, we have

---

Finding Dominant Eigenvalue and Associated Eigenvector of $\mathbf{x}' = \mathbf{A}\mathbf{x}$ ***by Iteration.*** For any starting vector $\mathbf{x}^{(0)}$, compute successive iterates $\mathbf{x}^{(k)}$ until the ratio $|\mathbf{x}^{(k)}|/|\mathbf{x}^{(k-1)}|$ converges to a fixed value. This value is the (absolute value) of the dominant eigenvalue and $\mathbf{x}^{(k)}$ is an associated eigenvector. If $\mathbf{x}^{(k)}$ becomes too large, "scale" it by dividing $\mathbf{x}^{(k)}$ by its largest entry.

Note that this iterative method was exactly how we found the stable probability distribution $\mathbf{p}^* = [.1, .2, .2, .2, .2, .1]$ for the frog Markov chain (implicitly, the largest eigenvalue was 1).

A geometrical illustration of the convergence of $\mathbf{A}^k\mathbf{x}$ to an eigenvector corresponding to the dominant eigenvalue is given in Figure 3.1. Using the matrix $\mathbf{A} = \begin{bmatrix} 3 & 1 \\ 2 & 2 \end{bmatrix}$ in the computer–dog growth model, we plot what happens to the *set* of vectors $\mathbf{x}$ in the first and third quadrants of the Cartesian plane when we iterate with $\mathbf{A}$: $\mathbf{x}$, $\mathbf{A}\mathbf{x}$, $\mathbf{A}^2\mathbf{x}$, $\mathbf{A}^3\mathbf{x}$. Figure 3.2 follows a particular vector $\mathbf{x}_0$ through such iteration.

We next try our iterative scheme on a 3-by-3 matrix. Note that for 3-by-3 matrices, there is no simple formula available for finding eigenvalues as roots of the characteristic equation $\det(\mathbf{A} - \lambda\mathbf{I})$.

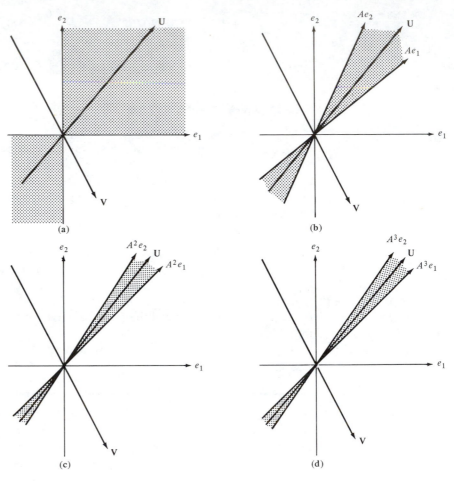

(a)

(b)

(c)

(d)

Figure 3.1

Figure 3.2

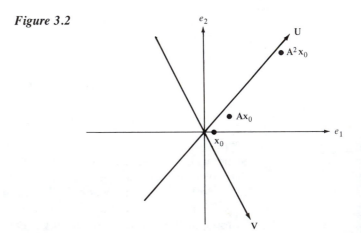

Example 3. **Dominant Eigenvalue in a 3-by-3 Matrix**

Let us expand our computer–dog growth model to include goats.

$$
\begin{aligned}
C' &= 3C + D + G \\
D' &= 2C + 2D \\
G' &= \quad\ \ + D + 2G
\end{aligned}
$$

We start iterating with $\mathbf{x}^{(0)} = [1, 1, 1]$.

$$\mathbf{x}^{(1)} = [5, 4, 3] \qquad \frac{\left|\mathbf{x}^{(1)}\right|_s}{\left|\mathbf{x}^{(0)}\right|_s} = 4$$

$$\mathbf{x}^{(2)} = [22, 18, 10] \qquad \frac{\left|\mathbf{x}^{(2)}\right|_s}{\left|\mathbf{x}^{(1)}\right|_s} = 4.17$$

$$\mathbf{x}^{(3)} = [94, 80, 38] \qquad \frac{\left|\mathbf{x}^{(3)}\right|_s}{\left|\mathbf{x}^{(2)}\right|_s} = 4.24$$

$$\mathbf{x}^{(4)} = [400, 348, 156] \qquad \frac{\left|\mathbf{x}^{(4)}\right|_s}{\left|\mathbf{x}^{(3)}\right|_s} = 4.26$$

Let us scale $\mathbf{x}^{(4)} = [400, 348, 156]$ by dividing by its largest entry:

$$\mathbf{x}^{*(4)} = [1.00, .87, .39]$$

$$\mathbf{x}^{(5)} = [4.26, 3.74, 1.65] \qquad \frac{\left|\mathbf{x}^{(5)}\right|_s}{\left|\mathbf{x}^{*(4)}\right|_s} = 4.26$$

$$\mathbf{x}^{(6)} = [18.17, 16.00, 7.04] \qquad \frac{\left|\mathbf{x}^{(6)}\right|_s}{\left|\mathbf{x}^{(5)}\right|_s} = 4.27$$

$$\mathbf{x}^{(7)} = [77.55, 68.34, 30.08] \qquad \frac{\left|\mathbf{x}^{(7)}\right|_s}{\left|\mathbf{x}^{(6)}\right|_s} = 4.27$$

So the dominant eigenvalue is 4.27. The scaled form of $\mathbf{x}^{(7)}$ is (rounded to the nearest hundred) $[1, .88, .38]$. ∎

Instead of taking the ratio of the sum norms of successive iterates to approximate the dominant eigenvalue λ^*, it is more accurate to use the following ratio, called the **Raleigh quotient.**

$$\lambda^* = \frac{\mathbf{x}^{(k)} \cdot \mathbf{x}^{(k+1)}}{\mathbf{x}^{(k)} \cdot \mathbf{x}^{(k)}} \qquad \text{or} \qquad = \frac{\mathbf{x}^{(k)} \cdot \mathbf{A}\mathbf{x}^{(k)}}{\mathbf{x}^{(k)} \cdot \mathbf{x}^{(k)}} \tag{3}$$

Applying the Raleigh quotient with just $k = 2$ in Example 3, we get

$$\lambda^* = \frac{\mathbf{x}^{(2)} \cdot \mathbf{x}^{(3)}}{\mathbf{x}^{(2)} \cdot \mathbf{x}^{(2)}} = \frac{3888}{908} = 4.282$$

Solving a System of Equations by Iteration

Our initial discussion will center around the Leontief economic model presented in Section 1.2. This model balanced supplies of a set of commodities against the demand for the commodities, demand by industry (as input for supplies produced), and demand by consumers. The sample model we introduced in Section 1.2 was

Supplies	Industrial Demands	Consumer Demand
Energy: $x_1 = .4x_1 + .2x_2 + .2x_3 + .2x_4 +$		100
Construct.: $x_2 = .3x_1 + .3x_2 + .2x_3 + .1x_4 +$		50
Transport.: $x_3 = .1x_1 + .1x_2 + \quad\quad + .2x_4 +$		100
Steel: $x_4 = \quad\quad + .1x_2 + .1x_3$		

$$(4)$$

or in matrix notation

$$\mathbf{x} = \mathbf{Dx} + \mathbf{c}$$

where

$$\mathbf{D} = \begin{bmatrix} .4 & .2 & .2 & .2 \\ .3 & .3 & .2 & .1 \\ .1 & .1 & 0 & .2 \\ 0 & .1 & .1 & 0 \end{bmatrix} \quad \text{and} \quad \mathbf{c} = \begin{bmatrix} 100 \\ 50 \\ 100 \\ 0 \end{bmatrix}$$

Recall that there was an input constraint that the sum of the coefficients in each column of $\mathbf{D}$ be < 1. This means that the sum norm of $\mathbf{D}$ (the sum norm is the largest column sum) is < 1:

$$\|\mathbf{D}\|_s < 1$$

We can rewrite $\mathbf{x} = \mathbf{Dx} + \mathbf{c}$ as $\mathbf{x} - \mathbf{Dx} = \mathbf{c}$ or

$$(\mathbf{I} - \mathbf{D})\mathbf{x} = \mathbf{c} \tag{5}$$

We can solve (5) algebraically with inverses to obtain

$$\mathbf{x} = (\mathbf{I} - \mathbf{D})^{-1}\mathbf{c} \tag{6}$$

We shall now give an algebraic formula for computing $(\mathbf{I} - \mathbf{D})^{-1}$.

Lemma. Let $\mathbf{A}$ be a matrix such that $\|\mathbf{A}\| < 1$ (any matrix norm can be used). Then

$$(\mathbf{I} - \mathbf{A})^{-1} = \sum_{k=0}^{\infty} \mathbf{A}^k = \mathbf{I} + \mathbf{A} + \mathbf{A}^2 + \mathbf{A}^3 + \cdots \qquad (7)$$

Here $\mathbf{A}^0 = \mathbf{I}$ just the way for any scalar r, $r^0 = 1$. Formula (7) is simply the matrix form for the geometric series

$$\frac{1}{1 - a} = \sum_{k=0}^{\infty} a^k = 1 + a + a^2 + a^3 + \cdots \qquad (8)$$

Recall that this series converges only when $|a| < 1$. The verification of (7) is similar to the way (8) was verified in high school—simply multiply $1 - a$ times the infinite series and show that the product equals 1 [equals $\mathbf{I}$ in (7)].

Since $\|\mathbf{D}\|_s < 1$, we can use (7) to compute $(\mathbf{I} - \mathbf{D})^{-1}$. Recall that $\|\mathbf{D}^k\|_s \le \|\mathbf{D}\|_s^k$, but $\|\mathbf{D}\|_s^k \to 0$ (since $\|\mathbf{D}\|_s < 1$). Then the sum norm (the largest column sum) of $\mathbf{D}^k$ approaches 0, so the individual entries of $\mathbf{D}^k$ approach 0. Thus we only need to calculate the sum $\Sigma \, \mathbf{D}^k$ up to, say, the twentieth power of $\mathbf{D}$—the remaining powers will be small enough to neglect.

Using the sum $\Sigma \, \mathbf{D}^k$ is not a very efficient way to compute $(\mathbf{I} - \mathbf{D})^{-1}$ for most matrices. However, the formula is simple and easy to program. The method also has the advantage of avoiding roundoff-error problems: The iterated multiplications will not magnify possible errors in values of the coefficients; instead, the errors shrink, since the entries in $\mathbf{D}^k$ all approach 0. Finally, this method guarantees that one can always solve a Leontief economic model for any $\mathbf{D}$ and any $\mathbf{c}$ (provided that $\|\mathbf{D}\|_s < 1$).

Theorem 1. Every Leontief supply–demand model $\mathbf{x} = \mathbf{Dx} + \mathbf{c}$ has a solution of nonnegative production levels for every nonnegative $\mathbf{c}$ and every nonnegative $\mathbf{D}$, provided that $\|\mathbf{D}\|_s < 1$.

The nonnegativity is very important, since a negative solution is essentially no solution. Nonnegativity follows from (7): All entries in the powers $\mathbf{D}^k$ will be ≥ 0 (since all entries in $\mathbf{D}$ are ≥ 0), so all entries in $\Sigma \, \mathbf{D}^k = (\mathbf{I} - \mathbf{D})^{-1}$ are ≥ 0. Also, $\mathbf{c} \ge 0$, so all entries in $(\mathbf{I} - \mathbf{D})^{-1}\mathbf{c}$ are ≥ 0.

Example 4. **Solution of a Leontief Model**

We solve the Leontief model in (4) by using formula (7) to solve $(\mathbf{I} - \mathbf{D})\mathbf{x} = \mathbf{c}$. We list below the first powers of $\mathbf{D}$ up to $\mathbf{D}^7$ plus $\mathbf{D}^{20}$, and the sum of the right-hand side of (7) up to $\mathbf{D}^{20}$ (using a computer program).

$$\mathbf{D}^2 = \begin{bmatrix} .24 & .18 & .14 & .14 \\ .23 & .18 & .13 & .13 \\ .07 & .07 & .06 & .03 \\ .04 & .04 & .02 & .03 \end{bmatrix} \qquad \mathbf{D}^3 = \begin{bmatrix} .164 & .13 & .098 & .094 \\ .159 & .126 & .095 & .09 \\ .055 & .044 & .031 & .033 \\ .03 & .025 & .019 & .016 \end{bmatrix}$$

$$\mathbf{D}^4 = \begin{bmatrix} .114 & .091 & .068 & .065 \\ .111 & .088 & .066 & .063 \\ .038 & .031 & .023 & .022 \\ .021 & .017 & .013 & .012 \end{bmatrix} \quad \mathbf{D}^5 = \begin{bmatrix} .080 & .064 & .048 & .046 \\ .077 & .062 & .046 & .044 \\ .027 & .021 & .016 & .015 \\ .015 & .012 & .009 & .009 \end{bmatrix}$$

$$\mathbf{D}^6 = \begin{bmatrix} .056 & .044 & .033 & .031 \\ .054 & .042 & .032 & .031 \\ .019 & .015 & .011 & .011 \\ .010 & .008 & .006 & .006 \end{bmatrix} \quad \mathbf{D}^7 = \begin{bmatrix} .039 & .031 & .023 & .022 \\ .038 & .030 & .022 & .022 \\ .013 & .010 & .008 & .007 \\ .007 & .006 & .004 & .004 \end{bmatrix}$$

$$\mathbf{D}^{20} = \begin{bmatrix} .0004 & .0003 & .0002 & .0002 \\ .0004 & .0003 & .0002 & .0002 \\ .0001 & .0001 & .0001 & .0001 \\ .0001 & .0001 & .0000 & .0000 \end{bmatrix} \tag{9}$$

Summing powers of $\mathbf{D}$ from $\mathbf{I}$ up through $\mathbf{D}^{20}$, we have

$$(\mathbf{I} - \mathbf{D})^{-1} \simeq \sum_{k=0}^{20} \mathbf{D}^k = \begin{bmatrix} 2.183 & .811 & .664 & .650 \\ 1.056 & 1.898 & .644 & .530 \\ .352 & .315 & 1.167 & .335 \\ .141 & .221 & .181 & 1.087 \end{bmatrix} \tag{10}$$

The entries in powers of $\mathbf{D}$ are decreasing quickly enough so that the numbers in (10) are accurate to the three decimal places shown.

   With (10) we can now solve the Leontief model for $\mathbf{x}$, the vector of the production levels for the four products.

$$\mathbf{x} = (\mathbf{I} - \mathbf{D})^{-1}\mathbf{c} = \begin{bmatrix} 2.183 & .811 & .664 & .650 \\ 1.056 & 1.898 & .644 & .530 \\ .352 & .315 & 1.167 & .335 \\ .141 & .221 & .181 & 1.087 \end{bmatrix} \begin{bmatrix} 100 \\ 50 \\ 100 \\ 0 \end{bmatrix}$$

$$= \begin{bmatrix} 2.183 \times 100 + & .811 \times 50 + & .664 \times 100 + & .650 \times 0 \\ 1.056 \times 100 + & 1.898 \times 50 + & .644 \times 100 + & .530 \times 0 \\ .352 \times 100 + & .315 \times 50 + & 1.167 \times 100 + & .335 \times 0 \\ .141 \times 100 & .221 \times 50 + & .181 \times 100 + & 1.087 \times 0 \end{bmatrix} \tag{11a}$$

$$= \begin{bmatrix} 218.3 + 40.55 + 66.4 + 0 \\ 105.6 + 94.9 + 64.4 + 0 \\ 35.2 + 15.75 + 116.7 + 0 \\ 14.1 + 11.05 + 18.1 + 0 \end{bmatrix} \tag{11b}$$

$$= \begin{bmatrix} 325 \\ 265 \\ 168 \\ 43 \end{bmatrix} \begin{array}{l} \text{units of energy} \\ \text{units of construction} \\ \text{units of transportation} \\ \text{units of steel} \end{array} \tag{11c}$$

The terms in $(\mathbf{I} - \mathbf{D})^{-1}$ allow us to see how the consumer demand affects the interindustry demands. For example, the total demand for energy, the first sum in (11a), is

$$2.183 \times 100 + .811 \times 50 + .664 \times 100 + .650 \times 0$$

This sum says that each of the 100 units of consumer demand for energy requires 2.183 units of energy to be produced, each of the 50 units of consumer demand for construction requires .811 unit of energy to be produced, and so on. ∎

If we do not need the inverse $(\mathbf{I} - \mathbf{D})^{-1}$ (to solve the Leontief system for many different consumer vectors) but just want the solution for one specific $\mathbf{c}$, we can shorten our effort by rewriting (11a) in terms of the powers of $\mathbf{D}$.

$$\mathbf{x} = (\mathbf{I} - \mathbf{D})^{-1}\mathbf{c} = \left(\sum_k \mathbf{D}^k \right)\mathbf{c} = \sum_k \mathbf{D}^k\mathbf{c} \tag{12}$$

Computing the sum of $\mathbf{D}^k\mathbf{c}$'s is faster than first computing the sum of $\mathbf{D}^k$'s and then multiplying by $\mathbf{c}$: We compute the vector $\mathbf{D}\mathbf{c}$; then by multiplying this vector by $\mathbf{D}$ we get $\mathbf{D}^2\mathbf{c}$, then $\mathbf{D}^3\mathbf{c}$, and so on, with each stage involving a matrix-vector product rather than matrix-matrix. It is left as an exercise for the reader to re-solve the Leontief problem using (12) (again stop at $\mathbf{D}^{20}\mathbf{c}$).

There is another way that we can recast the solution of $(\mathbf{I} - \mathbf{D})\mathbf{x} = \mathbf{c}$. The method is called **solution by iteration.** This is the method we used in Chapter 1 to get an approximate solution to the Leontief model. Iteration was also used to compute the stable distribution vector $\mathbf{p}^* = [.1, .2, .2, .2, .2, .1]$ of the frog Markov chain in Section 1.3. There we repeatedly computed the next-state distribution $\mathbf{p}^{(k)}$ using the transition equations

$$\mathbf{p}^{(k)} = \mathbf{A}\mathbf{p}^{(k-1)}$$

and $\mathbf{p}^{(k)}$ converged to $\mathbf{p}^*$. Let us recall how iteration with the Leontief system worked. We use the system of equations

$$\mathbf{x} = \mathbf{D}\mathbf{x} + \mathbf{c} \tag{13}$$

We guess values $\mathbf{x}^{(0)}$ for the vector $\mathbf{x}$, then substitute $\mathbf{x}^{(0)}$ in the right side of (13) and compute $\mathbf{D}\mathbf{x}^{(0)} + \mathbf{c}$. We check to see if $\mathbf{D}\mathbf{x}^{(0)} + \mathbf{c}$ equals the left side $\mathbf{x}^{(0)}$. If not, we set $\mathbf{x}^{(1)}$ equal to

$$\mathbf{x}^{(1)} = \mathbf{D}\mathbf{x}^{(0)} + \mathbf{c}$$

Suppose that we "guess" $\mathbf{x}^{(0)} = \mathbf{c}$; that is, we just produce enough to meet consumer demand. In our sample Leontief model in (4), $\mathbf{c}$ is the vector [100, 50, 100, 0]. Let us compute $\mathbf{D}\mathbf{x}^{(0)} + \mathbf{c}$ for this model, with $\mathbf{x}^{(0)} = \mathbf{c} = [100, 50, 100, 0]$.

$$\mathbf{D}\mathbf{x}^{(0)} + \mathbf{c} = \begin{bmatrix} .4\times100 + .2\times50 + .2\times100 + .2\times0 \\ .3\times100 + .3\times50 + .2\times100 + .1\times0 \\ .1\times100 + .1\times50 + \quad\quad + .2\times0 \\ .1\times50 + .1\times100 \end{bmatrix} + \begin{bmatrix} 100 \\ 50 \\ 100 \\ 0 \end{bmatrix}$$

$$= \begin{bmatrix} 70 + 100 \\ 65 + 50 \\ 15 + 100 \\ 15 + 0 \end{bmatrix} = \begin{bmatrix} 170 \\ 115 \\ 115 \\ 15 \end{bmatrix}$$

This vector does not equal $\mathbf{x}^{(0)}$, so we set $\mathbf{x}^{(1)} = [170, 115, 115, 5]$. This new estimate of the production levels equals consumer demands $\mathbf{c}$ plus the interindustrial demands $\mathbf{D}\mathbf{c}$ to meet the consumer demand.

We now compute $\mathbf{D}\mathbf{x}^{(1)} + \mathbf{c}$.

$$\mathbf{D}\mathbf{x}^{(1)} + \mathbf{c} = \begin{bmatrix} .4\times170 + .2\times115 + .2\times115 + .2\times15 \\ .3\times170 + .3\times115 + .2\times115 + .1\times15 \\ .1\times170 + .1\times115 + \quad\quad + .2\times15 \\ .1\times115 + .1\times115 \end{bmatrix} + \begin{bmatrix} 100 \\ 50 \\ 100 \\ 0 \end{bmatrix}$$

$$= \begin{bmatrix} 117 + 100 \\ 110 + 50 \\ 31.5 + 100 \\ 23 + 0 \end{bmatrix} = \begin{bmatrix} 217 \\ 160 \\ 131.5 \\ 23 \end{bmatrix}$$

Then we continue with the iteration, and we set

$$\mathbf{x}^{(2)} = \mathbf{D}\mathbf{x}^{(1)} + \mathbf{c} = [217, 160, 131.5, 23]$$

and in general we set

$$\mathbf{x}^{(k+1)} = \mathbf{D}\mathbf{x}^{(k)} + \mathbf{c} \tag{14}$$

In terms of a computer program, we are repeatedly performing the assignment statement

$$\mathbf{x} \leftarrow \mathbf{D}\mathbf{x} + \mathbf{c}$$

Every increase in production levels results in a further increase in the interindustry demand $\mathbf{Dx}$. The question is: Will this process converge to a solution $\mathbf{x}^*$ such that $\mathbf{x}^* = \mathbf{Dx}^* + \mathbf{c}$? In Chapter 1 we claimed that this iteration process did converge. Let us now give a theoretical justification for convergence.

Consider three successive iterates . . , $\mathbf{x}'$, $\mathbf{x}''$, $\mathbf{x}'''$, . . . , where

$$\mathbf{x}'' = \mathbf{Dx}' + \mathbf{c} \qquad \text{and} \qquad \mathbf{x}''' = \mathbf{Dx}'' + \mathbf{c}$$

Then

$$\mathbf{x}''' - \mathbf{x}'' = (\mathbf{Dx}'' + \mathbf{c}) - (\mathbf{Dx}' + \mathbf{c}) = \mathbf{Dx}'' - \mathbf{Dx}' = \mathbf{D}(\mathbf{x}'' - \mathbf{x}') \quad (15)$$

Since $\|\mathbf{D}\|_s < 1$, taking norms in (15), we have

$$|\mathbf{x}''' - \mathbf{x}''|_s \le \|\mathbf{D}\|_s |\mathbf{x}'' - \mathbf{x}'|_s < |\mathbf{x}'' - \mathbf{x}'|_s$$

This means that $\mathbf{D}$ lessens the change in $\mathbf{x}^{(k)}$ from iteration to iteration and the change will eventually shrink to zero.

If $\mathbf{x}^*$ is the solution so that $\mathbf{x}^* = \mathbf{Dx}^* + \mathbf{c}$ (Theorem 1 guarantees this solution exists), then replacing $\mathbf{x}''$ by $\mathbf{x}^*$ in (15), we have

$$\mathbf{x}''' - \mathbf{x}^* = (\mathbf{Dx}'' + \mathbf{c}) - (\mathbf{Dx}^* + \mathbf{c}) = \mathbf{Dx}'' - \mathbf{Dx}^* = \mathbf{D}(\mathbf{x}'' - \mathbf{x}^*) \quad (16)$$

and

$$|\mathbf{x}''' - \mathbf{x}^*|_s \le \|\mathbf{D}\|_s |\mathbf{x}'' - \mathbf{x}^*|_s < |\mathbf{x}'' - \mathbf{x}^*|_s \qquad (17)$$

so the iterates are getting closer and closer—that is, converging—to solution $\mathbf{x}^*$.

The following table gives the values we get in this iteration process (with the numbers rounded to integers).

$$\mathbf{x}^{(0)} = [100, 50, 100, 0]$$
$$\mathbf{x}^{(1)} = [170, 115, 115, 15]$$
$$\mathbf{x}^{(2)} = [217, 160, 132, 23]$$
$$\mathbf{x}^{(3)} = [250, 192, 142, 29]$$
$$\mathbf{x}^{(4)} = [273, 214, 150, 33]$$
$$\mathbf{x}^{(6)} = [300, 240, 159, 38]$$
$$\mathbf{x}^{(8)} = [313, 253, 163, 41]$$
$$\mathbf{x}^{(10)} = [319, 259, 166, 42]$$
$$\mathbf{x}^{(12)} = [322, 262, 167, 43]$$
$$\mathbf{x}^{(20)} = [325, 265, 168, 43]$$
$$\mathbf{x}^{(25)} = [325, 265, 168, 43]$$

All further $\mathbf{x}^{(n)}$, $n > 20$, equal $\mathbf{x}^{(20)}$. This is the same answer that we obtained previously in (11c) using the geometric series approach.

Note that we started with a very poor estimate $\mathbf{x}^{(0)} = \mathbf{c}$. Suppose that we had started with the more thoughtful guess of

$$\mathbf{x}^{(0)} = [300,\ 200,\ 200,\ 50]$$

Then iterating, we obtain

$$\mathbf{x}^{(1)} = [310,\ 245,\ 160,\ 40]$$
$$\mathbf{x}^{(2)} = [313,\ 253,\ 164,\ 41]$$
$$\mathbf{x}^{(3)} = [317,\ 256,\ 165,\ 42]$$

The first iteration is already quite close to the correct solution and, after three iterations, all entries are only 2% away from the true solution.

Now we show how this iterative approach is actually equivalent to the previous geometric series solution method. Recall $\mathbf{x}^{(0)} = \mathbf{c}$ and then

$$\mathbf{x}^{(1)} = \mathbf{D}\mathbf{x}^{(0)} + \mathbf{c} = \mathbf{D}(\mathbf{c}) + \mathbf{c}$$

Then

$$\mathbf{x}^{(2)} = \mathbf{D}\mathbf{x}^{(1)} + \mathbf{c} = \mathbf{D}(\mathbf{D}\mathbf{c} + \mathbf{c}) + \mathbf{c}$$
$$= \mathbf{D}^2\mathbf{c} + \mathbf{D}\mathbf{c} + \mathbf{c}$$

Continuing, we find that

$$\mathbf{x}^{(n)} = \mathbf{D}^n\mathbf{c} + \mathbf{D}^{n-1}\mathbf{c} + \cdots + \mathbf{D}\mathbf{c} + \mathbf{c} = \sum_{k=0}^{n} \mathbf{D}^k\mathbf{c} \qquad (18)$$

So this iterative method is just computing the partial sums in the geometric series for $(\mathbf{I} - \mathbf{D})^{-1}\mathbf{c}$ [see (12)].

A starting value $\mathbf{x}^{(0)}$ other than $\mathbf{c}$ speeds, or slows, convergence but it cannot prevent convergence. *We note that large real-world economic models are always solved by iterative methods, not by the elimination methods taught in standard mathematics books* (the reason is that iteration goes quickly in real-world problems where the matrix $\mathbf{D}$ is mostly 0's).

We now ask the question: Can we adapt this iterative technique to solving general systems of equations? The following theorem tells how to convert a system of equations into a form similar to a Leontief system.

Theorem 2. Given the system of equations $\mathbf{A}\mathbf{x} = \mathbf{b}$, let $\mathbf{D} = \mathbf{I} - \mathbf{A}$, so $\mathbf{A} = \mathbf{I} - \mathbf{D}$. Let the system be rewritten as

$$(\mathbf{I} - \mathbf{D})\mathbf{x} = \mathbf{b} \qquad \text{or} \qquad \mathbf{x} = \mathbf{D}\mathbf{x} + \mathbf{b} \qquad (19)$$

If $\|\mathbf{D}\| = \|\mathbf{I} - \mathbf{A}\| < 1$ (in any matrix norm), the iteration method

$$\mathbf{x}^{(k)} = \mathbf{D}\mathbf{x}^{(k-1)} + \mathbf{b} \qquad (20)$$

converges to the solution of $\mathbf{A}\mathbf{x} = \mathbf{b}$.

With this conversion, $\|\mathbf{D}\| < 1$ guarantees the iteration (20) converges to $(\mathbf{I} - \mathbf{D})^{-1}\mathbf{b}$, the required solution.

Example 5. Iteration Solution of an Oil Refinery Model

We return to our oil refinery model. Each refinery produces three petroleum-based products, heating oil, diesel oil, and gasoline, and x_i is the number of barrels of petroleum used by the ith refinery.

$$
\begin{aligned}
20x_1 + 4x_2 + 4x_3 &= 500 \\
10x_1 + 14x_2 + 5x_3 &= 850 \\
5x_1 + 5x_2 + 12x_3 &= 1000
\end{aligned}
\tag{21}
$$

Let $\mathbf{A}$ be the coefficient matrix in (21) and $\mathbf{b}$ be the vector of right-side demands. Theorem 2 does not apply to the system $\mathbf{Ax} = \mathbf{b}$, since our favorite norm, the sum norm, of $\mathbf{D}$ ($= \mathbf{I} - \mathbf{A}$) is 34 (the largest column sum). We want to rewrite the equations in (21) to make Theorem 2 apply.

To make the column sums or row sums of $\mathbf{I} - \mathbf{A}$ less than 1, we can divide each column (or row) of $\mathbf{A}$ by its largest entry. Dividing entries in the columns this way is equivalent to changing the units of the variables. That is, dividing the first column by 20 (its largest entry) is equivalent to replacing x_1 (the number of barrels of input to refinery 1) by $x_1' = 20x_1$ (input measured in $\frac{1}{20}$ of a barrel).

For all three columns we have

$$
x_1' = 20x_1, \qquad x_2' = 14x_2, \qquad x_3' = 12x_3
\tag{22}
$$

and hence

$$
x_1 = \frac{x_1'}{20}, \qquad x_2 = \frac{x_2'}{14}, \qquad x_3 = \frac{x_3'}{12}
$$

This change of variables divides the coefficients in the first column by 20, the coefficients in the second column by 14, and the coefficients in the third column by 12.

$$
\begin{aligned}
x_1' + \frac{4}{14} x_2' + \frac{4}{12} x_3' &= 500 \\
\frac{10}{20} x_1' + x_2' + \frac{5}{12} x_3' &= 850 \\
\frac{5}{20} x_1' + \frac{5}{14} x_2' + x_3' &= 1000
\end{aligned}
\tag{23}
$$

It is very important that in the new system the main-diagonal entries are all 1.

Now we can try to use Theorem 2. If $\mathbf{A}'$ is the new coefficient matrix in (23), the system for iteration $\mathbf{x}' = (\mathbf{I} - \mathbf{A}')\mathbf{x}' + \mathbf{b}$, or

$$
\begin{aligned}
x_1' &= & -\frac{4}{14}x_2' &- \frac{4}{12}x_3' &+ 500 \\
x_2' &= -\frac{10}{20}x_1' & &- \frac{5}{12}x_3' &+ 850 \\
x_3' &= -\frac{5}{20}x_1' &- \frac{5}{14}x_2' & &+ 1000
\end{aligned}
\tag{24}
$$

Note the nice form of (24): Each equation expresses one variable in terms of the other variables. The main-diagonal entries on the right side are 0 because the main-diagonal entries in (23) are 1.

In the matrix of coefficients $\mathbf{I} - \mathbf{A}'$ on the right side of (24), the sum of the (absolute values) in each column is < 1. Thus $\|\mathbf{I} - \mathbf{A}'\|_s < 1$, so Theorem 2 guarantees that iteration based on (24) will converge.

For simplicity we let $\mathbf{x}^{(0)} = \mathbf{0}$. Iterating with (24), we get (numbers are rounded to the nearest integer in this table)

$$
\begin{aligned}
\mathbf{x}^{(1)} &= [500, 850, 1000] \\
\mathbf{x}^{(2)} &= [-76, 183, 571] \\
\mathbf{x}^{(3)} &= [257, 650, 953] \\
\mathbf{x}^{(4)} &= [-3, 324, 704] \\
\mathbf{x}^{(5)} &= [173, 559, 757] \\
&\ \ \vdots \\
& \tag{25} \\
\mathbf{x}^{(10)} &= [84, 447, 797] \\
\mathbf{x}^{(11)} &= [106, 475, 819] \\
&\ \ \vdots \\
\mathbf{x}^{(20)} &= [97, 463, 810] \\
\mathbf{x}^{(21)} &= [98, 464, 810] \\
\mathbf{x}^{(22)} &= [\sim97.5, \sim463.75, \sim810] \quad \text{and no further change}
\end{aligned}
$$

Observe how our iterates oscillate above and below the final solution. This is due to the minus signs in (24). The reader should try to interpret the iteration process in terms of an iterative method a refinery manager might use to try and find the correct operating levels for the three refineries.

Converting $\mathbf{x}^{(22)}$ back into our original variables, we have

$$x_1 = \frac{x_1'}{20} = \frac{97.5}{20} = 4\frac{7}{8}$$

$$x_2 = \frac{x_2'}{14} = \frac{463.75}{14} = 33\frac{1}{8}$$

$$x_3 = \frac{x_3'}{12} = \frac{810}{12} = 67\frac{1}{2}$$

∎

In Example 2 we could also have rewritten the system of equation (21) by dividing each row (each equation) by its largest entry. This yields

$$x_1 + \frac{4}{20}x_2 + \frac{4}{20}x_3 = 25$$

$$\frac{10}{14}x_1 + x_2 + \frac{5}{14}x_3 = \frac{850}{14} \qquad (26)$$

$$\frac{5}{12}x_1 + \frac{5}{12}x_2 + x_3 = \frac{1000}{12}$$

Rewriting (26) in the form $\mathbf{x} = (\mathbf{I} - \mathbf{A}'')\mathbf{x} + \mathbf{b}''$, we have

$$x_1 = -\frac{4}{20}x_2 - \frac{4}{20}x_3 + 25$$

$$x_2 = -\frac{10}{14}x_1 - \frac{5}{14}x_3 + \frac{850}{14} \qquad (27)$$

$$x_3 = -\frac{5}{12}x_1 - \frac{5}{12}x_2 + \frac{1000}{12}$$

Observe that the sum of the second row of coefficients in (27), $\left|-\frac{10}{14}\right| + \left|-\frac{5}{14}\right|$ is > 1, so the max norm is not < 1 as required. The first column sum is also greater than 1, so Theorem 2 does not apply to (27).

Dividing the ith row by the coefficient of x_i, as in (26)–(27), has the advantage that it does not involve a change of variables. In (27), we are simply solving the first equation of the original system for x_1 (in terms of the other variables), solving the second equation for x_2, and solving the third equation for x_3. For a general system of equations

$$a_{11}x_1 + a_{12}x_2 + \cdots + a_{1n}x_n = b_1$$
$$a_{21}x_1 + a_{22}x_2 + \cdots + a_{2n}x_n = b_2 \qquad (28)$$
$$\vdots \qquad \vdots \qquad \vdots \qquad \vdots \qquad \vdots$$
$$a_{n1}x_1 + a_{n2}x_2 + \cdots + a_{nn}x_n = b_n$$

the row equations for iteration become

$$x_1 = \frac{-a_{12}x_2 - a_{13}x_3 - \cdots - a_{1n}x_n + b_1}{a_{11}}$$

$$x_2 = \frac{-a_{21}x_1 - a_{23}x_3 - \cdots - a_{2n}x_n + b_2}{a_{22}} \qquad (29)$$

$$\vdots \qquad \vdots$$

Iteration using this system is called **Jacobi iteration.**

It can be proven that iteration scheme (29) derived from row division converges if and only if the iteration scheme derived from column division [as in (24)] converges. The following theorem states what conditions must hold for this iteration scheme to work, that is, conditions so that after row or column division the max or sum norm will be < 1.

Theorem 3. Jacobi iteration using the system (29) converges if either of the following two conditions hold:

(i) For each row i, the coefficient a_{ii} of x_i is larger than the (absolute value) sum of the other coefficients in the row:

$$\sum_{\substack{j \\ i \neq j}} |a_{ij}| < |a_{ii}| \qquad \text{each } i \qquad (30)$$

or

(ii) For each column j, the coefficient a_{jj} of x_j is larger than the (absolute value) sum of the other coefficients in the column:

$$\sum_{\substack{i \\ i \neq j}} |a_{ij}| < |a_{jj}| \qquad \text{each } j \qquad (31)$$

The reader should check that condition (31) was satisfied in the refinery problem. It is a straightforward exercise to check that (31) guarantees that after column division [as in (23)–(24)] the resulting matrix $\mathbf{D}'' = \mathbf{I} - \mathbf{A}''$ will have sum norm < 1, and that (30) guarantees that after row division the max norm of $\mathbf{D}' = \mathbf{I} - \mathbf{A}'$ is < 1.

There are other iteration methods based on more advanced theory (see numerical analysis references). Exercise 17 mentions a simple way, called **Gauss–Seidel iteration,** to speed up the convergence of Jacobi iteration.

We conclude this section by linking the iteration, $\mathbf{x}^{(k)} = \mathbf{A}\mathbf{x}^{(k-1)}$, for the dominant eigenvector at the start of this section with the iteration in Theorem 2, $\mathbf{x}^{(k)} = \mathbf{D}\mathbf{x}^{(k-1)} + \mathbf{b}$, for solving the system $(\mathbf{I} - \mathbf{D})\mathbf{x} = \mathbf{b}$. The following trick converts the latter iteration into the former.

Define the $(n + 1)$-vector $\mathbf{x}^*$ and the $(n + 1)$-by-$(n + 1)$ matrix $\mathbf{D}^*$:

$$\mathbf{x}^* = [\mathbf{x}, 1] = [x_1, x_2, \ldots, x_n, 1] \tag{32}$$

$$\mathbf{D}^* = \begin{bmatrix} \mathbf{D} & \mathbf{b} \\ 0 & 1 \end{bmatrix}$$

For example, for the Leontief matrix **D** in (4), **D*** is

$$\mathbf{D}^* = \begin{bmatrix} .1 & .2 & .2 & .2 & 100 \\ .3 & .3 & .2 & .1 & 50 \\ .1 & .1 & 0 & .2 & 100 \\ 0 & .1 & .1 & 0 & 0 \\ 0 & 0 & 0 & 0 & 1 \end{bmatrix}$$

The reader should check that

$$\mathbf{x}^{*(k)} = \mathbf{D}^*\mathbf{x}^{*(k-1)} \quad \text{is equivalent to} \quad \mathbf{x}^{(k)} = \mathbf{D}\mathbf{x}^{(k-1)} + \mathbf{b}$$

Then the iteration scheme $\mathbf{x}^{(k)} = \mathbf{D}\mathbf{x}^{(k-1)} + \mathbf{b}$ of Theorem 2 will converge to a solution in which for large k,

$$\mathbf{x}^{(k)} \simeq \mathbf{x}^{(k-1)}$$

if and only if for large k,

$$\mathbf{x}^{*(k)} \simeq \mathbf{x}^{*(k-1)}$$

But the latter condition for the **x***'s means that the dominant eigenvalue of **D*** is 1.

Theorem 4. The iteration scheme $\mathbf{x}^{(k)} = \mathbf{D}\mathbf{x}^{(k-1)} + \mathbf{b}$ for solving $(\mathbf{I} - \mathbf{D})\mathbf{x} = \mathbf{b}$ converges to a solution if and only if the dominant eigenvalue of the augmented matrix **D*** [see (32)] is 1.

Section 3.4 **Exercises**

Summary of Exercises
Exercises 1–5 involve iterative methods for determining the largest eigenvalue and its eigenvector. Exercises 6–9 involve solutions by sum of powers. Exercises 10–15 deal with iterative methods to solve a system of equations. Exercises 16 and 17 introduce related iterative methods, and Exercise 16 introduces the Gauss–Seidel iteration.

1. Use iteration to determine the dominant eigenvalue and an associated eigenvector for the following systems of equations.

(a) $\begin{bmatrix} 1 & 1 \\ 0 & 2 \end{bmatrix}$ (b) $\begin{bmatrix} 1 & -2 \\ -2 & 1 \end{bmatrix}$ (c) $\begin{bmatrix} 1 & 3 \\ -2 & 6 \end{bmatrix}$ (d) $\begin{bmatrix} -4 & 4 & 4 \\ -1 & 1 & 2 \\ -3 & 2 & 4 \end{bmatrix}$

Use [1, 2] or [1, 2, 0] as your starting vector. This means that for part (a), you iterate the system

$$x_1' = 1x_1 + 1x_2$$
$$x_2' = 0 + 2x_2$$

Use the Raleigh quotient to refine your estimate of the dominant eigenvalue.

2. Repeat the iteration in Exercise 1 using the vector [1, 1] or [1, 1, 1] as a starting vector. How does this affect the speed of convergence to the dominant eigenvector? For one of the matrices, you do not converge to the dominant eigenvector—why?

3. (a) Use iteration to determine the dominant eigenvalue and an associated eigenvector for the following system of equations. Use [1, 0] as your starting vector.

$$x' = .707x - .707y$$
$$y' = .707x + .707y$$

   (b) Plot the successive iterates on *x-y* graph paper. Try other starting vectors. State in words the effect in *x-y* coordinates of this linear model.
   (c) Solve the characteristic equation $\det(\mathbf{A} - \lambda\mathbf{I}) = 0$ to determine the eigenvalues for this matrix of coefficients. You are finding out that imaginary eigenvalues correspond to rotations. Note that .707 = $\sin 45° = \cos 45°$.

4. (a) Use iteration to determine the dominant eigenvalue and an associated eigenvector for the following system of equations. Use [1, 1] as your starting vector.

$$x' = 2x - y$$
$$y' = 3x - 2y$$

   (b) Repeat the iteration starting with [0, 1].
   (c) Solve the characteristic equation $\det(\mathbf{A} - \lambda\mathbf{I}) = 0$ to determine the eigenvalues for this matrix of coefficients. Does this give you any hints about what was wrong in the iteration procedure in part (b)?

5. (a) Use iteration to determine the dominant eigenvalue and an associated eigenvector for the following system of equations. Use [1, 0, 0] as your starting vector. You have to iterate a long time to get the iterates to stabilize at an eigenvector.

$$x' = \qquad + 4y + z$$
$$y' = .4x$$
$$z' = \qquad .6y$$

(This is a population growth that we will study in Section 4.5; here x is number of babies, y number of adolescents, and z adults.)

(b) The other two eigenvalues of this system are -1.118 and $-.152$. How do these eigenvalues help explain the slow convergence of the iteration procedure?

Hint: See equation (2).

6. Suppose that we want to solve the same Leontief system as was solved in Example 4, but now the consumer demand vector **c** has been changed. Use the formula in equation (11a) with the following new **c**'s to determine the new vector **x** of production levels.
   (a) $\mathbf{c} = [50, 50, 50, 100]$ **(b)** $\mathbf{c} = [0, 100, 0, 0]$
   (c) $\mathbf{c} = [0, 0, 100, 0]$ **(d)** $\mathbf{c} = [0, 0, 50, 50]$
   (e) $\mathbf{c} = [100, 10, 10, 100]$

7. Use the sum-of-powers method in equations (10)–(11) to solve the following Leontief systems.
   (a) $x_1 = .1x_1 + .2x_2 + .2x_3 + 100$
   $\qquad x_2 = .2x_1 + .1x_2 \qquad\quad + 100$
   $\qquad x_3 = .2x_1 \qquad\quad + .1x_3 + 100$

   (b) $x_1 = .3x_1 + .1x_2 + .2x_3 + 100$
   $\qquad x_2 = .1x_1 + .1x_2 + .1x_3 + 100$
   $\qquad x_3 = .1x_1 + .1x_3 \qquad\quad + 100$

   Use computer programs for both systems; convergence is fast because the norm of **D** is small—you only need to go up to the sixth power of the coefficient matrix **D**.

8. Find the inverse of the following matrices by writing them in the form $\mathbf{I} - \mathbf{D}$ and using the sum-of-powers method on **D**. Check the accuracy of your answer by using the determinant-based formula for the inverse of a 2-by-2 matrix (see Section 3.3).

   (a) $\begin{bmatrix} .7 & -.2 \\ -.4 & .8 \end{bmatrix}$ **(b)** $\begin{bmatrix} .6 & .3 \\ .2 & .5 \end{bmatrix}$ **(c)** $\begin{bmatrix} .6 & 0 \\ 0 & .5 \end{bmatrix}$

9. Try using the sum-of-powers method to solve the following system of equations. Why does the method fail?

   $$x_1 = .4x_1 + .3x_2 + .4x_3 + 100$$
   $$x_2 = .3x_1 + .4x_2 + .6x_3 + 100$$
   $$x_3 = .7x_1 + .8x_2 + .5x_3 + 100$$

10. Use the iteration method in equation (14) to solve the Leontief systems in Exercise 7.

11. Consider the following systems of equations.

(i) $7x_1 + x_2 + 2x_3 = 30$ (ii) $6x_1 + 3x_2 + x_3 = 15$

    $x_1 + 5x_2 + 3x_3 = 10$ $2x_1 + 5x_2 + 2x_3 = 50$

   $2x_1 + 3x_2 + 8x_3 = 12$ $x_1 + x_2 + 4x_3 = 10$

(a) Use the formulation in (29) to rewrite the systems in the form $\mathbf{x} = \mathbf{Dx} + \mathbf{b}$ with $\|\mathbf{D}\| < 1$.

(b) Solve this system by iteration as described in Theorem 2, starting with $\mathbf{x}^{(0)} = [0, 0, 0]$.

(c) Repeat part (b) with starting vector $\mathbf{x}^{(0)} = [100, 100, 100]$.

12. In the two systems of equations in Exercise 11, divide each column by the main-diagonal entry and rewrite as $\mathbf{x}' = \mathbf{Dx}' + \mathbf{b}$, as done in Example 5. Then solve the systems by iteration, starting with $\mathbf{x}^{(0)} = [0, 0, 0]$. Are the iterates the same as in Exercise 11 (allowing for the changes of variable)?

13. Consider the system of equations $\mathbf{Ax} = \mathbf{b}$, where

$$\mathbf{A} = \begin{bmatrix} 9 & 4 & -3 \\ -3 & 3 & 10 \\ 4 & 8 & -3 \end{bmatrix}$$

and $\mathbf{b} = [10, 20, 30]$

(a) Rearrange the equations (rows) and divide each equation by appropriate numbers so that this system can be rewritten in the form $\mathbf{x} = \mathbf{Dx} + \mathbf{b}$ with $\|\mathbf{D}\| < 1$.

(b) Solve this system by iteration as described in Theorem 2.

14. (a) For which of the following systems of equations does Theorem 3 apply [is (30) or (31) satisfied]?

(i) $3x_1 - 4x_2 = 2$ (ii) $6x_1 + 2x_2 - x_3 = 4$

   $2x_1 + x_2 = 4$ $x_1 + 5x_2 + x_3 = 3$

                                 $2x_1 + x_2 + 4x_3 = 27$

(iii) $2x_1 + x_2 = 3$

    $4x_1 - x_2 = 5$

(b) In the systems where Theorem 3 does not apply directly, try to rearrange the rows and/or divide the rows or columns by the largest coefficient to make Theorem 3 apply.

(c) Try the iterative method for solving each system. Does the method work on a system where Theorem 3 could not be made to apply?

15. (a) Suppose that we use a starting vector $\mathbf{x}^{(0)} = \mathbf{w}$ in the iteration scheme $\mathbf{x}^{(k+1)} = \mathbf{D}\mathbf{x}^{(k)} + \mathbf{c}$. Using the same reasoning as led to equation (18), find formulas for $\mathbf{x}^{(1)}$, $\mathbf{x}^{(2)}$, and $\mathbf{x}^{(n)}$ in terms of $\mathbf{D}$ and $\mathbf{w}$.

(b) Use your formula for $\mathbf{x}^{(n)}$ in part (a) and the fact that $\|\mathbf{D}\| < 1$ to show that the starting vector does not influence the final values in the iteration process.

16. A well-known method to speed up the convergence of Jacobi iteration, called *Gauss–Seidel iteration,* is to use the *new value* of x_1 obtained from the first equation in the second and third equations (in place of the previous value of x_1); similarly, the new value for x_2 is used in the third equation. In the refinery problem, the first two equations in (24) are

$$
\begin{aligned}
x_1' &= \qquad\quad -\tfrac{4}{20}x_2' - \tfrac{4}{20}x_3' + 500 \\
x_2' &= -\tfrac{10}{14}x_1' \qquad\quad - \tfrac{5}{14}x_3' + 750
\end{aligned}
$$

Starting with $\mathbf{x}^{(0)} = [0, 0, 0]$, we would compute x_1' as $\tfrac{4}{20}(0) - \tfrac{4}{20}(0) + 500 = 500$. Then we use this value of 500 for x_1' in the second equation to compute x_2' as $\tfrac{4}{20}(500) - \tfrac{4}{20}(0) + 750 = 550$. The third equation would use the values for both x_1 and x_2 just computed.

(a) Use Gauss–Seidel iteration on the refinery problem [the three equations in (24)] starting with $\mathbf{x}^{(0)} = [0, 0, 0]$. How many iterations are required to attain the solution vector [135, 263, 868]?

(b) Use Gauss–Seidel iteration to solve the Leontief system in Example 4.

(c) Use Gauss–Seidel iteration to solve the system of equations in Exercise 11, part (i).

17. Another method of iteration is to average the two previous iterates. In an iteration process such as (25), where the iterates are oscillating above and below the final solution, such an averaging will increase convergence. On the other hand, when the iterates are increasing as in the Leontief system, this averaging method will slow down the convergence.

(a) Use this method of averaging the two previous iterates to re-solve the refinery equations [equations (24)] starting again with $\mathbf{x}^{(0)} = [0, 0, 0]$. How many iterations are required to attain the solution vector [135, 263, 868]?

(b) Use the method of averaging to re-solve the Leontief system in the text. How many iterations are required to attain the solution?

Numerical Analysis of Systems of Equations

Computational Complexity of Solving Systems of Linear Equations

In this section we look at some of the numerical difficulties and shortcuts that are possible during elimination computations. Numerical linear algebra is a large, growing field (see the References). We shall just touch on some of the basic results. Our discussion will concern systems of n equations in n unknowns, but generally our results also apply to systems of m equations in n unknowns.

The first issue we address is the computational complexity of elimination: How many arithmetic operations are required to solve a system of n equations in n unknowns (when the answer is unique)? We shall measure computation in terms of the number of multiplications required (division will be treated as equivalent to multiplication); the number of additions and subtractions is always about the same as the number of multiplications.

The fundamental computation in Gaussian elimination is subtracting a multiple l_{ki} of row i from row k, $k = i + 1, i + 2, \ldots, n$: this operation makes entry (k, i) zero. Each entry in row i must be multiplied by l_{ki} and then subtracted from the corresponding entry in row k, for $k > i$. There are $n - i + 1$ entries in row i (the entries to the right of the main diagonal plus the right side value) involved, and $n - i$ rows below row i. So there are approximately $(n - i + 1)(n - i)$ multiplications; for simplicity, we say about $(n - i)^2$ multiplications. To perform elimination of $x_1, x_2, \ldots, x_{n-1}$ requires

$$(n - 1)^2 + (n - 2)^2 + \cdots + (1)^2 \simeq \frac{n^3}{3} \text{ multiplications}$$

When elimination is finished, it takes one division to compute x_n, one multiplication and one subtraction (and one division) to compute x_{n-1}, and generally k multiplications and k subtractions to compute x_{n-k}. Altogether, back substitution requires about $n^2/2$ multiplications. For large n, $n^2/2$ is negligible beside $n^3/3$.

Now let us quickly go over the operation count for elimination by pivoting. The one difference is that a variable x_i is now eliminated from all the other $n - 1$ rows; in addition, every entry in the pivot row is divided by the pivot entry. It takes $(n - i + 1)n$ multiplications to eliminate x_i from all other rows. Summing over all i, we get a grand total of about $n^3/2$ multiplications. There is no back substitution.

If we want to compute the inverse by pivoting, then each row will require $(n - i + n)$ multiplications, for n (instead of 1) right-side terms. The total number of multiplications works out to about n^3. One can use Gaussian elimination with n right-hand sides and multiple back substitution, but the back substitution now is n times more complicated; the total process also requires about n^3 multiplications.

We summarize our discussion with a theorem.

Theorem 1. A system of n equations in n unknowns requires approximately $n^3/3$ multiplications (and subtractions) to solve by Gaussian elimination and $n^3/2$ multiplications to solve by pivoting. Either method requires approximately n^3 multiplications to invert an n-by-n matrix.

If we are solving $\mathbf{Ax} = \mathbf{b}$ for several different right-hand sides, then the best method is the **LU** method, presented in Section 3.2, of storing elimination multipliers in **L** and the final reduced matrix in **U**. Applying the multipliers in **L** to a new $\mathbf{b}^*$ will require about $n^2/2$ multiplications. As noted above, back substitution also requires $n^2/2$ multiplications. Thus, using **L** and **U**, we can solve the new system $\mathbf{Ax} = \mathbf{b}^*$ with just n^2 multiplications. This is the same number of multiplications required to compute the matrix-vector product $\mathbf{A}^{-1}\mathbf{b}^*$, the solution using the inverse (assuming that $\mathbf{A}^{-1}$ is known). However, the result with **L** and **U** will have less roundoff error: Using $\mathbf{A}^{-1}$ necessarily introduces some additional error.

In Section 3.4 we introduced the Jacobi iteration method for solving a system of equations. Each iteration requires a matrix-vector multiplication that takes n^2 operations. If the matrix is sparse, fewer operations are required (see Section 2.6). The problem is that we do not know how many iterations will be necessary to converge to the solution. When n is large and the coefficient matrix is sparse, an iteration method is likely to be much faster.

Solving Tridiagonal Systems

Let us next consider how much more quickly elimination can be performed for a well-structured sparse matrix. In particular, let us look at a triadiagonal matrix, whose only nonzero entries are on the main diagonal and just to the left and right of the main diagonal.

Look at the form of a tridiagonal matrix before and after the elimination step for x_i. A $*$ indicates a nonzero entry.

$$
\begin{array}{c}
\\
\\
B\\
E\\
F\\
O\\
R\\
E
\end{array}
\quad
\begin{array}{c}
\\
\\
\\
\\
i\\
i+1\\
\\

\end{array}
\quad
\begin{array}{cccccccc}
 & & & i & i+1 & & & \\
\left[\begin{array}{cccccccc}
* & * & 0 & 0 & & & & \\
0 & * & * & 0 & 0 & & & \\
0 & 0 & \cdot & \cdot & & & & \\
 & & \cdot & \cdot & & & & \\
0 & 0 & a_{ii} & * & 0 & 0 & & \\
 & 0 & * & * & * & 0 & 0 & \\
 & & 0 & * & * & * & 0 & 0 \\
 & & & & * & * & * &
\end{array}\right]
\end{array}
$$

$$
\begin{array}{c}
 \\
 \\
A \\
F \\
T \quad i \\
E \quad i+1 \\
R
\end{array}
\begin{array}{cc}
i & i+1 \\
\left[\begin{array}{cccccccc}
* & * & 0 & 0 & & & & \\
0 & * & * & 0 & 0 & & & \\
0 & 0 & \cdot & \cdot & & & & \\
& & \cdot & \cdot & & & & \\
& 0 & 0 & a_{ii} & * & 0 & 0 & \\
& & 0 & 0 & C & * & 0 & 0 \\
& & & 0 & * & * & * & 0 & 0 \\
& & & & & * & * & *
\end{array}\right]
\end{array}
$$

The only alteration was that entry $(i + 1, i)$ became 0, and entry $(i + 1, i + 1)$, marked with a C, changed to a new nonzero value—only entry C's new value must be computed, together with a change in the right side of row $i + 1$. This requires only two multiplications and two subtractions, plus one division to find the elimination multiplier.

Back substitution for each row in the reduced system will require one multiplication, one subtraction, and one division, since each row in the reduced system looks like $qx_i + rx_{i+1} = s$ (where x_{i+1} is already known), so $x_i = (s - rx_{i+1})/q$. Together, we have

Theorem 2. An n-by-n tridiagonal system of equations can be solved by Gaussian elimination in just $5n$ multiplications and $3n$ subtractions.

This is an incredibly fast result. Compared with the normal $n^3/3$ multiplications in Theorem 1, this means that solving a 50-by-50 tridiagonal matrix requires about 250 multiplications versus over 40,000 operations for a full 50-by-50 matrix. Savings are possible on any band matrix.

Let us illustrate the speed of elimination on a tridiagonal matrix we have seen frequently in this book.

Example 1. Computing the Stable Probability Distribution for the Frog Markov Chain

We return to the familiar frog Markov chain that has six states (representing different positions in the highway). The transition matrix is

$$
\mathbf{A} = \begin{bmatrix}
.50 & .25 & 0 & 0 & 0 & 0 \\
.50 & .50 & .25 & 0 & 0 & 0 \\
0 & .25 & .50 & .25 & 0 & 0 \\
0 & 0 & .25 & .50 & .25 & 0 \\
0 & 0 & 0 & .25 & .50 & .50 \\
0 & 0 & 0 & 0 & .25 & .50
\end{bmatrix}
\tag{1}
$$

By letting this Markov chain run for many iterations, we found in Section 1.3 that the probability distribution approached a stable distribution $\mathbf{p}*$ with the property $\mathbf{A}\mathbf{p}* = \mathbf{p}*$. In matrix algebra terminology, $\mathbf{p}*$ is an eigenvector of $\mathbf{A}$ with eigenvalue 1. Let us solve the matrix system

$$\mathbf{A}\mathbf{p} = \mathbf{p} \qquad \text{or, equivalently,} \qquad (\mathbf{A} - \mathbf{I})\mathbf{p} = 0$$

That is,

$$
\begin{aligned}
-.50p_1 + .25p_2 \qquad\qquad\qquad\qquad\qquad &= 0 \\
.50p_1 - .50p_2 + .25p_3 \qquad\qquad\qquad\quad &= 0 \\
.25p_2 - .50p_3 + .25p_4 \qquad\qquad\quad &= 0 \\
.25p_3 - .50p_4 + .25p_5 \qquad\quad &= 0 \\
.25p_4 - .50p_5 + .50p_6 &= 0 \\
.25p_5 - .50p_6 &= 0
\end{aligned}
\tag{2}
$$

Use Gaussian elimination. To eliminate p_1 from equation (2), we add the first equation in (2) to the second and obtain

$$
\begin{aligned}
-.50p_1 + .25p_2 \qquad\qquad\qquad\qquad\qquad &= 0 \\
-.25p_2 + .25p_3 \qquad\qquad\qquad\qquad\quad &= 0 \\
.25p_2 - .50p_3 + .25p_4 \qquad\qquad\quad &= 0 \\
.25p_3 - .50p_4 + .25p_5 \qquad\quad &= 0 \\
.25p_4 - .50p_5 + .50p_6 &= 0 \\
.25p_5 - .50p_6 &= 0
\end{aligned}
\tag{3}
$$

To eliminate p_2 from the third equation in (3), we add the second equation to the third.

$$
\begin{aligned}
-.50p_1 + .25p_2 \qquad\qquad\qquad\qquad\qquad &= 0 \\
-.25p_2 + .25p_3 \qquad\qquad\qquad\qquad\quad &= 0 \\
-.25p_3 + .25p_4 \qquad\qquad\qquad\quad &= 0 \\
.25p_3 - .50p_4 + .25p_5 \qquad\quad &= 0 \\
.25p_4 - .50p_5 + .50p_6 &= 0 \\
.25p_5 - .50p_6 &= 0
\end{aligned}
\tag{4}
$$

A simple pattern is emerging of simply adding the ith equation to the $(i + 1)$st equation. To eliminate p_3 from the fourth equation in (4), we add the third equation to the fourth and after that we eliminate p_4 from the fifth equation by adding the fourth equation to the fifth. After these two further elimination steps, we have

$$
\begin{aligned}
-.50p_1 + .25p_2 && = 0 \\
-.25p_2 + .25p_3 && = 0 \\
-.25p_3 + .25p_4 && = 0 \\
-.25p_4 + .25p_5 && = 0 \\
-.25p_5 + .50p_6 &= 0 \\
.25p_5 - .50p_6 &= 0
\end{aligned}
\tag{5}
$$

Now however, we are on the verge of a problem we had back in Section 3.2.: The last equation in (5) is just the negative of the fifth equation. When we add the fifth equation to the sixth equation to eliminate p_5, we obtain

$$
\begin{aligned}
-.50p_1 + .25p_2 && = 0 \\
-.25p_2 + .25p_3 && = 0 \\
-.25p_3 + .25p_4 && = 0 \\
-.25p_4 + .25p_5 && = 0 \\
-.25p_5 + .50p_6 &= 0 \\
0 &= 0
\end{aligned}
\tag{6}
$$

Multiplying all equations in (6) by 4 and bringing one term in each equation to the right side, we have the simple system

$$
\begin{aligned}
2p_1 &= p_2, & p_2 &= p_3, \\
p_3 &= p_4, & p_4 &= p_5, \\
p_5 &= 2p_6, & 0 &= 0
\end{aligned}
\tag{7}
$$

Our problem is that an eigenvector is really a family of eigenvectors: Any multiple of an eigenvector is again an eigenvector. We can give any value q to p_1, then from (7), $2p_1 = p_2$ implies that $p_2 = 2q$ and further that $p_5 = p_4 = p_3 = p_2 = 2q$. Then

$$
\mathbf{p} = [q, 2q, 2q, 2q, 2q, q]
\tag{8}
$$

But we want a special eigenvector, one that is a probability distribution—whose entries sum to 1. Requiring that the components in **p** sum to 1, we have the constraint

$$
q + 2q + 2q + 2q + 2q + q = 1 \quad \text{or} \quad 10q = 1
$$
$$
\rightarrow q = .1
$$

Thus our stable distribution is

$$
\mathbf{p}^* = [.1, .2, .2, .2, .2, .1]
$$

This is the same result we got by iteration in Section 1.3. ■

The simple nature of the elimination process in (2)–(6) shows that if the frog were on a 10-lane superhighway yielding a 12-by-12 transition matrix, the computations to find the stable distribution would still be easy (the n-lane problem is solved in Section 4.4). Tridiagonal systems arising from Markov chains and other real-world problems often have a simple elimination pattern. For example, in Section 4.7 we shall easily solve a 100-by-100 tridiagonal system.

One of the dangers in elimination in general sparse matrices (not band matrices) is **fill-in.** Fill-in is the creation of new nonzero entries during the elimination process, the loss of sparseness. Every nonzero entry created below the main diagonal will require additional computation to eliminate it later. In elimination by pivoting, nonzero entries above the main diagonal also cause extra work. To illustrate the trickiness of fill-in, observe what happens in elimination by pivoting when we re-solve the stable probability distribution in Example 1.

Example 2. Sparse Matrix Fill-in

After pivoting on entry $(1, 1)$, we have the same result as in (2) except that the first row is $[1, -.25, 0, 0, 0, 0]$, since we divide the pivot row by the pivot entry. After pivoting on entry $(2, 2)$, we obtain

$$
\begin{aligned}
&\text{(a)} & p_1 \quad &- .50p_3 & &= 0 \\
&\text{(b)} & p_2 - \ &p_3 & &= 0 \\
&\text{(c)} & &- .25p_3 + .25p_4 & &= 0 \\
&\text{(d)} & &.25p_3 - .50p_4 + .25p_5 & &= 0 \\
&\text{(e)} & &.25p_4 - .50p_5 + .50p_6 &&= 0 \\
&\text{(f)} & &.25p_5 - .50p_6 &&= 0
\end{aligned}
\tag{9}
$$

When we pivot on entry $(3, 3)$, we have to remove the nonzeros in entries $(1, 3)$ and $(2, 3)$. The result is

$$
\begin{aligned}
&\text{(a')} = \text{(a)} + .5\text{(c')} & p_1 \quad &- .50p_4 & &= 0 \\
&\text{(b')} = \text{(b)} + \text{(c')} & p_2 - \ &p_4 & &= 0 \\
&\text{(c')} = \text{(c)}/(-.25) & p_3 - \ &p_4 & &= 0 \\
&\text{(d')} = \text{(d)} & &- .25p_4 + .25p_5 & &= 0 \\
&\text{(e')} = \text{(e)} & &.25p_4 - .50p_5 + .50p_6 &&= 0 \\
&\text{(f')} = \text{(f)} & &.25p_5 - .50p_6 &&= 0
\end{aligned}
\tag{10}
$$

We just pushed the nonzero entries over one column and now will have to deal with them on the next pivot. So when it is time to pivot on entry (i, i) in this system, the entries above the main diagonal in column i will *all* be nonzero. When we used Gaussian elimination in Example 1, we never had such problems. ∎

Stable Elimination

We consider now the problem of the stability of computations during elimination. Will roundoff errors tend to grow and make the solution computed inaccurate, or will the errors stay small? The answer is that some systems of equations are inherently unstable, while others are very dependent on the order of the equations; that is, reordering the equations and variables can sometimes greatly reduce roundoff error. We shall discuss ways to choose a good arrangement and to estimate the underlying stability of the system of equations.

To see how computations with systems of equations can be stable or unstable depending on the order of the equations or variables, consider the following example.

Example 3. Roundoff Error in Elimination Computations

Gaussian elimination on the system

$$
\begin{aligned}
.0001x + y &= 1 \\
x + y &= 2
\end{aligned}
\tag{11}
$$

yields

$$
\begin{aligned}
.0001x + y &= 1 \\
-9999y &= -9998
\end{aligned}
\tag{12}
$$

from which we have $y \simeq .9999$, so back substitution yields

$$
.0001x + .9999 = 1 \rightarrow x = 1
$$

But suppose that roundoff error in the elimination had produced

$$
\begin{aligned}
.0001x + y &= 1 \\
-10{,}000y &= -10{,}000
\end{aligned}
\tag{13}
$$

yielding $y = 1$. Now back substitution gives

$$
.0001x + 1 = 1 \rightarrow x = 0
$$

Although the y value stays about the same, the difference in x values is very significant: (12) yields $x = 1$, and (13) yields $x = 0$.

The problem came from the small size .0001 of a coefficient of x in (11). This coefficient is the pivot entry in elimination by pivoting. The antidote is to avoid pivoting on small entries. In the case of system (11), we should pivot on the coefficient of x in the second equation or on the coefficient of y in the first equation. In Gaussian elimination terms, we should interchange either the equations or the variables.

Interchanging equations, we get

$$x + y = 2$$
$$.0001x + y = 1$$

(14)

and Gaussian elimination gives

$$x + \quad y = \quad 2$$
$$.9999y = .9999$$

(15)

yielding $y \simeq .9999$ and $x \simeq 1.0001$. Now there are no roundoff-error problems: A small error in the computations in (15) will yield only a small error in the values of x and y.

Interchanging the order of the variables in (11), we get

$$y + .0001x = 1$$
$$y + \quad x = 2$$

(16)

Now Gaussian elimination gives

$$y + .0001x = 1$$
$$.9999x = 1$$

(17)

again yielding $x \simeq 1.0001$ and $y \simeq .9999$. Again, a small error in computations in (17) has a small effect on x and y.

What a difference the rearrangements make! ■

The immediate conclusion would seem to be to pick an entry that is largest in its row and column for the pivot, and perform an exchange of equations and/or variables to get this entry up to the first coefficient in the first equations.

Unfortunately, the situation is more complicated than that. Suppose that we multiply the first equation of (11) by 10^4 and multiply the second equation by 10^{-4}. Further, let us replace x by $s = 10^{-4}x$ and y by $t = 10^4 y$ (so $x = 10^4 s$, $y = 10^{-4}t$). These scaling transformations convert (11) into

$$10^4\{.0001(10^4 s) + (10^{-4}t) = 1\}$$
$$10^{-4}\{(10^4 s) + (10^{-4}t) = 2\}$$

or

$$10^4 s + \quad t = 10^4$$
$$s + 10^{-8}t = 2 \times 10^{-4}$$

(18)

The coefficient of s in the first equation is now the largest coefficient in (18), but these scaling changes have not really changed the arithmetic. If we solve (18) with possible roundoff error in the fourth significant digit, as above in (13), the solution of (18) will be

$$s = 0, \quad t = 10^4 \;\rightarrow\; x = 10^4 s = 0, \quad y = 10^{-4}t = 1 \qquad (19)$$

the same error as before.

To undo the confusion caused by (18), it is important before making any pivot choices (i.e., rearrangement of equations or variables) to *scale* the coefficient matrix: Multiply equations and rescale variables by constants chosen to make the largest entry in each equation and each column the same size, say, equal to 1. Then pick an entry that is largest in its row and column for the pivot. After eliminating one variable from the other equations, repeat this process for the remaining $n - 1$ equations in $n - 1$ unknowns; and so on for each successive choice of pivot entry.

Example 4. Stable Elimination

Let us apply the preceding advice to system (18).

$$\begin{aligned} 10^4 s + \quad t &= 10^4 \\ s + 10^{-8}t &= 2 \times 10^{-4} \end{aligned} \qquad (18)$$

We divide the first equation by 10^4 to obtain

$$\begin{aligned} s + 10^{-4}t &= 1 \\ s + 10^{-8}t &= 2 \times 10^{-4} \end{aligned} \qquad (20)$$

The largest entry in each equation is 1, but the column of coefficients for t are all too small. Let us replace t by $y = 10^{-4}t$. Then (20) becomes

$$\begin{aligned} s + \quad y &= 1 \\ s + .0001y &= .0002 \end{aligned} \qquad (21)$$

Now we could pick y in the first equation or s in the second. Let us pivot on the y term in the first equation. Interchanging the order of the variables, we have

$$\begin{aligned} y + s &= 1 \\ .0001y + s &= .0002 \end{aligned} \qquad (22)$$

Gaussian elimination gives

$$\begin{aligned} y + \quad s &= 1 \\ .9999s &= .0001 \end{aligned} \qquad (23)$$

yielding $s \simeq .0001$ and $y \simeq .9999$. Recall that $x = 10^4 s$, so $x \simeq 1$.

The reader should check that small errors in (23) lead to a small change in the answer.   ∎

We repeat the rule of stable elimination. This rule is called **complete pivoting** in the numerical analysis literature.

Rule of Stable Elimination. First apply scaling to rows and columns as necessary to make the largest entry in each row and column equal to 1.

   An entry that is the largest (in absolute value) entry in its row and column should be chosen for the pivot. Interchange equations and variables to make the pivot the first entry in the first equation. Now eliminate the first variable from the remaining equations.

   Repeat this whole process for each round of elimination.

Let us look next at the question of inaccurate solutions from the view of the "person on the street" who needs to solve a system of equations. He or she will probably not worry about inaccuracy until it happens. So the key question is: How do you know if the solution $\mathbf{x}^*$ that you computed to the system of equations $\mathbf{Ax} = \mathbf{b}$ is accurate? The answer is simple. If $\mathbf{x}^*$ were the true solution, $\mathbf{Ax}^*$ would equal $\mathbf{b}$. A simple measure of error is the vector $\boldsymbol{\varepsilon}$:

$$\boldsymbol{\varepsilon} = \mathbf{Ax}^* - \mathbf{b} \qquad (\text{or } \mathbf{Ax}^* = \mathbf{b} + \boldsymbol{\varepsilon}) \qquad (24)$$

A true solution makes $\boldsymbol{\varepsilon}$ equal $\mathbf{0}$. If $\boldsymbol{\varepsilon}$ is unacceptably large, one should re-solve the system from scratch using a stable elimination method just given.

Optional

Maybe you already used this method, but the system was inherently unstable. Then the best way to proceed is to correct $\mathbf{x}^*$ as follows. If $\mathbf{x}^0$ is the true solution, our error in $\mathbf{x}^*$, $\mathbf{e} = \mathbf{x}^* - \mathbf{x}^0$, satisfies the equation

$$\begin{aligned}
\mathbf{Ae} = \mathbf{A}(\mathbf{x}^* - \mathbf{x}^0) &= \mathbf{Ax}^* - \mathbf{Ax}^0 \\
&= (\mathbf{b} + \boldsymbol{\varepsilon}) - \mathbf{b} \qquad (25) \\
&= \boldsymbol{\varepsilon}
\end{aligned}$$

So we should solve the equation $\mathbf{Ae} = \boldsymbol{\varepsilon}$ (by the stable elimination method) and subtract our solution $\mathbf{e}^*$ from $\mathbf{x}^*$ to get a corrected solution $\mathbf{x}^{**} = \mathbf{x}^* - \mathbf{e}^*$ for the original system.

Condition Number of a Matrix

Now we turn to the study of inherently unstable systems. We saw a case of instability back in Section 1.2 with the canoe-with-sail equations, which had the form

$$x_1 + kx_2 = b_1 \tag{26}$$
$$x_1 + x_2 = b_2$$

When k became close to 1, these two equations represented almost parallel lines. A small change in the value of k near 1 has a large effect on where these lines will intersect. The choice of pivots is not at issue here.

In Section 3.1 we obtained Cramer's rule, a determinant-based formula, for the solution to a system of equations. This formula involves division by the determinant of the coefficient matrix. For the coefficient matrix $\mathbf{A}$ in (26), $\det(\mathbf{A}) = 1 \cdot 1 - k \cdot 1 = 1 - k$. As $k \to 1$, $\det(\mathbf{A}) \to 0$. So in Cramer's rule we are almost dividing by 0 and problems will abound.

The critical issue is not just the size of the $\det(\mathbf{A})$ but the size of $\det(\mathbf{A})$ relative to the size of the entries in $\mathbf{A}$ (which are used in the numerator in Cramer's rule).

A more rigorous analysis of errors needs to use the norm of the coefficient matrix $\mathbf{A}$. Recall that

$$\|\mathbf{A}\| = \max \frac{|\mathbf{Ax}|}{|\mathbf{x}|}$$

here $|\mathbf{x}|$ and $|\mathbf{Ax}|$ are the sizes of these vectors measured by some norm (the euclidean norm, the sum norm, or the max norm, introduced in Section 2.5). $\|\mathbf{A}\|$ is the maximum magnifying effect that matrix multiplication can have on a vector. Recall that the sum norm $\|\mathbf{A}\|_s$ is simply the largest column sum and the max norm $\|\mathbf{A}\|_{mx}$ is the largest row sum. In both norms, the sums are of the absolute values of the entries.

For any vector x,

$$\|\mathbf{Ax}\| \le \|\mathbf{A}\| \cdot |\mathbf{x}| \tag{27}$$

Suppose that $\mathbf{E}$ represents a matrix of errors (either in recording data or roundoff errors): The true matrix $\mathbf{A}$ has become the matrix $\mathbf{A} + \mathbf{E}$. Then let us see how changing $\mathbf{A}$ to $\mathbf{A} + \mathbf{E}$ changes the solution to the matrix equation $\mathbf{Ax} = \mathbf{b}$. Suppose that $\mathbf{x}$ is the solution to the correct equation and $\mathbf{x} + \mathbf{e}$ represents the solution to the altered equation. Thus we have

$$\mathbf{Ax} = \mathbf{b} \quad \text{and} \quad (\mathbf{A} + \mathbf{E})(\mathbf{x} + \mathbf{e}) = \mathbf{b} \tag{28}$$

We now derive a bound on the relative size of $\mathbf{e}$ in terms of the relative size of $\mathbf{E}$, a bound of the form

$$\frac{|e|}{|x + e|} \leq c(A) \frac{\|E\|}{\|A\|} \tag{29}$$

We derive (29) by subtracting the first equation in (28) from the second to obtain

$$(A + E)(x + e) - Ax = b - b = 0 \qquad \text{or} \qquad Ae + E(x + e) = 0 \tag{30a}$$

or

$$Ae = -E(x + e) \tag{30b}$$

We assume that **A** is invertible (or else the solution is not unique). Then we can solve (30b) for **e**.

$$e = -A^{-1}E(x + e) \tag{31}$$

Taking norms in (31) and using (27), we have

$$|e| = |A^{-1}E(x + e)| \leq \|A^{-1}E\| \cdot |x + e| \tag{32}$$

Now we use the fact given in Section 2.5 that for any matrices **A**, **B**: $\|AB\| \leq \|A\| \cdot \|B\|$. With it, we have

$$\|A^{-1}E\| \leq \|A^{-1}\| \cdot \|E\| \tag{33}$$

Combining (33) with (32), we have

$$|e| \leq \|A^{-1}E\| \cdot |x + e| \leq \|A^{-1}\| \cdot \|E\| \cdot |x + e| \tag{34}$$

Dividing by $|x + e|$ yields the bound we were seeking.

$$\frac{|e|}{|x + e|} \leq \|A^{-1}\| \cdot \|E\| \tag{35}$$

Equation (35) can be rewritten as

$$\frac{|e|}{|x + e|} \leq (\|A^{-1}\| \cdot \|A\|) \frac{\|E\|}{\|A\|} \tag{36}$$

So the constant in (29) turns out to be $c(A) = \|A^{-1}\| \cdot \|A\|$. This product $c(A)$ is called the **condition number** of the matrix **A**. A small condition number means that the matrix is well behaved and yields stable computations during elimination, since by (36) small errors in **A** can only produce small errors in the solution vector.

The condition number can also be shown to bound the effects on **x** of an error in **b**, when $A(x + e) = b + e'$.

$$\frac{|e|}{|x|} \leq c(A) \frac{|e'|}{|b|} \tag{37}$$

The exact value of the condition number is dependent on which matrix norm we use.

Example 5. **Condition Number of Canoe-with-Sail Problem**

The canoe-with-sail equations in (26) have the coefficient matrix

$$A = \begin{bmatrix} 1 & k \\ 1 & 1 \end{bmatrix} \tag{38}$$

We want to study the sensitivity of a solution x of $Ax = b$ to small errors in the value of k when k is close to 1. From the preceding discussion, we compute the condition number $c(A) = \|A\|_s\|A^{-1}\|_s$, using the sum matrix norm. The sum norm of A is the largest column sum (remember absolute values). When k is close to 1, the sum of each column is 2 or about 2; so we say that $\|A\|_s = 2$. Computing the inverse of A (by the determinant-based inverse formula for a 2-by-2 matrix in Section 3.1), we have

$$A^{-1} = \begin{bmatrix} \dfrac{1}{1 - k} & \dfrac{-k}{1 - k} \\ \dfrac{-1}{1 - k} & \dfrac{1}{1 - k} \end{bmatrix} \tag{39}$$

Again with k close to 1, the sum of (the absolute values) of each column in A^{-1} is about $2/(1 - k)$. So let $\|A^{-1}\|_s = 2/(1 - k)$. Then

$$c(A) = \|A\|_s \cdot \|A^{-1}\|_s = 2 \cdot \frac{2}{1 - k} = \frac{4}{1 - k} \tag{40}$$

For $k = .75$, $c(A) = 4/(1 - .75) = 16$, so (where $\|A\|_s = 2$)

$$\frac{|e|}{|x + e|} \leq c(A) \frac{\|E\|}{\|A\|} = 16 \frac{\|E\|}{2} = 8\|E\| \tag{41}$$

Thus a small error of, say $\frac{1}{12}$ in the value of k, with $\|E\|_s = \frac{1}{12}$ (the error matrix E for A is all 0's except for $\frac{1}{12}$ in k's entry) could lead to a large percentage error in the solution $x + e$ of up to $8(\frac{1}{12}) = 67\%$. Recall that back in Section 1.2, changing k from $\frac{3}{4}$ to $\frac{2}{3}$ changed the solution of $Ax = [5, 7]$ from $x = [-1, 8]$ to $x' (= x + e) = [1, 6]$. So in this case, $|e|_s/|x + e|_s = (2 + 2)/(1 + 6) = 57\%$. ∎

For a geometric picture of why matrices with rows (or columns) that are almost equal are ill-conditioned (unstable) (see Exercise 4 of Section 3.3).

___

Example 6. A Well-Conditioned System

Consider the refinery system of equations introduced in Section 1.2 whose coefficient matrix is

$$\mathbf{A} = \begin{bmatrix} 20 & 4 & 4 \\ 10 & 14 & 5 \\ 5 & 5 & 12 \end{bmatrix}$$

In Example 5 of Section 3.3, we computed its inverse to be

$$\mathbf{A}^{-1} = \begin{bmatrix} .05958 & -.01166 & -.015 \\ -.03958 & .09167 & -.025 \\ -.00833 & -.03333 & .1 \end{bmatrix}$$

Taking the maximum (absolute value) of the column sums, we have $\|\mathbf{A}\|_s = 35$ (first column) and $\|\mathbf{A}^{-1}\|_s = .14$ (third column). Thus the condition number of $\mathbf{A}$ is

$$c(\mathbf{A}) = \|\mathbf{A}\|_s \|\mathbf{A}^{-1}\|_s = 35 \times .14 = 4.9$$

This is a reasonably well conditioned matrix. For the demand vector $\mathbf{b} = [500, 850, 1000]$ we have used in this refinery model, we found (in Section 3.2) that the solution $\mathbf{x}$ of $\mathbf{Ax} = \mathbf{b}$ was

$$x_1 = 4\tfrac{7}{8} \qquad x_2 = 33\tfrac{1}{8} \qquad x_3 = 67\tfrac{1}{2}$$

If we change entry $(2, 3)$ of $\mathbf{A}$ from 5 to 7 to get $\mathbf{A}'$, the error matrix $\mathbf{E}$ [a matrix of all 0's except entry $(2, 3)$ is 2] has $\|\mathbf{E}\|_s = 2$. The error bound (35) gives

$$\frac{|\mathbf{e}|_s}{|\mathbf{x} + \mathbf{e}|_s} \le c(\mathbf{A}) \frac{\|\mathbf{E}\|_s}{\|\mathbf{A}\|_s} = 4.9 \frac{2}{35} = .28 \qquad (42)$$

So a $\tfrac{2}{35}$ change in the norm of $\mathbf{A}$ can yield a 28% error in the norm of the solution. Solving $\mathbf{A}'\mathbf{x}' = \mathbf{b}$ for the same $\mathbf{b}$, one would obtain

$$x_1' = 6.5 \qquad x_2' = 19.7 \qquad x_3' = 72.3$$

Let $\mathbf{x}' = \mathbf{x} + \mathbf{e}$, where $|\mathbf{x}'| = |\mathbf{x} + \mathbf{e}| = 98.5$; then

$$\mathbf{e} = \mathbf{x}' - \mathbf{x} = [6.5 - 4.8, 19.7 - 33.2, 72.3 - 67.5]$$
$$= [1.7, -13.5, 4.8]$$

with $|e|_s = 20$. Thus $|e|_s/|x + e|_s = 20/98.5 = .20$, which is not far from the maximum percentage error of .28 given in (42). ∎

In the Exercises the reader has the chance to examine how the solution to $Ax = b$ is affected by small changes in a coefficient of A, for a variety of well-conditioned and ill-conditioned matrices.

Section 3.5 Exercises

Summary of Exercises

Exercises 1–12 concern the speed of elimination computations and elimination on tridiagonal matrices. Exercises 13–17 involve choice of pivots. Exercises 18–27 deal with the condition number of a matrix; it is assumed that the sum norm is being used. (Note that the word problems in Exercises 24–26 use inverses that would have been computed in Exercises 15–17 of Section 3.3.)

1. Let A be an 8-by-8 matrix. How many multiplications (approximately) are required to perform each of the following operations?
   (a) Compute A^2.   **(b)** Solve $Ax = b$.   **(c)** Compute A^{-1}.

2. Let A be an n-by-n matrix. How many multiplications (approximately) are required to perform each of the following operations?
   (a) Compute A^3.   **(b)** Solve $Ax = b$.
   (c) Iterate $x^{(k+1)} = Ax^{(k)}$ 10 times.   **(d)** Compute A^{-1}.

3. Let A be a 200-by-200 tridiagonal matrix. How large must k be so that squaring a k-by-k matrix takes as many multiplications as solving $Ax = b$?

4. Solve the following tridiagonal systems of equations.

   (a)
   $$
   \begin{aligned}
   x_1 - x_2 &= 1 \\
   -x_1 + 2x_2 - x_3 &= 0 \\
   -x_2 + 2x_3 - x_4 &= -1 \\
   -x_3 + 2x_4 - x_5 &= 1 \\
   -x_4 + 2x_5 - x_6 &= 1 \\
   -x_5 + 2x_6 &= 2
   \end{aligned}
   $$

   (b)
   $$
   \begin{aligned}
   x_1 + x_2 &= 1 \\
   2x_1 + x_2 + x_3 &= 0 \\
   2x_2 + x_3 + x_4 &= 0 \\
   3x_3 + 2x_4 + x_5 &= 0 \\
   2x_4 + 3x_5 + x_6 &= 0 \\
   2x_5 + 3x_6 &= -4
   \end{aligned}
   $$

5. Repeat Exercise 4, part (a) with the last equation changed to $-x_5 + x_6 = -2$. You now get a set of solutions. Express these solutions in terms of x_6.

6. Find the stable distribution (as done in Example 1) for Markov chains with the following transition matrices.

(a) $$\begin{bmatrix} \frac{1}{3} & \frac{1}{3} & 0 & 0 & 0 & 0 \\ \frac{2}{3} & \frac{1}{3} & \frac{1}{3} & 0 & 0 & 0 \\ 0 & \frac{1}{3} & \frac{1}{3} & \frac{1}{3} & 0 & 0 \\ 0 & 0 & \frac{1}{3} & \frac{1}{3} & \frac{1}{3} & 0 \\ 0 & 0 & 0 & \frac{1}{3} & \frac{1}{3} & \frac{2}{3} \\ 0 & 0 & 0 & 0 & \frac{1}{3} & \frac{1}{3} \end{bmatrix}$$

(b) $$\begin{bmatrix} \frac{2}{3} & \frac{2}{3} & 0 & 0 & 0 & 0 \\ \frac{1}{3} & \frac{1}{6} & \frac{2}{3} & 0 & 0 & 0 \\ 0 & \frac{1}{6} & \frac{1}{6} & \frac{2}{3} & 0 & 0 \\ 0 & 0 & \frac{1}{6} & \frac{1}{6} & \frac{2}{3} & 0 \\ 0 & 0 & 0 & \frac{1}{6} & \frac{1}{6} & \frac{2}{3} \\ 0 & 0 & 0 & 0 & \frac{1}{6} & \frac{1}{3} \end{bmatrix}$$

(c) $$\begin{bmatrix} \frac{2}{3} & \frac{2}{3} & 0 & 0 & 0 & 0 \\ \frac{1}{3} & \frac{1}{6} & \frac{2}{3} & 0 & 0 & 0 \\ 0 & \frac{1}{6} & \frac{1}{6} & \frac{1}{6} & 0 & 0 \\ 0 & 0 & \frac{1}{6} & \frac{1}{6} & \frac{1}{6} & 0 \\ 0 & 0 & 0 & \frac{2}{3} & \frac{1}{6} & \frac{1}{3} \\ 0 & 0 & 0 & 0 & \frac{2}{3} & \frac{2}{3} \end{bmatrix}$$

(d) $$\begin{bmatrix} 0 & \frac{1}{2} & 0 & 0 & 0 & 0 \\ 1 & 0 & \frac{1}{2} & 0 & 0 & 0 \\ 0 & \frac{1}{2} & 0 & \frac{1}{2} & 0 & 0 \\ 0 & 0 & \frac{1}{2} & 0 & \frac{1}{2} & 0 \\ 0 & 0 & 0 & \frac{1}{2} & 0 & 1 \\ 0 & 0 & 0 & 0 & \frac{1}{2} & 0 \end{bmatrix}$$

7. Expand the frog Markov chain from 6 states to 20 states (all the middle columns, like the middle columns of the current 6-state transition matrix). Solve for the stable distribution.

8. Expand the frog Markov chain from 6 states to n states (all the middle columns, like the middle columns of the current 6-state transition matrix). Solve for the stable distribution.

9. Suppose that you have an 10-by-10 matrix that has a tridiagonal form except for one row.
   (a) Explain why if this row is the last row, the speed of solving $\mathbf{Ax} = \mathbf{b}$ with a tridiagonal matrix is barely affected.
   (b) Explain why if this row is the first row, the speed of solving $\mathbf{Ax} = \mathbf{b}$ can become proportional to n^2 operations.

10. (a) Verify that finding the inverse of an n-by-n matrix will take about n^3 multiplications (or divisions) using elimination by pivoting.
   Hint: Because of special right sides, only $n^3/6$ steps are used on right sides during the elimination procedure.
   (b) Verify that finding the inverse of an n-by-n matrix will take about n^3 multiplications (or divisions) using Gaussian elimination with back substitution.

11. Show that for a band matrix with bandwidth w, Gaussian elimination takes approximately $w^2 n$ multiplications (or divisions).

12. (a) Find $\mathbf{A}^{-1}$ for the upper triangular matrix

$$\mathbf{A} = \begin{bmatrix} 1 & 2 & 3 & 4 \\ 0 & 1 & 1 & 2 \\ 0 & 0 & 2 & 4 \\ 0 & 0 & 0 & 1 \end{bmatrix}$$

using back substitution [or equivalently, pivoting on the augmented matrix [$\mathbf{A}$ $\mathbf{I}$], except start with entry (4, 4), then entry (3, 3), (2, 2), (1, 1)].

(b) Generalize the computation in part (a) to show that the inverse of an n-by-n upper triangular matrix requires about $n\frac{3}{6}$ multiplications (or divisions) to compute.

(c) Generalize the computation in part (a) to show that the inverse of an upper triangular matrix is upper triangular.

13. Solve by regular Gaussian elimination the following systems of equations with three significant digits (i.e., 2.002 becomes 2.00 and .9996 is rounded to 1.00).

(a) $.001x - y = 1$ **(b)** $.001x + 2y = 1$ **(c)** $.002x - 3y = 1$
      $3x + y = 0$ $-2x - y = 3$ $4x + y = 1$

14. Re-solve the problems in Exercise 13 using the stable elimination rules for rearranging rows and columns and scaling. Compare your solutions to the ones obtained in Exercise 13.

15. Solve by regular Gaussian elimination the following systems of equations with three significant digits.

(a) $.001x + y - z = 1$ **(b)** $x + .001y + z = 1$
       $x + 2y + z = 2$ $2x + .001y + z = 3$
    $-x - y + 2z = 3$ $-x + 3y + z = 0$

16. Re-solve the problems in Exercise 15 using the stable elimination rules for rearranging row and columns and scaling. Compare your solution to the one obtained in Exercise 15.

17. Take the wrong answer $\mathbf{x}^*$ you obtained for each problem $\mathbf{Ax} = \mathbf{b}$ in Exercise 13 and compute the right-side error $\boldsymbol{\varepsilon} = \mathbf{Ax}^* - \mathbf{b}$ (the right side obtained by $\mathbf{Ax}^*$ minus the true right side). Solve $\mathbf{Ax} = \boldsymbol{\varepsilon}$ (using the stable elimination rules) and subtract the solution $\mathbf{e}^*$ from $\mathbf{x}^*$ to get a more accurate answer $\mathbf{x}^{**} = \mathbf{x}^* - \mathbf{e}^*$. Compare the answer $\mathbf{x}^{**}$ with the answer you obtained from this problem in Exercise 14.

For Exercises 18–27 involving the condition number of a matrix, always use the sum norm.

18. What is the condition number of the n-by-n identity matrix?

19. Determine the condition number of the following matrices. Comment on whether or not small errors in data of each matrix **A** can result in large errors in the solution of **Ax** = **b**.

(a) $\begin{bmatrix} 1 & 3 \\ 2 & 4 \end{bmatrix}$ **(b)** $\begin{bmatrix} 2 & -3 \\ -2 & 1 \end{bmatrix}$ **(c)** $\begin{bmatrix} 1 & -3 \\ -2 & 6 \end{bmatrix}$ **(d)** $\begin{bmatrix} \frac{1}{3} & \frac{1}{4} & \frac{1}{5} \\ \frac{1}{4} & \frac{1}{5} & \frac{1}{6} \\ \frac{1}{5} & \frac{1}{6} & \frac{1}{7} \end{bmatrix}$

20. In the matrix **A** in Example 5, if k is changed from .8 to .85, how large a relative error in the solution to **Ax** = **b** is possible?

21. (a) In the refinery problem in Example 6, if entry (1, 1) is changed from 20 to 10 (yielding matrix **A'**), how large a relative error in the solution to **A'x** = **b** is possible?
   (b) Solve the system **A'x** = **b** for the **A'** in part (a) and compare the actual relative error with the relative error bound given in part (b).

22. (a) Compute the condition number of

$$\mathbf{A} = \begin{bmatrix} 1 & 1 & 1 \\ 2 & 4 & -1 \\ 1 & -1 & 2 \end{bmatrix}$$

   (b) Solve **Ax** = **b**, where **b** = [1, 2, 3].
   (c) Suppose that we change entry (2, 1) from 2 to 1 to get a new **A**. How large a relative change in solution of **Ax** = **b** is possible with this change in **A**? [use the condition number estimate (36)].
   (d) Solve **Ax** = **b** for this new **A**, and compare the observed relative change to the one predicted in part (c).
   (e) The large condition number of **A** in part (a) means that this matrix is close to being noninvertible, that is, that some combination of two rows of **A** almost equals a third row—show that $\frac{1}{3}$ of the sum of two of the rows almost equals the other row.

23. Answer parts (a) and (b) using equation (37) in the text.
   (a) If **b** = [2, 1] is changed to [2, 2], how large a change can this yield in the solution to **Ax** = **b** for **A** in Exercise 19 part (a)?
   (b) If **b** = [1, 2, 3] is changed to [2, 2, 3], how large a change can this yield in the solution to **Ax** = **b** for **A** in Exercise 22?
   (c) Derive the bound $|\mathbf{e}|/|\mathbf{x} + \mathbf{e}| \le c(\mathbf{A})\{|\mathbf{e}'|/|\mathbf{b}|\}$ in equation (37) by following the reasoning in equations (30) and (32) to obtain $|\mathbf{e}| \le \|\mathbf{A}^{-1}\| \cdot |\mathbf{e}'|$, then divide by $|\mathbf{x}|$ ($\ge |\mathbf{b}|/\|\mathbf{A}\|$).

24. (This is a continuation of Exercise 17 of Section 3.2 and Exercise 15 of Section 3.3.) The staff dietician at the California Institute of Trigonometry has to make up a meal with 600 calories, 20 grams of protein, and 200 milligrams of vitamin C. There are three food types to choose from: rubbery jello, dried fish sticks, and mystery meat. They have the following nutritional content per ounce.

	Jello	Fish Sticks	Mystery Meat
Calories	10	50	200
Protein	1	3	.2
Vitamin C	30	10	0

If there is at most a 5% error in (sum) norm of any column in this matrix of data **A**, how large a relative error can occur in solving the dietician's problem?

25. (This is a continuation of Exercise 18 in Section 3.2 and Exercise 16 in Section 3.3) A furniture manufacturer makes tables, chairs, and sofas. In one month, the company has available 300 units of wood, 350 units of labor, and 225 units of upholstery. The manufacturer wants a production schedule for the month that uses all of these resources. The different products require the following amounts of the resources.

	Table	Chair	Sofa
Wood	4	1	3
Labor	3	2	5
Upholstery	2	0	4

If the amount of wood needed to make a table was accidentally entered as 3 instead of 4, how large a relative error in the solution to this production problem is possible?

26. (This is a continuation of Exercise 20 of Section 3.2 and Exercise 17 of Section 3.3.) An investment analyst is trying to find out how much business a secretive TV manufacturer has. The company makes three brands of TV set: brand A, brand B, and brand C. The analyst learns that the manufacturer has ordered from suppliers 450,000 type 1 circuit boards, 300,000 type 2 circuit boards, and 350,000 type 3 circuit boards. Brand A uses 2 type-1 boards, 1 type-2 board, and 2 type-3 boards. Brand B uses 3 type-1 boards, 2 type-2 boards, and 1 type-3 board. Brand C uses 1 board of each type.

If there were a mistake in getting the type 1 circuit board orders and the analyst thought 350,000 boards were ordered instead of 450,000 boards, how large a relative error in the solution to this TV production problem is possible?

27. Show that for any invertible matrix **A** with condition number $c(\mathbf{A})$ (using the sum norm), $c(\mathbf{A}) \geq 1$.

Hint: Use the fact that $\|\mathbf{AB}\| \leq \|\mathbf{A}\| \cdot \|\mathbf{B}\|$.

4

A Sampling of Linear Models

Linear Transformations in Computer Graphics

In this chapter we discuss some linear models in greater detail. Three of these models were introduced in Chapter 1, Markov chains, linear programming, and population growth. Using matrix algebra and solution techniques learned in the previous chapters, we shall be able to analyze and solve these linear models.

A general solution for most of the models in this chapter requires one to solve a system of linear equations; in matrix form, solve $\mathbf{Ax} = \mathbf{b}$. In solving some of these systems of linear equations, various theoretical difficulties will arise. Those problems will motivate the theory of solutions to systems of linear equations, which is discussed in Chapter 5.

A common use of linear models is to predict the values of a set of variables in the future as a linear function of the variables' current values. A Markov chain is such a model, as was the rabbit–fox population model. Other models of this type are presented in this chapter. These models assume the matrix form

$$\mathbf{w}' = \mathbf{Aw} \tag{1}$$

or more generally (in future examples)

255

$$\mathbf{w}' = \mathbf{A}\mathbf{w} + \mathbf{b} \tag{2}$$

where $\mathbf{w}$ is the vector of current values and $\mathbf{w}'$ the vector of future values. In this section we give a geometric interpretation of (1) and (2).

A **linear transformation** T of the plane maps each point $\mathbf{w} = (x, y)$ into a point $\mathbf{w}' = T(\mathbf{w})$, where $T(\mathbf{w}) = \mathbf{A}\mathbf{w}$, for some 2-by-2 matrix $\mathbf{A}$. A linear transformation in n dimensions is defined the same way ($\mathbf{A}$ is then an n-by-n matrix). When linear transformations are programmed on a computer, they can be used to move figures about and create the special visual effects we have come to associate with computer graphics. In this section the reader will learn how to build up complicated graphics effects out of simple transformations.

It will sometimes be convenient in this section to drop matrix notation and write $\mathbf{w}' = T(\mathbf{w}) = \mathbf{A}\mathbf{w}$ to represent the pair of linear equations

$$\begin{aligned} x' &= ax + by \\ y' &= cx + dy \end{aligned} \tag{3}$$

A slightly more general transformation, corresponding to $\mathbf{w}' = \mathbf{A}\mathbf{w} + \mathbf{b}$, is an **affine linear transformation.**

$$\begin{aligned} x' &= ax + by + e \\ y' &= cx + dy + f \end{aligned} \tag{4}$$

Clearly, all linear transformations are affine linear transformations (with $e = f = 0$). Shortly, we will see that affine linear transformations lack some very important properties that linear transformations have.

Figure 4.1b, c, and d show the effect of transformations T_1, T_2, and T_3, respectively, on the square in Figure 4.1a whose corners are $A = (0, 0)$, $B = (1, 0)$, $C = (1, 1)$, $D = (0, 1)$.

$$T_1: \quad \begin{aligned} x' &= 2x + 4 \\ y' &= 3y + 2 \end{aligned} \tag{5}$$

$$T_2: \quad \begin{aligned} x' &= \cos 45°x - \sin 45°y = .707x - .707y \\ y' &= \sin 45°x + \cos 45°y = .707x + .707y \end{aligned} \tag{6}$$

$$T_3: \quad \begin{aligned} x' &= x + y \\ y' &= y \end{aligned} \tag{7}$$

Transformation T_1 doubles the width and triples the height of the square and also moves it 4 units to the right and 2 units up. Transformation T_2 has the effect of revolving the square 45° counterclockwise about the origin, but does not change the square's size. Transformation T_3 slants the y-axis and lines parallel to it by 45°. To help understand the effect of these transfor-

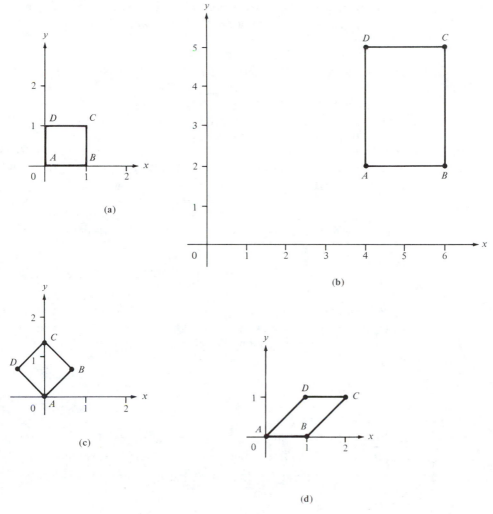

Figure 4.1 (a) Unit square. (b) Square transformed by T_1. (c) Square transformed by T_2. (d) Square transformed by T_3.

mations, readers should evaluate T_1, T_2, and T_3 at point C ($=[1, 1]$) of the square in Figure 4.1a. Exercise 31 shows that revolving a point about the origin always has the form (6).

There is an important computational question to ask about transforming a square or any figure built out of line segments. Is it sufficient to compute just the new coordinates of the corners, and then connect these new corners with straight lines to obtain the full transformed figure? Fortunately, the answer is yes. This result is a simple consequence of two basic laws of matrix algebra. We state these laws in linear transformation form as a theorem.

Theorem 1. Let T be a linear transformation with $\mathbf{w}' = T(\mathbf{w})$ and $\mathbf{v}' = T(\mathbf{v})$. Then for any scalar constants r and s,

    (i) $T(\mathbf{w} + \mathbf{v}) = T(\mathbf{w}) + T(\mathbf{v}) = \mathbf{w}' + \mathbf{v}'$

    (ii) $T(r\mathbf{w}) = rT(\mathbf{w}) = r\mathbf{w}'$

    (iii) $T(r\mathbf{w} + s\mathbf{v}) = rT(\mathbf{w}) + sT(\mathbf{v}) = r\mathbf{w}' + s\mathbf{v}'$

When Theorem 1, parts (i) and (ii) are rewritten in matrix form with $T(\mathbf{w}) = \mathbf{Aw}$, they become the familiar matrix laws $\mathbf{A}(\mathbf{w} + \mathbf{v}) = \mathbf{Aw} + \mathbf{Av}$ and $\mathbf{A}(r\mathbf{w}) = r(\mathbf{Aw})$. Theorem 1, part (iii) is just a combination of parts (i) and (ii). Theorem 1 is not true for affine linear transformations $T(\mathbf{w}) = \mathbf{Aw} + \mathbf{b}$—parts (ii) and (iii) fail (see Exercise 33 for counterexamples). Theorem 1, part (iii) generalizes to linear combinations of three or more points.

If $\mathbf{w}$ and $\mathbf{v}$ are the two endpoints of a line segment L, any point $\mathbf{t}$ on L can be written as a linear combination of $\mathbf{w}$ and $\mathbf{v}$ of the form

$$\mathbf{t} = r\mathbf{w} + (1 - r)\mathbf{v}, \qquad \text{for some } r, \quad 0 \leq r \leq 1 \qquad (8)$$

The constant r is the fraction of the distance $\mathbf{t}$ is from $\mathbf{w}$ to $\mathbf{v}$. For example, if $\mathbf{t}$ were halfway between $\mathbf{w}$ and $\mathbf{v}$, then $r = .5$. When $\mathbf{w}$ and $\mathbf{v}$ are mapped by some linear transformation T to points $\mathbf{w}'$ and $\mathbf{v}'$, the line segment between them will be all points of the form

$$\mathbf{t}' = r\mathbf{w}' + (1 - r)\mathbf{v}', \qquad \text{for some } r, \quad 0 \leq r \leq 1 \qquad (9)$$

By Theorem 1, part (iii), we see that if $\mathbf{t}$ is as in (8), then $\mathbf{t}'$ [$= T(\mathbf{t})$] is the expression in (9). So linear transformations map lines into lines. This result is also true for affine linear transformations [it is easily verified using matrix algebra (see Exercise 32)].

Theorem 2. An affine linear transformation T maps line segments into line segments.

Theorem 2 allows us to compute transformations of straight-line figures simply by transforming corners of figures and then drawing lines between the transformed corners.

In computer graphics applications, Theorem 1 says that if the coordinates of corners, or other critical points, in a figure can be expressed as linear combinations of the coordinates of some "key" points, then to transform the figure we only need to apply the linear transformation to the coordinates of these key points; the coordinates of other points can quickly be obtained from the coordinates of the transformed key points.

Theorem 1 restated the basic fact of matrix algebra that vector addition and scalar multiplication are preserved by matrix-vector multiplication. Another almost-as-easy consequence from matrix algebra is the following result.

Theorem 3. If T_1 and T_2 are two affine linear transformations, the composite transformation $\mathbf{w}'' = T_2(T_1(\mathbf{w}))$, obtained by mapping $\mathbf{w}$ to $\mathbf{w}' = T_1(\mathbf{w})$ and then mapping $\mathbf{w}'$ to $\mathbf{w}'' = T_2(\mathbf{w}')$, is also an affine linear transformation.

Proof. Let $T_1(\mathbf{w})$ be the mapping $\mathbf{w}' = \mathbf{A}_1\mathbf{w} + \mathbf{b}_1$ and $T_2(\mathbf{w})$ be the mapping $\mathbf{w}' = \mathbf{A}_2\mathbf{w} + \mathbf{b}_2$. Then using matrix algebra, $T_2(T_1(\mathbf{w}))$ can be written

$$T_2(T_1(\mathbf{w})) = \mathbf{A}_2(\mathbf{A}_1\mathbf{w} + \mathbf{b}_1) + \mathbf{b}_2 \tag{10}$$
$$= \mathbf{A}_2\mathbf{A}_1\mathbf{w} + \mathbf{A}_2\mathbf{b}_1 + \mathbf{b}_2$$

Then $T_2(T_1(\mathbf{w}))$ is the affine linear transformation $T_3(\mathbf{w}) = \mathbf{A}_3\mathbf{w} + \mathbf{b}_3$, where

$$\mathbf{A}_3 = \mathbf{A}_2\mathbf{A}_1 \quad \text{and} \quad \mathbf{b}_3 = \mathbf{A}_2\mathbf{b}_1 + \mathbf{b}_2 \tag{11} \blacksquare$$

This theorem was easy to prove using matrix algebra. Without it, we would have to substitute one system of equations for affine linear transformation T_1 into another system of equations for affine linear transformation T_2—a giant mess!

Theorem 3 lets us build up complicated transformations out of simpler transformations that revolve, expand distances along the x- or y-axis, move left (right) or up (down), slant axes, and other changes. Another use is in creating animated motion. If the rotation T_2 [in equation (6)] used an angle of $1°$, we would obtain a linear transformation T_2' that would revolve, say, a square around the origin by $1°$. If we repeatedly applied T_2' 360 times, we would obtain an "animated" sequence of figures that create one full revolution of the square around the origin.

Example 1. **Transforming a Set of Squares**

Draw a figure F consisting of a set of eight unit squares whose corners are at points (g, h), $g = 1, 2, 3$ and $h = -2, -1, 0, 1, 2$ (see Figure 4.2a). Observe that the coordinates of all corner points are linear combinations of the coordinates of the lower-left corner $(1, -2)$ and the "change in coordinates" points $(1, 0)$ and $(0, 1)$. For example, the point $(2, 2) = (1, -2) + 1(1, 0) + 4(0, 1)$. Now let us transform F first by applying the linear transformation $T_2(T_3(x, y))$ (T_3 followed by T_2), and second by applying the linear transformation $T_3(T_2(x, y))$. The transformed figures are drawn in Figure 4.2b and c. For convenience, we restate T_2 and T_3 here:

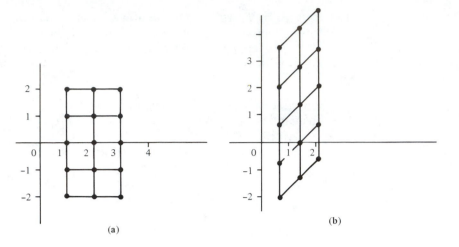

Figure 4.2 (a) Grid. (b) Grid transformed by T_3 followed by T_2.
(c) Grid transformed by T_2 followed by T_3.

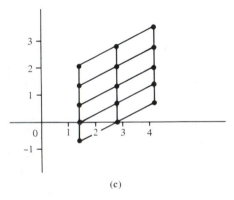

$$T_2: \quad \begin{aligned} x' &= \cos 45°x - \sin 45°y = .707x - .707y \\ y' &= \sin 45°x + \cos 45°y = .707x + .707y \end{aligned} \quad (6)$$

$$T_3: \quad \begin{aligned} x' &= x + y \\ y' &= y \end{aligned} \quad (7)$$

We successively apply T_2 and T_3 in both orders to the key points $(1, -2)$, $(1, 0)$, and $(0, 1)$.

$$\begin{aligned} T_2(T_3(1, -2)) &= (.707, -2.121) & T_3(T_2(1, -2)) &= (1.414, -.707) \\ T_2(T_3(1, 0)) &= (.707, .707) & T_3(T_2(1, 0)) &= (1.414, .707) \\ T_2(T_3(0, 1)) &= (0, 1.414) & T_3(T_2(0, 1)) &= (0, .707) \end{aligned}$$

We use Theorem 1, part (iii) to obtain the other corners as linear combinations of the transformed key points. For example, since the point $(2, 2) = (1, -2) + 1(1, 0) + 4(0, 1)$, then

$$\begin{aligned} T_2(T_3(2, 2)) &= T_2T_3(1, -2) + 1T_2T_3(1, 0) + 4T_2T_3(0, 1)) \\ &= (.707, -2.121) + 1(.707, .707) + 4(0, 1.414) \\ &= (1.414, 4.242) \end{aligned}$$

Note that changing the order of the transformations changes the composite transformation. Composing linear transformations is not commutative! This is because matrix multiplication is not commutative. ∎

There are many geometric properties of figures that we might hope affine linear transformations would preserve: an angle at a corner, the area of a square, and the distance of each point from the origin. Each of the properties is satisfied by some but not all of T_1, T_2, T_3, defined above. With a little thought, the reader should be able to guess conditions on affine linear transformations which make each of these properties true.

Another interesting property is reversibility—if $\mathbf{u}' = T(\mathbf{u})$, does there exist another transformation T^{-1} such that $\mathbf{u} = T^{-1}(\mathbf{u}')$? That is, does T have an inverse? T^{-1} should exist if the matrix used to define T has an inverse; details are left as an exercise.

Next let us consider transformations to represent three-dimensional figures in two dimensions. Any time one draws a three-dimensional figure on a piece of paper or displays it on a computer screen, one is performing such a transformation. A transformation $(x', y') = T(x, y, z)$ that maps a three-dimensional figure into two dimensions is really operating in three dimensions, but the z-coordinate always becomes zero [i.e., $(x', y', 0) = T(x, y, z)$].

The simplest type of linear transformation from three to two dimensions is a projection onto the x-y plane (or onto the x-z or y-z planes). This has the form $T(x, y, z) = (x, y)$—just delete the z-coordinate. A more general projection would project three dimensional space onto some plane that is not parallel to any pair of coordinate axes. Figure 4.3 illustrates such a projection of two dimensions onto one dimension. Here we have projected the x-y plane onto the (one-dimensional) line $x = y$ using the transformation

Figure 4.3 Projecting x-y plane onto line $y = x$.

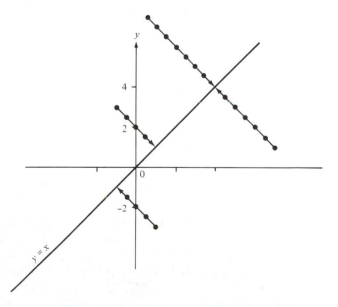

$$x' = .5x + .5y \tag{12}$$
$$y' = .5x + .5y$$

The arrows in Figure 4.3 show how sample points are projected onto this line. When three-dimensional points are projected onto a plane, the situation is similar to Figure 4.3, but in three dimensions it is harder to illustrate with a figure. In Chapter 5 we will learn more about projection mappings. (We note as an aside that there is no way to reverse a projection, that is, there is no affine linear transformation that maps a line onto the whole plane, since by Theorem 2 lines are always mapped onto lines; similarly, a plane cannot be mapped linearly onto all of three-dimensional space.)

The following two examples illustrate a standard way to map three dimensions into two, as well as a way to map two dimensions into three.

Example 2. **Projection of a Cube into the Plane**

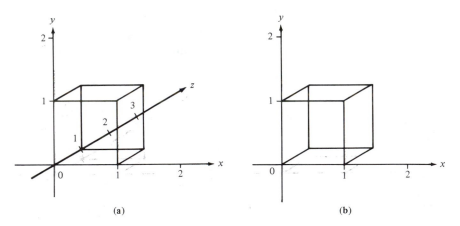

(a)   (b)

Figure 4.4 (a) Three-dimensional unit cube. (b) Projection of unit cube into *x-y* plane.

Devise a linear transformation that projects the three-dimensional unit cube shown in Figure 4.4a into the (x, y)-plane so that the cube looks just the way it is drawn on the (two-dimensional) page of this book in Figure 4.4a. In Figure 4.4a the *z*-axis is represented as a line at a 30° angle to the *x*-axis. Moreover, distances along the *z*-axis in Figure 4.4a are drawn with half the length of distances along the *x*- or *y*-axis. So the projection we want acts on a point (x, y, z) as follows: The *z*-coordinate should alter the *x*, *y* coordinates in the direction of a 30° angle above the *x*-axis and the distance of the displacement should be half the value of the *z*-coordinate.

$$T': \qquad x' = x + \tfrac{1}{2}\cos 30°z = x + .433z \qquad (13)$$
$$y' = y + \tfrac{1}{2}\sin 30°z = y + .25z$$

See Figure 4.4b.   ■

Example 3. Revolve a Letter Around the *x*-Axis

Devise a set of linear transformations that take the letter L (for **L**inear) in Figure 4.5a and revolve it 5° around the *x*-axis, then 10°, then 15°, and so on, around the *x*-axis, to make an animated movie of the L revolving around the *x*-axis. The revolution around the *x*-axis takes place in three dimensions. Let us agree to represent the transformed L in two dimensions using the projection T' given by (13). That is, we treat the L as a figure in three dimensions, then successively revolve it 5° around the *x*-axis and display the results at each stage in two dimensions using T'.

Figure 4.5 (a) Letter L. (b) Letter L revolved 50° about *x*-axis and projected back onto *x-y* plane. (c) Letter L revolved 50° about *x*-axis, then shrunk by T_{10} and projected onto *x-y* plane. (d) Letter L revolved 150° about *x*-axis, then shrunk by T_{30} and projected onto *x-y* plane.

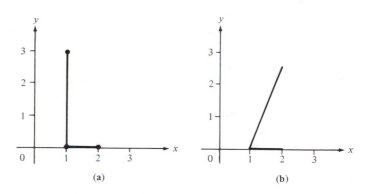
(a)   (b)

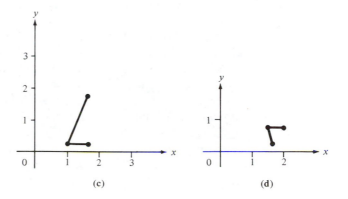
(c)   (d)

The corners of L are (1, 3), (1, 0), (2, 0), which in three dimensions become (1, 3, 0), (1, 0, 0), (2, 0, 0). Revolving L around the x-axis is similar to revolving a figure in the (x, y)-plane around the origin; this is what T_2 did, at the beginning of this section. We keep the x-coordinate fixed and revolve the y- and z-coordinates the way T_2 did. Thus the required linear transformation T in three dimensions is

$$T: \qquad x' = x$$
$$y' = \cos 5°y - \sin 5°z \qquad (14)$$
$$z' = \sin 5°y + \cos 5°z$$

First we apply T to the corners of L, (1, 3, 0), (1, 0, 0), (2, 0, 0), and then apply T again to the transformed corners and continue applying T. Each time we apply T, we display L in two dimensions by using the projection T'. The original L and the result after applying T 10 times, a 50° revolution (and then applying T'), are shown in Figure 4.5a and b. Note that T leaves corners (1, 0, 0) and (2, 0, 0) unchanged. ∎

The result $\mathbf{w}^*$ of applying T 10 times to an initial 3-vector $\mathbf{u}$ could also be computed by multiplying $\mathbf{u}$ by the tenth power of the matrix of coefficients in (14). Of course, the smart way to obtain a 50° rotation is just to substitute $\cos 50°$ and $\sin 50°$ in (14).

There are several variations on the transformation in Example 3 that are used in computer graphics. In practice, we would probably want to transform a whole set of letters that spell out a word. We might shrink or magnify the letters by an amount r as they revolve; we use the transformation $(x'', y'', z'') = (rx', ry', rz')$ composed with T. We might want the letters to recede back into the distance, away from the (x, y)-plane. This can be accomplished by increasing the z-coordinate a little more each time and shrinking the letters a little (in all coordinates). The following affine linear transformation could be used after k applications of T.

$$T_k: \qquad (x'', y'', z'') = (.95^k x', .95^k y', .95^k z' + .1k) \qquad (15)$$

Note that T_k has the undesirable affect of shrinking the x-coordinate toward the left (toward the origin). Figure 4.5c shows the result of applying T 10 times followed by T_{10}; Figure 4.5d shows the result of applying T 30 times followed by T_{30} (again T' is used to get planar depictions).

Let us add a warning about roundoff errors. A small error in computing (14) (to revolve the letter L) may become a noticeable error after dozens of iterations of (14). One easy way to eliminate this type of error is on every tenth iteration to compute the new coordinates of the transformed L directly from the original coordinates, (1, 3), (1, 0), (2, 0), by performing a 50° rotation [as noted above, do this by using $\cos 50°$ and $\sin 50°$ in (14)]. This method of eliminating an accumulation of errors by "updating" is used frequently in many different types of linear models.

Hopefully, Examples 1, 2, and 3 have given the reader a sense of how to generate a variety of transformations to move figures around in two and three dimensions. We have tried to provide the reader with the basic tools. We leave the creation of more interesting graphics transformations as projects for the readers (some simple graphics projects are suggested in the Exercises). Please note how extremely tedious animated graphics would be without computer programs to map sets of points repeatedly, as in Example 3, and to build a complex transformation from a sequence of simple transformations, as permitted by Theorem 3.

There is one very important problem in computer graphics that we have not discussed—the hidden surface problem. In Figure 4.4 we show all the corners and edges of the cube, but actually some are not visible because they lie at the back of the cube. The problem of determining which corners, edges, and surfaces of an object are visible is tricky and its solution relies heavily on linear algebra; even determinants are involved.

In earlier chapters we saw how eigenvectors can simplify the computation in iterating the system $\mathbf{x}' = \mathbf{A}\mathbf{x}$. We now give a geometric interpretation of how eigenvectors simplify a linear transformation $T(\mathbf{w}) = \mathbf{A}\mathbf{w}$. Recall that an eigenvector $\mathbf{u}$ of a matrix $\mathbf{A}$ has the property that multiplying $\mathbf{u}$ by $\mathbf{A}$ has the effect of multiplying $\mathbf{u}$ by a scalar. That is, there is some scalar λ, called an eigenvalue of $\mathbf{A}$, such that $\mathbf{A}\mathbf{u} = \lambda\mathbf{u}$. Since matrix $\mathbf{A}$ and the linear transformation $T(\mathbf{w}) = \mathbf{A}\mathbf{w}$ are really "the same thing," we will speak interchangeably about a vector $\mathbf{u}$ being an eigenvector of $\mathbf{A}$ or of T.

Eigenvectors allow us to break a linear transformation into simple parts. (This reverses our previous goal of building up complicated transformations from simple ones.) The idea is to change to a coordinate system based on the eigenvectors of $\mathbf{A}$, as was done in Example 8 of Section 2.5, and apply T in terms of this new coordinate system (in Section 3.3 we showed that converting to eigenvector coordinates was equivalent to writing $\mathbf{A}$ in the form $\mathbf{U}\mathbf{D}_\lambda\mathbf{U}^{-1}$).

Example 4. Eigenvector Coordinates to Simplify a Linear Transformation

Consider the linear transformation T,

$$\begin{aligned} x' &= x + y \\ y' &= \tfrac{1}{2}x + \tfrac{3}{2}y \end{aligned} \qquad (16)$$

The standard (x, y) coordinate system expresses a point as a linear combination of the point $(1, 0)$—the distance along the x-axis—and the point $(0, 1)$—the distance along the y-axis. That is, $(x, y) = x(1, 0) + y(0, 1)$. Now let us use a coordinate system in which points are expressed as a linear combination of two eigenvectors of T (these are magically provided by the author). They are $\mathbf{u}_1 = (1, 1)$ and $\mathbf{u}_2 = (-2, 1)$. Applying T to $\mathbf{u}_1$ and $\mathbf{u}_2$, we have

$$T(1, 1) = (2, 2) \quad = 2(1, 1)$$
$$T(-2, 1) = (-1, \tfrac{1}{2}) = .5(-2, 1)$$

We want to give an arbitrary point (x, y) new coordinates (r, s) such that

$$(x, y) = r(1, 1) + s(-2, 1) \tag{17}$$

Since T multiplies vector $(1, 1)$ by 2, then by Theorem 1, T also multiplies vector $r(1, 1)$ (for any r) by 2; similarly for multiples of $(-2, 1)$. In a coordinate system (r, s) based on $(1, 1)$ and $(-2, 1)$, T becomes

$$r' = 2r \tag{18}$$
$$s' = .5s$$

It will require some work to convert a point in standard (x, y) coordinates [based on points $(1, 0)$ and $(0, 1)$] to this new coordinate system based on $(1, 1)$ and $(-2, 1)$. But if T is to be applied repeatedly, the simple form (18) of T in the new coordinates is worth the effort of conversion. Writing (17) as a system of equations for the coordinates, we have

$$\begin{bmatrix} x \\ y \end{bmatrix} = r \begin{bmatrix} 1 \\ 1 \end{bmatrix} + s \begin{bmatrix} -2 \\ 1 \end{bmatrix} \tag{19}$$

or

$$\mathbf{w} = \mathbf{Et}, \quad \text{where } \mathbf{E} = \begin{bmatrix} 1 & -2 \\ 1 & 1 \end{bmatrix}, \quad \mathbf{t} = \begin{bmatrix} r \\ s \end{bmatrix}$$

Here $\mathbf{E}$ is the matrix whose columns are the eigenvectors.

Solving (19) for r and s, we obtain

$$\mathbf{t} = \mathbf{E}^{-1}\mathbf{w}: \quad r = \tfrac{1}{3}x + \tfrac{2}{3}y \tag{20}$$
$$s = -\tfrac{1}{3}x + \tfrac{1}{3}y$$

Let us consider the effect of repeatedly applying T to the point $(1, 0)$ [given in (x, y) coordinates]. Using (20), we convert $(1, 0)$ to $(\tfrac{1}{3}, -\tfrac{1}{3})$ in (r, s) coordinates. Now we repeatedly apply T to $(\tfrac{1}{3}, -\tfrac{1}{3})$ using (18). We get the sequence of points [in (r, s) coordinates]

$$(\tfrac{1}{3}, -\tfrac{1}{3}), (\tfrac{2}{3}, -\tfrac{1}{6}), (\tfrac{4}{3}, -\tfrac{1}{12}), (\tfrac{8}{3}, -\tfrac{1}{24}), \ldots$$

Each application of T doubles the first coordinate and halves the second coordinate (see Figure 4.6). So after k applications of T we get

$$(r_k, s_k) = \tfrac{1}{3}(2^k, -(\tfrac{1}{2})^k)$$

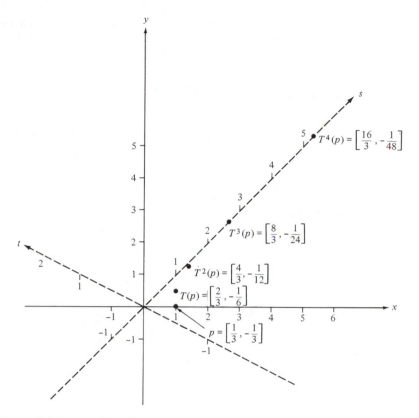

Figure 4.6 Repeated application of T in Example 4 to point $p = (0, 1)$. Points $T^k(p)$ using s, t coordinates are shown.

If we want to express the coordinates of this point (r_k, s_k) back in the original (x, y) coordinates, we simply convert back with (19) to obtain

$$x_k = r_k - 2s_k = \tfrac{1}{3}2^k + \tfrac{2}{3}(\tfrac{1}{2})^k$$
$$y_k = r_k + s_k = \tfrac{1}{3}2^k - \tfrac{1}{3}(\tfrac{1}{2})^k \tag{21}$$

For large k, 2^k is much much larger than $\tfrac{1}{3}(\tfrac{1}{2})^k$, so we can neglect the latter term (the effects of computer roundoff will eventually drop the smaller term for us).

In the long term, the largest eigenvalue always dominates the effects of all other eigenvalues (as was discussed in Section 2.5). In this case we have

$$(x_k, y_k) = \tfrac{1}{3}(2^k, 2^k) \tag{22}$$

This simple result would have been much harder to obtain without the change of coordinates. Clearly, using eigenvector-based coordinates can make calculations with linear transformations much easier.

∎

Note: The interested reader should consult one of the computer graphics references at the end of the book for more information about this fascinating application of linear algebra.

Section 4.1 Exercises

Summary of Exercises
Exercises 1–24 call for the construction of various affine linear transformations and plotting their effect on certain figures. Exercises 25–27 have eigenvector-coordinate computations. Exercises 28–33 involve associated theory.

1. Construct affine linear transformations to do the following to the square in Figure 4.1a.
   (a) Rotate the square 180° counterclockwise around the origin (in the plane).
   (b) Move the square 7 units to the right, 3 units up, and double its width.
   (c) Make the vertical lines of the square slant at a 45° angle (height unchanged):

   (d) Reflect the square about the *y*-axis.

2. Write out the affine linear transformations in Exercise 1 in the form **u′** = **Au** + **b**, giving **A** and **b**.

3. Compute the new coordinates of the corners of the square in Figure 4.1a for each of the transformations in Exercise 1.

4. If T is the triangle with corners at $(-1, 1)$, $(1, 1)$, $(1, -1)$, draw T after it is transformed by each of the transformations in Exercise 1.

5. Apply each of the transformations of Exercise 1 to the grid in Figure 4.2a and plot your answer.

6. The following exercise verifies which types of affine transformations are commutative. All transformations act on the *x-y* plane. Let

   T_a double the *x*-coordinate: $x' = 2x,\ y' = y$.
   T_b double the *y*-coordinate: $x' = x,\ y' = 2y$.
   T_c reflect about the *y*-axis: $x' = -x,\ y' = y$.
   T_d reflect about the *x*-axis: $x' = x,\ y' = -y$.
   T_e shift *x*-value 2 units: $x' = x + 2,\ y' = y$.
   T_f shift *y*-value 3 units: $x' = x,\ y' = y + 3$.
   T_g rotate 45° around the origin $[= T_2$ in (6)$]$.

    (a) Compute $T_a T_c$ and $T_c T_a$. Do these two transformations commute?
    (b) Compute $T_b T_e$ and $T_e T_b$. Do these two transformations commute?
    (c) Compute $T_b T_f$ and $T_f T_b$. Do these two transformations commute?
    (d) Compute $T_b T_g$ and $T_g T_b$. Do these two transformations commute?
    (e) Compute $T_d T_g$ and $T_g T_d$. Do these two transformations commute?
    (f) Compute $T_c T_f$ and $T_f T_c$. Do these two transformations commute?
    (g) Compute $T_e T_g$ and $T_g T_e$. Do these two transformations commute?

7. Construct affine linear transformations to do the following to the square in Figure 4.1a and plot the square after the transformation is performed.
    (a) Double the width of the square (double x coordinates) and rotate it 90° around the origin.
    (b) Reflect the square about the y-axis and then reflect about the x-axis.
    (c) Move the square 7 units to the right, 3 units up, and then rotate the square 180° counterclockwise around the origin.

8. Reflect the square in Figure 4.1a about the line $y = x$. Give your transformation.

9. Reflect the square in Figure 4.1a about the line $y = 2$. Give your transformation.

10. Rotate the square in Figure 4.1a 90° about the point $(\frac{1}{2}, 0)$ by first moving the square so that its center is at the origin, then rotate it 90°, and finally reverse the initial move. Give your transformation.

11. Rotate the grid in Figure 4.2a 90° counterclockwise about the point $(3, 1)$.

    Hint: See Exercise 10.
    Give your transformation.

12. Rotate the square in Figure 4.1a 45° about its center.

    Hint: See Exercise 10.
    Give your transformation.

13. By squaring the associated matrix, determine the transformation of repeating twice the transformations in Exercise 7, parts (a) and (b). Plot the square in Figure 4.1a after applying each squared transformation. Finally, describe in words the effect of the squared transformations.

14. Cube the matrix associated with the linear transformation in Exercise 1, part (a) and verify that the resulting linear transformation is the same as the original one (rotating 180° three times is the same as rotating 180° once).

15. Square the matrix associated with the projection linear transformation T' as (13) and verify that the square equals the original matrix. Explain this result in words.

16. Devise an affine linear transformation to project any (x, y) point onto the following lines.

    (a) $y = -x$ **(b)** $y = 2x$

    (c) $y = x + 2$ **(d)** $y = 3x - 8$

17. **(a)** Verify that $x' = 2x - y$, $y' = 2x - y$ projects any x-y point onto the line $y' = x'$ and that the projection is not perpendicular like (12), but, rather, the line segment from (x, y) to $(2x - y, 2x - y)$ has a slope of 2.

    (b) Construct a projection T that maps any x-y point $\mathbf{w}$ onto the line $y = 2x$ so that the line segment from $\mathbf{w}$ to $T(\mathbf{w})$ has a slope of 1.

18. Give the x-y coordinates of the corners of the following x-y-z figures after they are projected onto the x-y plane using projection T' in (13) and draw the figures (in the x-y plane).

    (a) A triangle with corners $(1, 1, 0)$, $(1, 0, 1)$, $(0, 1, 1)$.

    (b) A pyramid with base $(0, 0, 0)$, $(2, 0, 0)$, $(0, 0, 2)$, $(2, 0, 2)$ and top at $(1, 2, 1)$.

19. Construct a linear transformation to do the following to the letter L in Figure 4.5a.

    (a) Revolve it $30°$ around the y-axis.

    (b) Revolve it $10°$ around the z-axis.

    (c) Revolve it $30°$ around the y-axis and then $30°$ around the z-axis.

20. Give the composite linear transformation of performing the revolution T in Example 3 followed by the projection T' in (13).

21. Revolve the grid in Figure 4.2a $30°$ around the x-axis and project onto the x-y plane [using T' in (13)] (plot this).

22. Revolve the square in Figure 4.1a $60°$ around the x-axis (in three dimensions), then shrink all coordinates to half-size and project it onto the x-y plane using T' in (13).

23. **(a)** Construct a linear transformation to revolve an object $30°$ around the line of points $(x, 1, 0)$ (the line is parallel to the x-axis with $y = 1, z = 0$).
        Hint: See Exercise 8.

    (b) Apply the linear transformation in part (a) to the grid in Figure 4.2a and project the result onto the x-y plane with projection T' in (13).

24. Construct a linear transformation to make an animated movie in which in each successive frame the object

    (a) Revolves $30°$ around the y-axis and shrinks its x- and y-coordinates by 10%.

    (b) Revolves $10°$ around the y-axis, shrinks all coordinates by 10%, and then moves 2 units along the z-axis.

25. Repeat Example 4 to find the approximate point after k transformations using T in (16) if the initial point is
(a) $(3, 2)$ (b) $(-1, 1)$ (c) $(-2, -1)$

26. Find the eigenvalues and associated eigenvectors of the following linear transformations (use the method in Section 3.1).
(a) $x' = 3x + y, y' = 2x + 2y$ (b) T' in (13)
(c) T_2 in (6)
Hint: Eigenvalues are complex.
(d) T_3 in (7)   (e) T_1 in (5)

27. Use your results in Exercise 26, part (a) to find the result of applying that transformation 5 times to the point [5, 2] and the approximate value of applying it 20 times.

28. This exercise gives a "picture" of how, when two columns of **A** are almost the same, the inverse of **A** almost does not exist. For the following matrices **A**, solve the system $\mathbf{A} \begin{bmatrix} x_1 \\ x_2 \end{bmatrix} = \begin{bmatrix} 1 \\ 0 \end{bmatrix}$. Then plot $x_1 \mathbf{a}_1^C$ and $x_2 \mathbf{a}_2^C$ in a two-dimensional coordinate system and show geometrically how the sum of vectors $x_1 \mathbf{a}_1^C$ and $x_2 \mathbf{a}_2^C$ is $\begin{bmatrix} 1 \\ 0 \end{bmatrix}$ (here $\mathbf{a}_1^C$, $\mathbf{a}_2^C$ denote the two columns of **A**).

(a) $\begin{bmatrix} 2 & 3 \\ 1 & 2 \end{bmatrix}$ (b) $\begin{bmatrix} 2 & 3 \\ 2 & 2 \end{bmatrix}$ (c) $\begin{bmatrix} 5 & 6 \\ 6 & 7 \end{bmatrix}$ (d) $\begin{bmatrix} 8 & 9 \\ 7 & 8 \end{bmatrix}$.

29. A linear transformation $\mathbf{w}' = T(\mathbf{w}) = \mathbf{Aw}$ can be reversed if **A** is invertible. Then the reverse transformation is $T^*(\mathbf{w}) = \mathbf{A}^{-1}\mathbf{w}$. Find the reverse transformation, if possible, for
(a) T_3 in (7)   (b) T_2 in (6)   (c) T' in (13)

30. An affine linear transformation $\mathbf{w}' = T(\mathbf{w}) = \mathbf{Aw} + \mathbf{b}$ can be reversed if **A** is invertible.
(a) In matrix notation, what is the inverse transformation $T^*(\mathbf{w})$ for $T(\mathbf{u}) = \mathbf{Aw} + \mathbf{b}$ (so that T^*T is the identity transformation)?
(b) Find the inverse transformation for T_1 in (5).

31. Consider the point $\mathbf{u} = (1, 0)$ and suppose that we want to rotate it $\theta°$ counterclockwise around the origin. Then its new position will be distance 1 from the origin and at an angle of $\theta°$ (with respect to the x-axis). Using a similar argument for the point $\mathbf{v} = (0, 1)$, show that for $\mathbf{u}' = \mathbf{Au}$ and $\mathbf{v}' = \mathbf{Av}$ to have the right values, **A** must be

$$\mathbf{A} = \begin{bmatrix} \cos \theta° & -\sin \theta° \\ \sin \theta° & \cos \theta° \end{bmatrix}$$

32. Verify that affine linear transformations map lines into lines by showing if $\mathbf{u}' = T(\mathbf{u}) = \mathbf{A}\mathbf{u} + \mathbf{b}$, points of the form $\mathbf{w} = r\mathbf{u} + (1 - r)\mathbf{v}'$ are mapped into points of the form $\mathbf{w}' = r\mathbf{u}' + (1 - r)\mathbf{v}'$, where $\mathbf{u}' = T(\mathbf{u})$ and $\mathbf{v}' = T(\mathbf{v})$, and $0 \le r \le 1$.

33. Make up counterexamples to show that Theorem 1, parts (i) and (ii), are false for affine linear transformations (virtually any affine example will do).

Section 4.2 Linear Regression

One of the fundamental problems in building linear models is estimating coefficients and other constants in the linear equations. For example, in the refinery model introduced in Section 1.2 we stated that from 1 barrel of crude oil the first refinery would produce 20 gallons of heating oil, 10 gallons of diesel oil, and 5 gallons of gasoline. These production levels would vary from one batch of crude oil to another and might also vary depending on how much crude oil was being processed each day. The numbers given are estimates, not precise values. The first important work in linear algebra grew out of an estimation problem.

In 1818 the famous mathematician Karl Friedrich Gauss was commissioned to make a geodetic survey of the kingdoms of Denmark and Hanover (geodetic surveys create very accurate maps of a portion of the earth's spherical surface). In making estimates for the positions of different locations on a map, Gauss developed the least-squares theory of regression that we present in this section. This theory yields a system of linear equations that must be solved. To solve them, Gauss invented the algorithm we now call Gaussian elimination, presented in Section 3.2. This method is still the best way known to solve systems of linear equations. It should be noted that not only did this survey project cause Gauss to start the theories of statistics and linear algebra, but to compensate for the slightly nonspherical change of the earth, Gauss was also led to develop the theory of differential geometry! The following equation summarizes this paragraph.

$$\text{one good application} + \text{one genius} = \text{important new mathematics} \quad (*)$$

Let us return humbly to the problem of estimating coefficients. Recall the linear model from Example 2 of Section 1.4 for predicting C, the college grade average of a Scrooge High School graduate, in terms of the student's Scrooge High average S. The proposed model was

$$C = 1.1 \times S - .9 \tag{1}$$

The constants in (1) were chosen to "fit" as closely as possible data about eight graduates. The heart of these models is the choice of the constants.

Let us restate the problem of finding a linear model such as (1) in the following standardized form:

> ***Linear Regression Model.*** Given a set of points (x_1, y_1), (x_2, y_2), $\ldots$, (x_n, y_n), find constants q and r such that the linear relation
>
> $$\hat{y}_i = qx_i + r \qquad (2)$$
>
> gives the best possible fit for these points.

The point $(x_i, \hat{y}_i)$ is the estimate for (x_i, y_i). The name "regression," which means movement back to a less developed state, comes from the idea that our model recaptures a simple relationship between the x_i and the y_i which randomness has obscured (the variables regress to a linear relationship). A model involving several input variables is called **multiple linear regression**. If we try to fit the data to a more complex function, such as $\hat{y} = q_2 x^2 + q_1 x + r$ or $\hat{y} = e^{qx}$, the model is called **nonlinear regression**. We shall concentrate first on simple linear regression. Once this is well understood, we can extend our analysis to the multiple regression problem (using matrix algebra). We shall also show how some problems in nonlinear regression can be transformed into linear regression problems.

Example 1. Using the Model $\hat{y} = qx$

Let us consider a very simple regression problem. Suppose that we want to fit the three points $(0, 1)$, $(2, 1)$, and $(4, 4)$ to a line of the form

$$\hat{y} = qx \qquad (3)$$

(see Figure 4.7). The x-value might represent the number of semesters of college mathematics a student has taken and the y-value the student's score on some test. There are thousands of other settings that might give rise to these values. The estimate (3) would help us predict the

Figure 4.7 Regression estimates for points $(0, 1)$, $(2, 1)$, and $(4, 4)$.

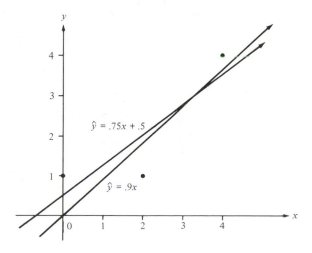

y-values for other x-values, for example, predict how other students might do on the test based on the amount of mathematics they have taken.

The three points in this problem are readily seen not to lie on a common line, much less a line through the origin [any line of the form (3) passes through the origin]. So we have to find a choice of q that gives the best possible fit, that is, a line $\hat{y} = qx$ passing as close to these three points as possible.

What do "best possible fit" and "as close as possible" mean? The most common approach used in such problems is to minimize the sum of the squares of the errors. The error at a point $(x_i, \hat{y}_i)$ will be $|\hat{y}_i - y_i| = |qx_i - y_i|$, the absolute difference between the value qx_i predicted by (3) and the true value y_i (the absolute value is needed so that a "negative" error and a "positive" error cannot offset each other). However, absolute values are not easy to use in mathematical equations. Taking the squares of differences yields positive numbers without using absolute values. There is also a geometric reason we shall give for using squares.

For the points $(0, 1)$, $(2, 1)$, $(4, 4)$, the expression $\Sigma \, (\hat{y}_i - y_i)^2$ for the sum of squares of the errors (SSE) is

$$\begin{aligned} SSE &= (0q - 1)^2 + (2q - 1)^2 + (4q - 4)^2 \\ &= 1 + (4q^2 - 4q + 1) + (16q^2 - 32q + 16) \qquad (4) \\ &= 20q^2 - 36q + 18 \end{aligned}$$

The geometric justification for using a sum of squares is based on the following interpretation of our estimation problem. Let $\mathbf{x}$ be the vector of our x-values and $\mathbf{y}$ be the vector of our corresponding y-values. In this case, $\mathbf{x} = [0, 2, 4]$ and $\mathbf{y} = [1, 1, 4]$. Further, let $\hat{\mathbf{y}}$ be the vector of estimates for $\mathbf{y}$. Equation (3) can now be rewritten

$$\hat{\mathbf{y}} = q\mathbf{x} \qquad (5)$$

That is, the estimates $\hat{y} = [\hat{y}_1, \hat{y}_2, \hat{y}_3]$ from (3) will be q times the x-values $[0, 2, 4]$.

Think of $\mathbf{x}$, $\mathbf{y}$, and $\hat{\mathbf{y}}$ as points in three-dimensional space, where $\hat{\mathbf{y}}$ is a multiple of $\mathbf{x}$. Then the obvious strategy is to pick the value of q that makes $q\mathbf{x}$ ($= \hat{\mathbf{y}}$) as close as possible to $\mathbf{y}$ (see Figure 4.8). That is, we want to minimize the distance $|q\mathbf{x} - \mathbf{y}|_e$ (in the euclidean norm) in three-dimensional space between $q\mathbf{x}$ and $\mathbf{y}$. This distance between $q\mathbf{x} = [0q, 2q, 4q]$ and $\mathbf{y} = [1, 1, 4]$ is simply

$$|q\mathbf{x} - \mathbf{y}|_e = \sqrt{(0q - 1)^2 + (2q - 1)^2 + (4q - 4)^2} \qquad (6)$$

Comparing (4) and (6), we see that $|q\mathbf{x} - \mathbf{y}|_e$ is the square root of SSE. So minimizing SSE will also minimize the distance $|q\mathbf{x} - \mathbf{y}|_e$. Recall that in vector notation $|\mathbf{a}|_e^2$ equals $\mathbf{a} \cdot \mathbf{a}$. So

$$SSE = |q\mathbf{x} - \mathbf{y}|_e^2 = (q\mathbf{x} - \mathbf{y}) \cdot (q\mathbf{x} - \mathbf{y}) \qquad (7)$$

Figure 4.8

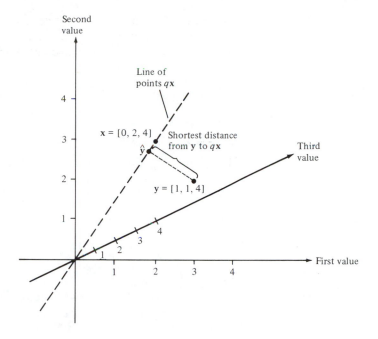

The value of q that minimizes $20q^2 - 36q + 18$ is found by differentiating this expression with respect to q and setting this derivative equal to 0:

$$\frac{d\text{SSE}}{dq} = 40q - 36 = 0 \rightarrow q = .9 \tag{8}$$

The desired regression equation is thus $\hat{y} = .9x$ (see Figure 4.7). Using our regression equation $\hat{y}_i = .9x_i$, our estimate for (0, 1) is (0, 0), for (2, 1) is (2, 1.8), and for (4, 4) is (4, 3.6) [the bad estimate for (0, 1) arose from the fact that any line $\hat{y} = qx$ must go through the origin]. So SSE $= 1^2 + .8^2 + .4^2 = 1.80$, and the distance in 3-space between our estimate vector $\hat{y}$ and the true $\mathbf{y}$ is $|\hat{y} - \mathbf{y}| = \sqrt{\text{SSE}} = \sqrt{1.80} \simeq 1.34$.   ■

Readers should pause a moment to get their geometric bearings. Example 1 started as a problem of estimating a relationship between some x- and y-values that we plotted in Figure 4.7 in x-y space. But then we considered a new geometric picture with three-dimensional vectors, formed by the x-values, the y-values, and the $\hat{y}$ estimates. Let us present the data in a matrix:

	x-value	y-value
First reading	0	1
Second reading	2	1
Third reading	4	4

In this new setting, our objective is to find a multiple q of the first column $\mathbf{x}$ that would estimate the second column $\mathbf{y}$ as closely as possible. Geometrically, we want to find a point $\mathbf{p}$ on the line formed by multiplies $\mathbf{x}$ ($= [0, 2, 4]$)—$\mathbf{p} = q\mathbf{x}$, for some q—such that $\mathbf{p}$ is as close to $\mathbf{y}$ ($= [1, 1, 4)$ as possible. Point $\mathbf{p}$ ($= \hat{\mathbf{y}}$) is the *projection* of $\mathbf{y}$ onto the line from the origin through $\mathbf{x}$.

Example 2. Using the Model $\hat{y} = qx + r$

Let us fit the points in Example 1, $(0, 1)$, $(2, 1)$, $(4, 4)$, using the full linear regression model (2): $\hat{y} = qx + r$. We can also write our regression model as

$$\hat{\mathbf{y}} = q\mathbf{x} + r\mathbf{1}: \qquad \begin{bmatrix} 1 \\ 1 \\ 4 \end{bmatrix} = q \begin{bmatrix} 0 \\ 2 \\ 4 \end{bmatrix} + r \begin{bmatrix} 1 \\ 1 \\ 1 \end{bmatrix} \qquad (9)$$

We want to find an estimate vector $\hat{\mathbf{y}}$ as close to the vector $\mathbf{y}$ of y-values as possible in 3-space. Now $\hat{\mathbf{y}}$ is formed from a linear combination of the vectors $\mathbf{x}$ and $\mathbf{1}$. (The set of all possible linear combinations of $\mathbf{x}$ and $\mathbf{1}$ will be a plane.)

Again we pick q and r by minimizing the sum of squares of errors,

$$\begin{aligned} \text{SSE} &= |\hat{\mathbf{y}} - \mathbf{y}|^2 = \Sigma\,(\hat{y}_i - y_i)^2 = \Sigma\,(qx_i + r - y_i)^2 \\ &= (0q + r - 1)^2 + (2q + r - 1)^2 + (4q + r - 4)^2 \\ &= (r^2 - 2r + 1) + (4q^2 + r^2 + 4qr - 4q - 2r + 1) \\ &\quad + (16q^2 + r^2 + 8qr - 32q - 8r + 16) \\ &= 20q^2 + 3r^2 + 12qr - 36q - 12r + 18 \qquad (10) \end{aligned}$$

To minimize (10) with respect to q and r, we differentiate with respect to q and r and set the partial derivatives equal to 0.

It is left as an exercise for the reader to verify that the partial derivatives of SSE in (10) with respect to q and r are

$$\frac{\partial \text{SSE}}{\partial q} = 40q + 12r - 36 = 0 \qquad \text{or} \qquad 40q + 12r = 36$$

$$\frac{\partial \text{SSE}}{\partial r} = 6r + 12q - 12 = 0 \qquad \text{or} \qquad 12q + 6r = 12 \qquad (11)$$

Solving the pair of equations in (11) for q and r, we obtain

$$q = .75, \qquad r = .5 \qquad (12)$$

So our regression equation is

$$\hat{y} = .75x + .5 \qquad (13)$$

(see Figure 4.7). This time our estimate for $(0, 1)$ is $(0, .5)$, for $(2, 1)$ is $(2, 2)$, and for $(4, 4)$ is $(4, 3.5)$, and $\text{SSE} = .5^2 + 1^2 + .5^2 = 1.50$. The distance $|\hat{\mathbf{y}} - \mathbf{y}|$ between our estimate vector $\hat{\mathbf{y}}$ and the true y-value vector $\mathbf{y}$ is $\sqrt{\text{SSE}} = \sqrt{1.50} = 1.22$. In Example 1 the distance was 1.34. Thus, for these data, our fuller model provided little improvement over the simple model $\hat{y} = qx$. ∎

If we were applying linear regression models (2) or (3) to n points, the x- and y-values would form n-vectors and the distance $|\hat{\mathbf{y}} - \mathbf{y}|$ would be calculated in n dimensions (the reasoning is the same). These calculations would be quite tedious. However, nowadays we have computers to handle the tedium. It is as easy to program a computer to do regression on n points as it is on three points with the following observation.

Proposition. The derivative of a sum of functions is the sum of the derivatives of each function.

This proposition greatly simplifies taking derivatives to find the minimum of SSE. In the model $\hat{y} = qx$, SSE has the following form for the set of points (x_i, y_i), $i = 1, 2, \ldots, n$:

$$\text{SSE} = \Sigma (qx_i - y_i)^2 = \Sigma (q^2 x_i^2 - 2qx_i y_i + y_i^2) \tag{14}$$

By the proposition, the derivative of SSE is the sum of the derivatives of the individual terms in (14):

$$\frac{d\text{SSE}}{dq} = \Sigma (2x_i^2 q - 2x_i y_i) = 2(\Sigma x_i^2)q - 2 \Sigma x_i y_i \tag{15}$$

Setting (15) equal to 0 (to find the minimizing value of q), we obtain

---

Formula for Regression Model $\hat{y} = qx$

$$q = \frac{\Sigma x_i y_i}{\Sigma x_i^2} = \frac{\mathbf{x} \cdot \mathbf{y}}{\mathbf{x} \cdot \mathbf{x}} \tag{16}$$

Let us informally rederive (16) using matrix algebra. Recall that $\text{SSE} = (q\mathbf{x} - \mathbf{y}) \cdot (q\mathbf{x} - \mathbf{y})$. In vector calculus, we treat $(q\mathbf{x} - \mathbf{y}) \cdot (q\mathbf{x} - \mathbf{y})$ like $(q\mathbf{x} - \mathbf{y})^2$, and SSE's derivative is

$$2\mathbf{x} \cdot (q\mathbf{x} - \mathbf{y}) = 2q\mathbf{x} \cdot \mathbf{x} - 2\mathbf{x} \cdot \mathbf{y}$$

If we set this expression equal to zero, we get

$$2q\mathbf{x} \cdot \mathbf{x} - 2\mathbf{x} \cdot \mathbf{y} = 0 \qquad \text{or} \qquad q\mathbf{x} \cdot \mathbf{x} = \mathbf{x} \cdot \mathbf{y}$$

Hence we have directly $q = \mathbf{x} \cdot \mathbf{y}/\mathbf{x} \cdot \mathbf{x}$.

The simple form of the formula (16) for q was not apparent in the calculations we did in Example 1. If we were working with 10 or more points, it would clearly be much easier to do the general calculations just performed to obtain (16) and then plug in the given x- and y-values, rather than to multiply out all the squared factors in SSE and collect terms, as was done in equation (4) in Example 1. General symbolic computations can make life a lot easier! This is what mathematics is all about. The details of an individual problem often hide a nice general structure for solutions. In programming terms, it is often easier to write a computer program to solve a general class of problems and use it to solve one specific problem, rather than write a specialized program for the single problem.

The calculations for the regression model $\hat{y}_i = qx_i + r$ are obtained similarly by generalizing the equations in Example 2. The partial derivatives of SSE can be shown to have the form (see Exercise 10)

$$
\begin{aligned}
\frac{\partial \text{SSE}}{\partial q} &= 2(\Sigma\, x_i^2)q + 2(\Sigma\, x_i)r - 2\,\Sigma\, x_iy_i \\
\frac{\partial \text{SSE}}{\partial r} &= 2(\Sigma\, x_i)q + 2nr - 2\,\Sigma\, y_i
\end{aligned}
\tag{17}
$$

We set the derivatives equal to 0 and solve this pair of equations for q and r. That is, we solve the equations

$$
\begin{aligned}
aq + br &= e \\
cq + dr &= f
\end{aligned}
\tag{18}
$$

where

$$
a = 2\,\Sigma\, x_i^2, \qquad b = 2\,\Sigma\, x_i, \qquad e = 2\,\Sigma\, x_iy_i,
$$
$$
c = 2\,\Sigma\, x_i, \qquad d = 2n, \qquad f = 2\,\Sigma\, y_i
$$

The solution is

Formula for Regression Model $\hat{y} = qx + r$

$$
q = \frac{n\,\Sigma\, x_iy_i - (\Sigma\, x_i)(\Sigma\, y_i)}{n\,\Sigma\, x_i^2 - (\Sigma\, x_i)^2}
$$

$$
r = \frac{(\Sigma\, y_i)(\Sigma\, x_i^2) - (\Sigma\, x_i)(\Sigma\, x_iy_i)}{n\,\Sigma\, x_i^2 - (\Sigma\, x_i)^2}
\tag{19}
$$

Again we note the advantage of generality. Solving a pair of linear equations in terms of constants a, b, and so on, and then substituting complex expressions for the constants of (18), is much easier than directly solving the specific system of equations arising from (17).

Although the formulas in (19) are certainly complicated, they are still quite simple to program. In fact, because linear regression is so widely used, many (nonprogrammable) hand-held calculators have built-in routines to calculate q and r. One simply enters successive (x_i, y_i) pairs. When a pair is entered, the calculator updates sums that it is computing for Σx_i, Σx_i^2, Σy_i, and $\Sigma x_i y_i$. After the last pair is entered, the user presses a Regression key and the calculator inserts these sums into the formulas in (19). The following BASIC program shows how the calculator works.

```
1 N=0: SX=0: SX2=0: SY=0: SXY=0
10 INPUT X,Y
20 N = N + 1
30 SX = SX + X
40 SX2 = SX2 + X*X
50 SY = SY + Y
60 SXY = SXY + X*Y
70 IF Regression key not pushed THEN GOTO 10
100 D = N*SX2 − SX*SX
120 PRINT "R ="; (N*SXY−SX*SY)/D
130 PRINT "Q = "; (SY*SX2−SX*SXY)/D
140 END
```

A word of warning about the formulas in (19). Roundoff errors in computing the terms in the denominators of these formulas can sometimes seriously affect the accuracy of the results. In line 100 of the BASIC program, if N*SX2 and SX*SX were large numbers that were nearly equal, then their difference D may be very inaccurate. Also, one data point quite different from the rest (caused by an unusual event or a recording error) can significantly affect the values of q and r; such points are called **outliers**. Exercise 5 illustrates the effect of outliers.

We now show a convenient shortcut for the linear regression model $\hat{y} = qx + r$. With a computer or calculator programmed to do regression, this shortcut does not save time, but it does eliminate roundoff-error difficulties.

We shall perform an elementary transformation of x-coordinates.

$$x_i' = x_i - \bar{x}, \qquad \text{where } \bar{x} = \frac{1}{n} \Sigma x_i \tag{20}$$

The term $\bar{x}$ we use to shift the x-coordinate is just the average of the x_i. Note that if the x-coordinates are integers (as is common) and are equally spaced along the x-axis, their average will be the middle integer, if n is odd (or midway between the middle two integers, if n is even). In the case in Example 2, the average of x_i is $(0 + 2 + 4)/3 = 2$. Since the y-values are unchanged, we are simply renumbering the x-axis (see Figure 4.9).

Figure 4.9 Figure 4.7 transformed so that $x' = 0$ is the mean of x'-values. The old regression line $\hat{y} = 0.75x + 0.5$ becomes $\hat{y} = 0.75x' + 2$, where the constant 2 is the mean of the y-values.

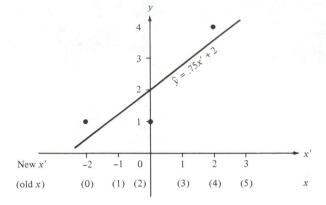

In the new coordinate system, the average of the x'_i will be 0—that was the whole idea of the shift, to center the x-values around the origin. If the average of the x'_i is 0, so is the sum of the x'_i (since the average is the sum divided by n). Now in (19) all products involving $\Sigma\, x_i$ are 0. The formulas for r and q in (19) simplify considerably, to become

$$q' = \frac{\Sigma\, x'_i y_i}{\Sigma\, x'^2_i}$$

$$r' = \frac{1}{n} \Sigma\, y_i$$

(21)

Here r' is just the average of the y_i, and q' is the same formula that we obtained in (16) for the regression model $\hat{y} = qx$. Roundoff error can no longer distort the denominator in (21) as was possible in (19).

Shifting the x-coordinate will not change the slope of a line, so q' equals q in the original model $\hat{y} = qx + r$. The reader can verify with a geometric argument that $r = r' - q\, \Sigma\, x_i/n$.

If we had also transformed the y-values by their average, then $r' = 0$. However, the regression formula for q' does not simplify further if we transform the y-values, so a y-transformation serves no purpose. Also, transforming just the y-values instead of the x-values will not simplify the denominator in (19) as happened in (21).

Example 3. Predicting Printing Costs

A copy center bases its fees on the number of (duplicate) units that have been ordered (a unit is 100 pages). Table 4.1 gives some sample fees. Based on these sample fees, what would be a reasonable charge for 15 units?

Let us fit a line $\hat{y} = qx + r$ to these five data points, (1, 6), (3, 5.5), (5, 5), (10, 3.5), (12, 3), and then determine $\hat{y}$ when $x =$

Table 4.1

Number of Units	Cost per Unit
1	$6
3	5.5
5	5
10	3.5
12	3

15. Using the shortcut described above, we transform the coordinates by subtracting $6.2 \ (= \Sigma \ x_i/n)$ from each x_i. Then (21) gives

$$q' = \frac{\Sigma \ x_i' y_i}{\Sigma \ x_i'^2}$$

$$= \frac{(-5.2) \times 6 + (-3.2) \times 5.5 + (-1.2) \times 5 + 3.8 \times 3.5 + 5.8 \times 3}{(-5.2)^2 + (-3.2)^2 + (-1.2)^2 + (3.8)^2 + (5.8)^2}$$

$$= -.3$$

$$(22)$$

and $r' = \Sigma \ y_i/5 = 4.6$. Then

$$q = q' = -.3 \quad \text{and} \quad r = r' - \frac{q \ \Sigma \ x_i}{n} = 4.6 - (-.3)(6.2) \approx 6.5$$

Thus our regression model is $\hat{y} = -.3x + 6.5$. And when $x = 15$, we obtain $\hat{y} = -.3 \times 15 + 6.5 = 2$. ∎

Next we give an example of a nonlinear regression problem and show how it can be converted into simple linear regression.

Example 4. **Nonlinear Regression**

Consider the following pairs of x- and y-values; the points are shown in Figure 4.10a. The x-values could be the age of a wine and the y-values ratings by expert wine tasters.

x	1	2	3	4	5	6	7
y	.4	.6	1.1	1.6	2.9	5	8

Suppose that by inspection and experience we believe that the relationship between x- and y-values is best explained by an exponential model

$$\hat{y} = re^{qx} \qquad (23)$$

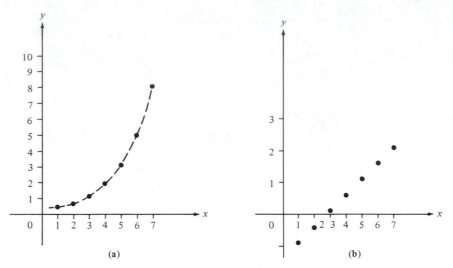

Figure 4.10 (a) Data points and curve $y = \frac{1}{4}e^{x/2}$. (b) Data transformation by $y' = \log y$.

Then we perform the following (nonlinear) transformation on the y-values.

$$y' = \log y \qquad (24)$$

The new data values are

x	1	2	3	4	5	6	7
y'	$-.9$	$-.5$	.1	.5	1.1	1.6	2.1

Clearly, we have a fairly good linear fit here (see Figure 4.10b). (One can also plot the original data points on log paper.) We let $\hat{y}'_i$ be the estimate for the transformed problem:

$$\hat{y}'_i = \log \hat{y}_i = \log re^{qx_i} = \log r + qx_i$$

Letting $r' = \log r$, we have the standard simple linear regression model

$$\hat{y}' = qx + r' \qquad (25)$$

That is, the logarithm function (24) transforms exponential curves into straight lines. We can apply the formulas in (19) to determine q and r' from the transformed data and insert these into (23) to get a model in the original coordinate system. We obtain

$$q = \tfrac{1}{2}, \qquad r' = -1.4 \quad (r = e^{-1.4} \simeq \tfrac{1}{4})$$

The curve $\hat{y} = \frac{1}{4}e^{x/2}$ is plotted in dashed lines in Figure 4.10a.   ∎

This technique can be used any time that there is a mapping that transforms the proposed regression model, such as (23), into a simple linear regression model. A statistician will test out several different regression models and settle on the one that gives the best fit (e.g., that minimizes the sum of the squares of the errors).

Another way to perform nonlinear regression is to fit the y values to a polynomial in x, such as $y_i = ax_i^3 + bx_i^2 + cx_i + d$, where we treat x_i^3, x_i^2, and x_i like the three distinct variables, v_i, w_i, and x_i. We cannot solve a problem with several variables on the right yet, but we will come back to least-squares polynomial fitting in Section 5.3.

Optional

We conclude this section with a vector calculus derivation of the general multivariable regression model in which we allow y to be a linear function of several input values. The same results will be obtained in Section 5.3 more simply using vector space techniques.

To be concrete, we consider the following model for $\hat{y}_i$:

$$\hat{y}_i = q_1 v_i + q_2 w_i + q_3 x_i + r \tag{26}$$

In matrix notation, we write

$$\hat{\mathbf{y}} = q_1 \mathbf{v} + q_2 \mathbf{w} + q_3 \mathbf{x} + r\mathbf{1} \tag{27a}$$

$$= \mathbf{X}\mathbf{q} \tag{27b}$$

where $\mathbf{q} = [q_1, q_2, q_3, r]$ and $\mathbf{X} = [\mathbf{v} \quad \mathbf{w} \quad \mathbf{x} \quad \mathbf{1}]$.

Now let us compute SSE and its derivative in terms of (27a). Later we do it in terms of (27b).

$$\text{SSE} = (q_1 \mathbf{v} + q_2 \mathbf{w} + q_3 \mathbf{x} + r\mathbf{1} - \mathbf{y}) \tag{28}$$
$$\cdot (q_1 \mathbf{v} + q_2 \mathbf{w} + q_3 \mathbf{x} + r\mathbf{1} - \mathbf{y})$$

By the informal vector calculus used previously,

$$\frac{d\text{SSE}}{dq_1} = 2\mathbf{v} \cdot (q_1 \mathbf{v} + q_2 \mathbf{w} + q_3 \mathbf{x} + r\mathbf{1} - \mathbf{y}) \tag{29}$$

The derivatives with respect to q_2 and q_3 and r will be similar. Multiplying $\mathbf{v}$ by the vectors in the parentheses in (29) and setting the result equal to 0, we obtain

$$q_1 \mathbf{v} \cdot \mathbf{v} + q_2 \mathbf{v} \cdot \mathbf{w} + q_3 \mathbf{v} \cdot \mathbf{x} + r\mathbf{v} \cdot \mathbf{1} = \mathbf{v} \cdot \mathbf{y} \tag{30}$$

Three other similar equations will be obtained from the other three derivatives. This gives us four equations in the four unknowns q_1, q_2, q_3, and r that can be solved by Gaussian elimination. (Recall that it was this system of equations of estimating regression unknowns that forced Gauss to develop Gaussian elimination.)

Let us indicate how (30) and its sister equations can be obtained as a matrix equation. We write the SSE using (27b):

$$\text{SSE} = (\hat{\mathbf{y}} - \mathbf{y}) \cdot (\hat{\mathbf{y}} - \mathbf{y}) = (\mathbf{Xq} - \mathbf{y}) \cdot (\mathbf{Xq} - \mathbf{y}) \tag{31}$$
$$= \mathbf{Xq} \cdot \mathbf{Xq} - 2\mathbf{Xq} \cdot \mathbf{y} + \mathbf{y} \cdot \mathbf{y}$$

By matrix algebra $\mathbf{Xq} \cdot \mathbf{y} = \mathbf{X}^T\mathbf{y} \cdot \mathbf{q}$ (see Exercise 14). Then by vector calculus, the derivative of $\mathbf{X}^T\mathbf{y} \cdot \mathbf{q}$ with respect to $\mathbf{q}$ is $\mathbf{X}^T\mathbf{y}$. By more advanced vector calculus, the derivative of $\mathbf{Xq} \cdot \mathbf{Xq}$ is $2(\mathbf{X}^T\mathbf{X})\mathbf{q}$. Thus

$$\frac{d\text{SSE}}{d\mathbf{q}} = 2(\mathbf{X}^T\mathbf{X})\mathbf{q} - 2\mathbf{X}^T\mathbf{y} \tag{32}$$

Setting (32) equal to 0, we obtain the famous normal equations of regression.

$$\mathbf{X}^T\mathbf{Xq} = \mathbf{X}^T\mathbf{y} \tag{33}$$

whose solution, using inverses, is

$$\mathbf{q} = (\mathbf{X}^T\mathbf{X})^{-1}\mathbf{X}^T\mathbf{y} \tag{34}$$

The matrix expression $(\mathbf{X}^T\mathbf{X})^{-1}\mathbf{X}^T$ is called the pseudoinverse of $\mathbf{X}$, since it allows us to solve (approximately) the system $\mathbf{Xq} = \mathbf{y}$. Pseudoinverses are discussed in Section 5.3.

Section 4.2 Exercises

Summary of Exercises
Exercises 1–8 involve regression models, Exercises 6–8 being nonlinear. Exercises 9–14 are theoretical.

1. Seven students earned the following scores on a test after studying the subject matter for different numbers of weeks:

Student	*A*	*B*	*C*	*D*	*E*	*F*	*G*
Length of Study	0	1	2	3	4	5	6
Test Score	3	4	7	6	10	6	10

(a) Fit these data with a regression model of the form $\hat{y} = qx$, where x is number of weeks studied and y is the test score. Plot the observed scores and the predicted scores. What is the sum of squares of errors?

(b) Fit these data with a regression model of the form $\hat{y} = qx + r$. Plot the observed scores and the predicted scores. What is the sum of squares of errors?

(c) Repeat the calculations in part (b) by first shifting the x-values to make the average x-value be 0 [see equations (21)].

2. The following data indicate the numbers of accidents that bus drivers had in one year as a function of the numbers of years on the job.

Years on Job	2	4	6	8	10	12
Accidents	10	8	3	8	4	5

(a) Fit these data with a regression model of the form $\hat{y} = qx$, where x is number of years experience and y is number of bus accidents. Plot the observed numbers of accidents and the predicted numbers. What is the sum of squares of errors?

(b) Fit these data with a regression model of the form $\hat{y} = qx + r$. Plot the observed numbers of accidents and the predicted numbers. What is the sum of squares of errors? Is this model significantly better than the model in part (a)?

(c) Repeat the calculations in part (b) by first shifting the x-values to make the average x-value be 0 [see equations (21)].

3. (a) Reverse the roles of y and x in Exercise 2—now y is number of years of experience—and fit the regression model $\hat{y} = qx + r$ to these data. Plot the observed years experience and the predicted numbers. What is the sum of squares of errors?

(b) Compare your results with those in Exercise 2, part (b) or (c)— why are the numbers not the same?

4. (a) The following data show the GPA and the job salary (5 years after graduation) of six mathematics majors from Podunk U.

GPA	2.3	3.1	2.7	3.4	3.7	2.8
Salary	25,000	38,000	28,000	35,000	30,000	32,000

Fit these data with a regression model of the form $\hat{y} = qx + r$. Plot the observed salaries and the predicted salaries. Is the regression fit reasonably good?

(b) Repeat the calculations in part (a) by first shifting the x-values to make the average x-value be 0 [see equations (21)].

5. Consider the following relationship between the height of a student's mother and the number of F's the student gets at Podunk U.

Student	Mother's Height (inches)	Number of F's
A	62	2
B	65	6
C	59	1
D	63	4
E	60	11
F	69	6
G	63	1
H	60	3

(a) Determine q and r in the regression model $\hat{y} = qx + r$ (where x is mother's height and y is number of F's).

(b) Delete student E from your study and repeat part (a). Does deleting E make much of a difference?

(c) Repeat the calculations in part (b) by first shifting the x-values to make the average x-value be 0 [see equations (21)].

6. Consider the following set of data, which are believed to obey (approximately) a relation of the form $y = qx^2$:

x-value	1	2	3	4	5	6	7
y-value	.5	1	2	3.5	7	9	12

Perform a transformation $y' = f(y)$ on y so that the regression model $\hat{y}' = q'x$ is fairly accurate. Then determine q', reverse the transformation to determine q, and plot the curve $\hat{y} = qx^2$.

7. Consider the following set of data, which are believed to obey an inverse relation of the form $y = q(1/x)$.

Experience (x)	1	2	3	4	5	6
Number of Accidents (y)	10	5	4	3	2	2

Perform a transformation $y' = f(y)$ on y so that the regression model $y' = q'x$ is fairly accurate. Then determine q', reverse the transformation to determine q, and plot the curve $\hat{y} = q(1/x)$.

8. Consider the following set of data, which are believed to obey a square root law $y = q\sqrt{x}$.

Age	10	15	20	25	30	35	40
Strength	7	12	13	16	17	19	20

Perform a transformation $y' = f(y)$ on y so that the regression model $y' = q'x$ is fairly accurate. Then determine q', reverse the transformation to determine q, and plot the curve $\hat{y} = q\sqrt{x}$.

9. Verify the calculation of the partial derivatives in (11).

10. Verify (17).

11. Verify the expression for $r = r' - q \Sigma x_i/n$, where r' and q $(= q')$ are the regression coefficients in the transformed problem [see (21)].

12. Show that the formula for q and r makes the regression line $y = qx + r$ go through the point $(\bar{x}, \bar{y})$, where $\bar{x}$ is the average x-value and $\bar{y}$ is the average y-value.

    Hint: First shift the x-values so that $\bar{x} = 0$.

13. In vector notation, the sum of squares to be minimized in the model $y = qx + r$ is SSE $= (q\mathbf{x} + r\mathbf{1} - \mathbf{y}) \cdot (q\mathbf{x} + r\mathbf{1} - \mathbf{y})$. Compute the vector derivative of this expression with respect to q and with respect to r [see (29)]. Show that the two derivatives are the same as the expressions in (17).

14. Verify that $\mathbf{Xq} \cdot \mathbf{y} = \mathbf{X}^T\mathbf{y} \cdot \mathbf{q}$.

Section 4.3	# Linear Models in the Physical Sciences and Differential Equations

The examples in this section deal with physical-science applications. Until two decades ago, the physical sciences were almost the only disciplines that used mathematics. In those days everyone who studied calculus also studied physics. Today the majority of American students who study calculus and related mathematics take little physics. For them, physical-science applications of mathematics are very hard to follow, since these applications usually depend on a general familiarity with the physical problem being modeled. Further, because linear models play such a large role in the physical sciences, the right place to study them is in a physical science course where the mathematics and science are naturally integrated. On the other hand, students experienced with physical-science linear models can learn much by seeing how similar models are used in other disciplines.

For this reason, none of the models examined in depth in this book will come from the physical sciences. The linear models, such as Markov chains and growth models, that we shall repeatedly use to illustrate concepts are "neutral" models that are easily comprehended by all students. However, for completeness in this section a few models based on basic physical laws will be presented. (Social scientists should consider this section as the book's "College of Arts and Letters' distribution requirement in science.")

The same reasoning applies to differential equations. Students who will use differential equations in courses in their major should have a full course in differential equations. We will only sample the subject to see some basic ways that linear systems of equations arise in differential equations.

Example 1. **Balancing Chemical Equations**

In a chemical reaction, a collection of molecules are brought together in the proper setting (e.g., in boiling water) and they rearrange themselves into new molecules. In this process the number of atoms of each element is conserved. If the molecules put into the reaction have a total of 12 hydrogen (H) atoms, the resulting set of molecules must also contain 12 H's. Consider the reaction in which permanganate (MnO_4) and hydrogen (H) ions combine to form manganese (Mn) and water (H_2O):

$$MnO_4 + H \rightarrow Mn + H_2O \tag{1}$$

where O represents oxygen. Let x_1 be the number of permanganate ions, x_2 the number of hydrogen ions, x_3 the number of manganese atoms, and x_4 the number of water molecules. To have the same number of atoms in the molecules on each side of the reaction, we obtain the system of equations.

$$
\begin{aligned}
\text{H:} \quad & x_2 = 2x_4 \\
\text{Mn:} \quad & x_1 = x_3 \\
\text{O:} \quad & 4x_1 = x_4
\end{aligned}
\tag{2a}
$$

or

$$
\begin{aligned}
x_2 \quad\quad\quad - 2x_4 &= 0 \\
x_1 \quad - x_3 \quad\quad\quad &= 0 \\
4x_1 \quad\quad\quad - x_4 &= 0
\end{aligned}
\tag{2b}
$$

Notice that we have four unknowns but only three equations.

Let us solve this system using elimination by pivoting. We pivot on entry (2, 1) to obtain

$$
\begin{aligned}
x_2 \quad\quad\quad - 2x_4 &= 0 \\
x_1 \quad - x_3 \quad\quad\quad &= 0 \\
4x_3 \quad - x_4 &= 0
\end{aligned}
$$

We do not need to pivot in column 2. Next we pivot on entry $(3, 3)$

$$
\begin{aligned}
x_2 \quad &- 2x_4 = 0 \\
x_1 \quad &- \tfrac{1}{4}x_4 = 0 \qquad \text{or} \\
x_3 &- \tfrac{1}{4}x_4 = 0
\end{aligned}
\qquad
\begin{aligned}
x_2 &= 2x_4 \\
x_1 &= \tfrac{1}{4}x_4 \\
x_3 &= \tfrac{1}{4}x_4
\end{aligned}
$$

As vectors, the solutions in (3) have the form

$$
[\tfrac{1}{4}x_4, \, 2x_4, \, \tfrac{1}{4}x_4, \, x_4] \qquad \text{or} \qquad x_4[\tfrac{1}{4}, \, 2, \, \tfrac{1}{4}, \, 1] \tag{3}
$$

For example, if $x_4 = 4$, then $x_2 = 8$, $x_1 = x_3 = 1$, and the reaction equation becomes

$$
MnO_4 + 8H \rightarrow Mn + 4H_2O
$$

The solution we obtain makes the amounts of the first three types of molecules fixed ratios of the amount of the fourth type, which are free to give any value (i.e., x_4 is a free variable). This makes sense in physical terms, since doubling our chemical "recipe" of inputs should just double the output.

In another series of pivots, say at entries $(1, 2)$, $(2, 3)$, and $(3, 4)$, we get x_1 as the free variable.

$$
\begin{aligned}
-8x_1 + x_2 \quad\quad &= 0 \\
-x_1 \quad\ + x_3 \quad &= 0 \qquad \text{or} \\
-4x_1 \quad\quad\ + x_4 &= 0
\end{aligned}
\qquad
\begin{aligned}
x_2 &= 8x_1 \\
x_3 &= x_1 \\
x_4 &= 4x_1
\end{aligned}
$$

yielding solution vectors

$$
[x_1, \, 8x_1, \, x_1, \, 4x_1] \qquad \text{or} \qquad x_1[1, \, 8, \, 1, \, 4] \tag{4}
$$

In this solution the values of the last three variables are fixed ratios of the first variable. ∎

Example 2. Currents in an Electrical Network

In this example we compute the current in different parts of the electrical network in Figure 4.11a. There are three basic laws that are used to analyze simple electrical networks. The following review of elementary physics summarizes the concepts behind these laws. A battery or other source of electrical power "forces" electricity through electrical devices, such as a light or a doorbell. The force applied to a device depends on two factors: (i) the *resistance* of the device—a measure of how hard it is to push electricity through the device; and (ii) the *current*, the rate at which electricity flows through the device. The fundamental law of electricity, due to Ohm, says

Ohm's Law: force = resistance × current.

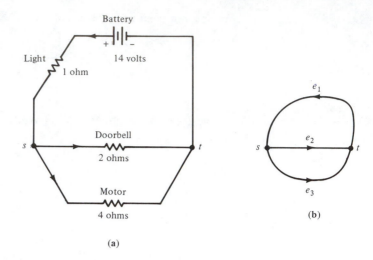

Figure 4.11 (a) Electrical network. (b) Associated graph.

Force is measured in volts, current in amperes, and resistance in ohms. In terms of these units, Ohm's law is

$$\text{volts} = \text{ohms} \times \text{amperes}$$

A battery supplies a fixed force into a network. Batteries send their electricity out from a positive terminal and receive it back at a negative terminal. All the voltage (i.e., force) provided by a battery is used up by the time the electricity returns to the terminal. This property of voltage is called

Kirchhoff's Voltage Law. In any cycle (closed path) in a network, the sum of the voltages used by resistive devices equals the voltage from the battery(ies).

The final law says that current is conserved at any branch node.

Kirchhoff's Current Law. The sum of the currents flowing into any node is equal to the sum of the current flowing out of the node.

Let us use these three laws to derive a set of linear equations modeling the behavior of currents in the network in Figure 4.11a. Later we shall express each of Kirchhoff's laws in the form of a matrix equation.

Assume that the battery delivers 14 volts. Further let c_1 be the current flowing through the section of the network with the battery and the light, whose resistance is 1 ohm; let c_2 be the current through the section with the doorbell, whose resistance is 2 ohms; and let c_3 be the current through the section with the motor, whose resistance is 4 ohms. Currents flow in the direction of the arrows along these edges. By the current law, at node s we have the following equation:

$$c_1 = c_2 + c_3 \tag{5}$$

Note that at node t we get the same equation. Next we use the voltage law. Following the cycle, battery to light to doorbell to battery, we have

$$\text{voltage at light} + \text{voltage at doorbell} = \text{battery voltage} \tag{6a}$$

We use Ohm's law (voltage = resistance $\times$ current) to determine the voltages at the devices. The battery voltage is 14. So (6a) becomes

$$1c_1 + 2c_2 = 14 \tag{6b}$$

Following a second cycle, battery to light to motor to battery, we have

$$1c_1 + 4c_3 = 14 \tag{7}$$

There is still a third cycle that we could use, node s to doorbell to node t to motor (going against the current flow) to node s. If we go against the current flow, the voltage is treated as negative. So the voltage law for this cycle is

$$2c_2 - 4c_3 = 0 \tag{8}$$

Note that equation (8) is simply what is obtained when we subtract (7) from (6b). Intuitively, the third cycle is the net result of going forward on the first cycle and then backward on the second cycle.

We need to solve the three equations (5), (6), and (7) in the three unknown currents, which we write as

$$\begin{aligned} c_1 - c_2 - c_3 &= 0 \\ c_1 + 2c_2 &= 14 \\ c_1 + 4c_3 &= 14 \end{aligned} \tag{9}$$

Solving by Gaussian elimination, we find

$$c_1 = 6 \text{ amperes}, \qquad c_2 = 4 \text{ amperes}, \qquad c_3 = 2 \text{ amperes} \tag{10}$$

■

A general analysis of currents in networks involves a combination of physics and mathematics. The critical mathematical problem is proving that there will always be enough different equations to determine uniquely all the currents.

Before leaving this problem, let us show how this problem can be cast in matrix notation. Kirchhoff's current law can be restated in matrix form using the incidence matrix $\mathbf{M}(G)$ of the underlying graph (when batteries and resistive devices are ignored). Figure 4.11b shows the graph G associated with the network in Figure 4.11a. G contains two nodes s, t and three edges e_1, e_2, e_3 joining s and t. In this graph each edge has a direction; the directions for G are shown in Figure 4.11b. The current c_i in edge e_i is positive if it flows in the direction of e_i and negative if it flows in the opposite direction.

Recall from Section 2.3 that $\mathbf{M}(G)$ has a row for each node of G and a column for each edge. In the case of directed edges, entry $m_{ij} = +1$ if the jth edge is directed into ith node, $= -1$ if the jth edge is directed out from the ith node, and $= 0$ if the jth edge does not touch the ith node. For the graph G in Figure 4.11b, $\mathbf{M}(G)$ is

$$\mathbf{M}(G) = \begin{array}{c} \\ s \\ t \end{array} \begin{array}{ccc} e_1 & e_2 & e_3 \\ \left[\begin{array}{ccc} +1 & -1 & -1 \\ -1 & +1 & +1 \end{array}\right] \end{array} \tag{11}$$

Kirchhoff's current law says that the flow into a node equals the flow out of the node, or in other words, the net current flow is zero. At node s, this means that

$$+c_1 - c_2 - c_3 = 0 \tag{12}$$

Observe that currents going into s are associated with edges that have $+1$ in row s of $\mathbf{M}(G)$, and currents going out are associated with edges that have -1 in row s of $\mathbf{M}(G)$. If $\mathbf{m}_s$ denotes row s of $\mathbf{M}(G)$ and $\mathbf{c}$ is the vector of currents (c_1, c_2, c_3), then (12) can be rewritten as

$$\mathbf{m}_s \cdot \mathbf{c} = 0 \tag{13}$$

This equation is true for all rows of $\mathbf{M}(G)$. Thus (13) generalizes to

Kirchhoff's current law: $\mathbf{M}(G)\mathbf{c} = 0$ $\qquad\qquad\qquad$ (14)

This result is true for any associated graph G.

We can also define a special cycle matrix $\mathbf{K}(G)$ for G with a row for each cycle (closed path) of G and a column for each edge. Let r_j be the resistance in the jth edge. Then define entry k_{ij} of $\mathbf{K}(G) = +r_j$ if the ith cycle uses the jth edge traversing this edge in the direction of its arrow, $-r_j$ if the ith circuit uses the jth edge in the opposite direction of its arrow, and $= 0$ otherwise. For example, $\mathbf{K}(G)$ in Example 2 would be, with cycles listed in the order they were discussed,

$$\mathbf{K}(G) = \begin{bmatrix} 1 & 2 & 0 \\ 1 & 0 & 4 \\ 0 & 2 & -4 \end{bmatrix}$$

As occurred in Example 1, some circuits in a graph are always redundant and there is a simple rule for picking a minimal set of cycles for $\mathbf{K}(G)$. If $\mathbf{k}_i$ is the ith row of $\mathbf{K}(G)$, then $\mathbf{k}_i \cdot \mathbf{c}$ will be the voltages used on the ith cycle. If f_i is the voltage force of batteries on the ith cycle, then Kirchhoff's voltage law becomes

$$\mathbf{k}_i \cdot \mathbf{c} = f_i \tag{15}$$

and if $\mathbf{f}$ is the vector of f_i's, we have

Kirchhoff's voltage law: $\mathbf{K}(G)\mathbf{c} = \mathbf{f}$ (16)

Our current problem can now be stated: Solve the system of equations (14) and (16) for $\mathbf{c}$.

The next four examples involve differential equations. A common mathematical model for many dynamic systems, such as falling objects, vibrating strings, or economic growth, is a differential equation of the form

$$y''(t) = a_1 y'(t) + a_0 y(t) \tag{17}$$

where $y(t)$ is a function that measures the "position" of the quantity, $y'(t)$ denotes the first derivative with respect to t (representing time), $y''(t)$ denotes the second derivative, and a_1 and a_0 are constants. The differential equation is called *linear* because the right side is a linear combination of the function and its derivative. Solutions of (17) are functions of the form

$$y(t) = Ae^{kt} \tag{18}$$

where e is Euler's constant, and A and k are constants that depend on the particular problem. This form of solution also works if higher derivatives are involved in the linear differential equation.

Example 3. Differential Equation for Instantaneous Interest

The simple differential equation

$$y'(t) = .10y(t) \tag{19}$$

describes the amount of money $y(t)$ in a savings account after t years when the account earns 10% interest compounded instantaneously. Recall that $y'(t)$ is the instantaneous rate of change, or graphically, the slope, of $y(t)$. Thus (19) says that the instantaneous growth rate of the savings account is 10% of the account's current value.

Recall that the derivative of e^{kt} is ke^{kt}. Let us try setting $y(t) = Ae^{kt}$ [as given in (18)]. Now (19) becomes

$$kAe^{kt} = .10Ae^{kt} \tag{20}$$

Dividing both sides of (20) by Ae^{kt}, we obtain $k = \ ^.10$. So the solution to (19) is

$$y(t) = Ae^{.10t} \tag{21}$$

The constant A is still to be determined because to know how much money we shall have after t years, we must know how much we started with. Suppose that we started with 1000 dollars. The starting time is $t = 0$. Thus we have (using the fact $e^0 = 1$)

$$1000 = y(0) = Ae^0 = A \tag{22}$$

Then the solution of (19) with $y(0) = 1000$ is

$$y(t) = 1000e^{.10t} \tag{23}$$

∎

Example 4. **Solving Second-Order Linear Differential Equations**

Consider the following differential equation that might describe the height of a falling particle in a special force field.

$$y''(t) = 6y'(t) - 8y(t) \tag{24}$$

This differential equation is called a *second-order* equation because it involves the second derivative. To solve this equation, we also need to know the starting conditions, what are the initial height $y(0)$ and the initial speed $y'(0)$. Here the derivative $y'(t)$ measures speed, that is, the rate of change of the height. Suppose in this problem that

$$f(0) = 100 \quad \text{and} \quad f'(0) = -20 \tag{25}$$

We solve this problem in two stages. First we substitute (18) for $y(t)$ in (24).

$$\frac{d^2}{dt^2} Ae^{kt} = 6\frac{d}{dt} Ae^{kt} - 8Ae^{kt} \tag{26}$$

Recall that the second derivative of e^{kt} is $k^2 e^{kt}$. So (26) becomes

$$k^2 Ae^{kt} = 6kAe^{kt} - 8Ae^{kt} \tag{27}$$

If we divide by Ae^{kt}, (27) becomes

$$k^2 = 6k - 8 \quad \text{or} \quad k^2 - 6k + 8 = 0 \tag{28}$$

Equation (28) is called the **characteristic equation** of the linear differential equation (24). The roots of (28) are easily verified to be 2 and 4. So we have two possible types of solutions to (24).

$$y(t) = Ae^{4t} \qquad \text{and} \qquad y(t) = A'e^{2t}$$

The constants A and A' can have any value and these solutions will still satisfy (24). In fact, it can readily be checked [see Exercise 16, part (b)] that any linear combination of the basic solutions e^{2t} and e^{4t} is a solution. Thus

$$y(t) = Ae^{4t} + A'e^{2t} \tag{29}$$

is the general form of a solution to (24). The constants A and A' depend on the starting values. From (25) we have

$$
\begin{aligned}
100 &= y(0) = Ae^0 + A'e^0 \\
-20 &= y'(0) = 4Ae^0 + 2A'e^0
\end{aligned} \tag{30a}
$$

which simplifies to

$$
\begin{aligned}
A + A' &= 100 \\
4A + 2A' &= -20
\end{aligned} \tag{30b}
$$

In (30a), we obtain $y'(0)$ by differentiating (29) and setting $t = 0$. Now we have our old "friend," a system of two equations in two unknowns. We solve (30b) and obtain

$$A = -110 \qquad \text{and} \qquad A' = 210 \tag{31}$$

Substituting these values in (29), we obtain the required solution

$$y(t) = -110e^{4t} + 210e^{2t} \tag{32}$$

∎

The calculations for any other second-order differential equation would proceed in a similar fashion: First substitute (18) in the differential equation to obtain the characteristic equation [as in (28)] and solve for its roots; then determine A and A' from the pair of equations for starting values [as in (30)]. This method generalizes to kth-order differential equations; then the characteristic equation has k roots, we need k initial values, and we have to solve k equations in k unknowns.

For completeness, we note that if the two roots of the characteristic equation (28) were the same, such as 2 and 2, then the starting value equations cannot be solved, since the two equations of (30a) will be the same. In the case of identical roots of the characteristic equation, $y(t)$ instead has the form

$$y(t) = Ae^{rt} + A'te^{rt} \tag{33}$$

where r is the double root. It is an exercise to check that in the case of a multiple root (and only then), te^{rt} is also a solution to the differential equation.

Example 5. A System of Differential Equations

Let us consider a pair of first-order differential equations which describe motion of an object in x-y space with one equation governing the x-coordinate and one the y-coordinate.

$$
\begin{aligned}
x'(t) &= 2x(t) - y(t) \\
y'(t) &= -x(t) + 2y(t)
\end{aligned} \tag{34}
$$

The starting values are $x(0) = y(0) = 1$. Let $\mathbf{u}(t)$ be the vector function $\mathbf{u}(t) = (x(t), y(t))$. Then (34) can be written in matrix notation as

$$
\mathbf{u}'(t) = \mathbf{B}\mathbf{u}(t), \qquad \text{where } \mathbf{B} = \begin{bmatrix} 2 & -1 \\ -1 & 2 \end{bmatrix} \tag{35}
$$

and $\mathbf{u}(0) = [1, 1] = \mathbf{1}$. In Example 3 we saw that

$$
y'(t) = by(t), \quad y(0) = A \;\rightarrow\; y(t) = Ae^{bt} \tag{36}
$$

Substituting $\mathbf{B}$ for b and $\mathbf{1}$ for A in (36), we obtain the solution to (35):

$$
\mathbf{u}(t) = e^{\mathbf{B}t}\mathbf{1} \tag{37}
$$

■

A matrix in the exponent looks strange. But one definition of e^x is in terms of the power series.

$$
e^x = 1 + x + \frac{x^2}{2!} + \frac{x^3}{3!} + \cdots + \frac{x^k}{k!} + \cdots \tag{38}
$$

Similarly, $e^{\mathbf{B}}$ is defined

$$
e^{\mathbf{B}} = \mathbf{I} + \mathbf{B} + \frac{\mathbf{B}^2}{2!} + \frac{\mathbf{B}^3}{3!} + \cdots + \frac{\mathbf{B}^k}{k!} + \cdots \tag{39}
$$

The power series (39) is well defined for all matrices. Although this power series of matrices may look forbidding, it is easy to use if we work in eigenvector-based coordinates so that $\mathbf{B}$ and its powers act like scalar multipliers (as in $\mathbf{B}\mathbf{u} = \lambda\mathbf{u}$). Recall Theorem 5 of Section 3.3, which said that if $\mathbf{U}$ is a matrix whose columns were different eigenvectors of $\mathbf{B}$ and if $\mathbf{D}_\lambda$ is a diagonal matrix of associated eigenvalues, then

$$
\mathbf{B} = \mathbf{U}\mathbf{D}_\lambda\mathbf{U}^{-1} \tag{40}
$$

Substituting with (40) for **B** in (39), we obtain

$$e^{\mathbf{B}} = \mathbf{I} + \mathbf{UD}_\lambda \mathbf{U}^{-1} + \frac{\mathbf{UD}_\lambda^2 \, \mathbf{U}^{-1}}{2!} + \frac{\mathbf{UD}_\lambda^3 \, \mathbf{U}^{-1}}{3!} + \cdots$$

$$+ \frac{\mathbf{UD}_\lambda^k \, \mathbf{U}^{-1}}{k!} + \cdots \tag{41a}$$

$$= \mathbf{U}\left(\mathbf{I} + \mathbf{D}_\lambda + \frac{\mathbf{D}_\lambda^2}{2!} + \frac{\mathbf{D}_\lambda^3}{3!} + \cdots + \frac{\mathbf{D}_\lambda^k}{k!} + \cdots\right)\mathbf{U}^{-1} \tag{41b}$$

$$= \mathbf{U}e^{\mathbf{D}_\lambda}\mathbf{U}^{-1} = \mathbf{U}
\begin{bmatrix}
e^{\lambda_1} & 0 & 0 & \cdots \\
0 & e^{\lambda_2} & 0 & \cdots \\
0 & 0 & e^{\lambda_3} & \cdots \\
\vdots & \vdots & \vdots & \vdots
\end{bmatrix}
\mathbf{U}^{-1} \tag{41c}$$

The reason that $e^{\mathbf{D}_\lambda}$ turns out to be simply a diagonal matrix with diagonal entries e^{λ_1}, e^{λ_2}, . . . , e^{λ_n} is that in (41b), the matrices $\mathbf{D}_\lambda^k/k!$ are diagonal with entries $\lambda_1^k/k!$, $\lambda_2^k/k!$, . . . and summing these matrices we get a matrix whose entry $(1, 1)$ is $1 + \lambda_1 + \lambda_1^2/2! + \lambda_1^3/3! + \ldots$, which equals e^{λ_1}; and similarly for the other diagonal entries.

Example 6. **Converting a Second-Order Differential Equation into a Pair of First-Order Differential Equations**

Let us consider again the second-order equation from Example 4:

$$y''(t) = 6y'(t) - 8y(t) \tag{42}$$

We convert (42) into a pair of first-order equations by introducing a second function $x(t)$ defined

$$x(t) = y'(t) \quad \text{and thus} \quad x'(t) = y''(t) \tag{43}$$

Now (42) can be written as the pair of the first-order equations

$$x'(t) = 6x(t) - 8y(t) \tag{44}$$
$$y'(t) = x(t)$$

Defining the vector function $\mathbf{u}(t) = [x(t), y(t)]$, we have

$$\mathbf{u}'(t) = \mathbf{B}\mathbf{u}(t), \quad \text{where } \mathbf{B} =
\begin{bmatrix}
6 & -8 \\
1 & 0
\end{bmatrix} \tag{45}$$

with initial conditions from Example 4 of $\mathbf{u}(0) = [-20, 100]$.

As in Example 5, the solution to (45) should be

$$\mathbf{u}(t) = e^{\mathbf{B}t}\mathbf{u}(0) \tag{46}$$

where $e^{\mathbf{B}t}$ is defined by the power series

$$e^{\mathbf{B}t} = \mathbf{I} + t\mathbf{B} + \frac{t^2\mathbf{B}^2}{2!} + \frac{t^3\mathbf{B}^3}{3!} + \cdots + \frac{t^k\mathbf{B}^k}{k!} + \cdots \tag{47}$$

Remember that we already know from Example 4 the solution of (42), so $\mathbf{u}(t)$ in (46) must equal

$$\mathbf{u}(t) = \begin{bmatrix} x(t) \\ y(t) \end{bmatrix} = \begin{bmatrix} -440e^{4t} + 420e^{2t} \\ -110e^{4t} + 210e^{2t} \end{bmatrix} \tag{48}$$

where $x(t) = y'(t) = -440e^{4t} + 420e^{2t}$ is obtained by differentiating the solution for $y(t)$.

We now use the eigenvector-coordinates approach from Sections 2.5 and 3.3 to show how the intimidating formula for $\mathbf{u}(t)$ in (46) is the same as (48). Since by (47) $e^{\mathbf{B}t}$ involves powers of $\mathbf{B}$, the computation of multiplying $e^{\mathbf{B}t}$ times $\mathbf{u}(0)$ will be simplified if we express $\mathbf{u}(0)$ in terms of $\mathbf{B}$'s eigenvectors.

In Section 3.1 we learned how to find the eigenvalues of a matrix $\mathbf{B}$—they are the roots of the characteristic polynomial det $(\mathbf{B} - \lambda\mathbf{I})$—and from them, the associated eigenvectors. The characteristic polynomial for $\mathbf{B}$ is $\lambda^2 - 6\lambda + 8$ and its roots are 4 and 2. Eigenvectors $\mathbf{u}_1$, $\mathbf{u}_2$ of $\mathbf{B}$ associated with the eigenvalues 4 and 2 are (Exercise 13):

$$\mathbf{u}_1 = [4, 1] \quad \text{for } \lambda_1 = 4 \qquad \mathbf{u}_2 = [2, 1] \quad \text{for } \lambda_2 = 2$$

Writing $\mathbf{u}(0) = [-20, 100]$ as a linear combination of $\mathbf{u}_1$ and $\mathbf{u}_2$ (we must solve the system $\mathbf{u}(0) = a\mathbf{u}_1 + b\mathbf{u}_2$ for a and b (see Section 2.5 for details), we obtain

$$\mathbf{u}(0) = [-20, 100] = -110\mathbf{u}_1 + 210\mathbf{u}_2 \tag{49}$$

We now can compute $e^{\mathbf{B}t}\mathbf{u}(0)$, which we rewrite using (47) as

$$\begin{aligned} \mathbf{u}(t) &= e^{\mathbf{B}t}\mathbf{u}(0) \\ &= \mathbf{I}\mathbf{u}(0) + t\mathbf{B}\mathbf{u}(0) + \frac{t^2}{2!}\mathbf{B}^2\mathbf{u}(0) + \cdots \\ &\quad + \frac{t^k}{k!}\mathbf{B}^k\mathbf{u}(0) + \cdots \end{aligned} \tag{50}$$

Substituting $\mathbf{u}(0) = -110\mathbf{u}_1 + 210\mathbf{u}_2$ in (50), we have

$$\mathbf{u}(t) = \mathbf{I}(-110\mathbf{u}_1 + 210\mathbf{u}_2) + t\mathbf{B}(-110\mathbf{u}_1 + 210\mathbf{u}_2)$$

$$+ \frac{t^2}{2!} \mathbf{B}^2(-110\mathbf{u}_1 + 210\mathbf{u}_2) + \cdots$$

$$+ \frac{t^k}{k!} \mathbf{B}^k(-110\mathbf{u}_1 + 210\mathbf{u}_2) + \cdots$$

$$= -110\left\{ \mathbf{Iu}_1 + t\mathbf{Bu}_1 + \frac{t^2}{2!} \mathbf{B}^2\mathbf{u}_1 + \cdots + \frac{t^k}{k!} \mathbf{B}^k\mathbf{u}_1 + \cdots \right\}$$

$$+ 210\left\{ \mathbf{Iu}_2 + t\mathbf{Bu}_2 + \frac{t^2}{2!} \mathbf{B}^2\mathbf{u}_2 + \cdots + \frac{t^k}{k!} \mathbf{B}^k\mathbf{u}_2 + \cdots \right\}$$

$$\tag{51}$$

But since $\mathbf{u}_1$ and $\mathbf{u}_2$ are eigenvectors, the term $\mathbf{B}^k\mathbf{u}_1$ equals $4^k\mathbf{u}_1$, and $\mathbf{B}^k\mathbf{u}_2 = 2k\mathbf{u}_2$. So (51) becomes

$$\mathbf{u}(t) = -110\left\{ \mathbf{u}_1 + t4\mathbf{u}_1 + \frac{t^2 4^2}{2!} \mathbf{u}_1 + \cdots + \frac{t^k 4^k}{k!} \mathbf{u}_1 + \cdots \right\}$$

$$+ 210\left\{ \mathbf{u}_2 + t2\mathbf{u}_2 + \frac{t^2 2^2}{2!} \mathbf{u}_2 + \cdots + \frac{t^k 2^k}{k!} \mathbf{u}_2 + \cdots \right\}$$

$$= -110\left\{ 1 + t4 + \frac{t^2 4^2}{2!} + \cdots + \frac{t^k 4^k}{k!} + \cdots \right\}\mathbf{u}_1$$

$$+ 210\left\{ 1 + t2 + \frac{t^2 2^2}{2!} + \cdots + \frac{t^k 2^k}{k!} + \cdots \right\}\mathbf{u}_2$$

$$= -110e^{4t}\mathbf{u}_1 + 210e^{2t}\mathbf{u}_2$$

$$= -110e^{4t}\begin{bmatrix} 4 \\ 1 \end{bmatrix} + 210e^{2t}\begin{bmatrix} 2 \\ 1 \end{bmatrix}$$

$$= \begin{bmatrix} -440e^{4t} + 420e^{2t} \\ -110e^{4t} + 210e^{2t} \end{bmatrix} \tag{52}$$

Observe that for a different starting vector $\mathbf{u}^*(0)$, we would get $\mathbf{u}^*(0) = a'\mathbf{u}_1 + b'\mathbf{u}_2$, for some a', b' and then the result in (52) would be $\mathbf{u}(t) = a'e^{4t}\mathbf{u}_1 + b'e^{2t}\mathbf{u}_2$.

We now give a shorter derivation of this result using the matrix formula in (41) to handle the exponential series. For $e^{\mathbf{B}t}$, (41) becomes

$$e^{\mathbf{B}t} = \mathbf{U}e^{\mathbf{D}_\lambda t}\mathbf{U}^{-1} \tag{53}$$

where $e^{\mathbf{D}_\lambda t}$ is a diagonal matrix with diagonal entries $e^{\lambda_i t}$. Recall that $\mathbf{U}$ has eigenvectors $\mathbf{u}_1$ and $\mathbf{u}_2$ as its columns. Thus

$$\mathbf{U} = \begin{bmatrix} 4 & 2 \\ 1 & 1 \end{bmatrix} \qquad \text{and we compute} \qquad \mathbf{U}^{-1} = \begin{bmatrix} \frac{1}{2} & -1 \\ -\frac{1}{2} & 2 \end{bmatrix} \tag{54}$$

Using (53) to substitute for $e^{\mathbf{B}t}$ in $\mathbf{u}(t) = e^{\mathbf{B}}\mathbf{u}(0)$, we obtain

$$
\begin{aligned}
\mathbf{u}(t) = e^{\mathbf{B}t}\mathbf{u}(0) &= \mathbf{U}e^{\mathbf{D}_\lambda t}\mathbf{U}^{-1}\mathbf{u}(0) \\
&= \begin{bmatrix} 4 & 2 \\ 1 & 1 \end{bmatrix} \begin{bmatrix} e^{4t} & 0 \\ 0 & e^{2t} \end{bmatrix} \begin{bmatrix} \frac{1}{2} & -1 \\ -\frac{1}{2} & 2 \end{bmatrix} \begin{bmatrix} -20 \\ 100 \end{bmatrix} \\
&= \begin{bmatrix} 4 & 2 \\ 1 & 1 \end{bmatrix} \begin{bmatrix} e^{4t} & 0 \\ 0 & e^{2t} \end{bmatrix} \begin{bmatrix} -110 \\ 210 \end{bmatrix} \\
&= \begin{bmatrix} 4e^{4t} & 2e^{2t} \\ e^{4t} & e^{2t} \end{bmatrix} \begin{bmatrix} -110 \\ 210 \end{bmatrix} \\
&= \begin{bmatrix} -440e^{4t} + 420e^{2t} \\ -110e^{4t} + 210e^{2t} \end{bmatrix}
\end{aligned} \tag{55}
$$

∎

The conversion of (42) to a system of first-order differential equations can be applied to any linear higher-order differential equation.

Example 7. **Converting a Third-Order Differential Equation into a System of Three First-Order Differential Equations**

Consider the third-order linear differential equation

$$
y'''(t) = y''(t) + 2y'(t) + 3y(t) \tag{56}
$$

We introduce the two new functions $w(t)$ and $z(t)$:

$$
w(t) = y'(t) \quad \text{and} \quad z(t) = w'(t) \quad [= y''(t)] \tag{57}
$$

Then (56) can be written

$$
z'(t) = z(t) + 2w(t) + 3y(t) \tag{58}
$$

Defining the vector function $\mathbf{u}(t) = [y(t), w(t), z(t)]$, we can rewrite (57) and (58) as the matrix equation

$$
\mathbf{u}'(t) = \mathbf{B}\mathbf{u}(t): \qquad \begin{bmatrix} z'(t) \\ w'(t) \\ y'(t) \end{bmatrix} = \begin{bmatrix} 1 & 2 & 3 \\ 1 & 0 & 0 \\ 0 & 1 & 0 \end{bmatrix} \begin{bmatrix} z(t) \\ w(t) \\ y(t) \end{bmatrix} \tag{59}
$$

The solution to (59) is $\mathbf{u}(t) = e^{\mathbf{B}t}\mathbf{u}(0)$, which we would evaluate using eigenvectors, as discussed in Example 6. ∎

Section 4.3 **Exercises**

Summary of Exercises

Exercises 1–4 are chemical reaction balancing problems. Exercises 5–7 are electrical circuit problems. Exercises 8–18 involve differential equations, with Exercises 16–18 being theory questions.

1. Write out a system of equations required to balance the following chemical reactions and solve. Here C represents carbon, N represents nitrogen, H represents hydrogen, and O represents oxygen.
   (a) $N_2H_4 + N_2O_4 \rightarrow N_2 + H_2O$
   (b) $C_6H_6 + O_2 \rightarrow CO_2 + H_2O$

2. Write out a system of equations required to balance the following chemical reaction and solve.

$$SO_2 + NO_3 + H_2O \rightarrow H + SO_4 + NO$$

   where S represents sulfur, N represents nitrogen, H represents hydrogen, and O represents oxygen.

3. Write out a system of equations required to balance the following chemical reaction and solve.

$$PbN_6 + CrMn_2O_8 \rightarrow Cr_2O_3 + MnO_2 + Pb_3O_4 + NO$$

   where Pb represents lead, N represents nitrogen, Cr represents chromium, Mn represents manganese, and O represents oxygen.

4. Write out a system of equations required to balance the following chemical reaction and solve.

$$H_2SO_4 + MnS + As_2Cr_{10}O_{35} \rightarrow HMnO_4 + AsH_3 + CrS_3O_{12} + H_2O$$

   where H represents hydrogen, S represents sulfur, O represents oxygen, Mn represents manganese, As represents arsenic, and Cr represents chromium.

5. Determine the currents in each branch of the following circuit.

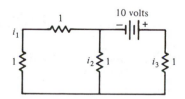

6. Determine the currents in each branch of the following circuits, in which the incoming amperage (on the left) is given.

(a)   (b)

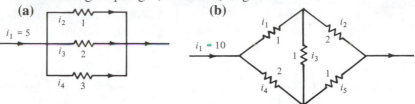

7. Determine the currents in each branch of the following circuit. The voltage in the battery is 19.

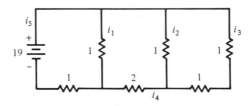

8. Solve the following first-order differential equations, with given initial values.
   (a) $y'(t) = .5y(t)$, $y(0) = 100$
   (b) $y'(t) - 4y(t) = 0$, $y(0) = 10$

9. Suppose that a population of bacteria is continuously doubling its size every unit of time. Write a differential equation for $y(t)$, the size of the population.

10. Solve the following second-order differential equations, with given initial values. Use the method based on the characteristic equation (see Example 4).
    (a) $y''(t) = 5y'(t) - 4y(t)$, $y(0) = 20$, $y'(0) = 5$
    (b) $y''(t) = -5y'(t) + 6y(t)$, $y(0) = 1$, $y'(0) = 15$
    (c) $y''(t) = 2y'(t) + 8y(t)$, $y(0) = 2$, $y'(0) = 0$

11. Convert the following differential equations into systems of simultaneous first-order differential equations. *Do not solve*.
    (a) $y''(t) = 5y'(t) - 4y(t)$
    (b) $y''(t) = -5y'(t) - 6y(t)$
    (c) $y'''(t) = 4y''(t) + 3y'(t) - 2y(t)$
    (d) $y'''(t) = 2y'(t) + y(t)$

12. Check that 3 and 1 are the eigenvalues for $\mathbf{B} = \begin{bmatrix} 2 & -1 \\ -1 & 2 \end{bmatrix}$ in Example 5 and that associated eigenvectors are $[1, -1]$ and $[1, 1]$. Solve the system of differential equations in Example 5 using the method in Example 6.

13. Check that 4 and 2 are the eigenvalues for $\mathbf{B} = \begin{bmatrix} 6 & -8 \\ 1 & 0 \end{bmatrix}$ in Example 6 and that associated eigenvectors are [4, 1] and [2, 1]. Also verify (49).

14. Re-solve the second-order differential equations in Exercise 10 by converting them to a pair of first-order differential equations and solving by the method in Example 6 (you must find the eigenvalues and eigenvectors).

15. Solve the following pairs of first-order differential equations by using the solution technique in Example 6 (you must find the eigenvalues and eigenvectors). The initial condition is $x(0) = y(0) = 10$.
    (a) $x'(t) = 4x(t)$, $y'(t) = 2x(t) + 2y(t)$
    (b) $x'(t) = 2x(t) + y(t)$, $y'(t) = 2x(t) + 3y(t)$
    (c) $x'(t) = x(t) + 4y(t)$, $y'(t) = 2x(t) + 3y(t)$

16. (a) Show that any multiple $ry^*(t)$ of a solution $y^*(t)$ to a second-order differential equation $y''(t) = ay'(t) + by(t)$ is again a solution.
    (b) Show that any linear combination $ry^*(t) + sy^0(t)$ of solutions $y^*(t)$, $y^0(t)$ to $y''(t) = ay'(t) + by(t)$ is again a solution.

17. Suppose that $y^0(t)$ is some solution to $y''(t) - ay'(t) - by(t) = f(t)$ and $y^*(t)$ is a solution to $y''(t) - ay'(t) - by(t) = 0$. Then show that for any r, $y^0(t) + ry^*(t)$ is also a solution to $y''(t) - ay'(t) - by(t) = f(t)$.

18. Verify that $y(t) = te^{\lambda t}$ is a solution to the differential equation $y''(t) = cy'(t) + dy(t)$ whose characteristic equation $k^2 - ck - d = 0$ has λ as its double root.

    Note: If $k^2 - ck - d = 0$ has λ as a double root, the characteristic equation can be factored as $(k - \lambda)^2 = 0$. This means that $c = 2\lambda$ and $d = -\lambda^2$. Use these values for c and d in verifying that $te^{\lambda t}$ is a solution.

Section 4.4 Markov Chains

Markov chains were introduced in Section 1.3. They are probability models for simulating the behavior of a system that randomly moves among different "states" over successive periods of time. If a Markov chain is currently in state S_j, there is a transition probability a_{ij} that 1 unit of time later it will be in state S_i. The matrix $\mathbf{A}$ of transition probabilities completely describes the Markov chain. If $\mathbf{p} = [p_1, p_2, \ldots, p_n]$ is the vector giving the probabilities p_i that S_i is the current state of the chain and $\mathbf{p}' = [p'_1, p'_2, \ldots, p'_n]$ is the vector of probabilities p_i' that S_i is the next state of the chain, then we have

$$\mathbf{p}' = \mathbf{Ap} \qquad (1a)$$

For a particular p_i', this is

$$p_i' = a_{i1}p_1 + a_{i2}p_2 + \cdots + a_{in}p_n \qquad (1b)$$

Example 1. Frog in Highway Revisited

In Section 1.3 we considered the Markov chain for a frog wandering across a highway that was divided into six states. The transition matrix **A** was

$$
\mathbf{A} = \begin{array}{c} \\ \text{Next} \\ \text{State} \end{array}
\begin{array}{c} \\ 1 \\ 2 \\ 3 \\ 4 \\ 5 \\ 6 \end{array}
\overset{\begin{array}{cccccc} 1 & 2 & 3 & 4 & 5 & 6 \end{array}}{
\begin{bmatrix}
.50 & .25 & 0 & 0 & 0 & 0 \\
0 & .50 & .25 & 0 & 0 & 0 \\
0 & .25 & .50 & .25 & 0 & 0 \\
0 & 0 & .25 & .50 & .25 & 0 \\
0 & 0 & 0 & .25 & .50 & .50 \\
0 & 0 & 0 & 0 & .25 & .50
\end{bmatrix}} \qquad (2)
$$

We started with probability vector $\mathbf{p} = [1, 0, 0, 0, 0, 0]$, that is, the frog started in state 1. We computed a table of the probability distributions after varying numbers of periods. In matrix notation, we computed

$$\mathbf{p}^{(k)} = \mathbf{A}^k\mathbf{p}$$

for increasing values of k. We found that as k got large, $\mathbf{p}^{(k)}$ converged to the probability vector

$$\mathbf{p}^* = [.1, .2, .2, .2, .2, .1] \qquad (3)$$

which satisfied the equation

$$\mathbf{p}^* = \mathbf{Ap}^* \qquad (4)$$

Thus $\mathbf{p}^*$ is a stable (unchanging) probability distribution for this Markov chain. In matrix terminology, $\mathbf{p}^*$ is an eigenvector of **A** with associated eigenvalue 1.

In Section 3.5 we solved the eigenvector equations (4)—actually, we solved $(\mathbf{A} - \mathbf{I})\mathbf{p} = \mathbf{0}$—and obtained a general solution of the form

$$[q, 2q, 2q, 2q, 2q, q] \qquad (5)$$

Making the components in (5) sum to 1 (to be a probability distribution), we obtained $\mathbf{p}^* = [.1, .2, .2, .2, .2, .1]$. So this $\mathbf{p}^*$ is the unique stable distribution of the frog Markov chain. The fact was brought out in the Exercises of Section 1.3 that if the starting probability vector $\mathbf{p}$ had been different, we still would have found that $\mathbf{p}^{(k)}$ converged to this $\mathbf{p}^*$.

Suppose that the starting probability vector were the jth unit vector $\mathbf{e}_j$, with a 1 in the jth position and 0's elsewhere (the original $\mathbf{p}$ was $\mathbf{e}_1$). It was noted in Section 2.4 that for any matrix $\mathbf{B}$,

$$\mathbf{B}\mathbf{e}_j = \mathbf{b}_j^C \qquad (\mathbf{b}_j^C = \text{the } j\text{th column of } \mathbf{B})$$

Then

$$\mathbf{p}^{(k)} = \mathbf{A}^k\mathbf{e}_j = (\mathbf{A}^k)_j^C \qquad \text{(the } j\text{th column of } \mathbf{A}^k) \qquad (6)$$

Since $\mathbf{p}^{(k)}$ converges to $\mathbf{p}^*$, we conclude that the jth column of $\mathbf{A}^k$ approaches $\mathbf{p}^*$, for k large,

$$\mathbf{A}^k \rightarrow \begin{bmatrix} .1 & .1 & .1 & .1 & .1 & .1 \\ .2 & .2 & .2 & .2 & .2 & .2 \\ .2 & .2 & .2 & .2 & .2 & .2 \\ .2 & .2 & .2 & .2 & .2 & .2 \\ .2 & .2 & .2 & .2 & .2 & .2 \\ .1 & .1 & .1 & .1 & .1 & .1 \end{bmatrix} \qquad (7)$$

Does this property of any starting probability vector converging to a stable probability distribution hold true for all Markov chains? The answer is no.

Example 2. Markov Chain Not Converging to Stable Distribution

Consider the simple two-state Markov chain with transition matrix and starting vector

$$\mathbf{A} = \begin{bmatrix} 0 & 1 \\ 1 & 0 \end{bmatrix} \qquad \text{and} \qquad \mathbf{p} = [1, 0] \qquad (8)$$

It is easy to check that $\mathbf{p}^{(k)} = [0, 1]$ for k odd, and $\mathbf{p}^{(k)} = [1, 0]$ for k even. More generally, for a starting vector of $\mathbf{p} = [r, 1 - r]$ for any r, $0 \leq r \leq 1$, we have $\mathbf{p}^{(k)} = [1 - r, r]$ for k odd, and $\mathbf{p}^{(k)} = [r, 1 - r]$ for k even. Note that $\mathbf{p}^0 = [.5, .5]$ is a stable vector (an eigenvector with eigenvalue 1), but this Markov chain will not converge to $\mathbf{p}^0$; if we do not start at $\mathbf{p}^0$, we never get to $\mathbf{p}^0$.

The powers of $\mathbf{A}$ have a similar odd–even cyclic pattern, with $\mathbf{A}^k = \mathbf{A}$ for k odd, and $\mathbf{A}^k = \mathbf{I}$ for k even. ∎

So now the question is: Under what conditions does a Markov transition matrix **A** have a stable probability vector to which any starting vector will converge, or equivalently, when will the columns in powers of **A** all converge to a given probability vector as in (7)? The question is not tied to the existence of an eigenvector with eigenvalue 1, since in Example 2, [.5, .5] was such an eigenvector, but there was no convergence. Instead, the answer depends on the absence of any cyclic or other nonrandom pattern, as seen in Example 2.

Definition. A Markov chain with transition matrix **A** is **regular** if for some positive integer h, the matrix $\mathbf{A}^h$ has all positive entries.

The matrix in Example 2 was not regular. A regular Markov chain mixes, or randomizes, patterns so as to eliminate any cyclic behavior. If a Markov chain is regular, then every column of $\mathbf{A}^h$ has all positive entries, meaning that starting from state j it is possible after h periods to be in any of the states. The following theorem requires a lengthy, but not advanced, proof that may be found in any of the texts on Markov chains listed in the References.

Theorem 1. Every regular Markov chain with transition matrix **A** has a stable probability vector **p*** to which $\mathbf{p}^{(k)} = \mathbf{A}^k\mathbf{p}$ converges, for any probability vector **p**. All the columns of $\mathbf{A}^k$ also converge to **p***.

One way to find the stable distribution of a regular Markov chain is, as done in Section 3.5, by solving $(\mathbf{A} - \mathbf{I})\mathbf{p} = \mathbf{0}$ and then picking the constant in the solution to make the components sum to 1 [see equation (5)]. Another approach is to add the additional constraint $\mathbf{1} \cdot \mathbf{p} \ (= \Sigma \ p_i) = 1$.

$$(\mathbf{A} - \mathbf{I})\mathbf{p} = 0 \qquad\qquad (9)$$
$$\mathbf{1} \cdot \mathbf{p} = 1$$

This is a set of $n + 1$ equations in n unknowns.

Example 3. Solving for Stable Distribution

Consider a simpler Markov chain which involves just two states that represent two islands, isle 1 and isle 2, in an isolated country. We are interested in the flow of money between these two islands. Assume that no money enters or leaves the country. Then a Markov chain should provide a reasonable model for currency flow. We shall perform a general analysis of this model rather than use specific values for the transition probabilities a_{ij}. Since columns must sum to 1, the transition matrix **A** can be written in terms of the off-diagonal entries thus:

$$
\begin{array}{cc}
 & \text{Current State} \\
 & \begin{array}{cc} 1 & \quad 2 \end{array} \\
\mathbf{A} = \begin{array}{c} \text{Next 1} \\ \text{State 2} \end{array} & \begin{bmatrix} 1 - b & a \\ b & 1 - a \end{bmatrix}
\end{array}
$$

We require $0 < a, b < 1$ so that the Markov chain will be regular. Let us solve (9) for this **A**.

$$(\mathbf{A} - \mathbf{I})\mathbf{p} = \mathbf{0}: \qquad -bp_1 + ap_2 = 0$$
$$bp_1 - ap_2 = 0 \qquad (10)$$
$$\mathbf{1} \cdot \mathbf{p} = 1: \qquad p_1 + p_2 = 1$$

The second equation in (10) is just the first equation multiplied by -1. So the second equation is redundant, and we are back to the standard situation of two equations in two unknowns. When solved, they yield the stable distribution

$$p_1 = \frac{a}{a + b}, \qquad p_2 = \frac{b}{a + b} \qquad (11)$$

Note that (11) will always be well defined unless a and b are both 0, in which case we get the trivial Markov chain: $p_1' = p_1, p_2' = p_2$. ■

For any Markov transition matrix, the system $(\mathbf{A} - \mathbf{I})\mathbf{p} = \mathbf{0}$ always has redundancy because the sum of the right-hand sides of all the equations is 0 (**A** has column sums of 1, but the $-\mathbf{I}$ term makes the column sums of $\mathbf{A} - \mathbf{I}$ equal to 0) or, eqivalently, the last row is minus the sum of all the preceding rows. When such redundancy exists, the last row will be zeroed out in Gaussian elimination (the reasons for this are discussed in Section 5.2). Thus the last row can be replaced by the constraint $\mathbf{1} \cdot \mathbf{p} = 1$, as implicitly happened in Example 3.

However, Gaussian elimination is so simple in tridiagonal systems, like the frog Markov chain, that adding this new row creates as much trouble as it saves. We illustrate the advantage of a tridiagonal matrix with the following large-scale example.

Example 4. **An *n*-State Frog Markov Chain**

Let us generalize the frog Markov chain to a chain with n states, where n is an arbitrary number. The system of equations $(\mathbf{A} - \mathbf{I})\mathbf{p} = \mathbf{0}$ is

$$-.50p_1 + .25p_2 \qquad\qquad = 0$$
$$.50p_1 - .50p_2 + .25p_3 \qquad\qquad = 0$$
$$.25p_2 - .50p_3 + .25p_4 \qquad\qquad = 0$$
$$\vdots$$
$$.25p_{n-3} - .50p_{n-2} + .25p_{n-1} \qquad = 0$$
$$.25p_{n-2} - .50p_{n-1} + .50p_n = 0$$
$$.25p_{n-1} - .50p_n = 0$$
$$(12)$$

Performing elimination in the first column requires us simply to add the first equation to the second.

$$
\begin{aligned}
-.50p_1 + .25p_2 &= 0 \\
-.25p_2 + .25p_3 &= 0 \\
.25p_2 - .50p_3 + .25p_4 &= 0 \\
\vdots \\
.25p_{n-3} - .50p_{n-2} + .25p_{n-1} &= 0 \\
.25p_{n-2} - .50p_{n-1} + .50p_n &= 0 \\
.25p_{n-1} - .50p_n &= 0
\end{aligned}
$$

Similarly, for elimination in the second column we add the second equation to the third.

$$
\begin{aligned}
-.50p_1 + .25p_2 &= 0 \\
-.25p_2 + .25p_3 &= 0 \\
-.25p_3 + .25p_4 &= 0 \\
.25p_3 - .50p_4 + .25p_5 &= 0 \\
\vdots \\
.25p_{n-3} - .50p_{n-2} + .25p_{n-1} &= 0 \\
.25p_{n-2} - .50p_{n-1} + .50p_n &= 0 \\
.25p_{n-1} - .50p_n &= 0
\end{aligned}
$$

The situation when we come to perform elimination in the third column is the same as in the second column and again involves adding the third equation to the fourth. This situation will stay the same for every column from the second through the $(n-1)$st. After elimination in the first $n-1$ columns, we have

$$
\begin{aligned}
-.50p_1 + .25p_2 &= 0 \\
-.25p_2 + .25p_3 &= 0 \\
-.25p_3 + .25p_4 &= 0 \\
\vdots \\
-.25p_{n-2} + .25p_{n-1} &= 0 \\
-.25p_{n-1} + .50p_n &= 0 \\
0 &= 0
\end{aligned}
\tag{13}
$$

Note that the $(n-1)$st equation in (13) is the negative of the original last equation, so the last equation is zeroed out when we perform elimination in the $(n-1)$st column. For concreteness, the reader may want to refer back to Example 1 of Section 3.5, where we solved this system for the original frog Markov chain, where $n = 6$.

Letting $p_n = q$, we perform back substitution and find from the $(n - 1)$st equation in (13) that $.25p_{n-1} = .50p_n \; (= .50q)$, so $p_{n-1} = 2q$. Equations 2 through $n - 2$ in (13) say that successive p_i, p_{i+1} pairs from p_2 through p_{n-1} are equal. Since $p_{n-1} = 2q$, all these p_i's equal $2q$. Finally, we see that $p_1 = q$. So our solution has the form

$$[q, 2q, 2q, \ldots, 2q, 2q, q]$$

The sum of the entries in this general solution is $(2n - 2)q$. For this sum to equal 1, we require that $q = 1/(2n - 2)$. So our stable distribution is

$$\mathbf{p}^* = \left[\frac{1}{2n - 2}, \frac{2}{2n - 2}, \frac{2}{2n - 2}, \ldots, \frac{2}{2n - 2}, \frac{2}{2n - 2}, \frac{1}{2n - 2} \right]$$

The effect of replacing the last equation in (12) by $\mathbf{1} \cdot \mathbf{p} = 1$ is discussed in the Exercises. ∎

Next we consider an important type of nonregular Markov chain, called an absorbing Markov chain. A state S_i in a Markov chain is called an **absorbing state** if $a_{ii} = 1$, that is, once you enter state S_i you never leave it. A Markov chain with one or more absorbing states is called an **absorbing Markov chain**.

Example 5. A Gambling Model with Absorbing States

Absorbing states complicate the behavior of a Markov chain and lead to a variety of different stable probabilities. Consider the following Markov chain for gambling, with states representing the gambler's winnings. Each round, the gambler has a probability .3 of winning $1, .33 of losing $1, and .37 of staying the same. The gambler stops if he or she loses all of the money, and also stops if the winnings reach $6. So 0 and 6 will be absorbing states in this Markov chain.

Current State

		0	1	2	3	4	5	6
	0	1	.33	0	0	0	0	0
	1	0	.37	.33	0	0	0	0
	2	0	.30	.37	.33	0	0	0
Next	3	0	0	.30	.37	.33	0	0
State	4	0	0	0	.30	.37	.33	0
	5	0	0	0	0	.30	.37	0
	6	0	0	0	0	0	.30	1

$$\mathbf{A} = \qquad\qquad (14)$$

After playing a very long time, a person is certain to have stopped, either having gone broke or having won. Thus the stable probabilities should only involve the absorbing states 0 and 6, that is, the stable vector $\mathbf{p}^*$ has the form

$$p_0^* = p, \quad p_6^* = 1 - p, \quad \text{and} \quad p_1^* = p_2^* = p_3^* = p_4^* = p_5^* = 0$$
(15)

By looking at the transition probabilities for states 0 and 6 alone,

		Current State	
		0	6
Next	0	1	0
State	6	0	1

it is easy to see that *any* $\mathbf{p}^*$ of the form in (15), with $0 \leq p \leq 1$, is a stable vector.

Before doing any mathematical analysis of absorbing Markov chains, let us explore the behavior of (14) by letting a computer pro-

Table 4.2

	Probability Distribution After k Rounds						
Rounds	**0**	**1**	**2**	**3**	**4**	**5**	**6**
0	0	0	0	1	0	0	0
1	0	0	.33	.37	.30	0	0
2	0	.109	.244	.335	.222	.09	0
3	.036	.121	.234	.270	.212	.100	.027
4	.076	.128	.212	.240	.193	.101	.057
5	.116	.115	.194	.216	.177	.095	.087
6	.154	.107	.178	.196	.161	.088	.116
8	.222	.090	.149	.164	.135	.074	.166
10	.278	.075	.125	.137	.113	.062	.209
15	.383	.048	.080	.088	.072	.040	.288
20	.451	.031	.051	.056	.047	.026	.336
25	.494	.020	.033	.036	.030	.016	.371
50	.563	.002	.003	.004	.003	.002	.423
75	.570	~0	~0	~0	~0	~0	.428
100	.571	~0	~0	~0	~0	~0	.429

gram produce a table of probability distributions over many rounds when we start with $3. Since $3 is halfway between losing and winning, we could expect the chances of losing to be close to .5, but a little above .5, since there is always .33-versus-.30 bias toward losing a dollar in any state.

So if we start with $3, the probability of eventually losing (before we reach $6) is $p = .571$. Instead of repeating this computer simulation for other starting values, we shall now develop a theory that lets us calculate directly the probability of losing or winning, when we start with different amounts of money. ∎

The first step in our development is to divide the states of an absorbing Markov chain into two groups, the absorbing states and the nonabsorbing states. Assume that there are r absorbing states and s nonabsorbing states. If the absorbing states are listed first, the transition matrix **A** can be partitioned into the form

$$
\mathbf{A} = \begin{array}{cc} & \begin{array}{cc} Ab & NAb \end{array} \\ \begin{array}{c} Ab \\ NAb \end{array} & \begin{bmatrix} \mathbf{I} & \mathbf{R} \\ \mathbf{O} & \mathbf{Q} \end{bmatrix} \end{array} \tag{16}
$$

where **I** is an r-by-r identity matrix, **O** is an s-by-r matrix of 0's, **R** is an r-by-s matrix with entry r_{ij} giving the probability of going from nonabsorbing state j to absorbing state i, and **Q** is the s-by-s transition matrix among the nonabsorbing states. The transition matrix (14) in Example 5 becomes

$$
\mathbf{A} = \begin{array}{c} 0 \\ 6 \\ 1 \\ 2 \\ 3 \\ 4 \\ 5 \end{array}
\begin{array}{c} \begin{array}{ccccccc} 0 & 6 & 1 & 2 & 3 & 4 & 5 \end{array} \\
\begin{bmatrix}
1 & 0 & .33 & 0 & 0 & 0 & 0 \\
0 & 1 & 0 & 0 & 0 & 0 & .30 \\
0 & 0 & .37 & .33 & 0 & 0 & 0 \\
0 & 0 & .30 & .37 & .33 & 0 & 0 \\
0 & 0 & 0 & .30 & .37 & .33 & 0 \\
0 & 0 & 0 & 0 & .30 & .37 & .33 \\
0 & 0 & 0 & 0 & 0 & .30 & .37
\end{bmatrix}
\end{array} \tag{17}
$$

Using the rule for matrix multiplication of a partitioned matrix from Section 2.6 (just treat the submatrices like individual entries), we have

$$
\begin{aligned}
\mathbf{A}^2 &= \begin{bmatrix} \mathbf{I} & \mathbf{R} \\ \mathbf{O} & \mathbf{Q} \end{bmatrix}\begin{bmatrix} \mathbf{I} & \mathbf{R} \\ \mathbf{O} & \mathbf{Q} \end{bmatrix} = \begin{bmatrix} \mathbf{II} + \mathbf{RO} & \mathbf{IR} + \mathbf{RQ} \\ \mathbf{IO} + \mathbf{OQ} & \mathbf{OR} + \mathbf{QQ} \end{bmatrix} \\
&= \begin{bmatrix} \mathbf{I} & \mathbf{R} + \mathbf{RQ} \\ \mathbf{O} & \mathbf{Q}^2 \end{bmatrix}
\end{aligned} \tag{18}
$$

and multiplying **A** times $\mathbf{A}^2$, we find that

$$\mathbf{A}^3 = \begin{bmatrix} \mathbf{I} & \mathbf{R} + \mathbf{RQ} + \mathbf{RQ}^2 \\ \mathbf{O} & \mathbf{Q}^3 \end{bmatrix} = \begin{bmatrix} 1 & 0 & .53 & .19 & .04 & 0 & 0 \\ 0 & 1 & 0 & 0 & .03 & .16 & .48 \\ 0 & 0 & .16 & .20 & .12 & .04 & 0 \\ 0 & 0 & .18 & .27 & .23 & .12 & .04 \\ 0 & 0 & .10 & .21 & .27 & .23 & .12 \\ 0 & 0 & .03 & .10 & .21 & .27 & .20 \\ 0 & 0 & 0 & .03 & .10 & .18 & .16 \end{bmatrix}$$

It is left as an exercise to check that for higher powers of $\mathbf{A}$, the partitioned form is

$$\mathbf{A}^k = \begin{bmatrix} \mathbf{I} & \mathbf{R}_k^* \\ \mathbf{O} & \mathbf{Q}^k \end{bmatrix},$$

where $\mathbf{R}_k^* = \mathbf{R} + \mathbf{RQ} + \mathbf{RQ}^2 + \cdots + \mathbf{RQ}^{k-1}$ (19)

Note that $\mathbf{Q}^k$ is the standard kth power of the nonabsorbing transition matrix $\mathbf{Q}$, for play among the nonabsorbing states.

As k gets large, the entries in $\mathbf{Q}^k$ will approach 0, since over time the probability of not getting absorbed approaches 0. The important submatrix in (19) is $\mathbf{R}_k^*$. Entry (i, j) of $\mathbf{R}_k^*$ is the probability of being in absorbing state i after k rounds if we start in nonabsorbing state j. Let us explain what this probability is in detail. To go from nonabsorbing state j to absorbing state i after k rounds, we can either go immediately on the first round from j to i—with probability r_{ij}—(and remain in absorbing state i), or we can wander among the nonabsorbing states for several rounds, ending up after w rounds in nonabsorbing state h—with probability given by entry (j, h) in $\mathbf{Q}^w$—and then go from state h to absorbing state i—with probability r_{ih} (and thereafter remaining in state i). The total probability of starting in a nonabsorbing state j, wandering among nonabsorbing states for w rounds, and then going from some nonabsorbing state to absorbing state i is given by entry (i, j) in $\mathbf{RQ}^w$. Since the number w can range up to $k - 1$, we obtain the sum for $\mathbf{R}_k^*$ given in (19).

The limiting matrix $\mathbf{R}^*$ for $\mathbf{R}_k^*$ as k approaches infinity will give the probabilities r_{ij}^* that starting in nonabsorbing state j we eventually end up in absorbing state i.

$$\mathbf{R}^* = \mathbf{R} + \mathbf{RQ} + \mathbf{RQ}^2 + \cdots = \mathbf{R}(\mathbf{I} + \mathbf{Q} + \mathbf{Q}^2 + \cdots) \quad (20)$$

$\mathbf{R}^*$ is the matrix that would tell us in the gambling model the probability of eventually losing or winning, when we start with different amounts.

There are two ways to compute $\mathbf{R}^*$. The first way is to compute $\mathbf{Q}^k$ for all k up to some large number, say 50, and sum these matrices and multiply by $\mathbf{R}$ to obtain $\mathbf{R}^*$ as in (20). The other way rewrites $\mathbf{R}^*$ as

$$\mathbf{R}^* = \mathbf{R}(\mathbf{I} + \mathbf{Q} + \mathbf{Q}^2 + \cdots) = \mathbf{R}(\mathbf{I} - \mathbf{Q})^{-1} \quad (21)$$

using the geometric series identity introduced in equation (7) of Section 3.4.

We call $(\mathbf{I} - \mathbf{Q})^{-1}$ the **fundamental matrix of an absorbing Markov chain** and use the matrix $\mathbf{N}$ to denote it.

$$\mathbf{N} = \mathbf{I} + \mathbf{Q} + \mathbf{Q}^2 + \cdots = (\mathbf{I} - \mathbf{Q})^{-1} \qquad \text{and} \qquad \mathbf{R}^* = \mathbf{RN} \quad (22)$$

We can calculate this inverse using elimination by pivoting. $\mathbf{N}$ is given in (24).

The geometric series identity used in (21) required that $\|\mathbf{Q}\| < 1$ for some matrix norm. The sum norm $\|\mathbf{Q}\|_s$ (largest column sum) of $\mathbf{Q}$ in (17) is 1, and the max norm is > 1. However, $\|\mathbf{Q}\| < 1$ in the euclidean norm.

The matrix $\mathbf{N}$ contains some very useful information by itself. It tells us the expected number of times we will visit (nonabsorbing) state i if we start in (nonabsorbing) state j. The reasoning is as follows. The average number of times we visit state i starting from state j after *exactly* one round is simply $0(1 - q_{ij}) + 1q_{ij} = q_{ij}$—the weighted average of visiting state i zero times and of visiting state i one time. The average number of times we visit state i starting from state j after exactly two rounds is $0(1 - q_{ij}^{(2)}) + 1q_{ij}^{(2)} = q_{ij}^{(2)}$, where $q_{ij}^{(2)}$ denotes entry (i, j) in $\mathbf{Q}^2$. The average number of visits after exactly k rounds is entry (i, j) in $\mathbf{Q}^k$. Probability theory states that the average number of visits from state j to state i totaled over all rounds is simply the sum of the average number of visits on each specific round. So the expected number of times we visit nonabsorbing state i starting from nonabsorbing state j is the sum of the (i, j) entries in $\mathbf{Q}^k$ for all $\mathbf{Q}^k$, that is, entry (i, j) in $\mathbf{N}$.

Furthermore, if we sum the entries of the jth column of $\mathbf{N}$—the expected number of times, starting from state j, that we visit state 1 plus the expected number of times we visit state 2, and so on—we obtain the expected number of rounds until we are absorbed. The vector-matrix product $\mathbf{1N}$ computes the sum of each column of $\mathbf{N}$.

We summarize this wealth of information about absorbing Markov chains we can get from $\mathbf{N}$ with the following theorem. The term **absorption** is used in this theorem to mean going to an absorbing state.

Theorem 2. Let $\mathbf{N}$ be the fundamental matrix of an absorbing Markov chain [$\mathbf{N}$ is defined in (22)]. Then the following are true.

  (i) Entry n_{ij} of $\mathbf{N}$ is the expected number of times we visit the nonabsorbing state i (before absorption) when we start in nonabsorbing state j.

  (ii) The jth entry in the vector $\mathbf{1N}$ gives the expected number of rounds before absorption when we start in nonabsorbing state j.

  (iii) Entry (i, j) in $\mathbf{RN}$ is the probability of eventually ending up in absorbing state i when we start in nonabsorbing state j.

Example 5 (continued). A Gambling Model

With Theorem 2 we can answer a variety of interesting questions about this model. We must compute the matrix $\mathbf{N}$ by finding the inverse of $\mathbf{I} - \mathbf{Q}$, where $\mathbf{Q}$ is

$$\mathbf{Q} = \begin{array}{c} \\ 1 \\ 2 \\ 3 \\ 4 \\ 5 \end{array} \begin{array}{ccccc} 1 & 2 & 3 & 4 & 5 \\ \left[\begin{array}{ccccc} .37 & .33 & 0 & 0 & 0 \\ .30 & .37 & .33 & 0 & 0 \\ 0 & .30 & .37 & .33 & 0 \\ 0 & 0 & .30 & .37 & .33 \\ 0 & 0 & 0 & .30 & .37 \end{array} \right] \end{array} \tag{23}$$

Using elimination by pivoting (described in Section 3.3), we obtain

$$\mathbf{N} = (\mathbf{I} - \mathbf{Q})^{-1} = \begin{array}{c} \\ 1 \\ 2 \\ 3 \\ 4 \\ 5 \end{array} \begin{array}{ccccc} 1 & 2 & 3 & 4 & 5 \\ \left[\begin{array}{ccccc} 2.64 & 2.21 & 1.73 & 1.21 & .63 \\ 2.01 & 4.21 & 3.30 & 2.31 & 1.21 \\ 1.43 & 3.00 & 4.73 & 3.30 & 1.73 \\ .91 & 1.91 & 3.00 & 4.21 & 2.21 \\ .43 & .91 & 1.43 & 2.01 & 2.64 \end{array} \right] \end{array} \tag{24}$$

From (24) and Theorem 2, we see that if we started with \$3, there would be 3.3 rounds during an average gambling session when we would be in state 2 (when we would have \$2).

Next we sum the columns of $\mathbf{N}$:

$$\mathbf{1N} = [7.42, 12.24, 14.19, 13.04, 8.42] \tag{25}$$

The third entry in (25) tells us that if we start with \$3, we get to play about 14 rounds, on average, before the game ends.

Finally, we compute $\mathbf{R}^* = \mathbf{RN}$, where we see from (17) that $\mathbf{R}$ is

$$\mathbf{R} = \begin{array}{c} \\ 0 \\ 6 \end{array} \begin{array}{ccccc} 1 & 2 & 3 & 4 & 5 \\ \left[\begin{array}{ccccc} .33 & 0 & 0 & 0 & 0 \\ 0 & 0 & 0 & 0 & .3 \end{array} \right] \end{array} \tag{26}$$

$$\mathbf{R}^* = \mathbf{RN} = \begin{array}{c} \\ 0 \\ 6 \end{array} \begin{array}{ccccc} 1 & 2 & 3 & 4 & 5 \\ \left[\begin{array}{ccccc} .87 & .73 & .57 & .40 & .21 \\ .13 & .27 & .43 & .60 & .79 \end{array} \right] \end{array}$$

Entry (0, 3) of $\mathbf{R}^*$ confirms our earlier simulation result that the probability of going broke when we start with \$3 is .57.   ■

We close this section by noting that some of these results about absorbing Markov chains can be applied to regular Markov chains with the following trick. Let $\mathbf{A}$ be the transition matrix of a regular Markov chain. We convert one state, say state p, into an absorbing state by replacing the pth column $\mathbf{a}_p^C$ of $\mathbf{A}$ by the pth unit vector $\mathbf{e}_p$. Now whenever we come to state p, we stay there. Our theory of absorbing Markov chains can be applied to this modified transition matrix to determine the expected number of rounds it takes to get to state p if we start from any other state j (and also the expected number of visits in any third state on the journey from j to p).

If we convert two states p and r into absorbing states with unit vector columns, then we compute the relative probability of reaching p or r first when we start from some other state j. These calculations are requested for Example 1 in the Exercises.

Example 6. **Ice Cream Selection**

We surveyed a group of students eating blueberry, mint, and strawberry ice cream about which flavor they would choose next time. Suppose that $\frac{3}{4}$ of those eating blueberry would choose blueberry the next time, while the remaining quarter would choose strawberry. Responses from others yielded the following transition matrix:

$$
\begin{array}{c}
\\
\\
\text{Next}\\
\text{Time}
\end{array}
\begin{array}{c}
\\
\text{Blueberry}\\
\text{Mint}\\
\text{Strawberry}
\end{array}
\overset{\displaystyle \begin{array}{ccc}\text{Current Flavor}\end{array}}{\overset{\displaystyle\begin{array}{ccc}\text{Blueberry} & \text{Mint} & \text{Strawberry}\end{array}}{\left[\begin{array}{ccc}
\frac{3}{4} & \frac{1}{2} & 0\\
0 & \frac{1}{2} & \frac{1}{3}\\
\frac{1}{4} & 0 & \frac{2}{3}
\end{array}\right]}} \qquad (27)
$$

We treat the selection of flavors as a Markov process and pose the question: How many rounds does it take on average for a person to switch from strawberry to blueberry?

To answer this question, we change the transition matrix in (27) by making blueberry an absorbing state. The modified transition matrix is

$$
\mathbf{A} = \begin{array}{c}b\\m\\s\end{array}\overset{\displaystyle\begin{array}{ccc}b & m & s\end{array}}{\left[\begin{array}{ccc}
1 & \frac{1}{2} & 0\\
0 & \frac{1}{2} & \frac{1}{3}\\
0 & 0 & \frac{2}{3}
\end{array}\right]} = \left[\begin{array}{cc}\mathbf{I} & \mathbf{R}\\ \mathbf{O} & \mathbf{Q}\end{array}\right] \quad \text{with } \mathbf{Q} = \left[\begin{array}{cc}\frac{1}{2} & \frac{1}{3}\\ 0 & \frac{2}{3}\end{array}\right] \qquad (28)
$$

Then we find that

$$
\mathbf{N} = (\mathbf{I} - \mathbf{Q})^{-1} = \left[\begin{array}{cc}\frac{1}{2} & -\frac{1}{3}\\ 0 & \frac{1}{3}\end{array}\right]^{-1} = \left[\begin{array}{cc}2 & 2\\ 0 & 3\end{array}\right]
$$

and

$$
\mathbf{1N} = [2, 5]
$$

By Theorem 2, part (ii), the second entry, 5, in $\mathbf{1N}$ is the average number of rounds until absorption (blueberry) when starting from strawberry. Moreover, by Theorem 2, part (i), the second column of

N tells us that on average a person starting with strawberry would choose strawberry three times (including the initial time) and choose mint twice before choosing blueberry.   ■

With modest effort (see any textbook on Markov chains in the References), one can also prove the following interesting result.

Theorem 3. Let $\mathbf{p}^* = [p_1^*, p_2^*, \ldots, p_n^*]$ be the stable probability vector of a regular Markov chain. Then, if we start in state i, the expected number of rounds before we return to i again is $1/p_i^*$.

Section 4.4 **Exercises**

Summary of Exercises
Exercises 1–14 concern regular Markov chains, their stable distributions, and long-term behavior. Exercises 15–25 involve analysis of absorbing Markov chains.

1. Describe the behavior of the Markov chain

$$\begin{bmatrix} 0 & 1 & 0 \\ 0 & 0 & 1 \\ 1 & 0 & 0 \end{bmatrix}$$

with starting vector $[1, 0, 0]$. Are there any stable vectors?

2. Which of the following transition matrices belong to regular Markov chains? Find a stable distribution for each chain.

(a) $\begin{bmatrix} 0 & \frac{1}{2} \\ 1 & \frac{1}{2} \end{bmatrix}$ (b) $\begin{bmatrix} \frac{1}{2} & 0 \\ \frac{1}{2} & 1 \end{bmatrix}$ (c) $\begin{bmatrix} \frac{1}{2} & 1 & 0 \\ 0 & 0 & 1 \\ \frac{1}{2} & 0 & 0 \end{bmatrix}$

3. Compute the stable distribution for the weather Markov chain introduced in Section 1.3 with transition matrix

$$\begin{array}{cc} & \begin{array}{cc} \text{Sunny} & \text{Cloudy} \end{array} \\ \begin{array}{c} \text{Sunny} \\ \text{Cloudy} \end{array} & \begin{bmatrix} \frac{3}{4} & \frac{1}{2} \\ \frac{1}{4} & \frac{1}{2} \end{bmatrix} \end{array}$$

4. The printing press in a newspaper has the following pattern of breakdowns. If it is working today, tomorrow it has 90% chance of working (and 10% chance of breaking down). If the press is broken today, it has a 60% chance of working tomorrow (and 40% chance by being broken again). Compute the stable distribution for this Markov chain.

5. If the local professional basketball team, the Sneakers, wins today's game, they have a $\frac{2}{3}$ chance of winning their next game. If they lose this game, they have a $\frac{1}{2}$ chance of winning their next game. Compute the stable distribution for this Markov chain and give the approximate values of entries in $\mathbf{A}^{100}$, where $\mathbf{A}$ is this Markov chain's transition matrix.

6. If the stock market went up today, historical data show that it has a 60% chance of going up tomorrow, a 20% chance of staying the same, and a 20% chance of going down. If the market was unchanged today, it has a 20% chance of being unchanged tomorrow, a 40% chance of going up, and a 40% chance of going down. If the market goes down today, it has a 20% of going up tomorrow, a 20% chance of being unchanged, and a 60% chance of going down. Compute the stable distribution for the stock market.

7. Write down a Markov chain to model the following situation: Assume that there are three types of voters in Texas: Republicans, Democrats, and Independent. From one (national) election to the next, 60% of Republicans remain Republican and similarly for the two other groups; among the 40% who change parties, 30% become Independent and 10% go to the other major party, except that the Independents who change all become Republicans. Determine the stable distribution among the three parties and from it give the approximate values of entries in $\mathbf{A}^{1000}$.

8. **(a)** Make a Markov chain model for a rat wandering through the following maze if, at the end of each period, the rat is equally likely to leave its current room through any of the doorways. (It never stays where it is.)

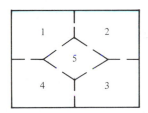

   (b) What is the stable distribution?

9. Repeat the questions in Exercise 8 for the following maze.

10. Find the stable distribution for Markov chains with the following transition matrices.

(a) $\begin{bmatrix} \frac{1}{3} & \frac{1}{3} & 0 & 0 & 0 & 0 \\ \frac{2}{3} & \frac{1}{3} & \frac{1}{3} & 0 & 0 & 0 \\ 0 & \frac{1}{3} & \frac{1}{3} & \frac{1}{3} & 0 & 0 \\ 0 & 0 & \frac{1}{3} & \frac{1}{3} & \frac{1}{3} & 0 \\ 0 & 0 & 0 & \frac{1}{3} & \frac{1}{3} & \frac{2}{3} \\ 0 & 0 & 0 & 0 & \frac{1}{3} & \frac{1}{3} \end{bmatrix}$

(b) $\begin{bmatrix} \frac{2}{3} & \frac{2}{3} & 0 & 0 & 0 & 0 \\ \frac{1}{3} & \frac{1}{6} & \frac{2}{3} & 0 & 0 & 0 \\ 0 & \frac{1}{6} & \frac{1}{6} & \frac{2}{3} & 0 & 0 \\ 0 & 0 & \frac{1}{6} & \frac{1}{6} & \frac{2}{3} & 0 \\ 0 & 0 & 0 & \frac{1}{6} & \frac{1}{6} & \frac{2}{3} \\ 0 & 0 & 0 & 0 & \frac{1}{6} & \frac{1}{3} \end{bmatrix}$

(c) $\begin{bmatrix} \frac{2}{3} & \frac{2}{3} & 0 & 0 & 0 & 0 \\ \frac{1}{3} & \frac{1}{6} & \frac{2}{3} & 0 & 0 & 0 \\ 0 & \frac{1}{6} & \frac{1}{6} & \frac{1}{6} & 0 & 0 \\ 0 & 0 & \frac{1}{6} & \frac{1}{6} & \frac{1}{6} & 0 \\ 0 & 0 & 0 & \frac{2}{3} & \frac{1}{6} & \frac{1}{3} \\ 0 & 0 & 0 & 0 & \frac{2}{3} & \frac{2}{3} \end{bmatrix}$

(d) $\begin{bmatrix} 0 & \frac{1}{2} & 0 & 0 & 0 & 0 \\ 1 & 0 & \frac{1}{2} & 0 & 0 & 0 \\ 0 & \frac{1}{2} & 0 & \frac{1}{2} & 0 & 0 \\ 0 & 0 & \frac{1}{2} & 0 & \frac{1}{2} & 0 \\ 0 & 0 & 0 & \frac{1}{2} & 0 & 1 \\ 0 & 0 & 0 & 0 & \frac{1}{2} & 0 \end{bmatrix}$

11. Repeat Exercise 10, parts (a) and (d) with the number of states expanded from six to n, as done in Example 4.

12. Determine the two eigenvalues and associated eigenvectors for the Markov chain in the following exercises. Give the distribution after six periods for the given starting distribution $\mathbf{p}$ by representing $\mathbf{p}$ as a linear combination of the eigenvectors as was done in the end of Section 3.1.
    (a) Exercise 3, starting $\mathbf{p}$: Sunny 0, Cloudy 1.
    (b) Exercise 4, starting $\mathbf{p}$: Working 1, Broken 0.
    (c) Exercise 5, starting $\mathbf{p}$: Winning $\frac{1}{2}$, Losing $\frac{1}{2}$.

13. Show that if $\mathbf{A}$ is a tridiagonal Markov transition matrix, then in solving $(\mathbf{A} - \mathbf{I})\mathbf{p} = \mathbf{0}$ by Gaussian elimination, the L in LU decomposition of $\mathbf{A}$ is a matrix with 1's on the main diagonal and -1 just below the diagonal entries. That is, in Gaussian elimination one always adds the current row (times 1) to the next row.

14. Re-solve the stable distribution problems in Exercise 10 with the last row of the matrix equation $(\mathbf{A} - \mathbf{I})\mathbf{p}$ replaced by the constraint $\mathbf{1} \cdot \mathbf{p} = 1$ (the last row always drops out—becomes 0—and the additional constraint that the probabilities sum to 1 can be put in its place).

15. The following questions refer to the gambling Markov chain in Example 5. If you started with \$4, what is the expected number of rounds that you have \$3, and what is the expected number of rounds until the game ends?

16. The following model for learning a concept over a set of lessons identifies four states of learning: I = ignorance, E = exploratory thinking, S = superficial understanding, and M = mastery. If now in state I,

after one lesson you have $\frac{1}{2}$ probability of still being in I and $\frac{1}{2}$ probability of being in E. If now in state E, you have $\frac{1}{4}$ probability of being in I, $\frac{1}{2}$ in E, and $\frac{1}{4}$ in S. If now in state S, you have $\frac{1}{4}$ probability of being in E, $\frac{1}{2}$ in S, and $\frac{1}{4}$ in M. If in M, you always stay in M (with probability 1).

(a) Write out the transition matrix for this Markov chain with the absorbing state as the first state.

(b) Compute the fundamental matrix **N**.

(c) What is the expected number of rounds until mastery is attained if currently in the state of ignorance?

17. In the following maze suppose that a rat has a 20% chance of going into the middle room, which is an absorbing state, and a 40% chance each of going to the room on the left or on the right.

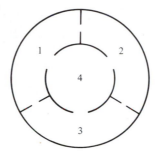

(a) What is the expected number of times a rat starting in room 1 enters room 2?

(b) What is the expected number of rounds until the rat goes to the middle room?

18. (a) Make a Markov chain model for a rat wandering through the following maze if, at the end of each period, the rat is equally likely to leave its current room through any of the doorways. The center room is an absorbing state. (It never stays in the same room.)

(b) If the rat starts in room 4, what is the expected number of times it will be in room 2?

(c) If the rat starts in room 4, what is the expected rounds until absorption?

19. Consider the game of Ping-Pong with the following states:

A: Player 1 is hitting the ball.
B: Player 2 is hitting the ball.
C: Play is dead because 1 hit the ball out or in the net.
D: Play is dead because 2 hit the ball out or in the net.

The transition matrix is

$$
\begin{array}{c}
\text{(A hitting ball) 1} \\
\text{(B hitting ball) 2} \\
\text{(A hit ball out) 3} \\
\text{(B hit ball out) 4}
\end{array}
\begin{array}{cccc}
1 & 2 & 3 & 4 \\
\left[\begin{array}{cccc}
0 & .9 & 0 & 0 \\
.4 & 0 & 0 & 0 \\
.6 & 0 & 1 & 0 \\
0 & .1 & 0 & 1
\end{array}\right]
\end{array}
$$

If we start play with player *A* hitting the ball (in state 1)
(a) What is the expected number of times player *A* hits the ball (before the point is over)?
(b) What is the expected number of hits by *A* and *B* (before the point is over)?
(c) What is the probability that player *A* hits the ball out (i.e., that player *B* wins the point)?

20. Repeat Exercise 18 for the following maze, in which rooms 1 and 5 are absorbing. Start in room 2.

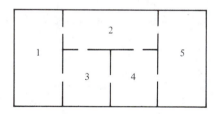

21. Repeat the Markov chain model of a poker game given in Example 5 but now with probability $\frac{1}{3}$ that a player wins 1 dollar in a period, with probability $\frac{1}{3}$ a player loses 1 dollar, and with probability $\frac{1}{3}$ a player stays the same. The game ends if the player loses all his or her money *or* if the player has 6 dollars. Compute **N, 1N,** and **RN** for this problem.

22. Three tanks *A, B,* and *C* are engaged in a battle. Tank *A*, when it fires, hits its target with hit probability $\frac{1}{2}$. *B* hits its target with hit probability $\frac{1}{3}$, and *C* with hit probability $\frac{1}{6}$. Initially (in the first period), *B* and *C* fire at *A* and *A* fires at *B*. Once one tank is hit, the remaining tanks aim at each other. The battle ends when there is one or no tank left. The transition matrix for this game is (the states are the subsets of tanks surviving)

$$
\begin{array}{c}
\quad\quad\quad\;\; ABC \quad AC \quad BC \quad A \quad B \quad C \quad \text{None} \\
\begin{array}{c}
ABC \\ AC \\ BC \\ A \\ B \\ C \\ \text{None}
\end{array}
\left[
\begin{array}{ccccccc}
\frac{5}{18} & 0 & 0 & 0 & 0 & 0 & 0 \\
\frac{5}{18} & \frac{5}{12} & 0 & 0 & 0 & 0 & 0 \\
\frac{4}{18} & 0 & \frac{10}{18} & 0 & 0 & 0 & 0 \\
0 & \frac{5}{12} & 0 & 1 & 0 & 0 & 0 \\
0 & 0 & \frac{5}{18} & 0 & 1 & 0 & 0 \\
\frac{4}{18} & \frac{1}{12} & \frac{2}{18} & 0 & 0 & 1 & 0 \\
0 & \frac{1}{12} & \frac{1}{18} & 0 & 0 & 0 & 1
\end{array}
\right]
\end{array}
$$

(a) Determine the expected number of rounds that the battle lasts (starting from state ABC).

(b) What are the chances of the different tanks winning (being the sole surviving tank)?

23. Compute $\mathbf{A}^3$ and $\mathbf{A}^4$, in partitioned form, for the partitioned matrix $\mathbf{A}$ in (17).

24. Modify the frog Markov chain in Example 1 by making states 1 and 6 absorbing states. Compute the probability, when started in state 3 of being absorbed in state 1. Also compute the expected number of periods until absorption (in state 1 or 6).

25. Modify the frog Markov chain in Example 1 by making state 1 an absorbing state. Starting from state 5, compute the expected number of periods until absorption and the expected number of visits to state 6.

Section 4.5 Growth Models

In this section we examine three models for growing populations. We have already seen a simple linear model, introduced in Section 1.3, for the growth of two competing species, rabbits and foxes. Here we will study models for the growth of one species that is subdivided into different age groups. For simplicity we again let rabbits be the object of study in the models. However, our models apply to any renewable natural resource, from animals to forests, and to many human enterprises, be they economic or social. The first model has been applied to human populations to predict population cycles and to set insurance rates.

Example 1. Age-Specific Population Model

We want a model that breaks down a population into different age groups. Human population models commonly have about 20 age groups, with each age group spanning 5 years, plus a special group for the first year of life (since mortality rates for newborns are different from other young children) and a last group consisting of everyone past some advanced age, say 90 years. Each age group is really two groups, one for men and one for women. The study of the sizes of human populations is called *demography*.

To make matters simple, we will content ourselves here with a three-age-group model of rabbits.

$$y = \text{young rabbits, up to 2 years old}$$
$$m = \text{midlife rabbits, between 2 and 4 years old}$$
$$o = \text{old rabbits, 4 to 6 years old}$$

Let one period of time equal 2 years. Our model for the next period's population vector $\mathbf{a}' = [y', m', o']$ in terms of the current population $\mathbf{a} = [y, m, o]$ is

$$
\begin{aligned}
y' &= 4m + o \\
m' &= .4y \\
o' &= .6m
\end{aligned}
\tag{1}
$$

The first equation in (1) says that each midlife rabbit gives birth to 4 young each period and that each old rabbit gives birth to 1 young each period (of course, only females have babies, but in this initial model we are not differentiating between sexes). The second equation says that 40% of all young rabbits survive through their first 2 years (one period). The third equation says that 60% of midlife rabbits live through a period to become old rabbits. Finally, assume that all old rabbits die within 2 years. If $\mathbf{L}$ is the matrix of coefficients in (1),

$$
\mathbf{L} = \begin{bmatrix} 0 & 4 & 1 \\ .4 & 0 & 0 \\ 0 & .6 & 0 \end{bmatrix}
\tag{2}
$$

then (1) has the matrix form

$$
\mathbf{a}' = \mathbf{L}\mathbf{a}
\tag{3}
$$

This population model is called a **Leslie model.** If there were more age groups, the matrix $\mathbf{L}$ of coefficients would have the form

$$
\mathbf{L} = \begin{bmatrix}
0 & b_2 & b_3 & b_4 & \cdots & & b_n \\
p_1 & 0 & 0 & 0 & \cdots & & 0 \\
0 & p_2 & 0 & 0 & \cdots & & 0 \\
0 & 0 & p_3 & 0 & \cdots & & 0 \\
0 & 0 & 0 & p_4 & \cdots & & 0 \\
\vdots & \vdots & \vdots & \vdots & & & \vdots \\
0 & 0 & 0 & 0 & \cdots & p_{n-1} & 0
\end{bmatrix}
\tag{4}
$$

where b_i is the number of offspring per individual in group i and p_i is the probability that an individual in group i survives one period to become a member of group $i + 1$.

The model is somewhat like a Markov chain, except that the numbers b_i are not probabilities. Rather than summing to 1, the three variables y, m, and o will grow larger or smaller over time. We want to know the behavior of this model over many periods. Will the total number of rabbits increase or decrease? Will there be any cyclic patterns in the population, such as a surge in the young one year followed a period later by a surge in midlifes, then the next period a surge in the young, continuing back and forth? Or after several periods, will there be a steady distribution of the population; for example, will the fractions of rabbits that are young, are midlife, and are old remain the same from period to period?

The answers to these questions depend on the eigenvalues and eigenvectors of **L**. As we saw in Sections 2.5 and 3.4, *the long-term population distribution* **L**^{k}**a**, *for large k, will be a multiple of the dominant eigenvector of* **L** *(the eigenvector associated with the largest eigenvalue), and the long-term growth rate will be the dominant (largest) eigenvalue.* Whether the model converges quickly to the dominant eigenvalue depends on how much the largest eigenvalue dominates the second largest eigenvalue.

Table 4.3

Period	Young	Midlife	Old	Total
0	100	50	30	180
1	230	40	30	300
2	190	92	24	306
3	392	76	55	523
4	359	157	47	563
5	673	144	94	916
6	669	269	86	1024
7	1162	266	161	1589
8	1232	465	160	1857
9	2021	493	279	2793
10	2250	808	295	3353
11	3529	900	485	4914
⋮				
14	7382	2474	980	11836
⋮				
19	33873	9553	4602	48028
20	42815	13549	5732	62096

To start, let us examine the behavior of our model with a computer simulation. Starting with an initial population of 100 young, 50 midlife, and 30 old, we compute a table of populations in successive periods using (1); we have rounded values to whole numbers in Table 4.3. Initially, we see a very pronounced cycling behavior between young and midlife rabbits, and this in turn leads to an uneven growth in the total population—there is little growth between periods 1 and 2, between 3 and 4, or between 5 and 6. The cycling is much smaller after 20 periods but still present. If we run the model a little longer, we see that the population stabilizes with a distribution and growth multiplier

Long-term distribution: 70% young, 21% midlifes, and 9% old

Growth multiplier: 1.334   (33.4% growth rate)   (5)

That is, the population vector in the next period is about 1.334 times this period's population vector.   ■

Using more advanced techniques introduced in the Appendix to Section 5.5 (or using the appropriate mathematical software), we find that the eigenvalues of $\mathbf{L}$ in decreasing absolute size are

$$\lambda_1 = 1.334, \qquad \lambda_2 = -1.118, \qquad \lambda_3 = -.152$$

and the dominant eigenvector (associated with λ_1) is

$$\mathbf{u}_1 = [.697, .209, .094]$$

The fact that λ_2 is close to λ_1 in absolute size is why the simulation took a long time to stabilize at $\mathbf{u}_1$.

In mathematics, the best way to understand a property of interest is often to study cases where the property fails to be true. We shall now take this approach and look at a Leslie model whose group percentages do not converge to the dominant eigenvector.

Example 2. A Cyclic Leslie Model

Consider the following Leslie model:

$$\begin{aligned} y' &= & 4o \\ m' &= .5y \\ o' &= & .5m \end{aligned}$$ (6)

with the Leslie matrix

$$\mathbf{L} = \begin{bmatrix} 0 & 0 & 4 \\ .5 & 0 & 0 \\ 0 & .5 & 0 \end{bmatrix}$$ (7)

Let the initial population be 100 young and no midlife or old rabbits. Then the next period's population is easily seen to have just 50 midlifes, the next following period just 25 old rabbits, and the third following period just 100 young—returning to the initial population. This cycle will repeat over and over again, and there will never be a stable distribution of age groups—we have pure cycling. No matter what the starting population is, it will repeat every three periods. Such cycling is very unusual and results from properties of the eigenvalues of this problem.

Let us compute the eigenvalues of **L** in (7). Recall from Section 3.1 that the eigenvalues λ are the zeros of the determinant of the matrix $(\mathbf{L} - \lambda \mathbf{I})$. In this example, $\det(\mathbf{L} - \lambda \mathbf{I})$ has a simple form that is easy to work with.

$$\mathbf{L} - \lambda \mathbf{I} = \begin{bmatrix} -\lambda & 0 & 4 \\ .5 & -\lambda & 0 \\ 0 & .5 & -\lambda \end{bmatrix} \tag{8}$$

so

$$\begin{aligned} \det(\mathbf{L} - \lambda \mathbf{I}) &= (-\lambda)(-\lambda)(-\lambda) + (4)(.5)(.5) \\ &= -\lambda^3 + 1 \end{aligned} \tag{9}$$

Although the determinant of a 3-by-3 matrix involves taking six diagonal products (see Section 3.1), the three 0's in (8) eliminate all but two of these products.

Setting the determinant $-\lambda^3 + 1$ equal to 0, we get

$$\lambda^3 = 1 \tag{10}$$

One obvious root of (10) is $\lambda = 1$. But all cubic equations must have three roots. The other two roots, called roots of unity, involve complex numbers. They are $-\frac{1}{2} \pm (\sqrt{\frac{3}{2}})i$, where $i = \sqrt{-1}$. These complex numbers have absolute value 1 also. So there are three dominant eigenvalues of size 1, instead of a single one as is usually the case. This is why there is not a single dominant long-term effect as occurred in Example 1. ∎

Perpetual cycling occurs if there are several largest eigenvalues (in absolute size). If the largest eigenvalue is complex, we must get cyclic behavior—since complex zeros of a polynomial always come in conjugate pairs ($c + id$ and $c - id$) of the same absolute value. The cyclic Markov chain in Example 2 of Section 4.4 had two largest eigenvalues. Recall that its transition matrix was

$$\mathbf{A} = \begin{bmatrix} 0 & 1 \\ 1 & 0 \end{bmatrix} \tag{11}$$

for which $\det(\mathbf{A} - \lambda\mathbf{I}) = \lambda^2 - 1$. So its eigenvalues were 1 and -1.

The following theorem, whose proof is beyond the scope of this book, gives a condition for assuring noncyclic behavior in a Leslie model.

Theorem. A Leslie growth model with matrix $\mathbf{L}$ given in (4) will have a unique largest eigenvalue that is a real number, and hence a stable long-term population distribution, if some consecutive pair b_i, b_{i+1} of entries in the first row of $\mathbf{L}$ are both positive, that is, if two consecutive age groups give birth to offspring.

The condition in this theorem is satisfied in Example 1 but not in Example 2.

Example 3. **Harvesting a Renewable Resource**

This time we shall grow rabbits for profit, to sell some of the rabbit population every year. The goal will be to determine the proper population size and distribution among age groups so that we can "harvest" a given number of rabbits each year without depleting the population. That is, we want a minimal-size collection of rabbits that will sustain a given harvest forever. This time we shall use a model that differentiates between females and males. We shall again use three different age categories, but now the age spans in the groups will vary. The groups are

$$
\begin{aligned}
bm &= \text{baby male rabbits (less than 1 year old)} \\
bf &= \text{baby female rabbits} \\
ym &= \text{yearling male rabbits (between 1 and 2 years old)} \\
yf &= \text{yearling female rabbits} \\
am &= \text{adult male rabbits (2 or more years old)} \\
af &= \text{adult female rabbits}
\end{aligned}
\tag{12}
$$

The time period is 1 year. In this model adults do not all die at the end of one time period but rather survive to the next year with probability .75. We shall only harvest adults, hm male rabbits harvested, and hf female rabbits harvested. Let us try using the following equations.

$$
\begin{aligned}
bm' &= & & & 2af \\
bf' &= & & & 2af \\
ym' &= .6bm \\
yf' &= & .6bf \\
am' &= & & .6ym & + .75am & & - hm \\
af' &= & & & .6yf & + .75af & - hf
\end{aligned}
\tag{13}
$$

If $\mathbf{A}$ is the matrix of coefficients on the right-hand side of (13), excluding terms $-$hm and $-$hf, **x** and **x'** are the 6-entry vectors of current and next-period age-group sizes, and **h** is the harvesting vector (0, 0, 0, 0, hm, hf), we have

$$\mathbf{x'} = \mathbf{Ax} - \mathbf{h} \tag{14}$$

For stable harvesting, we seek values for the variables in (12) such that specified amounts can be harvested from each group in a year [in (13) only am and af are harvested] and in the next year all the groups will be the same sizes, ready to be harvested again. Mathematically, this means that in (14), $\mathbf{x'} = \mathbf{x}$. So (14) becomes

$$\mathbf{x} = \mathbf{Ax} - \mathbf{h}$$

With matrix algebra, we have

$$(\mathbf{A} - \mathbf{I})\mathbf{x} = \mathbf{h} \tag{15}$$

or

$$
\begin{array}{rclcrcl}
-\text{bm} & & & & 2\text{af} &=& 0 \\
& -\text{bf} & & & 2\text{af} &=& 0 \\
.6\text{bm} & & -\text{ym} & & &=& 0 \\
& .6\text{bf} & & -\text{yf} & &=& 0 \\
& & .6\text{ym} & & -.25\text{am} &=& \text{hm} \\
& & & .6\text{yf} & -.25\text{af} &=& \text{hf}
\end{array}
\tag{16}
$$

Let us also compare, in a very general way, the difference between solving this model and the Leslie growth model in Example 1. The Leslie model is a dynamic growing model whose solution involves an eigenvalue problem, while here we have a static model (one period is like the next period) whose solution involves solving a standard system of n equations in n unknowns. A Leslie model like (4) in Example 1 will converge to a constant distribution of ages in all but the most exceptional cases, whereas system (15) can easily have no solution of constant population with harvesting. For example, if without harvesting the population naturally decreases (i.e., $\|\mathbf{A}\| < 1$), then with harvesting it will decrease even faster and eventually become extinct.

We cannot find a solution to (15) by iterating (13) and hoping for the variables to converge after many periods to the stable harvest distribution. The reverse happens: If you do not start with the right population, the populations over successive periods will move farther away from the desired answer. Such divergence means that one has to be very careful about roundoff errors in solving (16).

Let us choose the values hm $=$ hf $=$ 100, and solve (16) by Gaussian elimination. We obtain (with answers rounded to whole numbers)

$$\text{bm} = 426, \qquad \text{bf} = 426, \qquad \text{ym} = 255,$$
$$\text{yf} = 255, \qquad \text{am} = 213, \qquad \text{af} = 213 \qquad (17)$$

Observe that while only females produce offspring, it appears that this model really groups males and females together, in that there are the same numbers of males and females in each age group. So far we have seen only that this equality occurs when hm = hf = 100. To see that it is true for hm = hf = k, for any $k > 0$, write (16) in matrix form as

$$(\mathbf{A} - \mathbf{I})\mathbf{x} = k\mathbf{h}^*, \qquad \text{where } \mathbf{h}^* = [0, 0, 0, 0, 100, 100] \qquad (18)$$

In (17), we have the solution $\mathbf{x}^* = [426, 426, 255, 255, 213, 213]$ to (18) when $k = 1$: that is, $(\mathbf{A} - \mathbf{I})\mathbf{x}^* = \mathbf{h}^*$. Then by linearity,

$$\mathbf{A}(k\mathbf{x}^*) = k(\mathbf{A}\mathbf{x}^*) = k\mathbf{h}^* \qquad (19)$$

So when hm = hf = 100k, the stable population in our harvesting model will be $k\mathbf{x}^* = [426k, 426k, 255k, 255k, 213k, 213k]$.

We see from (17) that the sex differentiation can be dropped from the original model in (13) when hm = hf, yielding the simpler model

$$b' = \qquad\qquad 2a$$
$$y' = .6b \qquad\qquad\qquad (20)$$
$$a' = \qquad .6 + .75a + h$$

where b = baby rabbits, y = yearlings, a = adults, h = harvest. It was not at all obvious in advance that these two models would be equivalent.

Suppose that hm ≠ hf. If we harvest twice as many of one sex as of the other, we obtain the results shown in Table 4.4 by re-solving (16) for these new values of hm and hf.

Table 4.4

Harvest		Stable Population					
hm	hf	x_1	x_2	x_3	x_4	x_5	x_6
50	100	426	426	255	255	413	213
100	50	212	212	127	127	**−93**	106

Surprise! Our model gives us a negative answer when we try to harvest twice as many adult males as females. Why did this happen? What is the smallest ratio of harvested males to females that can occur?

To get a fuller understanding of our harvesting model, let us compute the inverse of the coefficient matrix in (16).

$$(A - I)^{-1} = \begin{bmatrix} -1 & 1.53 & 0 & 2.55 & 0 & 4.25 \\ 0 & .53 & 0 & 2.55 & 0 & 4.25 \\ -.6 & .91 & -1 & 1.53 & 0 & 2.55 \\ 0 & .32 & 0 & .53 & 0 & 2.55 \\ -1.44 & 2.21 & -2.4 & 3.68 & -4 & 6.13 \\ 0 & .77 & 0 & 1.28 & 0 & 2.13 \end{bmatrix} \tag{21}$$

Notice how the sparsity of the original coefficient matrix is lost in the inverse. This inverse gives an explicit formula for x in terms of h that can be invaluable. For example, if only adult males and females are harvested, then $h = [0, 0, 0, 0, hm, hf]$ and multiplying $(A - I)^{-1}$ times this h yields a formula for the stable population vector x.

$$\begin{aligned} x &= (A - I)^{-1}h \\ &= [4.25hf, 4.25hf, 2.55hf, 2.55hf, -4hm + 6.13hf, 2.13hf] \end{aligned} \tag{22}$$

The form of the solution vector in (22) answers all questions about the behavior of this particular model. In particular, for the fifth component to be ≥ 0, hf should be at least $\frac{2}{3}$ of hm. ∎

We now consider a simplified model for rabbit growth that results in a single equation. However, this equation will involve the population in the current period *and* the previous period.

Example 4. **Recurrence Model for Rabbit Growth**

First we consider an extremely simplified model, in which the population doubles in each successive period. If r_n is the rabbit population in the nth period, we have

$$r_n = 2r_{n-1} \tag{23}$$

If we started with $r_0 = A$ rabbits in period 0, then in period 1 we would have $r_1 = 2A$ rabbits, and in the next period $r_2 = 4A$ rabbits. It is not hard to see that the formula for r_n is

$$r_n = 2^n A$$

For the general problem,

$$r_n = cr_{n-1} \qquad \text{has solution} \qquad r_n = c^n r_0 \tag{24}$$

Now let us consider a model with adults and young rabbits. We suppose that once a pair of rabbits are 1 year old, they have one pair of offspring every year for the rest of their lives. Assume that all pairs consist of one female and one male. We make a two-group mathematical model of the rabbit population.

Let m_n denote the number of mature pairs (at least 1 year old) during period n, y_n denote the number of young pairs (under 1 year old) during period n, and r_n denote the total number of rabbit pairs during period n. If we assume that no rabbits die (the model will not be valid for long periods), then from one period of time to the next, these quantities obey the equations

$$\begin{aligned} y_n &= m_{n-1} \\ m_n &= m_{n-1} + y_{n-1} \end{aligned} \tag{25}$$

and

$$r_n = m_n + y_n \tag{26}$$

Observe that r_n can also be expressed as r_{n-1} plus the number of new pairs born in the nth year (i.e., y_n).

$$r_n = r_{n-1} + y_n \tag{27}$$

Comparing the right-hand sides of (26) and (27), we conclude that

$$m_n = r_{n-1} \tag{28}$$

In words, we explain (28) by the fact that any rabbit, young or mature, alive one period ago will be an adult this period. And restating (28) for the previous year, we have $m_{n-1} = r_{n-2}$. When this identity is combined with $y_n = m_{n-1}$ [equation (25)], we have

$$y_n = m_{n-1} = r_{n-2} \tag{29}$$

Substituting (29) in the equation for y_n in (27), we obtain the following simple relation for r_n:

$$r_n = r_{n-1} + r_{n-2} \tag{30}$$

Equations such as (30) that tell how to compute the next number in a sequence $r_0, r_1, r_2, \ldots$ are called **recurrence relations.** Equation (30) is called a second-order relation because the right-hand side goes back 2 years. Recurrence relations are the discrete counterpart to differential equations, that is, when time is measured in discrete units rather than continuously.

Equation (30) is called the **Fibonacci relation** (named after the thirteenth-century Italian mathematician Fibonacci, who first studied this growth model). For example, if we started with one young pair of rabbits (i.e., $r_0 = 1$), after one period we would have one adult pair ($r_1 = 1$) and thereafter we could use (30) to get the following sequence of population sizes:

$$r_2 = 2, \qquad r_3 = 3, \qquad r_4 = 5, \qquad r_5 = 8,$$
$$r_6 = 13, \qquad r_7 = 21, \qquad r_8 = 34$$

and so on. The numbers in this sequence are called the **Fibonacci numbers.** Fibonacci numbers arise in many settings in nature and mathematics. Perhaps the most famous example is the spiral pattern of leaves around a blossom: On apple or oak stems, there are $5 (= r_4)$ leaves for every two $(= r_2)$ spiral turns; on pear stems, $8 (= r_5)$ leaves for every three $(= r_3)$ spiral turns; on willow stems, $13 (= r_6)$ leaves for every five $(= r_4)$ spiral turns. There are biological reasons involving the Fibonacci relation for these numbers.

The Leslie model in Example 1 was a system of recurrence relations [as was our original model (25)]. That is, (1) could have been written

$$
\begin{aligned}
y_n &= & 4m_{n-1} + o_{n-1} \\
m_n &= .4y_{n-1} \\
o_n &= & .6m_{n-1}
\end{aligned}
\tag{31}
$$

Harvesting equations (13) in Example 3 are recurrence relations (but there we are not interested in growth of the model over many periods; instead, we seek a starting distribution which will remain constant with annual harvesting). The transition equations of a Markov chain are also recurrence relations.

[Note that the matrix generalization of the solution for a first-order recurrence relation, given in (24), tells us that the solution of the system $\mathbf{a}_n = \mathbf{L}\mathbf{a}_{n-1}$ in (31) is

$$\mathbf{a}_n = \mathbf{L}^n \mathbf{a}_0$$

Unfortunately, this is not new information.]

Let us return to the second-order recurrence relation for r_n in (30). The theory of recurrence relations says that any *linear* recurrence relation for r_n of the form (where k and the c_i are constants)

$$r_n = c_1 r_{n-1} + c_2 r_{n-2} + \cdots + c_k r_{n-k} \tag{32}$$

has solutions of the form

$$r_n = b\alpha^n \tag{33}$$

where b and α are values to be determined. Note the similarity with the form of solution for linear differential equations given in Section 4.3. We can determine α by substituting (33) into equation (32). For the relation $r_n = r_{n-1} + r_{n-2}$, (33) yields

$$b\alpha^n = b\alpha^{n-1} + b\alpha^{n-2}$$

We can simplify this equation by dividing both sides by $b\alpha^{n-2}$ to obtain

$$\alpha^2 = \alpha + 1 \quad \text{or} \quad \alpha^2 - \alpha - 1 = 0 \tag{34}$$

Paralleling the terminology for differential equations, equation (34) is called the **characteristic equation** of the recurrence relation. The left-side polynomial in (34) is called its **characteristic polynomial.**

The two roots of the characteristic equation in (34) are found by the quadratic formula

$$ax^2 + bx + c = 0 \quad \text{has roots} \quad \frac{-b \pm \sqrt{b^2 - 4ac}}{2a} \tag{35}$$

In our case, the solutions are $(1 \pm \sqrt{5})/2$, or approximately 1.681 and $-.618$. It is quite surprising that the simple sequence formed by the Fibonacci numbers should turn out to be a function of an irrational number such as $\sqrt{5}$.

As with differential equations, the general solution to (24) is a linear combination of the two solutions we have found:

$$r_n = b_1\left(\frac{1 + \sqrt{5}}{2}\right)^n + b_2\left(\frac{1 - \sqrt{5}}{2}\right)^n \tag{36}$$

For simplicity, let $\alpha_1 = (1 + \sqrt{5})/2 \ (\approx 1.618)$ and $\alpha_2 = (1 - \sqrt{5})/2 \ (\approx -.618)$. So (36) becomes

$$r_n = b_1\alpha_1^n + b_2\alpha_2^n \tag{37}$$

As with differential equations, we solve for b_1 and b_2 by inserting the starting conditions in (37). Suppose that $r_0 = r_1 = 1$.

$$\begin{aligned} 1 = r_0 = b_1\alpha_1^0 + b_2\alpha_2^0 = b_1 + b_2' \\ 1 = r_1 = b_1\alpha_1 + b_2\alpha_2 = \alpha_1b_1 + \alpha_2b_2 \end{aligned} \tag{38}$$

Solving these two equations in two unknowns, we obtain

$$\begin{aligned} b_1 = \frac{\alpha_2 - 1}{\alpha_2 - \alpha_1} = \frac{1 + \sqrt{5}}{2\sqrt{5}} = \frac{\alpha_1}{\sqrt{5}} \\ b_2 = \frac{1 - \alpha_1}{\alpha_2 - \alpha_1} = -\frac{1 - \sqrt{5}}{2\sqrt{5}} = \frac{-\alpha_2}{\sqrt{5}} \end{aligned} \tag{39}$$

Remember that $\alpha_2 \approx -.618$, so $|\alpha_2^n|$ will always be $< \frac{1}{2}$. Thus the formula for the Fibonacci numbers is

$$r_n = \text{closest integer to } \frac{1}{\sqrt{5}}\left[\frac{1 + \sqrt{5}}{2}\right]^{n+1} \tag{40}$$

Unfortunately, this formula is much harder to use than the original recurrence relation (30). That is, iteratively computing successive Fibonacci numbers by summing the previous two numbers is much easier than using (40). The one important fact we do get from (40) is that r_n *grows at an exponential rate* (faster than any polynomial in n). ∎

There is one important similarity to note between Example 1 and Example 4. The largest root of the characteristic polynomial equation (35) is the growth rate of this model, just as the largest eigenvalue of the Leslie model is that model's growth rate. To illustrate this link further, let us treat our original pair of recurrence relations for young and adults as a Leslie model:

$$y_n = m_{n-1} \tag{41}$$
$$m_n = y_{n-1} + m_{n-1}$$

The matrix of coefficients in (41) is

$$\mathbf{L} = \begin{bmatrix} 0 & 1 \\ 1 & 1 \end{bmatrix} \tag{42}$$

The eigenvalues of this Leslie matrix $\mathbf{L}$ are the roots of $\det(\mathbf{L} - \lambda\mathbf{I})$.

$$\det(\mathbf{L} - \lambda\mathbf{I}) = \begin{vmatrix} -\lambda & 1 \\ 1 & 1-\lambda \end{vmatrix} = \lambda^2 - \lambda - 1 \tag{43}$$

But (43) is just the characteristic polynomial (35) for our second-order recurrence relation. So the roots of (43) are again $(1 \pm \sqrt{5})/2$.

Iterating the growth model $\mathbf{x}' = \mathbf{Lx}$ is equivalent to iterating the recurrence relation. Using the eigenvector coordinate approach from Section 2.5, let us write $\mathbf{x}$ as a linear combination of the eigenvectors $\mathbf{u}_1, \mathbf{u}_2$ of $\mathbf{L}$:

$$\mathbf{x} = a_1\mathbf{u}_1 + a_2\mathbf{u}_2$$

Then

$$\mathbf{x}^{(n)} = \mathbf{A}^n\mathbf{x} = a_1\mathbf{A}^n\mathbf{u}_1 + a_2\mathbf{A}^n\mathbf{u}_2$$

$$\text{or} \quad \begin{bmatrix} x_1^{(n)} \\ x_2^{(n)} \end{bmatrix} = a_1\alpha_1^n \begin{bmatrix} u_{11} \\ u_{12} \end{bmatrix} + a_2\alpha_2^n \begin{bmatrix} u_{21} \\ u_{22} \end{bmatrix} \tag{44}$$

$$= a_1\alpha_1^n\mathbf{u}_1 + a_2\alpha_2^n\mathbf{u}_2$$

The sum $x_1^{(n)} + x_2^{(n)}$ of the components of $\mathbf{x}^{(n)}$ is r_n, the total number of rabbits. From (44), this sum is

$$r_n = x_1^{(n)} + x_2^{(n)} = a_1\alpha_1^n(u_{11} + u_{12}) + a_2\alpha_2^n(u_{21} + u_{22}) \tag{45}$$

If we define constants b_1, b_2 as follows:

$$b_1 = a_1(u_{11} + u_{12}) \quad \text{and} \quad b_2 = a_2(u_{21} + u_{22})$$

then (45) becomes

$$r_n = b_1\alpha_1^n + b_2\alpha_2^n \tag{46}$$

Formula (46) is the same answer we got in (37). Then the b_1 and b_2 in (46) and in (37) must be the same constants. Thus the equations (38) for determining initial-condition constants b_1, b_2 are closely related to the equations for determining the weights a_1, a_2 in writing the initial vector $\mathbf{x}$ as a linear combination of the eigenvectors, $\mathbf{x} = a_1\mathbf{u}_1 + a_2\mathbf{u}_2$.

We close by noting that any recurrence relation can be recast as a system of first-order equations, the way the Fibonacci relation was in (41). For example,

$$r_n = a_1 r_{n-1} + a_2 r_{n-2} + a_3 r_{n-3} + a_4 r_{n-4} \tag{47}$$

becomes

$$
\begin{aligned}
r_n &= a_1 r_{n-1} + a_2 s_{n-1} + a_3 t_{n-1} + a_4 u_{n-1} \\
s_n &= \quad r_{n-1} \\
t_n &= \qquad\qquad s_{n-1} \\
u_n &= \qquad\qquad\qquad t_{n-1}
\end{aligned}
\qquad \text{or} \quad \mathbf{r}_n = \mathbf{Ar}_{n-1}
\tag{48}
$$

One can check that for the $\mathbf{A}$ in (48), the characteristic polynomial $p(\mathbf{A}, \lambda)$ is $\lambda^4 - a_1\lambda^3 - a_2\lambda^2 - a_3\lambda - a_4$.

In summary, the study of linear recurrence relations, when viewed as (48), is a special case of the study of linear growth models.

Section 4.5 Exercises

Summary of Exercises

Exercises 1 and 2 concern Leslie population growth models. Exercise 3 looks at eigenvalues of cyclic Markov chains. Exercises 4–10 deal with the harvesting model and variations. Exercises 11–21 involve recurrence relations, with Exercises 11–13 about building recurrence relations.

1. Find the long-term annual growth rate for the following Leslie growth models (use iteration; this growth rate is the size of the largest eigenvalue). Also find the long-term population distribution (percentages in each age group).

   (a) $\begin{aligned} y' &= \quad m + 2o \\ m' &= y \\ o' &= \quad m \end{aligned}$

   (b) $\begin{aligned} y' &= \qquad m + 2o \\ m' &= .5y \\ o' &= \qquad .5m \end{aligned}$

(c)
$$y' = \quad m + 4o$$
$$m' = .5y$$
$$o' = \quad .5m$$

(d)
$$y' = \quad\quad 4a + 4o$$
$$m' = .5y$$
$$a' = \quad\quad .5m$$
$$o' = \quad\quad\quad .5a$$

(e)
$$y' = \quad\quad 4o$$
$$m' = .5y$$
$$a' = \quad .5m$$
$$o' = \quad .5a$$

2. Consider the three-age-group Leslie matrix

$$\mathbf{L} = \begin{bmatrix} 0 & 0 & b \\ d_1 & 0 & 0 \\ 0 & d_2 & 0 \end{bmatrix}$$

The characteristic equation, $\det(\mathbf{L} - \lambda\mathbf{I}) = 0$, for this matrix is $\lambda^3 - bd_1d_2 = 0$, or

$$\lambda^3 = bd_1d_2$$

Briefly describe the behavior of the system $\mathbf{p}' = \mathbf{Lp}$ over time for the cases

(a) $bd_1d_2 < 1$ **(b)** $bd_1d_2 = 1$ **(c)** $bd_1d_2 > 1$

3. What are the eigenvalues for the Markov chain in Exercise 1 of Section 4.4? Explain that Markov chain's behavior in terms of the eigenvalues.

4. For the harvesting model in Example 3:
   (a) If hf $= 50$, how large can hm be (without negative herd values, as happened with hf $= 50$, hm $= 100$)?
   (b) Suppose that we harvest 100 yearling males and 100 yearling females. What is the stable herd vector now?
   (c) Suppose that we harvest 100 yearling males and 50 yearling females. What is the stable herd vector? Does it make sense?

5. Solve the rabbit harvesting system of equations (16) yourself by Gaussian elimination with $h_m = h_f = 100$.

6. In the harvesting model in Example 3, explain in words why the number of females and males is the same when we harvest the same number of adult males and adult females? Will this also be true if in addition we harvest equal numbers of yearling males and yearling females?

7. In the harvesting model in Example 3, explain in words why when hm $= 100$ and hf $= 50$ we get an impossible solution (involving a negative number of adult males).

8. Re-solve (16), using Gaussian elimination, with the survival probabilities of .6 changed to .5; again let hm = hf = 100. Explain in words the difference between your answer and the answer in (17) obtained with survival probabilities of .6.

9. (a) Suppose that we want the rabbit population to grow by 10% during the year. Rewrite the matrix equation (15) to reflect that fact that $x' = 1.1x$. Solve the new version of (16) with hm = hf = 100.
   (b) Find the inverse of the new coefficient matrix $(A - 1.1I)$ in (16).

   Note: Requires a computer program for inverses.

10. Try to solve our harvesting model by iteration with hm = hf = 100. That is, guess an initial value for the population vector x and insert that vector of values on the right side of (13). Use the resulting left-side values as your next estimated x and continue iterating. Try several different starting vectors. Do you ever get convergence to a solution?

11. Suppose that a_n, the level of radioactivity after n years from the element linearium, decreases by 20% a year. Write a recurrence relation that expresses this decay rate.

12. Let a_n be the number of dollars in a savings account after n years. Suppose that money earns 10% interest a year.
    (a) Write a recurrence relation to represent this interest rate.
    (b) If $a_0 = 100$, calculate a_5.

13. Let a_n = the number of different ways for an elf to climb a sequence of n stairs with steps of size 1 or 2. Explain why $a_n = a_{n-1} + a_{n-2}$. What are a_1 and a_2? Determine a_8.

14. Solve the following recurrence relations, given the initial values.
    (a) $a_n = 3a_{n-1} - 2a_{n-2}$, $a_0 = a_1 = 2$
    (b) $a_n = 6a_{n-1} - 8a_{n-2}$, $a_0 = 0, a_1 = 1$
    (c) $a_n = 3a_{n-1} + 4a_{n-2}$, $a_0 = a_1 = 1$
    (d) $a_n = 2a_{n-1} - a_{n-2}$, $a_0 = a_1 = 1$

    Hint: See Exercise 21.

15. Give an approximate formula for a_{20} for each recurrence relation in Exercise 14 (use just the largest root of the characteristic equation).

16. Convert each recurrence relation in Exercise 14 into a pair of first-order recurrence relations $x_n = Ax_{n-1}$ for $x_n = [a_n, b_n]$.
    Hint: Let $b_n = a_{n-1}$; see (48).
    Recast the initial values from Exercise 14 into a initial-value vector x_1. Check that the eigenvalues of each A equal the roots of the characteristic equation for the corresponding original recurrence relation in Exercise 14.

17. Solve the recurrence relations you got in Exercise 16 using the method at the end of Section 3.1 to get a formula for a_n.

18. Convert the following recurrence relations into a system of first-order recurrence relations.
    (a) $a_n = 2a_{n-1} + a_{n-2} - a_{n-3}$ (b) $a_n = a_{n-2} + a_{n-4}$

19. Show that $a_n = \alpha^n$, where $\alpha^2 = c_1\alpha + c_2$, will always be a solution to the recurrence relation $a_n = c_1 a_{n-1} + c_2 a_{n-2}$.

20. Show that any linear combination of solutions to the recurrence relation $a_n = c_1 a_{n-1} + c_2 a_{n-2}$ is again a solution.

21. Verify that $a_n = n\lambda^n$ is a solution to the recurrence relation $a_n = c_1 a_{n-1} + c_2 a_{n-2}$ whose characteristic equation $k^2 - c_1 k - c_2 = 0$ has λ as a double root.

    Note: If $k^2 - c_1 k - c_2 = 0$ has λ as a double root, the characteristic equation can be factored as $(k - \lambda)^2 = 0$. This means that $c_1 = 2\lambda$ and $c_2 = -\lambda^2$. Use these values for c_1 and c_2.

Section 4.6 Linear Programming

Studies have shown that about 25% of all scientific computing is devoted to solving linear progams. Linear programming is the principal tool of management science. The object of a linear program is to optimize—maximize or minimize—some linear function subject to a system of linear constraints. There are hundreds of different real-world problems that can be posed as linear programs. We start with a simple linear problem presented to maximize sales from furniture production.

Example 1. Production of Chairs and Tables

A factory can manufacture chairs and tables. Let x_1 be the number of chairs produced and x_2 the number of tables. Chairs sell for $40 a piece and tables for $200 a piece. The production of x_1 chairs and x_2 tables requires various amounts of raw materials whose supplies are limited. The following inequalities describe the requirements and supplies.

$$
\begin{aligned}
\text{Wood:} \quad & x_1 + 4x_2 \le 1400 \\
\text{Labor:} \quad & 2x_1 + 3x_2 \le 2000 \\
\text{Braces:} \quad & x_1 + 12x_2 \le 3600 \\
\text{Upholstery:} \quad & 2x_1 \quad\quad\ \le 1800
\end{aligned}
\tag{1}
$$

Subject to these constraints we want to pick x_1 and x_2 so as to maximize the objective function of the total sales. Our model is

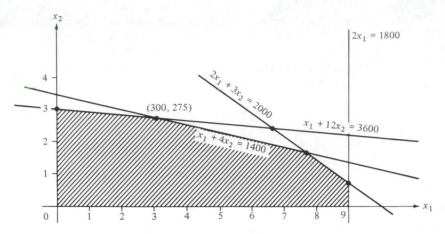

Figure 4.12 Feasible region of linear program.

$$\text{Maximize } 40x_1 + 200x_2 \tag{2}$$
$$\text{subject to } x_1 \geq 0, \, x_2 \geq 0 \text{ and } (1)$$

Linear programming models of energy usage in the American economy have used thousands of variables and constraint equations and inequalities. Linear models with close to 1,000,000 variables and over 100,000 constraints have been developed and solved in private industry. Such a system of equations has 10,000,000,000 (10 billion) coefficient terms. Of course, in these large problems almost all coefficients are zero. A large mathematical theory about linear programs has been developed with simplifying shortcuts that make it possible to solve huge linear programs.

In Figure 4.12, we have marked (the shaded area) the **feasible region** of $x_1 - x_2$ points that satisfy (1).

The key to solving a linear program is the following theorem.

Theorem. A linear objective function assumes its maximum and minimum values on the boundary of the feasible region (assuming that the feasible region is bounded). In fact, the optimal value is achieved at a corner point of this boundary.

Proof. Although true for all linear programs, we verify this proposition in the case of two variables (as in Example 1). Consider any line crossing through the feasible region. If we compute the values of a linear function along this line, we observe that as we move in one direction on the line, the linear function constantly increases and in the other direction constantly decreases. (There is one exception—the linear function could be constant along the line.) To find the maximum value of this linear function along the line, we should go as far as possible along the line in the direction of increasing values, that is, go to an end of the line segment where it meets the boundary. Thus the maximum of a linear objective function occurs at the boundary of the feasible region.

Next repeat this argument using a boundary line of the feasible region. Again it pays to go to an end of the boundary line, that is, to a corner of the feasible region. ∎

The theorem tells us to look at the corners of the feasible region for the optimal (x_1, x_2)-value. In theory, any pair of constraint lines could intersect to form a corner of the feasible region, but by plotting the constraint lines, as in Figure 4.12, we can see which pairs of lines intersect along the boundary of the feasible region. To find the intersection point of two constraint lines, we solve for an (x_1, x_2) point that lies on both lines—the same old problem of two equations in two unknowns. Table 4.5 lists the coordinates of the corners and the associated objective function values.

Table 4.5

Corner Coordinates	Intersecting Constraints	Objective Function
(0, 0)	$x_1 \geq 0$ and $x_2 \geq 0$	0
(0, 300)	$x_1 \geq 0$ and braces	60,000
(300, 275)	**Braces and wood**	**67,000*****
(760, 160)	Wood and labor	62,400
(900, 66.6)	Labor and upholstery	39,333
(900, 0)	Upholstery and $x_2 \geq 0$	36,000

So the optimal production schedule is to make 300 chairs and 275 tables, whose sales value will be $67,000. ∎

Before giving a general procedure for solving linear programs, we present some examples that show how to build linear programming models. The reader may also want to refer back to the crop planting linear program presented in Section 1.4.

Example 2. Dietician's Problem

Suppose that a meal must contain at least 500 units of vitamin A, 1000 units of vitamin C, 100 units of iron, and 50 grams of protein. A dietician has two foods for the meal, meat and fruit. Meat costs 50 cents a unit and fruit costs 40 cents a unit.

Each unit of meat has 20 units of vitamin A, 30 units of vitamin C, 10 units of iron, and 15 grams of protein. Each unit of fruit has 50 units of vitamin A, 100 units of vitamin C, 1 unit of iron, and 2 units of protein.

The dietician wants to have the cheapest meal that satisfies the four nutritional constraints. Let us formulate the dietician's problem as a linear program. Make x_1 be the number of units of meat used and x_2 the number of units of fruit used. Then the objective is to minimize the cost of the meal

Minimize $.5x_1 + .4x_2$

subject to the nutritional lower-bound constraints

Vitamin A:	$20x_1 +$	$50x_2 \geq$	500
Vitamin C:	$30x_1 +$	$100x_2 \geq$	1000
Iron:	$10x_1 +$	$1x_1 \geq$	100
Protein:	$15x_1 +$	$2x_2 \geq$	50

$$x_1 \geq 0, \quad x_2 \geq 0 \qquad \blacksquare$$

We should note that the constraints in linear programs do not have to be inequalities. Later in this section we shall see how to convert inequalities to equations and equations to inequalities. The next example is an important type of linear programming problem in which the constraints are equations.

Example 3. A Transportation Problem

Warehouses 1, 2, and 3 have 20, 30, and 15 tons, respectively, of chicken wings. Colleges 1 and 2 need 25 and 40 tons, respectively, of chicken wings (to serve to students). The following table indicates the cost of shipping a ton from a given warehouse to a given college.

		College	
		A	B
Warehouses	1	80	45
	2	60	55
	3	40	65

Since the overall demand at both colleges is $25 + 40 = 65$ and the overall supply of all three warehouses is also $20 + 30 + 15 = 65$, all the supplies of each warehouse must be used. The constraints are that the total amount of chicken wings shipped from warehouse 1 must equal 20 tons; from warehouse 2, 30 tons; from warehouse 3, 15 tons; and the total amount shipped to college A must equal 25 tons and to college B 40 tons. If x_{ij} is the number of tons shipped from warehouse i to college j, then these constraints are

Warehouse   equations

$$
\begin{aligned}
x_{11} + x_{12} &= 20 \\
x_{21} + x_{22} &= 30 \\
x_{31} + x_{32} &= 15
\end{aligned}
\tag{3}
$$

College equations:
$$x_{11} + x_{21} + x_{31} = 25$$
$$x_{12} + x_{22} + x_{32} = 40$$

The objective is to minimize the transportation costs. In terms of the x_{ij}, the transportation costs are

$$80x_{11} + 45x_{12} + 60x_{21} + 55x_{22} + 40x_{31} + 65x_{32} \qquad (4)$$

Thus our linear program is to minimize (4) subject to (3) and $x_{ij} \geq 0$. ∎

Example 4. **Running a Chicken Farm**

We have a farm with 5000 chickens. Each year for 3 years we must decide how many of the chickens should lay eggs to be sold and how many chickens should be hatching eggs to produce more chickens next year (we assume that all chickens are hens; roosters are ignored). At the end of 3 years, all the chickens are sold for slaughter at $2 per bird (this includes chickens hatched during the third year). A chicken can hatch 30 eggs in a year. The eggs from one chicken in 1 year earn $7. It costs $2 a year in feed for chickens (no charge for chickens born during the year). The objective is to maximize income from eggs and from the final sale of the chickens. State this maximization problem as a linear program.

Let x_i and y_i be the number of chickens hatching and laying eggs for sale, respectively, in the ith year, $i = 1, 2, 3$. Then the following equations represent the fact that $x_i + y_i$ equals the total number of chickens each year (the original number 5000 plus the numbers of new chickens born thus far).

$$x_1 + y_1 = 5000$$
$$x_2 + y_2 = 5000 + 30x_1 \qquad (5)$$
$$x_3 + y_3 = 5000 + 30x_1 + 30x_2$$

Let us rewrite (5) with all the variables on the left side and each variable in a different column.

$$x_1 + y_1 = 5000$$
$$-30x_1 + x_2 + y_2 = 5000 \qquad (6)$$
$$-30x_1 - 30x_2 + x_3 + y_3 = 5000$$

As usual, all variables must be nonnegative.

$$x_1 \geq 0, \quad x_2 \geq 0, \quad x_3 \geq 0, \quad y_1 \geq 0,$$
$$y_2 \geq 0, \quad y_3 \geq 0 \qquad (7)$$

The objective function to be maximized is

$$\text{Maximize} \quad 2(5000 + 30x_1 + 30x_2 + 30x_3)$$
$$+ 7(y_1 + y_2 + y_3) \qquad (8)$$
$$- 2(15{,}000 + 60x_1 + 30x_2)$$

The first factor in (8) represents the sale of all chickens after 3 years, the second factor the income from eggs, and the last factor the feeding cost [the total number of chicken-years of feeding is the sum of the right-hand sides in (7)].

The required linear program is to maximize (8) subject to (6) and (7). ∎

We shall now develop a general procedure for solving linear programs. The discussion will be couched in terms of the chair–table linear program in Example 1. Our presentation will necessarily be sketchy. Readers interested in a more extensive treatment of linear programming should turn to any of the dozens of books on the subject (most colleges have several courses about linear programming, offered by mathematics, economics, and business departments).

Note that the method used in Example 1 of graphing and then checking the corner points of the feasible region for an optimal value is not feasible for larger problems. An n-variable problem with m constraints can have up to m^n corner points to check.

The theory of linear programming is centered about the following method, now called the **simplex algorithm**, which was developed over 30 years ago when the advent of digital computers first made it possible to try to solve moderate-sized linear programs. Intuitively, the simplex algorithm starts at the origin (a corner of the feasible region) and moves along a boundary edge of the feasible region to a better corner, where the objective function is larger, and continues in this way until it reaches an optimal corner.

We outline the simplex algorithm and then describe it in detail using Example 1 to illustrate the calculations.

Simplex Algorithm

Part 1. Let x_h be the variable with the largest positive coefficient in the objective function. Starting at the origin, increase x_h as much as possible until an inequality constraint is reached (while the other variables remain equal to 0).

Part 2. Make a partial change of coordinates by replacing x_h with x_h' so that the new corner, which we reached by increasing x_h, becomes the origin in the new coordinates. If any coefficient in the new objective function is positive, go back to part 1; otherwise, the new origin is an optimal corner.

To find a better corner, the simplex algorithm looks at the objective function and checks which coefficient c_j is largest (most positive). Let x_h be the variable with the largest coefficient in the objective function. Then increasing x_h yields the greatest rate of increase in the objective function; in economic terms, activity h is the most profitable.

The simplex algorithm increases the value of x_h as much as possible, that is, until increasing x_h any more would violate one of the inequality constraints. If all $c_j \leq 0$, so that increasing any x_j cannot increase the objective function, the current origin must be an optimal corner.

In the chair–table problem, the simplex algorithm would start at the origin $(0, 0)$. The coefficient of x_2 is greater than the coefficient of x_1 in the objective function ($c_2 = 200$ versus $c_1 = 40$), so x_2 would be increased (while x_1 is kept fixed $= 0$). In words, we produce as many tables (variable x_2) as possible because they are more profitable than chairs.

Recall the objective function and inequalities in this problem,

$$\text{Maximize } 40x_1 + 200x_2$$
$$x_1 \geq 0, \quad x_2 \geq 0$$

Wood:	$x_1 + 4x_2 \leq 1400$	
Labor:	$2x_1 + 3x_2 \leq 2000$	(9)
Braces:	$x_1 + 12x_2 \leq 3600$	
Upholstery:	$2x_1 \quad\quad\quad \leq 1800$	

Looking at Figure 4.12, we see that x_2 can be increased to $x_2 = 300$, where the objective function is 60,000. Any greater value of x_2 would violate the braces constraint.

Let us show how the new corner can be found algebraically (since in larger problems, we will not be able to draw a picture of the feasible region). We want to increase x_2 while keeping $x_1 = 0$. Substituting $x_1 = 0$ into the inequalities of (9), we obtain

Wood:	$4x_2 \leq 1400$	
Labor:	$3x_2 \leq 2000$	(10)
Braces:	$12x_2 \leq 3600$	
Upholstery:	$0 \leq 1800$	

We can ignore the upholstery inequality $0 \leq 1800$—it does not contain x_2 and will always be true. The other three inequalities in (10) are easily solved in terms of x_2 to become

Wood:	$x_2 \leq 350$	
Labor:	$x_2 \leq 666\frac{2}{3}$	(11)
Braces:	$x_2 \leq 300$	

The smallest of the bounds on x_2, namely 300, is the amount x_2 can be increased without violating any inequality. So again we find that $(0, 300)$ is the new corner point we reach by increasing x_2 as much as possible. This corner (0, 300) is formed by the intersection of constraints $x_1 = 0$ and $x_1 + 12x_2 = 3600$.

Next comes part 2 of the simplex algorithm. We perform a linear transformation that changes our coordinate system by replacing x_2 with a different coordinate variable, call it x_2'. The current corner $x_1 = 0$, $x_2 = 300$ will be transformed into the origin $x_1 = 0$, $x_2' = 0$ in this new coordinate system.

To motivate this change of variables, we first need to show how our system of constraint inequalities can be recast into a system of equations. For this recasting, we must introduce additional variables. A **slack variable** equals the difference between the left- and right-hand sides of an inequality. With slack variables, the linear program in (9) becomes

Maximize $40x_1 + 200x_2$

subject to $x_1 \geq 0$, $x_2 \geq 0$, $x_3 \geq 0$, $x_4 \geq 0$, $x_5 \geq 0$, $x_6 \geq 0$ and

$$
\begin{array}{llr}
\text{Wood:} & x_1 + 4x_2 + x_3 & = 1400 \\
\text{Labor:} & 2x_1 + 3x_2 + x_4 & = 2000 \\
\text{Braces:} & x_1 + 12x_2 + x_5 & = 3600 \\
\text{Upholstery:} & 2x_1 + x_6 & = 1800
\end{array}
\tag{12}
$$

where x_3, x_4, x_5, x_6 are the *slack variables*. We call x_1, x_2 **independent variables**—they are the original coordinate variables from (9). Like x_1, x_2, a slack variable must be ≥ 0.

A slack variable is what the simplex algorithm uses as x_2', the variable to replace x_2 in the change of coordinates mentioned above. In particular, we want to use x_5, the slack variable in the braces constraint $x_1 + 12x_2 + x_5 = 3600$. Remember that the corner (0, 300) is the intersection point of $x_1 + 12x_2 = 3600$ with axis line $x_1 = 0$. Forcing the slack variable x_5 to be 0 is the same as forcing x_1, x_2 to satisfy the equation $x_1 + 12x_2 = 3600$. The corner $x_1 = 0$, $x_2 = 300$ where lines $x_1 = 0$ and $x_1 + 12x_2 = 3600$ meet is in x_1, x_5-coordinates the origin, $x_1 = 0$, $x_5 = 0$. This is exactly what we were looking for (see Figure 4.13).

We must rewrite the system of equations in (12) so that x_5 replaces x_2 as an independent variable, that is, as a coordinate variable. We do this by using the elimination-by-pivoting process to eliminate x_2 from other equations in (12). We subtract the appropriate multiple of the braces equation in (12), $x_1 + 12x_2 + x_5 = 3600$, from the other equations to eliminate x_2.

The wood equation $x_1 + 4x_2 + x_3 = 1400$ has an x_2-coefficient of 4 while the braces equation has an x_2-coefficient of 12. So we subtract $\frac{4}{12}$, or $\frac{1}{3}$, times the braces equation from the wood equation

$$
\begin{array}{llr}
\text{Wood:} & x_1 + 4x_2 + x_3 & = 1400 \\
-\frac{1}{3}(\text{Braces:} & x_1 + 12x_2 + x_5 & = 3600) \\
\hline
\text{New wood:} & \frac{2}{3}x_1 + x_3 - \frac{1}{3}x_5 & = 200
\end{array}
\tag{13a}
$$

The labor constraint becomes

Figure 4.13 Feasible region with x_1, x_5 as independent variables.

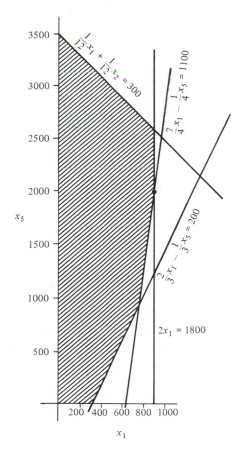

$$
\begin{array}{lllll}
\text{Labor:} & 2x_1 + 3x_2 + x_4 & = 2000 & \\
-\tfrac{1}{4}(\text{Braces:} & x_1 + 12x_2 & + x_5 = 3600) & \text{(13b)} \\
\hline
\text{New labor:} & \tfrac{7}{4}x_1 & + x_4 - \tfrac{1}{4}x_5 = 1100 &
\end{array}
$$

The upholstery equation is unchanged, since x_2 does not occur in it.

$$
\text{New upholstery:} \quad 2x_1 \qquad + x_6 = 1800 \qquad \text{(13c)}
$$

We also need to eliminate x_2 from the objective function. To do this, we write the braces equation as $x_1 + 12x_2 + x_5 - 3600 = 0$.

$$
\begin{array}{lll}
\text{Objective functions:} & 40x_1 + 200x_2 & \\
-\tfrac{50}{3}(\text{Braces:} & x_1 + 12x_2 + x_5 - 3600) & \text{(14)} \\
\hline
\text{New objective function:} & \tfrac{70}{3}x_1 \qquad - \tfrac{50}{3}x_5 + 60{,}000 &
\end{array}
$$

We must also rewrite the original braces constraint $x_1 + 12x_2 + x_5 = 3600$ to make x_2 look like a slack variable; that is, x_2 should have a coefficient of 1. Dividing by 12, we have

$$
\text{New braces:} \quad \tfrac{1}{12}x_1 + x_2 \qquad + \tfrac{1}{12}x_5 = 300 \qquad \text{(13d)}
$$

Collecting (13a)–(13d) and (14), we have the desired transformed linear program.

Maximize $\frac{70}{3} x_1 - \frac{50}{3} x_5 + 60{,}000$

subject to $x_1 \geq 0, \quad x_2 \geq 0, \quad x_3 \geq 0, \quad x_4 \geq 0, \quad x_5 \geq 0, \quad x_6 \geq 0$, and

$$
\begin{aligned}
\tfrac{2}{3}x_1 \quad\quad + x_3 \quad\quad\quad - \tfrac{1}{3}x_5 \quad\quad &= 200 \\
\tfrac{7}{4}x_1 \quad\quad\quad\quad + x_4 - \tfrac{1}{4}x_5 \quad\quad &= 1100 \\
\tfrac{1}{12}x_1 + x_2 \quad\quad\quad\quad \tfrac{1}{12}x_5 \quad\quad &= 300 \\
2x_1 \quad\quad\quad\quad\quad\quad\quad + x_6 &= 1800
\end{aligned}
\tag{15}
$$

Observe that (15) has the same general form as (12), except that x_2 and x_5 have interchanged roles as independent and slack variables. Since we have only restated the problem, a maximum for this problem is a maximum for our original problem (12). The feasible region for our linear program in the restated form is the "hashed" region in Figure 4.13; note that the coordinate axes are labeled x_1 and x_5.

This sequence of computations to interchange the independent and slack variable roles of x_2 and x_5 is called a **pivot exchange**. Recall from Section 3.2 about elimination by pivoting that we used the term *pivot on entry a_{ij}* to denote the process of using equation i to eliminate x_j from all other equations (and making the coefficient of x_j be 1 in equation i). If independent variable x_h is exchanged with the slack variable in equation g, then we are pivoting on the coefficient of x_h in equation g.

Using the concept of a pivot, we restate the simplex algorithm as follows.

Simplex Algorithm

Part 1. Let x_h be the (independent) variable with the largest positive coefficient in the current objective function. Increase x_h as much as possible until an inequality constraint is reached; call this the gth inequality.

Part 2. Perform a pivot exchange between x_h and the slack variable in the gth constraint. If any coefficient in the new objective function is positive, go back to part 1; otherwise, the new origin is an optimal corner.

We can convert the constraint equations in (15) back to inequality constraints by dropping the slack variables x_2, x_3, x_4, and x_6:

$$
\begin{aligned}
\text{Maximize } & \tfrac{70}{3} x_1 - \tfrac{50}{3} x_5 + 60{,}000 \\
& \tfrac{2}{3}x_1 - \tfrac{1}{3}x_5 \leq 200 \\
& \tfrac{7}{4}x_1 - \tfrac{1}{4}x_5 \leq 1100 \\
& \tfrac{1}{12}x_1 + \tfrac{1}{12}x_5 \leq 300 \\
& 2x_1 \quad\quad\quad \leq 1800
\end{aligned}
\tag{16}
$$

(see Figure 4.13).

Now we apply the two steps of the simplex algorithm to our problem in its new form (16). In step 1 we increase x_1, which has the only positive coefficient in the objective function, while keeping $x_5 = 0$. (As an aside, we note that back in Figure 4.12, keeping $x_5 = 0$ means that we move along the brace constraint line $x_1 + 12x_2 = 3600$). When we set $x_5 = 0$ in (16), we obtain the inequalities

$$\begin{array}{rcll}
\tfrac{2}{3}x_1 &\leq& 200 & \text{or} \quad x_1 \leq 300 \\
\tfrac{7}{4}x_1 &\leq& 1100 & \text{or} \quad x_1 \leq \tfrac{4400}{7} \\
\tfrac{1}{12}x_1 &\leq& 300 & \text{or} \quad x_1 \leq 3600 \\
2x_1 &\leq& 1800 & \text{or} \quad x_1 \leq 900
\end{array} \tag{17}$$

The smallest of these constraints is the first. So we can increase x_1 to 300, and *our new corner is at* $x_1 = 300$, $x_5 = 0$, where the first constraint $2x_1/3 - x_5/3 = 200$ and constraint $x_5 = 0$ meet (see Figure 4.13). To see where we are in the original problem, we can use the third constraint equation in (15) to compute x_2's value when $x_1 = 300$ and $x_5 = 0$—we get $x_2 = 275$; so our new corner corresponds to the point $x_1 = 300$, $x_2 = 275$ in Figure 4.12.

We return to the equation form (15) of the constraints. The first constraint is $2x_1/3 + x_3 - x_5/3 = 200$. Then x_3 is the slack variable for this constraint, so in part 2 of the simplex algorithm, we perform a pivot exchange between independent variable x_1 and slack variable x_3. [Note that the computations in (17) could be done with the constraints in equations form, as in (15): we simply divide the right-side value in each equation by the coefficient of x_1 and pick the smallest positive value.]

Using the first constraint to eliminate x_1 from the other equations in (15) and from the objective function, we obtain the new linear program.

$$\begin{array}{l}
\text{Maximize } - 35x_3 - 5x_5 + 67{,}000 \\
\text{subject to } x_i \geq 0, \quad i = 1, 2, 3, 4, 5, 6, \text{ and}
\end{array}$$

$$\begin{array}{rcll}
x_1 + \tfrac{3}{2}x_3 \quad\quad - \tfrac{1}{2}x_5 &=& 300 \\
- \tfrac{21}{8}x_3 + x_4 + \tfrac{5}{8}x_5 &=& 575 \\
x_2 - \tfrac{1}{8}x_3 \quad\quad + \tfrac{1}{8}x_5 &=& 275 \\
- 3x_3 \quad\quad + x_5 + x_6 &=& 1200
\end{array} \tag{18}$$

Since all coefficients in the current objective function are negative, the current origin is the maximum corner. The value of the objective function at the origin $x_3 = x_5 = 0$ is simply the constant term in the objective function, 67,000. By setting $x_3 = x_5 = 0$ in (18), we can also directly read off the values of $x_1 (= 300)$ and $x_2 (= 275)$ as well as the values of the other slack variables.

This finishes our example of how the simplex algorithm works. Any linear program of the general form

$$\text{Maximize } \mathbf{c} \cdot \mathbf{x} + d \qquad \text{subject to} \qquad \mathbf{Ax} \le \mathbf{b}, \quad \mathbf{x} \ge \mathbf{0} \qquad (19)$$

can be solved by the method we have just presented.

The key decision in the simplex algorithm is the choice of the pivot entry (part 1). *We pick the variable x_h with the largest coefficient in the objective function. Then we divide the right-side value in each constraint equation by the coefficient of x_h and pick the equation with the smallest quotient. The coefficient of x_h in that equation is the pivot entry.*

Just as in Gaussian elimination, we can display the computations in the simplex algorithm in matrix notation (omitting the variables). For the linear program in (19) we use the augmented matrix

$$\begin{bmatrix} \mathbf{c} & -d \\ \mathbf{A} & \mathbf{b} \end{bmatrix} \qquad (20)$$

Note that the constant d is written with a minus sign, $-d$. This is a technicality based on the fact that $\mathbf{b}$ is the vector of right sides of equations, while d is on the left side (if fictitiously d were put in the "right side" of the objective function, the constant would become $-d$).

Let us repeat the stages of the simplex algorithm for our chair–table program using matrix notation. At the outset we have (where x_1, x_2 are the initial independent variables).

	x_1	x_2	x_3	x_4	x_5	x_6	
Objective Function	40	200	0	0	0	0	0
Wood	1	4	1	0	0	0	1400
Labor	2	3	0	1	0	0	2000
Braces	1	(12)	0	0	1	0	3600
Upholstery	2	0	0	0	0	1	1800

$$(21)$$

We pick x_2 (200 is largest coefficient in first row) and then divide each positive entry in x_2's column into the corresponding entry in the last column. The smallest quotient is in the braces equation, so we pivot on entry (4, 2), which is circled in (21). After pivoting on entry (4, 2), we obtain

	x_1	x_2	x_3	x_4	x_5	x_6	
Objective Function	$\frac{70}{3}$	0	0	0	$-\frac{50}{3}$	0	$-60{,}000$
Wood	$\left(\frac{2}{3}\right)$	0	1	0	$-\frac{1}{3}$	0	200
Labor	$\frac{7}{4}$	0	0	1	$-\frac{1}{4}$	0	1,100
Braces	$\frac{1}{12}$	1	0	0	$\frac{1}{12}$	0	300
Upholstery	2	0	0	0	0	1	1,800

$$(22)$$

Next we pick x_1 and then the wood equation. So entry (2, 1) is the second pivot. After pivoting on entry (2, 1), we obtain

$$\begin{array}{c} \quad\quad x_1 \quad x_2 \quad x_3 \quad x_4 \quad x_5 \quad x_6 \\ \begin{array}{r} \text{Objective Function} \\ \text{Wood} \\ \text{Labor} \\ \text{Braces} \\ \text{Upholstery} \end{array} \left[\begin{array}{cccccc|c} 0 & 0 & -35 & 0 & -5 & 0 & -67,000 \\ \hline 1 & 0 & +\frac{3}{2} & 0 & -\frac{1}{2} & 0 & 300 \\ 0 & 0 & -\frac{21}{8} & 1 & \frac{5}{8} & 0 & 575 \\ 0 & 1 & -\frac{1}{8} & 0 & \frac{1}{8} & 0 & 275 \\ 0 & 0 & -3 & 0 & 1 & 1 & 1,200 \end{array}\right] \end{array} \quad (23)$$

Since the objective function has no positive coefficients, we now have an optimum with objective function value 67,000. The values of the current independent variables, x_3, x_5, are zero and from (23) we read off the values of the other variables to be

$$x_1 = 300, \qquad x_2 = 275, \qquad x_4 = 575, \qquad x_6 = 1200$$

The name given to these augmented matrices, (21), (22), and (23), is **simplex tableaus**.

Let us now go quickly through the solution of another linear program.

Example 5. **Simplex Algorithm Applied to a Linear Program**

Consider the following linear program for the production of sugar (x_1), syrup (x_2), and molasses (x_3):

$$\begin{aligned} &\text{Maximize } 3x_1 + 4x_2 + 2x_3 \\ &\text{subject to } x_1 \geq 0, \quad x_2 \geq 0, \quad x_3 \geq 0 \text{ and} \\ &\text{Transportation:} \quad 2x_1 + x_2 + x_3 \leq 6 \\ &\text{Labor:} \quad x_1 + 2x_2 + 3x_3 \leq 7 \qquad (24) \\ &\text{Machinery:} \quad 3x_1 + 2x_2 + 4x_3 \leq 15 \end{aligned}$$

In simplex tableau with slack variables added, we have (where x_1, x_2, x_3 are the initial independent variables)

$$\begin{array}{c} \quad\quad\quad x_1 \quad x_2 \quad x_3 \quad x_4 \quad x_5 \quad x_6 \\ \begin{array}{r} \text{Objective Function} \\ \text{Transportation} \\ \text{Labor} \\ \text{Machinery} \end{array} \left[\begin{array}{cccccc|c} 3 & 4 & 2 & 0 & 0 & 0 & 0 \\ \hline 2 & 1 & 1 & 1 & 0 & 0 & 6 \\ 1 & ② & 3 & 0 & 1 & 0 & 7 \\ 3 & 2 & 4 & 0 & 0 & 1 & 15 \end{array}\right] \end{array} \quad (25)$$

The pivot will be in x_2's column. Picking the minimum of $\frac{6}{1}$, $\frac{7}{2}$, $\frac{15}{2}$, we take the second, from labor's row. So the pivot entry is (3, 2). After pivoting there, we obtain

$$\begin{array}{c} \\ \text{Objective Function} \\ \text{Transportation} \\ \text{Labor} \\ \text{Machinery} \end{array} \begin{array}{cccccc} x_1 & x_2 & x_3 & x_4 & x_5 & x_6 \\ \left[\begin{array}{cccccc|c} 1 & 0 & -4 & 0 & -2 & 0 & -14 \\ \hline \tfrac{3}{2} & 0 & -\tfrac{1}{2} & 1 & -\tfrac{1}{2} & 0 & \tfrac{5}{2} \\ \tfrac{1}{2} & 1 & \tfrac{3}{2} & 0 & \tfrac{1}{2} & 0 & \tfrac{7}{2} \\ 2 & 0 & 1 & 0 & -1 & 1 & 8 \end{array}\right] \end{array} \qquad (26)$$

The next pivot will be in x_1's column. Picking the minimum of $(\tfrac{5}{2})/(\tfrac{3}{2})$, $(\tfrac{7}{2})/(\tfrac{1}{2})$, $\tfrac{8}{4}$, we take the first, from transportation's row. So the pivot entry is $(2, 1)$. After pivoting there, we obtain

$$\begin{array}{c} \\ \text{Objective Function} \\ \text{Transportation} \\ \text{Labor} \\ \text{Machinery} \end{array} \begin{array}{cccccc} x_1 & x_2 & x_3 & x_4 & x_5 & x_6 \\ \left[\begin{array}{cccccc|c} 0 & 0 & -\tfrac{11}{3} & -\tfrac{4}{3} & -\tfrac{5}{3} & 0 & -\tfrac{47}{3} \\ \hline 1 & 0 & -\tfrac{1}{3} & \tfrac{2}{3} & -\tfrac{1}{3} & 0 & \tfrac{5}{3} \\ 0 & 1 & \tfrac{5}{3} & -\tfrac{1}{3} & \tfrac{2}{3} & 0 & \tfrac{8}{3} \\ 0 & 0 & \tfrac{5}{3} & -\tfrac{4}{3} & -\tfrac{1}{3} & 1 & \tfrac{14}{3} \end{array}\right] \end{array}$$

$$(27)$$

Now there are no positive coefficients in the objective function, so we have a maximum, where the objective function equals $\tfrac{47}{3}$, the independent variables x_3, x_4, x_5 are zero, and the other variables are

$$x_1 = \tfrac{5}{3}, \qquad x_2 = \tfrac{8}{3}, \qquad x_6 = \tfrac{14}{3} \qquad \blacksquare$$

The simplex algorithm was invented in the earliest days of linear programming by G. Dantzig as an intuitive scheme for "walking" along the boundary of the feasible regions in search of better and better corners. Many more sophisticated methods have been proposed—in the 1970s Khachian's algorithm made the front pages of major newspapers for its theoretical advantages over the simplex algorithm, and in 1984 Karmarkar's algorithm gained publicity for being faster than the simplex algorithm for some linear programs—but the simplex algorithm is still the best general-purpose way to solve a linear program. The explanation of why it works so well requires advanced mathematical analysis.

There are many important variations in the basic theory of the simplex algorithm. We mention a few here and give some more in the Exercises and in an appendix to this section.

First, if the problem involves minimization, we can convert it to a maximization problem by multiplying the objective function by -1 (maximizing $-\mathbf{c} \cdot \mathbf{x}$ is the same as minimizing $\mathbf{c} \cdot \mathbf{x}$).

Second, if an inequality constraint is of the form $\mathbf{a} \cdot \mathbf{x} \geq b$, convert it to a constraint with a $\leq$ sign by multiplying both sides by -1 (to get $-\mathbf{a} \cdot \mathbf{x} \leq -b$).

Third, we consider linear programs whose constraints are equations. We converted the inequalities in the chair–table program to equations by introducing slack variables [see system (12)]. For a system of equations to be in slack variable form, we require that each equation should have a slack variable which has a coefficient of $+1$ and occurs only in that equation. If a system of constraint equations is not in slack variable form, we can use pivoting to get the system in this form. We illustrate this process with the transportation problem in Example 3.

Example 6. **Putting Transportation Constraint Equations in Slack Variable Form**

In Example 3 we obtained the following mathematical formulation of the problem of minimizing the cost of shipping chicken wings from three warehouses to two colleges.

Minimize $80x_{11} + 45x_{12} + 60x_{21} + 55x_{22} + 40x_{31} + 65x_{32}$

subject to $x_{ij} \geq 0$ and

$$
\begin{array}{ll}
\text{Warehouse} & x_{11} + x_{12} \hspace{5.5cm} = 20 \\
\text{equations} & \hspace{1.7cm} x_{21} + x_{22} \hspace{3.2cm} = 30 \\
& \hspace{4.4cm} x_{31} + x_{32} = 15 \hspace{1cm} (28) \\
\text{College} & x_{11} \hspace{1.5cm} + x_{21} \hspace{1.2cm} + x_{31} \hspace{1.4cm} = 25 \\
\text{equations} & \hspace{1.5cm} x_{12} \hspace{1.2cm} + x_{22} \hspace{1.3cm} + x_{32} = 40
\end{array}
$$

We want to use pivoting to convert the five constraint equations in (28) into an equivalent system of equations with slack variables (variables that each occur in just one equation).

Let us make x_{12} the slack variable for the first equation. To eliminate x_{12} from the fifth equation, we simply subtract the first equation from the fifth equation [i.e., we pivot on entry (1, 2), the coefficient of x_{12} in the first equation]. In a similar fashion we make x_{22} the slack variable for the second equation by pivoting on entry (2, 4), and we make x_{32} the slack variable for the third equation by pivoting on entry (3, 6). After these three pivots (which involve subtracting the first, second and third equations from the fifth equation), we have

$$
\begin{array}{ll}
x_{11} + x_{12} \hspace{5.5cm} = \hspace{0.5cm} 20 \\
\hspace{1.7cm} x_{21} + x_{22} \hspace{3.2cm} = \hspace{0.5cm} 30 \\
\hspace{4.4cm} x_{31} + x_{32} = \hspace{0.5cm} 15 \hspace{1cm} (29) \\
x_{11} \hspace{1.5cm} + x_{21} \hspace{1.2cm} + x_{31} \hspace{1.2cm} = \hspace{0.5cm} 25 \\
-x_{11} \hspace{1.4cm} - x_{21} \hspace{1.1cm} - x_{31} \hspace{1.2cm} = -25
\end{array}
$$

In (29), x_{12}, x_{22}, and x_{32} have the form of slack variables as required. These pivots had the effect of converting the last equation into the negative of the fourth equation—the last equation is redundant and can be eliminated (the reason for this redundancy is explained in Exercise 20). Let us make x_{31} the slack variable for the fourth equation. Pivoting on entry (4, 5), that is, subtracting the fourth equation from the third equation, we obtain

$$x_{11} + x_{12} \qquad\qquad\qquad\qquad = \quad 20$$
$$+ x_{21} + x_{22} \qquad\qquad\quad = \quad 30 \qquad (30)$$
$$-x_{11} \qquad - x_{21} \qquad\qquad + x_{32} = \; -10$$
$$x_{11} \qquad + x_{21} \qquad + x_{31} \qquad = \quad 25$$

Now each equation has a slack variable, and we can rewrite (30) as the following family of inequalities:

$$x_{11} \qquad\qquad\qquad \leq \quad 20$$
$$x_{21} \leq \quad 30$$
$$-x_{11} \quad - x_{21} \qquad \leq \; -10 \qquad (\leftrightarrow x_1 + x_{21} \geq 10) \qquad (31)$$
$$x_{11} \quad + x_{21} \qquad \leq \quad 25$$

In the pivoting process, we also have to restate the objective function in terms of just x_{11} and x_{21} (see Exercise 21 for details). Note that the origin in x_{11}, x_{21} coordinates is not in the feasible region.

We have reduced this problem in five equations and six unknowns to an easy problem of four inequalities in two variables. ∎

Sensitivity Analysis

We conclude this section with a brief discussion of the sensitivity of the solution of our linear program to changes in the input values in the constraints. In many economics applications, this sensitivity analysis to changes in the input is almost as important as solving the linear program.

Example 7. Sensitivity Analysis in Chair–Table Production

The final (slack-variable) form of our linear program when an optimum was obtained by the simplex algorithm was

$$\text{Maximize } -35x_3 - 5x_5 + 67{,}000$$
$$\text{subject to } x_i \geq 0, \; i = 1, 2, 3, 4, 5, 6, \text{ and}$$
$$x_1 + \tfrac{3}{2}x_3 \qquad\quad - \tfrac{1}{2}x_5 \qquad\qquad = \quad 300$$
$$- \tfrac{21}{8}x_3 + x_4 + \tfrac{5}{8}x_5 \qquad\qquad = \quad 575$$
$$x_2 - \tfrac{1}{8}x_3 \qquad + \tfrac{1}{8}x_5 \qquad\qquad = \quad 275 \qquad (32)$$
$$- 3x_3 \qquad + \; x_5 + x_6 = 1200$$

Here x_3 and x_5 are the current independent variables whose origin $x_3 = x_5 = 0$ is the optimal corner. These are the slack variables for the original wood and braces constraints. So $x_3 = x_5 = 0$ in the optimal solution means that the optimal production schedule will use all the wood and all the braces. To determine the values of the original independent variables and other slack variables, we simply set $x_3 = x_5 = 0$ in (32) and read off the values of the remaining variables in each equation:

$$x_1 = 300, \qquad x_2 = 275, \qquad x_4 = 575, \qquad x_6 = 1200 \quad (33)$$

As we had found earlier, we make 300 chairs (x_1) and 275 tables (x_2). The labor slack variable (x_4) is 575 and the upholstery slack variable (x_6) is 1200. These slack-variable values are the amounts of these two inputs that are *unused* in the optimal solution. We have an excess supply of labor and upholstery. Thus moderate decreases in the amount of labor or upholstery available will not affect our solution.

What about changes in the input materials we do use, wood and braces? This is a place where the simplex algorithm really shines. First it is convenient to solve for x_1 and x_2 in the first and third equations, which involve x_1 and x_2, respectively.

$$
\begin{aligned}
x_1 &= 300 - \tfrac{3}{2}x_3 + \tfrac{1}{2}x_5 \\
x_2 &= 275 + \tfrac{1}{8}x_3 - \tfrac{1}{8}x_5
\end{aligned}
\quad (34)
$$

Having 1 less unit of wood (1399 units instead of 1400) is equivalent to increasing the wood slack variable x_3 from 0 to 1. To determine the effects of 1 less unit of wood, we simply set $x_3 = 1$, while $x_5 = 0$, in (34). From (34) we have $x_1 = 300 - \tfrac{3}{2}$ and $x_2 = 275 + \tfrac{1}{8}$. The new value of the objective function with 1 less unit of wood is also obtained by setting $x_3 = 1$ in the objective function (while $x_5 = 0$). We have $-35 + 67,000$.

In summary, the coefficients of x_3 in (34) and in the objective function give the effect of 1 less unit of wood: chair production will decrease by $\tfrac{3}{2}$, table production x_2 will increase by $\tfrac{1}{8}$, and profit will decrease by \$35. If we had 1 *more* unit of wood ($x_3 = -1$), the opposite occurs. Chair production increases by $\tfrac{3}{2}$, table production decreases by $\tfrac{1}{8}$, and profit increases by \$35. It is left as an exercise to the reader to evaluate the effects of changing the number of braces.

In economics, the increase in the profit caused by using 1 more unit of an input is called the **marginal value** of the input. In this case it means that we should be willing to pay \$35 for 1 additional unit of wood because that is the value of wood to us in increasing our sales. ∎

Section 4.6 Exercises

Summary of Exercises
Exercises 1–12 involve converting a "word problem" into a linear program (some of these problems previously appeared in Section 1.4); solutions, if requested, are to be obtained by graphing. Exercises 13–17 require one to solve linear programs. Exercises 18–21 ask for transportation problems to be converted into slack-variable form. Exercises 22–25 involve sensitivity analysis. Exercise 26 illustrates duality theory.

1. Suppose that a building supervisor must hire 18-year-olds and 35-year-olds to do the following jobs: clean 500 windows, empty 800 waste-paper baskets, and mop 8000 square feet of floors. In 1 day, an 18-year-old can clean 50 windows, empty 100 baskets, and mop 500 square feet of floors. A 35-year-old can clean 100 windows, empty 100 baskets, and mop 700 square feet of floors. An 18-year-old gets $40 a day and a 35-year-old gets $55 a day. Formulate the problem of minimizing the cost of hiring workers to do the required work as a linear program. (*Set up; do not solve.*)

2. The Arizona tile company manufactures three types of tiles, plain, regular, and fancy, in two different factories. Factory *A* produces 3000 plain, 2000 regular, and 1000 fancy tiles a day and costs $2000 a day to operate. Factory *B* produces 2000 plain, 4000 regular, and 2000 fancy tiles a day and costs $3000 a day to operate. Write down, but do not solve, a linear program for determining the least-cost way to produce at least 20,000 plain, at least 30,000 regular, and at least 10,000 fancy tiles.

3. A farmer has 400 acres on which he can plant any combination of two crops, barley and rye. Barley requires 5 worker-days and $15 of capital for each acre planted, while rye requires 3 worker-days and $20 of capital for each acre planted. Suppose that barley yields $40 per acre and rye $30 per acre. The farmer has $4000 of capital and 500 worker-days of labor available for the year. He wants to determine the most profitable planting strategy.
   (a) Formulate this problem as a linear program.
   (b) Plot the feasible region and find the corner point that maximizes profit.

4. Suppose that a Bored Motor Company factory requires 7 units of metal, 20 units of labor, 3 units of paint, and 8 units of plastic to build a car, while it requires 10 units of metal, 24 units of labor, 3 units of paint, and 4 units of plastic to build a truck. A car sells for $6000 and a truck for $8000. The following resources are available: 2000 units of metal, 5000 units of labor, 1000 units of paint, and 1500 units of plastic.
   (a) State the problem of maximizing the value of the vehicles produced with these resources as a linear program.
   (b) Plot the feasible region of this linear program and solve by checking the objective function at the corners (by looking at the objective function, you should be able to tell which corners are good candidates for the maximum).

5. Consider the two-refinery problem.

$$
\begin{array}{lrcl}
\text{Heating oil:} & 20x_1 + 6x_2 & = & 500 \\
\text{Diesel oil:} & 8x_1 + 15x_2 & = & 750 \\
\text{Gasoline:} & 4x_1 + 6x_2 & = & 1000
\end{array}
$$

Suppose that it costs $30 to refine a barrel in refinery 1 and $25 a barrel in refinery 2. What is the production schedule (i.e., values of x_1, x_2) that minimizes the cost while producing *at least* the amounts demanded of each product (i.e., at least 500 gallons of heating oil, etc.)? Solve by the method in Exercise 3.

6. Plot the boundary of the feasible region for the linear program in Example 2 and solve this linear program as discussed in Exercise 3.

7. Formulate the following transportation problem as a linear program (see Example 3); *do not solve*. There are three factories *A*, *B*, and *C* that ship motors to three stores 1, 2, and 3. Factory *A* makes 1000 motors, *B* makes 2000 motors, and *C* makes 3000 motors. Store 1 needs 1500 motors, store 2 needs 2000 motors, and store 3 needs 2500 motors. The following matrix gives the costs of shipping a motor from a given factory to a given store.

$$
\begin{array}{c}
\text{Factory} \\
\begin{array}{ccc}
A & B & C
\end{array} \\
\text{Store}\ \begin{array}{c} 1 \\ 2 \\ 3 \end{array}
\begin{bmatrix}
1 & 2 & 3 \\
1 & 3 & 2 \\
2 & 4 & 3
\end{bmatrix}
\end{array}
$$

8. There are four boys and four girls and we wish to pair them off in a fashion that minimizes the sum of the personality conflicts in the matches. Entry (i, j) in the following matrix gives a measure of conflict when girl i is matched with boy j. Set up this problem as a linear problem (see the hint at the end of Exercise 9).

$$
\begin{array}{c}
\text{Boys} \\
\begin{array}{cccc}
1 & 2 & 3 & 4
\end{array} \\
\text{Girls}\ \begin{array}{c} A \\ B \\ C \\ D \end{array}
\begin{bmatrix}
2 & 3 & 1 & 3 \\
0 & 3 & 7 & 9 \\
2 & 2 & 3 & 3 \\
3 & 1 & 5 & 1
\end{bmatrix}
\end{array}
$$

9. We wish to assign each person to a different job so as to minimize the total amount of time that must be spent to get all the jobs done. Entry (i, j) in the following matrix tells how many hours it takes person i to do job j (a dash "—" means that the person cannot do the job). Set up this problem as a linear program.

Hint: This is a special form of transportation problem in which the "demands" of jobs and "supplies" of people are all equal to 1.

$$
\begin{array}{c}
\text{Jobs} \\
\begin{array}{ccccc}
1 & 2 & 3 & 4 & 5
\end{array} \\
\text{People}\quad
\begin{array}{c}
A \\ B \\ C \\ D \\ E
\end{array}
\left[
\begin{array}{ccccc}
9 & 4 & - & - & 7 \\
- & 4 & 6 & 2 & 4 \\
6 & 5 & 4 & - & - \\
8 & 3 & - & 5 & 6 \\
- & - & 6 & 4 & 5
\end{array}
\right]
\end{array}
$$

10. We have a farm with 200 cows and capital of $5000 with the following idealized conditions. Cows produce milk for sale or milk to nurse two yearlings (which in a year become cows). A cow can generate $500 worth of milk in a year if not nursing. It costs $300 to feed a cow for a year (no matter what its milk is used for). Write a linear program to maximize the total income over 3 years for the farm (be sure not to spend more money in a year than you currently have).

11. An investor has money-making activities A and B available at the start of each of the next 5 years. Each dollar invested at the start of a year in A returns $1.40 *two* years later (in time for immediate reinvestment). Each dollar invested in B returns $1.70 *three* years later. There are in addition activities C and D that are only available once. Each dollar invested in C at the start of the second year returns $2.00 four years later, and each dollar invested in D at the start of the fifth year returns $1.30 in 1 year. The investor begins with $10,000 and she wants an investment plan that maximizes the gain at the end of 5 years. Give a linear program model for this problem.

12. The Expando Manufacturing Co. wishes to enlarge its capacity over the next six periods to produce umbrellas so as to maximize available capacity at the beginning of the seventh period. Each umbrella produced in a period requires d dollars input and one unit of plant capacity; an umbrella yields r dollars revenue at the start of the next period. In each period, Expando can expand capacity using two construction methods, A and B. A requires b dollars per unit and takes one period; B requires c dollars per unit and takes two periods. Expando has D dollars initially to finance production and expansion (in no period can more money be spent than is available). The capacity initially is K. Formulate a linear program to maximize *production capacity* in period 7.

13. Work through the simplex algorithm for the linear program in Example 5 to verify systems (26) and (27).

14. Solve the following linear programs using the simplex algorithm.

(a) Maximize $3x_1 + 2x_2$
   subject to
$$x_1 \geq 0, \quad x_2 \geq 0$$
$$3x_1 + 4x_2 \leq 12$$
$$4x_1 + 3x_2 \leq 12$$
$$x_1 + 2x_2 \leq 8$$

(b) Maximize $4x_1 + 6x_2$
   subject to
$$x_1 \geq 0, \quad x_2 \geq 0$$
$$3x_1 + 4x_2 \leq 12$$
$$x_1 + 2x_2 \leq 8$$
$$3x_1 + x_2 \leq 6$$

(c) Maximize $x_1 + x_2$
   subject to
$$x_1 \geq 0, \quad x_2 \geq 0$$
$$3x_1 + x_2 \leq 10$$
$$2x_1 + 3x_2 \leq 15$$
$$2x_1 + 2x_2 \leq 12$$

(d) Maximize $3x_1 + 2x_2$
   subject to
$$x_1 \geq 0, \quad x_2 \geq 0$$
$$2x_1 - x_2 \leq 6$$
$$x_1 + x_2 \leq 4$$
$$-3x_1 + x_2 \leq 3$$

15. Solve the following linear programs using the simplex algorithm.

(a) Maximize $2x_1 + 3x_2 + 4x_3$
   subject to
$$x_1 \geq 0, \quad x_2 \geq 0, \quad x_3 \geq 0$$
$$x_1 + 2x_2 + x_3 \leq 10$$
$$3x_1 - x_2 + 2x_3 \leq 12$$
$$x_1 + x_2 + x_3 \leq 6$$

(b) Maximize $3x_1 + x_2 + 2x_3$
   subject to
$$x_1 \geq 0, \quad x_2 \geq 0, \quad x_3 \geq 0$$
$$2x_1 + 3x_2 + x_3 \leq 15$$
$$x_1 + x_2 + 2x_3 \leq 9$$
$$x_1 + 2x_2 + x_3 \leq 10$$

(c) Maximize $3x_1 + 4x_2 + 2x_3$
   subject to
$$x_1 \geq 0, \quad x_2 \geq 0, \quad x_3 \geq 0$$
$$3x_1 + 2x_2 + 4x_3 \leq 15$$
$$x_1 + 2x_2 + 3x_3 \leq 7$$
$$2x_1 + x_2 + x_3 \leq 6$$

(d) Maximize $2x_1 + x_2 + x_3$
   subject to
$$x_1 \geq 0, \quad x_2 \geq 0, \quad x_3 \geq 0$$
$$4x_1 + 3x_2 + 3x_3 \leq 12$$
$$x_1 + 2x_2 - x_3 \leq 4$$
$$x_1 + x_2 + 2x_3 \leq 6$$

16. Solve the following linear program using the simplex algorithm.

$$\text{Maximize } 2x_1 + 4x_2 + x_3 + x_4$$
   subject to
$$x_1 \geq 0, \quad x_2 \geq 0, \quad x_3 \geq 0, \quad x_4 \geq 0$$
$$2x_1 + x_2 + 2x_3 + 3x_4 \leq 12$$
$$2x_2 + x_3 + 2x_4 \leq 20$$
$$2x_1 + x_2 + 4x_3 \leq 16$$

17. Solve the following linear program using the simplex algorithm.

$$\text{Maximize } 15x_1 + 28x_2 + 19x_3 + 24x_4 + 34x_5$$
   subject to

$$x_1 \geq 0, \quad x_2 \geq 0, \quad x_3 \geq 0, \quad x_4 \geq 0, \quad x_5 \geq 0$$
$$x_1 + 2x_2 + x_3 + 3x_4 + 2x_5 \leq 90$$
$$2x_1 + x_2 + x_3 + 4x_4 + x_5 \leq 70$$
$$2x_1 + x_2 + x_3 + x_4 + 2x_5 \leq 80$$
$$3x_2 + 2x_3 + x_4 + 3x_5 \leq 150$$

18. Put the transportation problem in Exercise 7 in slack variable form as shown in Example 6. Also express the problem in inequality constraints.

19. Put the dating problem in Exercise 8 in slack-variable form as shown in Example 6. Also express the problem in inequality constraints.

20. (a) Show that the last equation in the transportation problem in Example 6 is redundant by verifying that the sum of the warehouse equations equals the sum of the college equations; hence the sum of the warehouse equations minus the first college equation will equal the second college equation.
    (b) Explain in words why any nonnegative solution to the first four equations in (28) would also have to satisfy the fifth equation.

21. (a) Use the equations in (30) to rewrite the objective function for the transportation problem in Example 6 in terms of just x_{11} and x_{22}.
    (b) Now solve the transportation problem graphically using the linear program in (31) with the objective function from part (a).

22. Repeat the sensitivity analysis in Example 7 of the chair–table production problem for braces: How would profit change with 1 less unit of braces, how would the number of chairs and number of tables change?

23. Perform sensitivity analysis on the farming linear program in Example 4 of Section 1.4. What are the effects on profit and on amounts of corn and wheat planted if 1 less acre is planted? If 1 less dollar of capital is available?

24. Solve the farming problem in Exercise 3 using the simplex algorithm and perform a sensitivity analysis on all constraints that are fully used. For example, determine the affect of having 1 less dollar of capital.

25. Solve the car–truck production problem in Exercise 4 using the simplex algorithm and perform a sensitivity analysis on all constraints that are fully used.

26. Consider the following two linear programs.
    (i) Maximize $3x_1 + 3x_2$ (ii) Minimize $10x_1 + 8x_2$
    subject to   subject to
    $$x_1 \geq 0, \quad x_2 \geq 0$$ $$x_1 \geq 0, \quad x_2 \geq 0$$
    $$3x_1 + 2x_2 \leq 10$$ $$3x_1 + 4x_2 \geq 3$$
    $$4x_1 + x_2 \leq 8$$ $$2x_1 + x_1 \geq 3$$

Solve them by graphing and show that the optimum values of these two objective functions are the same. What relations link the input data in the two problems?

Appendix to Section 4.6: **Linear Programming Details**

Simplex Algorithm for a General Linear Program

We use the following notation involving n variables and m inequality constraints.

General Linear Program—Inequality Form

$$\text{Maximize } c_1x_1 + c_2x_2 + \cdots + c_nx_n + d$$
$$\text{subject to } x_1 \geq 0, \, x_2 \geq 0, \ldots, x_n \geq 0$$

$$
\begin{array}{l}
a_{11}x_1 + a_{12}x_2 + \cdots + a_{1h}x_h + \cdots + a_{1n}x_n \leq b_1 \\
a_{21}x_1 + a_{22}x_2 + \cdots + a_{2h}x_h + \cdots + a_{2n}x_n \leq b_2 \\
\quad \vdots \qquad \vdots \qquad \vdots \qquad \vdots \qquad \vdots \qquad \vdots \\
a_{g1}x_1 + a_{g2}x_2 + \cdots + a_{gh}x_h + \cdots + a_{gn}x_n \leq b_g \\
\quad \vdots \qquad \vdots \qquad \vdots \qquad \vdots \qquad \vdots \qquad \vdots \\
a_{m1}x_1 + a_{m2}x_2 + \cdots + a_{mh}x_h + \cdots + a_{mn}x_n \leq b_m
\end{array}
\tag{1a}
$$

or in matrix notation,

$$\text{Maximize } \mathbf{c} \cdot \mathbf{x} + d$$
$$\text{subject to}$$
$$\mathbf{Ax} \leq \mathbf{b}, \qquad \mathbf{x} \geq \mathbf{0}$$
$$\tag{1b}$$

Upon restating the inequality constraints in (1a) as equations with slack variables, the system is

General Linear Program—Equation Form

$$\text{Maximize } c_1x_1 + c_2x_2 + \cdots + c_hx_h + \cdots + c_nx_n + d$$
$$\text{subject to } x_1 \geq 0, \, x_2 \geq 0, \ldots, x_n \geq 0, \, x_{n+1} \geq 0, \ldots, x_{n+m} \geq 0, \text{ and}$$

$$
\begin{array}{l}
a_{11}x_1 + a_{12}x_2 + \cdots + a_{1h}x_h + \cdots + a_{1n}x_n + x_{n+1} \qquad\qquad\qquad = b_1 \\
a_{21}x_1 + a_{22}x_2 + \cdots + a_{2h}x_h + \cdots + a_{2n}x_n \qquad + x_{n+2} \qquad\qquad = b_2 \\
\quad \vdots \qquad \vdots \qquad \vdots \qquad \vdots \qquad \vdots \qquad \vdots \qquad\qquad \vdots \\
a_{g1}x_1 + a_{g2}x_2 + \cdots + a_{gh}x_h + \cdots + a_{gn}x_n \qquad\qquad + x_{n+g} \qquad = b_g \\
\quad \vdots \qquad \vdots \qquad \vdots \qquad \vdots \qquad \vdots \qquad \vdots \qquad\qquad \vdots \\
a_{m1}x_1 + a_{m2}x_2 + \cdots + a_{mh}x_h + \cdots + a_{mn}x_n \qquad\qquad\qquad + x_{n+m} = b_m
\end{array}
\tag{2a}
$$

The first n variables are our independent variables from (1a) and the last m variables are the slack variables. *Keep in mind that after iterations of the simplex algorithm, the positions of variables that play the role of slack variables become totally scrambled.* (In linear programming texts, the set of slack variables is called the *basis* of a linear program in equation form.)

If we let $\mathbf{x}^* = [\mathbf{x}, \mathbf{x}']$, where $\mathbf{x}' = [x_{n+1}, \ldots, x_{n+m}]$ is the vector of slack variables, and $\mathbf{A}^* = [\mathbf{A} \quad \mathbf{I}]$, then (2a) can be written

$$\text{Maximize } \mathbf{c} \cdot \mathbf{x} + d$$
$$\text{subject to} \tag{2b}$$
$$\mathbf{A}^*\mathbf{x}^* = \mathbf{b}, \qquad \mathbf{x}^* \geq 0$$

It is important to note that the simplex algorithm assumes that the origin of the current independent variables is a feasible corner. This is the solution obtained by setting all independent variables equal to 0 and setting the slack variable for row i equal to b_i. If the origin is not feasible, see the section "Finding an Initial Feasible Solution."

Now we state the general simplex algorithm. In part I we find the independent variable x_h with the largest positive coefficient c_h in the objective function. This is the variable whose increase provides the greatest rate of increase in the objective function. If no coefficient c_j is positive, the current corner is optimal.

Next we determine how much we can increase x_h (while all other independent variables remain $= 0$). The inequalities in (1a) reduce to

$$a_{1h}x_h \leq b_1 \rightarrow x_h \leq \frac{b_1}{a_{1h}}$$

$$a_{2h}x_h \leq b_2 \rightarrow x_h \leq \frac{b_2}{a_{2h}}$$

$$\vdots \qquad \vdots \qquad \vdots \qquad \vdots \tag{3}$$

$$a_{ih}x_h \leq b_i \rightarrow x_h \leq \frac{b_i}{a_{ih}}$$

$$\vdots \qquad \vdots \qquad \vdots \qquad \vdots$$

$$a_{mh}x_h \leq b_m \rightarrow x_h \leq \frac{b_m}{a_{mh}}$$

Then we can increase x_h by the minimum of these bounds. Let t_h be this amount. If some $a_{ih} \leq 0$, then x_h can increase indefinitely without violating the ith constraint. So we only want to examine constraints with $a_{ih} > 0$. If there is no positive a_{ih}, then x can increase indefinitely without violating any constraint—this "pathological" situation is mentioned later. Summarizing, we have

Part 1 of Simplex Algorithm

1. Let x_h be the independent variable with the largest positive coefficient c_h in the objective function. If all $c_j \leq 0$, the current corner is optimal.
2. Determine the row index i that achieves the minimum value for the ratio b_i/a_{ih} when $a_{ih} > 0$. Let g be the minimizing i.
3. (a) If all $a_{ih} \leq 0$, x_h can be increased infinitely and the problem is unbounded. See Abnormal Possibilities II below.
   (b) If the minimizing ratio b_g/a_{gh} equals 0, that is, $b_g = 0$ (and every other row i has $b_i \leq 0$ or $a_{ih} \leq 0$), special steps must be taken. See Abnormal Possibilities III below.

The intersection of the gth constraint with the constraints $x_j = 0$, $j \neq h$, forms the new corner $(0, 0, \ldots, t_h, \ldots, 0)$.

Part 2 of the simplex algorithm requires us to rewrite the equations in (2) to make this new corner become the origin of the new independent variables. We do this by performing a pivot exchange between independent variable x_h and the slack variable x'_g in equation g; that is, we pivot on entry a_{gh}.

Part 2 of Simplex Algorithm: Pivoting on a_{gh}

1. The old gth constraint equation is rewritten to make x_h become a slack variable, that is, make x_h have coefficient 1. This is accomplished by dividing the equation by a^*.
2. Set the column of coefficients for x_h equal to $\mathbf{e}_g$ (all 0's except in the gth equation). This is accomplished by the standard elimination by pivoting process.
3. The objective function undergoes the same change as in step 2 (x_h drops out and x'_h comes in).

The following diagram illustrates steps 1, 2, and 3. Suppose that $a^* = a_{gh}$ is again the pivot entry, $p = a_{gj}$ is another coefficient in the pivot row, $q = a_{ih}$ is another coefficient in the pivot column, and $r = a_{ij}$ is a oefficient in p's row and q's column (possibly q and r are in the objective function, or possibly p and r are b_i's). Then part 2 of the simplex algorithm has the form

$$\text{Columns}$$

		h	j	g'	
Before pivoting		$\vdots$			
	g	$\cdots\cdots a^* \cdots\cdots p \cdots\cdots 1$			(4)
Rows		$\vdots$	$\vdots$	$\vdots$	
	i	$q \cdots\cdots\cdots r \cdots\cdots 0$			
		$\vdots$			

Columns

After pivoting		h	j	g'
			$\vdots$	

$$g \quad \cdots\cdots 1 \cdots\cdots \frac{p}{a^*} \cdots\cdots \frac{1}{a^*} \qquad (5)$$

Rows

$$i \qquad\qquad 0 \cdots\cdots r - \frac{pq}{a^*} \cdots \frac{-q}{a^*}$$

If r is the constant $-d$ in the objective function, pivoting changes $-d$ to $-d - pq/a^*$ (the objective function constant increases from d to $d + pq/a^*$). In the display above we have listed column g', the column of the new independent variable x_g', which previously was the slack variable in row g.

After pivoting, we return to part 1 of the simplex algorithm for the linear program in this new form and continue doing part 1 and part 2 until no improvement in the objective function is possible in part 1. Then we shall have found the maximum.

Abnormal Possibilities

No Feasible Points. The feasible region for a linear program may be empty. That is, no point lies on the correct side of all the inequalities, for example, to satisfy $x_1 + x_2 \leq -1$, either x_1 or x_2 must be negative—but $x_j \geq 0$ is required in all linear programs. A linear program is called *infeasible* in this case.

Unbounded Feasible Region. When we increase x_h in part 1, perhaps x_h can be increased without limit (to infinity). See step 3(a) of part 1 in the simplex algorithm. This means that the feasible region is unbounded along the x_h-axis and the maximum value of the objective function is infinite. If a practical problem has been misformulated (or data not entered correctly), this difficulty can arise. When this happens, the linear program is called *unbounded*.

Degeneracy. In very rare cases, it can happen that in part 1 replacing an independent variable by a slack variable does not increase the objective function because some constraint has a zero right-hand side (see step 3(b) of part 1 in the simplex algorithm). This can happen without being at an optimal point. This phenomenon is called *degeneracy*. Since it is rare and requires special methods to handle, we shall ignore degeneracy in this book.

Finding an Initial Feasible Solution

How do we start the simplex algorithm if the origin [setting all the independent variables in (1a)] is not a corner of the feasible region? For example, consider a two-variable problem with constraints

$$2x_1 + x_2 \geq 4$$
$$x_1 - 3x_2 \leq -6$$

(6)

or, multiplying the first inequality by -1,

$$-2x_1 - x_2 \leq -4$$
$$x_1 - 3x_2 \leq -6$$

(7)

Neither of these inequalities is satisfied by $x_1 = x_2 = 0$. Let us put the inequalities of (7) in equation form with slack variables x_3, x_4.

$$-2x_1 - x_2 + x_3 \qquad = -4$$
$$x_1 - 3x_2 \qquad + x_4 = -6$$

(8)

As in (7), setting $x_1 = x_2 = 0$ cannot yield a solution of (8), since we require $x_3 \geq 0$, $x_4 \geq 0$.

There is a standard "trick" for finding a starting feasible corner (and associated set of independent variables) when the origin, $x_1 = x_2 = 0$, is infeasible. We add a new equality-violation variable on the left-hand side of each constraint with a negative right-hand side. For the system (8), we introduce x_5, x_6.

$$-2x_1 - x_2 + x_3 \qquad - x_5 \qquad = -4$$
$$x_1 - 3x_2 \qquad + x_4 \qquad - x_6 = -6$$

(9)

where $x_5 \geq 0$, $x_6 \geq 0$. Let us multiply (9) by -1 to get

$$2x_1 + x_2 - x_3 \qquad + x_5 \qquad = 4$$
$$-x_1 + 3x_2 \qquad - x_4 \qquad + x_6 = 6$$

(10)

Now $x_1 = x_2 = x_3 = x_4 = 0$ yields a feasible solution to (10), with $x_5 = 4$ and $x_6 = 6$. We can apply the simplex algorithm to (10).

Next we define a contrived objective function that makes us look for a corner where the equality-violation variables become zero, that is, a corner satisfying our original equations (8). The linear program we want to solve, with the simplex algorithm, is

$$\text{Maximize} - x_5 - x_6 \qquad (11)$$
$$\text{subject to (10) and } x_i \geq 0$$

Since x_5, x_6 are nonnegative, the best possible maximum for (11) would be 0, occurring when $x_5 = x_6 = 0$. The values of x_1, x_2, x_3, and x_4 when $x_5 = x_6 = 0$ will be the starting feasible solution we need to use the simplex algorithm on the original problem. If we solve the linear problem (11) and do not get a maximum of 0, then the original system of constraints was infeasible.

Note that since x_5 and x_6 are slack variables, we must rewrite the objective function in (11) in terms of the independent variables x_1, x_2, x_3, x_4. Reading off expressions for x_5 and x_6 from (10), we have the linear program

$$\text{Maximize} \ (2x_1 + x_2 - x_3 - 4) + (-x_1 + 3x_2 - x_4 - 6)$$
$$= x_1 + 4x_2 - x_3 - x_4 - 10 \qquad (12)$$
$$\text{subject to (10) and } x_i \geq 0$$

Matrix Representation of Pivoting and Revised Simplex Method

The pivoting operation of the simplex algorithm can be expressed as a matrix product. Let the linear program be written in equation form with slack variables as in (2):

$$\text{Maximize } \mathbf{c} \cdot \mathbf{x} + d$$
$$\text{subject to } \mathbf{A}^*\mathbf{x}^* = \mathbf{b}, \ \mathbf{x}^* \geq 0$$

We can obtain the new coefficient matrix after pivoting on entry a_{gh} as the matrix product $\mathbf{PA}^*$, where $\mathbf{P}$ is the m-by-m "pivot" matrix. Since $\mathbf{A}^* = [\mathbf{A} \quad \mathbf{I}]$, where $\mathbf{I}$ is the m-by-m identity matrix, we can write

$$\mathbf{PA}^* = \mathbf{P}[\mathbf{A} \quad \mathbf{I}] = [\mathbf{PA} \quad \mathbf{P}] \qquad (13)$$

From (13) we see that the matrix $\mathbf{P}$, assuming that $\mathbf{P}$ exists, is just the m-by-m submatrix formed by the last m columns (the columns of the slack variables) *after* pivoting is performed. As noted earlier, the columns in the slack-variable submatrix are unchanged by pivoting except for column $n + g$ [which is as shown in (5)]. Thus if $a^* = a_{gh}$ is the pivot entry, then

Column
g

$$
\mathbf{P} =
\begin{bmatrix}
1 & 0 & 0 & \cdots & 0 & \dfrac{-a_{1h}}{a^*} & 0 & \cdots & 0 \\[2ex]
0 & 1 & 0 & \cdots & 0 & \dfrac{-a_{2h}}{a^*} & 0 & \cdots & 0 \\[2ex]
0 & 0 & 1 & \cdots & 0 & \dfrac{-a_{3h}}{a^*} & 0 & \cdots & 0 \\[2ex]
& \ddots & & & & \vdots & & & \\[1ex]
& & & \ddots & & \vdots & & & \\[1ex]
& & & & \ddots & \dfrac{1}{a^*} & & & \\[2ex]
& & & & & \vdots & \ddots & & \\[1ex]
& & & & & \vdots & & \ddots & \\[1ex]
0 & 0 & 0 & \cdots & & \dfrac{-a_{mh}}{a^*} & 0 & \cdots & 0
\end{bmatrix}
\tag{14}
$$

$\mathbf{P}$ is the identity matrix with column g replaced by the hth column $\mathbf{a}'_h$ of $\mathbf{A}^*$ divided by a^*, except that entry (g, g) of $\mathbf{P}$ is $1/a^*$, the inverse of the pivot value $a^* = a_{gh}$.

As a check, let us compute entry (i, j), $j \neq g$, in the product $\mathbf{PA}$. This entry will be the scalar product $\mathbf{p}_i \cdot \mathbf{a}_j$ (where $\mathbf{p}_i$ denotes the ith row of $\mathbf{P}$ and $\mathbf{a}_j$ the jth column of $\mathbf{A}^*$). Row vector $\mathbf{p}_i$ has just two nonnegative entries, $p_{ii} = 1$ and $p_{ig} = -a_{ih}/a^*$. Thus

$$
\text{entry } (i, j) \text{ of } \mathbf{PA} = \mathbf{p}_i \cdot \mathbf{a}_j = p_{ii}a_{ij} + p_{ig}a_{gj}
$$

$$
= 1a_{ij} - \left(\frac{a_{ih}}{a^*}\right) a_{gj}
\tag{15}
$$

$$
= a_{ij} - \frac{a_{gj}a_{ih}}{a^*}
$$

This expression corresponds to the value of entry (i, j) in table (5), since $a_{ij} = r$, $a_{gj} = p$ and $a_{ih} = q$. So $\mathbf{P}$ is, as advertised, the desired pivot matrix.

As mentioned above for table (5), the column vector $\mathbf{b}$ and objective row vector $\mathbf{c}$ are changed in pivoting the same way $\mathbf{A}^*$ is. So we can expand $\mathbf{A}^*$ into an $(m + n + 1)$-by-$(m + 1)$ matrix $\mathbf{A}^+$ of all the data.

$$
\mathbf{A}^+ =
\begin{bmatrix}
\mathbf{c} & \mathbf{0} & -d \\
\mathbf{A} & \mathbf{I} & \mathbf{b}
\end{bmatrix}
\tag{16}
$$

(The minus sign for d is a technicality based on the fact that the constant in the objective function is not on the right-hand side of an equation, the way the b_i are.) The $(m + 1)$-by-m pivot matrix $\mathbf{P}^+$ for $\mathbf{A}^+$ has a corresponding additional top row with $-ch/a*$ in its gth entry and 0's elsewhere.

Let us use the notation $\mathbf{P}_{gh}$ to denote the pivot matrix when we pivot on entry (g, h) (to exchange independent variable x_h and the slack variable for row g). Then a sequence of k pivots on entries (g_1, h_1), (g_2, h_2), . . . , (g_k, h_k) would produce the new linear program with matrix $\mathbf{A}_k^+$:

$$\mathbf{A}_k^+ = \mathbf{P}_{g_1 h_1} \mathbf{P}_{g_2 h_2} \cdots \mathbf{P}_{g_k h_k} \mathbf{A}^+ \tag{17}$$

The revised simplex algorithm uses (17) to compute just the cost coefficients c_j in $\mathbf{A}_k^+$, needed to select the pivot column. We determine which column has the largest c_j; call its index h'. Next (17) is used to compute $\mathbf{b}$ and the h'th column of the constraint matrix in $\mathbf{A}_k^+$, needed to select the pivot row. *The other entries in* $\mathbf{A}$ *are never computed.* The pivot row is the row that achieves the minimum positive value of $b_i/a_{ih'}$; call it row g'. The next pivot matrix $\mathbf{P}_{g'h'}$ can be constructed using the entries of the h'th column [see (14)]. This completes one iteration of the revised simplex algorithm.

The pivot matrices are easy to store and multiply. Only column g' needs to be stored, and even this vector is likely to be very sparse, since the coefficient matrices in large linear programs are very sparse. Because we only calculate one row and two columns of the current linear program, a tremendous savings in time is achieved over the standard simplex algorithm.

Section 4.7 Linear Models for Differentiation and Integration

Although calculus deals with highly nonlinear functions, both the theory and computations associated with calculus are built on linear models. It is the computation that will be our primary interest in this section. We will show how linear approximations to arbitrary functions allow one to solve numerically almost any calculus or differential equation problem.

We first point out the central role of linearity in calculus. The derivative of a function is the essence of a linear model. The derivative $f'(x)$ of a function $y = f(x)$ at a point (x_0, y_0) is the slope of $f(x)$ at (x_0, y_0). That is, $f'(x_0)$ gives the slope of a line through (x_0, y_0) that coincides with $f(x)$ when x is very close to x_0 (see Figure 4.14). In other words, the derivative gives the slope of a linear approximation to $f(x)$ at a point.

An equally important aspect of linearity in calculus is the fact that differentiation is a *linear operation*, in the sense that if $f(x) = ag(x) + bh(x)$, where a, b are constants, then $f'(x) = ag'(x) + bh'(x)$ [the same way that $\mathbf{A}(a\mathbf{u} + b\mathbf{v}) = a\mathbf{A}\mathbf{u} + b\mathbf{A}\mathbf{v}$]. For example, we compute the derivative of $f(x) = 5x^3 + 6x^2$ by knowing that $3x^2$ is the derivative of x^3 and $2x$ is the derivative of x_2 and then $f'(x) = 5(3x^2) + 6(2x)$. Integration

Figure 4.14 Slope of line tangent to $y = f(x)$ is the derivative.

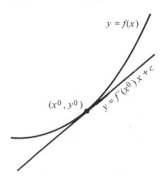

is also a linear operation. Without this linearity, computing the derivative or integral of a polynomial would be a very complicated process.

We now show how to use the linear approximation of a function given by the derivative to build an iterative scheme to find zeros of a function. The scheme is called **Newton's method.** For x-values close to $x = x_0$, we have

$$f(x) \simeq f'(x_0)x + c \qquad \text{for an appropriate constant } c \qquad (1)$$

Pretend that (1) is a good approximation for $f(x)$ over a wide interval around $x = x_0$. We shall use (1) to approximate where $f(x)$ has a zero [where $f(x) = 0$]. If $f(x)$ has value $f(x_0)$ at x_0 and has slope $f'(x_0)$, then the linear approximation (1) decreases by $f'(x_0)$ for each unit we decrease x and will be zero if we decrease x by $f(x_0)/f'(x_0)$. Thus we estimate the x-value x_1 that makes $f(x) = 0$ to be

$$x_1 = x_0 - \frac{f(x_0)}{f'(x_0)} \qquad (2)$$

(see Figure 4.15a). Note that if $f(x_0)/f'(x_0) < 0$, then $x_1 > x_0$.

Normally, $f(x_1) \neq 0$, because the derivative $f'(x_0)$ only approximates the slope of $f(x)$ when x is very near x_0. However, there is a fair chance that x_1 is close to a zero of $f(x)$. Let us use the approximation (1) again, now at the point x_1, to estimate where $f(x)$ is zero. Similar to (2), we get

$$x_2 = x_1 - \frac{f(x_1)}{f'(x_1)}$$

We continue in this method to approximate the true zero of $f(x)$. The general formula is

$$x_{n+1} = x_n - \frac{f(x_n)}{f'(x_n)} \qquad (3)$$

We stop when $f(x_n)$ is very close to 0. Once we get an x-value close to a zero of $f(x)$, the approximation (1) is quite accurate and our method converges quickly to a zero of $f(x)$.

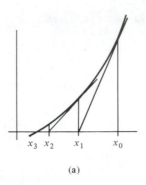

(a)

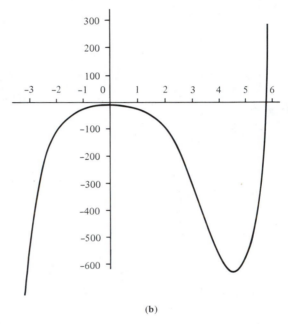

(b)

Figure 4.15 (a) Newton's method. (b) Function $y = x^5 - 5x^4 - 4x^3 - 3x^2 - 2x - 1$.

We should mention that if one cannot compute $f'(x)$ by a differentiation formula, then $f'(x)$ must be estimated by the approximation $[f(x + h) - f(x)]/h$ (for some small h).

Example 1. **Newton's Method**

Consider the function $f(x) = x^2 - 5x + 4$. We use Newton's method to find a zero of this function. First we note that this type of problem arises in determining eigenvalues. Suppose that we want to find the eigenvalues of a matrix $\mathbf{A} = \begin{bmatrix} 3 & 2 \\ 1 & 2 \end{bmatrix}$. Then we must find the zeros of $\det(\mathbf{A} - \lambda\mathbf{I})$,

$$\det(\mathbf{A} - \lambda \mathbf{I}) = \begin{vmatrix} 3 - \lambda & 2 \\ 1 & 2 - \lambda \end{vmatrix}$$

$$= (3 - \lambda)(2 - \lambda) - 2 \cdot 1 = \lambda^2 - 5\lambda + 4$$

For this $f(x)$, $f'(x) = 2x - 5$. Suppose that we start with $x_0 = 10$. We calculate that $f(10) = 54$ and $f'(10) = 15$. So Newton's method estimates the zero to be at

$$x_1 = x_0 - \frac{f(x_0)}{f'(x_0)} = 10 - \frac{54}{15} = 6.4$$

Next we calculate $f(6.4) = 12.96$ and $f'(6.4) = 7.8$. Then

$$x_2 = x_1 - \frac{f(x_1)}{f'(x_1)} = 6.4 - \frac{12.96}{7.8} \simeq 4.74$$

Calculating $f(4.74) \simeq 2.77$ and $f'(4.74) \simeq 4.48$ gives us

$$x_3 = x_2 - \frac{f(x_2)}{f'(x_2)} = 4.74 - \frac{2.77}{4.48} = 4.12$$

and

$$x_4 = x_3 - \frac{f(4.12)}{f'(4.12)} = 4.12 - \frac{.37}{3.24} = 4.006$$

$$x_5 = x_4 - \frac{f(4.006)}{f'(4.006)} = 4.006 - \frac{.018}{3.012} = 4.000024$$

Clearly, we are converging to 4. That is, $f(4) = 0$.   ■

Example 2. **Bad Performance by Newton's Method**

Use Newton's method to find a zero of the function $y = x^5 - 5x^4 - 4x^3 - 3x^2 - 2x - 1$. We have plotted this function in Figure 4.15b with the y-axis magnified near the origin. This curve gets close to a zero at $x = -.4$ but only has a relative maximum with $y \simeq -.56$. It decreases awhile and then increases sharply with $f(5.5) = -310$, $f(6) \simeq 300$, and $f(7) \simeq 3000$. If we start Newton's method with $x_0 = 0$, we get the sequence of points shown in Table 4.6.

We see that our sequence of points gets caught around the relative minimum at $x \simeq -.4$ and tends to swing back and forth from one side of $-.4$ to the other. It finally gets very near the minimum on the 107th iteration. Here the slope $f'(x_{107})$ is almost zero, so dividing by $f'(x_{107})$ $(= .04)$ in (3) brings a large change in x that gets us away from this

Table 4.6

i	x_i	$f(x_i)$	$f'(x_i)$
0	0	-1	-2
1	$-.5$	$-.594$	.812
2	.231	-1.683	-4.25
3	$-.165$	$-.738$	-1.24
4	$-.758$	-1.36	6.02
5	$-.531$	$-.644$	1.19
6	$-.008$	$-.980$	-1.95
7	$-.512$	$-.604$	.960
8	.117	-1.28	-2.90
9	$-.324$	$-.589$	$-.577$
10	-1.134	-14.7	49
11	-1.046	-4.85	20
12	$-.804$	-1.67	7.54
13	$-.581$	$-.703$	1.93
14	$-.218$	$-.675$	-1.04
15	$-.868$	-2.2	10.1
$\vdots$	$\vdots$	$\vdots$	$\vdots$
107	$-.420$	$-.56$	.04
108	13.14	233,700	101,763
109	10.84	75,585	42,266
110	9.06	24,144	17,789
111	7.70	7,505	7,710
112	6.73	2,183	3,578
113	6.12	530	1,943
114	5.848	76.5	1,400
115	5.793	2.65	1,304
116	5.791324	.0036	1,300
117	5.791321	~ 0	1,300

relative minimum—$x_{108} = 13.14$. Now Newton's method homes in on the minimum.

If we had started with $x_0 > 5$, the procedure would have converged quickly. Although it did finally escape from the region of the relative minimum, a better scheme would have been to pick a new starting value when the method had not converged after 20 iterations. ∎

Next let us consider the definition of the integral of a function $f(x)$. The integral of $f(x)$ is the area under a curve from 0 to x (sometimes 0 is replaced by another x-value x_0). The formal definition of an integral is the limit of the sum of areas of approximating rectangles

$$\int_0^x f(x)\,dx = \lim_{n\to\infty} \sum_{i=1}^n (t_i - t_{i-1})f(t_i) \qquad (4)$$

where $t_i = i(x/n)$. The t_i subdivide the interval $[0, x]$ into equal subintervals of length x/n, with $t_0 = 0$ and $t_n = x$ (see Figure 4.16a). We are approximating the area under the curve $f(x)$ with the area under the piecewise approximation to $f(x)$:

$$f_n(x) = f(t_i) \qquad \text{when } t_{i-1} < x \le t_i, \qquad i = 1, 2, \ldots, n \qquad (5)$$

That is, the sum on the right-hand side of (4) is the area under $f_n(x)$. The function $f_n(x)$ is a *piecewise* constant function, since it is made up of constant functions with a different constant on each subinterval of $[0, x]$.

Figure 4.16 (a) Set of rectangles whose areas approximate the area under the curve. (b) Piecewise linear approximation to curve $f(x)$ in part (a). Set of trapezoids approximate the area under $f(x)$.

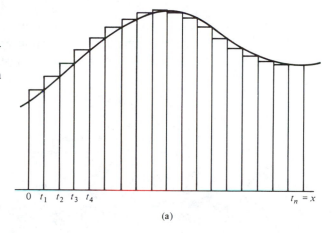

(a)

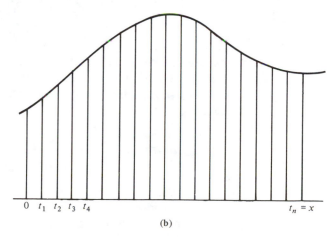

(b)

The points t_i in (4) are called **mesh points.** The collection of mesh points and the subintervals they form is called the **mesh.** As the mesh becomes finer (as n increases), it is intuitively clear that the area under $f_n(x)$ will approach the area under the curve $f(x)$.

Let us consider how we might try to calculate the area under a function $f(x)$ from 0 to x° when we do not know how to integrate exactly [i.e., no integration formulas apply to $f(x)$, and integration by parts, etc., all fail]. One obvious approach is to use the definition of the integral given in (4). That is, we pick some number of mesh points, say $n = 50$, for the interval $(0, x^\circ)$. Then we calculate the value of $f(x)$ at the mesh points and compute the sum on the right-hand side of (4). A better estimate for the area under $f(x)$ should be obtained by using linear approximations to $f(x)$ in each subinterval that agree with the values of $f(x)$ at the subinterval endpoints (see Figure 4.16b). It is left to the reader to check that the following piecewise linear approximation does this.

$$g_n(x) = \frac{f(t_i) - f(t_{i-1})}{t_i - t_{i-1}} (x - t_{i-1}) + f(t_{i-1})$$

$$\text{for } t_{i-1} < x \le t_i, \quad i = 1, 2, \ldots, n \tag{6}$$

By looking at Figure 4.16b, one sees that the region in each subinterval using (6) is a trapezoid. The area of this trapezoid is the same as the area of a rectangle with height halfway between $f(t_{i-1})$ and $f(t_i)$. That is, using (6) gives the same area as using the piecewise constant function

$$g_n^*(x) = \frac{f(t_i) + f(t_{i-1})}{2} \qquad \text{for } t_{i-1} < x \le t_i \tag{7}$$

Let $h = t_i - t_{i-1} = x^\circ/n$. Then the integral of $f(x)$ from 0 to x° is approximated by the area under $g_n^*(x)$,

$$\int g_n^*(x)\,dx = h \sum_{i=1}^{n} \frac{f(t_i) + f(t_{i-1})}{2}$$

$$= \frac{hf(t_0)}{2} + h \sum_{i=1}^{n} f(t_i) + \frac{hf(t_n)}{2} \tag{8}$$

Using (8) to approximate an integral is an integration scheme called the **trapezoidal rule** (named after the trapezoids in Figure 4.16b). Note that this rule weights the two endpoint values $f(t_0)$ and $f(t_n)$ half as much as the other $f(t_i)$. Various schemes have been developed that give different sets of weights to the $f(t_i)$.

Example 3. Piecewise Approximation of Area Under Curve

Consider the function $f(x) = 1/(x^2 + 1)$. We want to compute the area under $f(x)$ on the interval $[0, 2]$ using the piecewise approximations given in (5) and (6). To make calculations easy, let the mesh points be $t_0 = 0$, $t_1 = .5$, $t_2 = 1$, $t_3 = 1.5$, $t_4 = 2$. Then we get the table of function values

t_i	0	.5	1	1.5	2
$f(t_i)$	1	.8	.5	~.3	.2

The approximation of $f(x)$ by the function in (5),

$$f_4(x) = f(t_i) \qquad \text{for } t_{i-1} \le x \le t_i, \quad i = 1, 2, 3, 4$$

gives

$$f_4(x) = \begin{cases} .8 \text{ on } [0, .5] \\ .5 \text{ on } [.5, 1] \\ .3 \text{ on } [1, 1.5] \\ .2 \text{ on } [1.5, 1] \end{cases} \qquad (9)$$

and the approximation of $f(x)$ by the function in (6),

$$g_4(x) = \frac{f(t_i) - f(t_{i-1})}{t_i - t_{i-1}} (x - t_{i-1}) + f(t_{i-1})$$

$$\text{for } t_{i-1} < x \le t_i, \quad i = 1, 2, 3, 4$$

gives

$$g_4(x) = \begin{cases} -.4x + 1 \text{ on } [0, .5] \\ -.6x + .8 \text{ on } [.5, 1] \\ -.4x + .5 \text{ on } [1, 1.5] \\ -.2x + .3 \text{ } [1.5, 2] \end{cases} \qquad (10)$$

The area under $f_4(x)$ equals $.5(.8 + .5 + .3 + .2) = .9$ and the area under $g_4(x)$ equals [using (8)] $.5\{1/2 + (.8 + .5 + .3) + .2/2\} = 1.2$. The actual area under $1/(x^2 + 1)$ from 0 to 2 is about 1.11. So, as expected, $g_4(x)$ led to a better estimate of the area. ∎

In (5) we approximated $f(x)$ with a piecewise constant function $f_n(x)$, and in (6) we approximated $f(x)$ with a piecewise linear function $g_n^*(n)$. A better approximation to the area under $f(x)$ can be sought by using piecewise approximations to $f(x)$ that are quadratic or cubic functions. It turns out that cubic functions have several good properties that quadratics lack, so piecewise cubic functions are frequently used to approximate functions in integration and other calculations. The cubic function in the ith mesh interval would be

$$s_i(x) = a_i x^3 + b_i x^2 + c_i x + d_i \qquad \text{for } t_{i-1} \le x \le t_i \qquad (11)$$

We want each cubic piece to equal $f(x)$ at the mesh points t_{i-1} and t_i, that is,

$$s_i(t_{i-1}) = f(t_{i-1}) \quad \text{and} \quad s_i(t_i) = f(t_i), \quad i = 1, 2, \ldots, n \quad (12)$$

Another desirable property for an approximating function is to be smooth at the mesh points t_i, where the pieces meet. By this we mean that the first and second derivatives of two successive cubics coincide at their common mesh point. Then we require

$$\begin{aligned} s_i'(t_i) &= s_{i+1}'(t_i) \\ s_i''(t_i) &= s_{i+1}''(t_i) \quad \text{for } i = 1, 2, \ldots, n - 1 \end{aligned} \quad (13)$$

The term **spline** is the name given to a smooth piecewise function. Cubic splines are used to approximate curves in thousands of applications. For example, automotive designers use splines to approximate car contours in computer models that simulate a car's wind resistance. Splines are widely used in computer graphics to generate the forms of complex figures; they are used to join points made with a light pen in tracing out a figure on a terminal screen. Once a figure is represented by mathematical equations, it is an easy matter to rotate, shrink, and perform other transformations of the figure, as discussed in Section 4.1.

The problem with splines is determining the coefficients in each of the cubic pieces. However, the system of equations (12) and (13) determining the coefficients in the splines can be collapsed to a tridiagonal system of $n - 1$ equations in $n - 1$ unknowns. See the Appendix to this section for details. Such tridiagonal systems can be solved very quickly (see Section 3.5), so spline approximations can be computed very quickly.

Another important approach, called functional approximation, for approximating a function for integration and other purposes is to use one linear combination of nice functions (e.g., whose integrals are easily computed) to approximate $f(x)$ over the entire interval from 0 to $x°$. This approach is discussed in Section 5.4.

We next consider linear approximations to solutions of differential equations. The derivative $f'(x)$, the slope of function $f(x)$ at x, is approximated by

$$f'(x) \simeq \frac{f(x + h) - f(x)}{h} \quad (14)$$

Indeed, the limit of (14) as h goes to zero is by definition $f'(x)$. First and second derivatives of complicated functions are often needed in differential equations computations.

Example 4. Discrete Approximation to a Differential Equation

Consider $y(x)$, the temperature of a rod, as a function of x, the distance from the left end of the rod. If a heat source is applied to the rod with a temperature $f(x)$ at position x, the following differential equation describes how $f(x)$ affects $y(x)$.

$$\frac{d^2y}{dx^2} = -f(x) \qquad 0 \le x \le d \qquad (d = \text{length of rod}) \qquad (15)$$

For example, $f(x)$ might be zero everywhere except at two short segments of the rod. It is too difficult to solve this differential equation analytically for most interesting $f(x)$. Instead, one seeks an approximate solution by computing values of $y(x)$ at a set of mesh points x_0, $x_1, \ldots, x_n$ on the interval $(0, d)$. If $h = x_i - x_{i-1}$, then at $x = x_i$, (14) becomes

$$y'(x_i) \simeq \frac{y(x_{i+1}) - y(x_i)}{h} \qquad (16)$$

We could just as well use

$$y'(x_i) \simeq \frac{y(x_i) - y(x_{i-1})}{h} \qquad (16')$$

Next, using the fact that $y''(x)$ is the derivative of the derivative, we can estimate $y''(x_i)$ using (16) or (16$'$) twice. To make the result symmetric about x_i, we use (16) for $y'(x_i)$ and (16$'$) to get $y''(x_i)$.

$$
\begin{aligned}
y''(x_i) &\simeq \frac{\{y'(x_i) - y'(x_{i-1})\}}{h} \\
&\simeq \frac{\{[y(x_{i+1}) - y(x_i)]/h - [y(x_i) - y(x_{i-1})]/h\}}{h} \qquad (17) \\
&\simeq \frac{\{y(x_{i+1}) - 2y(x_i) + y(x_{i-1})\}}{h^2}
\end{aligned}
$$

Substituting (17) into (15), we obtain a system of equations for the values $y(x_i)$. Letting $y_i = y(x_i)$ and $f_i = f(x_i)$, we have

$$\frac{y_{i+1} - 2y_i + y_{i-1}}{h^2} = -f_i$$

or

$$y_{i+1} - 2y_i + y_{i-1} = -h^2 f_i, \qquad i = 1, 2, \ldots, n - 1 \qquad (18)$$

We have $n - 1$ equations for the $n + 1$ unknowns $y_0, y_1, \ldots, y_n$. Like the differential equations in Section 3.3, we need to specify two starting, or boundary, conditions. Typically, these have the form $y_0 = a$ and $y_n = b$. Suppose, for later reference, that we choose the boundary conditions

$$y_0 = y_n = 0 \tag{19}$$

The physical interpretation of (19) is that both ends of the rod are attached to objects that dissipate away all the heat at the ends. Now we drop y_0 and y_n from (18), since by (19) they are 0. This reduces (18) to a system of $n - 1$ equations in $n - 1$ unknowns.

In matrix terms, (18) becomes

$$\mathbf{Dy} = \mathbf{f}^* \tag{20}$$

where $\mathbf{y} = (y_1, y_2, \dots, y_{n-1})$—remember that y_0 and y_n are dropped, since they are 0—$\mathbf{f}^* = (h^2 f_1, h^2 f_2, \dots, h^2 f_{n-1})$, and $\mathbf{D}$ is

$\mathbf{D} =$

$$\begin{bmatrix}
-2 & 1 & 0 & 0 & 0 & & & & & & & \\
1 & -2 & 1 & 0 & 0 & & & & & & & \\
0 & 1 & -2 & 1 & 0 & & & & & & & \\
0 & 0 & 1 & -2 & 1 & & & & & & & \\
& & & 1 & -2 & 1 & & & & & & \\
& & & & 1 & -2 & 1 & & & & & \\
& & & & & \cdot & \cdot & \cdot & & & & \\
& & & & & & \cdot & \cdot & \cdot & & & \\
& & & & & & & \cdot & \cdot & \cdot & & \\
& & & & & & & 1 & -2 & 1 & 0 & 0 \\
& & & & & & & 0 & 1 & -2 & 1 & 0 \\
& & & & & & & 0 & 0 & 1 & -2 & 1 \\
& & & & & & & 0 & 0 & 0 & 1 & -2
\end{bmatrix} \tag{21}$$

Observe that $\mathbf{D}$ is a tridiagonal matrix, so $\mathbf{Dy} = \mathbf{f}$ can be solved very quickly by Gaussian elimination (see Section 3.5). Let us use (20) to get an approximate solution to our differential equation (15) on the interval [0, 1] with 100 mesh points. So $n = 100$ and $h = 1/n = .01$. Suppose that $f(x)$ represents a heat source applied to a point at the middle of the rod:

$$f_{50} = 10{,}000 \quad \text{and} \quad \text{all other } f_i = 0 \tag{22}$$

Then, if $\mathbf{f}^*, -h^2 f_{50} = -(.01)^2 \times 10{,}000 = -1$. So $\mathbf{f}^*$ is all 0's except -1 in its fiftieth entry.

Applying Gaussian elimination to (21) plus the right-side vector $\mathbf{f}^*$, we obtain the following matrix.

$$\left[\begin{array}{cccccccc}
-2 & 1 & 0 & 0 & & & & \bigm| \quad 0 \\
0 & -\frac{3}{2} & 1 & 0 & & & & \quad 0 \\
0 & 0 & -\frac{4}{3} & 0 & & & & \bigm| \quad 0 \\
0 & 0 & 0 & 1 & & & & \quad 0 \\
& & & \cdot & \cdot & \cdot & & \quad \cdot \\
& & & \cdot & \cdot & \cdot & & \quad \cdot \\
& & & \cdot & \cdot & \cdot & & \quad \cdot \\
& & -\frac{51}{50} & 1 & & & & \bigm| \; -1 \\
& & & -\frac{52}{51} & 1 & & & \; -\frac{50}{51} \\
& & & & \cdot & \cdot & \cdot & \quad \cdot \\
& & & & \cdot & \cdot & \cdot & \quad \cdot \\
& & & & \cdot & \cdot & \cdot & \quad \cdot \\
& & & & -\frac{98}{97} & 1 & 0 & \bigm| \; -\frac{50}{97} \\
& & & & 0 & -\frac{99}{98} & 1 & \; -\frac{50}{98} \\
& & & & 0 & 0 & -\frac{100}{99} & \bigm| \; -\frac{50}{99}
\end{array}\right] \quad (23)$$

Back substitution yields

$$y_k = \frac{k}{2} \qquad k = 1, 2, \ldots, 49$$

$$y_k = \frac{100 - k}{2} \qquad k = 50, 51, \ldots, 99 \qquad (24)$$

The continuous function $y(x)$ that (24) approximates is clearly

$$y(x) = \begin{cases} 50x, & 0 \le x \le .5 \\ 50 - 50x, & .5 \le x \le 1 \end{cases} \qquad (25)$$

The solution (25) says that if the middle of the rod is heated and the ends are kept at temperature 0, the temperature will decrease at a uniform rate along the rod toward the ends (as opposed to an exponential decay or some other nonlinear decrease). This uniform decrease is indeed what occurs in nature. ∎

By letting the number of mesh points grow very large, our solution vector **y** in (24) becomes a better and better approximation to the true solution $y(x)$. In the limit, **y** becomes an infinite vector that "is" $y(x)$. Thus any continuous function can be thought of as an infinite-length vector. In Section 5.4 we show another way to view continuous functions as vectors.

Most methods of solving differential equations by discrete approximation are called *finite difference* schemes. The other basic approach to approximating solutions to differential equations is *finite element* methods. Here the function $f(x)$ is approximated as a sum of special functions for which the differential equation is easily solved.

Section 4.7 Exercises

Summary of Exercises

Exercises 1–3 involve finding zeros with Newton's method. Exercises 4–9 involve approximating areas under curves. Exercises 10 and 11 involve finite difference methods for solving differential equations. Problems about splines appear in the Exercises in the appendix to this section.

1. Use Newton's method to find a zero of the following functions; let $x_0 = 2$.

    (a) $f(x) = 3x - 5$
    (b) $f(x) = x^2 + 6x + 5$
    (c) $f(x) = x^3 - 3x^2 + 3x - 1$
    (d) $f(x) = x^3 + x^2 - 2x$
    (e) $f(x) = \cos(x)$ (x in radians)
    (f) $f(x) = x^2 - 2x + 8$
    (g) $f(x) = x^4 - 4x^3 + 5x^2 - x - 15$
    (h) $f(x) = x^5 + 4x^3 - 8x - 3$

2. Use Newton's method to find all zeros of the functions in Exercise 1 in the interval from -2 to 6.

    Hint: If α is a zero of $f(x)$, the other zeros of $f(x)$ are also zeros of $f(x)/(x - \alpha)$.

3. Use Newton's method to find all eigenvalues of the following matrices. (See the hint in Exercise 2 and see Section 3.1 for determinant formulas.)

    (a) $\begin{bmatrix} 4 & 0 \\ 2 & 2 \end{bmatrix}$
    (b) $\begin{bmatrix} 4 & -1 & 0 \\ 1 & 0 & 2 \\ 0 & 2 & 1 \end{bmatrix}$
    (c) $\begin{bmatrix} 0 & 4 & 1 \\ .4 & 0 & 0 \\ 0 & .6 & 0 \end{bmatrix}$

    (d) $\begin{bmatrix} 0 & 0 & 4 & 8 \\ .5 & 0 & 0 & 0 \\ 0 & .5 & 0 & 0 \\ 0 & 0 & .5 & 0 \end{bmatrix}$

4. (a) Use the trapezoidal rule, equation (8), to estimate the area under the curve $f(x) = x^2 - 2x + 1$ from 0 to 4 with the mesh $\{0, 1, 2, 3, 4\}$. Determine the area exactly by integration. Also plot this function and the approximation given by the piecewise linear function in (6).
    (b) See how accurate your answer gets with a denser mesh. Compute with the trapezoidal rule again with meshes.
    (i) $\{0, .5, 1, 1.5, 2, 2.5, 3, 3.5, 4.0\}$
    (ii) $\{0, .25, .5, .75, 1, 1.25, 1.5, 1.75, 2, 2.25, 2.5, 2.75, 3, 3.25, 3.5, 3.75, 4\}$.

5. **(a)** Use the trapezoidal rule, equation (8), to estimate the area under the curve $f(x) = x^3 - 2x + 4$ from 0 to 4 with the mesh $\{0, 1, 2, 3, 4\}$. Determine the area exactly by integration. Also plot this function and the approximation given by the piecewise linear function in (6).

   (b) See how accurate your answer gets with a denser mesh. Compute with the trapezoidal rule again with the mesh $\{0, .5, 1, 1.5, 2, 2.5, 3, 3.5, 4\}$.

6. **(a)** Use the trapezoidal rule, equation (8), to estimate the area under the curve $f(x) = e^x$ from 0 to 4 with the mesh $\{0, 1, 2, 3, 4\}$. Determine the area exactly by integration. Also plot this function and the approximation given by the piecewise linear function in (6).

   (b) See how accurate your answer gets with a denser mesh. Compute with the trapezoidal rule again with meshes.

   (i) $\{0, .5, 1, 1.5, 2, 2.5, 3. 3.5, 4\}$

   (ii) $\{0, .25, .5, .75, 1. 1.25, 1.5, 1.75, 2, 2.25, 2.5, 2.75, 3, 3.25, 3.5, 3.75, 4\}$

7. **(a)** Use the trapezoidal rule, equation (8), to estimate the area under the curve $f(x) = 10/(x + 1)^2$ from 0 to 4 with the mesh $\{0, 1, 2, 3, 4\}$. Determine the area exactly by integration. Also plot this function and the approximation given by the piecewise linear function in (6).

   (b) See how accurate your answer gets with a denser mesh. Compute with the trapezoidal rule again with meshes:

   (i) $\{0, .5, 1, 1.5, 2, 2.5, 3, 3.5, 4\}$

   (ii) $\{0, .25, .5, .75, 1, 1.25, 1.5, 1.75, 2, 2.25, 2.5, 2.75, 3, 3.25, 3.5, 3.75, 4\}$

8. **(a)** Use the trapezoidal rule, equation (8), to estimate the area under the curve $f(x) = \sin (2\pi/x)$ from $\frac{1}{4}$ to 1 with the mesh $\{.25, .5, .75, 1\}$. Also plot this function and the approximation given by the piecewise linear function in (6). This function cannot be integrated by any standard integration technique.

   (b) See how much your answer changes with a denser mesh. Compute with the trapezoidal rule with the variable mesh $\{.25, .26, .27, .28, .29, .3, .32, .34, .36, .39, .42, .45, .5, .6, .7, .8, .9, 1\}$.

9. Simpson's rule approximates an integral by

$$\int_a^b f(x)\, dx \simeq \frac{h}{3} \{f(a) + 4f(t_1) + 2f(t)_2) + 4f(t_3) + 2f(t_4) + \cdots$$

$$+ 4f(t_{n-1}) + f(b)\}$$

where $t_i = a + hi$, $i = 1, 2, \ldots, n - 1$, n even, and $h = (b - a)/n$.

(a) Simpson's rule finds the exact integral of quadratic functions. Verify this by using Simpson's rule to approximate the given integrals and check by formal integration.
   (i) $f(x) = x^2 + 1$ from 0 to 1 with $n = 4$.
   (ii) $f(x) = x^2 - 2x + 4$ from 1 to 3 with $n = 2$.

(b) Repeat Exercise 5, part (a) with Simpson's rule. Is the Simpson's rule approximation more accurate?

(c) Repeat Exercise 6, part (a) with Simpson's rule. Is the Simpson's rule approximation more accurate?

(d) Repeat Exercise 7, part (a) with Simpson's rule. Is the Simpson's rule approximation more accurate?

10. Repeat the method for solving $d^2y/dx^2 = -f(x)$, $0 \le x \le 1$ approximately in Example 4 with $y(0) = y(1) = 0$ using the following $f(x)$.
   (a) $f(x) = \{f_{25} = -10{,}000$ and all other $f_i = 0\}$
   (b) $f(x) = \{f_1 = -10{,}000$ and all other $f_i = 0\}$

11. Use the method for solving a differential equation in Example 4 to solve approximately

$$\frac{d^2y}{dx^2} = -10{,}000, \qquad 0 \le x \le 1$$

with $h = .01$ and $y(0) = y(1) = 0$.

Appendix to Section 4.7: Computing Cubic Spline Approximations

In this appendix we show how the equations for determining the coefficients in a cubic spline can be greatly simplified and reduced to a tridiagonal system of $n - 1$ equations in $n - 1$ unknowns, where n is the number of subintervals in the cubic spline.

Suppose that we divide the interval $[a, b]$, into n subintervals. Let $a = t_0 < t_1 < \cdots < t_n = b$ be the mesh points. Recall that a **cubic spline** $s(x)$ for a function $f(x)$ is a piecewise cubic function such that

(a) $s(t_i)$ equals the function $f(t_i)$ at each t_i.
(b) The first and second derivatives $s'(x)$ and $s''(x)$ are continuous.

We only know the values of $f(x)$ at the mesh points $t_0, t_1, \ldots, t_n$ and want to interpolate values for $f(x)$ over the whole interval.

Let $s_i(x)$ be the cubic polynomial in subinterval $[t_i, t_{i+1}]$, $i = 0, 1, \ldots, n - 1$. Then condition (a) becomes

$$s_i(t_i) \quad = f(t_i), \qquad i = 0, 1, \ldots, n - 1 \qquad (1)$$
$$s_i(t_{i+1}) = f(t_{i+1}), \qquad i = 0, 1, \ldots, n - 1 \qquad (2)$$

and condition (b) becomes

$$s_i'(t_{i+1}) = s_{i+1}'(t_{i+1}), \qquad i = 0, 1, \ldots, n - 2 \qquad (3)$$
$$s_i''(t_{i+1}) = s_{i+1}''(t_{i+1}), \qquad i = 0, 1, \ldots, n - 2 \qquad (4)$$

This gives us $4n - 2$ equations in $4n$ unknowns. As in the discrete approximation of a differential equation in Example 4 of Section 4.7, we need two extra constraints. The ones that work out the best are

$$s_0''(t_0) = 0 \qquad \text{and} \qquad s_{n-1}''(t_n) = 0 \qquad (5)$$

Let us assume that the mesh points are equally spaced with $t_{i+1} - t_i = h$. The first step is to express $s_i(x)$ in the translated form

$$s_i(x) = a_i + b_i(x - t_i) + c_i(x - t_i)^2 \qquad (6)$$
$$+ d_i(x - t_i)^3, \qquad i = 1, \ldots, n$$

As a result of (6), when $x = t_i$, then $s_i(t_i) = a_i$, $s_i'(t_i) = b_i$ and $s_i''(t_i) = 2c_i$. Observe also that

$$s_i(t_{i+1}) = a_i + b_i h + c_i h^2 + d_i h^3$$
$$s_i'(t_{i+1}) = b_1 + 2c_i h + 3d_i h^2 \qquad (7)$$
$$s_i''(t_{i+1}) = 2c_i + 6d_i h$$

Using (7), we rewrite conditions (1)–(5) as

$$a_i = f(t_i), \qquad\qquad\qquad\qquad i = 0, 1, \ldots, n \qquad (1')$$
$$a_{i+1} = a_i + b_i h + c_i h^2 + d_i h^3, \quad i = 0, 1, \ldots, n - 1 \qquad (2')$$
$$b_{i+1} = b_i + 2c_i h + 3d_i h^2, \qquad i = 0, 1, \ldots, n - 2 \qquad (3')$$
$$2c_{i+1} = 2c_i + 6d_i h \text{ or}$$
$$c_{i+1} = c_i + 3d_i h, \qquad\qquad\qquad i = 0, 1, \ldots, n - 2 \qquad (4')$$
$$c_0 = 0 \text{ and } c_n = 0 \qquad\qquad\qquad\qquad\qquad\qquad\qquad (5')$$

Here we have "invented" $a_n[= f(t_n)]$ in (1') and c_n ($= 0$) in (5'). There is no $s_n(x)$ but the terms a_n and c_n are a useful way to represent conditions on the spline at t_n.

From (1') we see that the a_i's can be determined immediately from the values $f(t_i)$. We shall now proceed to show how to express the d_i's and b_i's in terms of the c_i's and the a_i's, and then we obtain a tridiagonal system to solve for the c_i's.

Solving (4') for d_i, we have

$$d_i = \frac{c_{i+1} - c_i}{3h}, \qquad i = 0, 1, \ldots, n - 2 \qquad (8)$$

Substituting this value for d_i in (2') and (3'), we obtain

$$a_{i+1} = a_i + hb_i + \frac{h^2}{3}(2c_i + c_{i+1}), \qquad i = 0, 1, \ldots, n-1 \quad (2'')$$

$$b_{i+1} = b_i + h(c_{i+1} + c_i), \qquad\qquad i = 0, 1, \ldots, n-2 \quad (3'')$$

Solving (2'') for b_i, we obtain

$$b_i = \frac{a_{i+1} - a_i}{h} - \frac{h(2c_i + c_{i+1})}{3}, \qquad i = 0, 1, \ldots, n-1 \quad (9)$$

Reducing the subscript by 1, (9) becomes

$$b_{i-1} = \frac{a_i - a_{i-1}}{h} - \frac{h(2c_{i-1} + c_i)}{3}, \qquad i = 1, 2, \ldots, n \quad (9')$$

Let us also rewrite (3'') with reduced subscript

$$b_i = b_{i-1} + h(c_i + c_{i-1}), \qquad i = 1, 2, \ldots, n-1 \quad (3''')$$

Now we substitute (9) for b_i and (9') for b_{i-1} in (3''') to obtain

$$\frac{a_{i+1} - a_i}{h} - \frac{h(2c_i + 1\,c_{i+1})}{3} = \frac{a_i - a_{i-1}}{h}$$
$$- \frac{h(2c_{i-1} + c_i)}{3} + h(c_i + c_{i-1}) \quad (10)$$

Multiplying by 3, dividing by h, and collecting the a_i's on the right side, we have

$$(2c_i + c_{i+1}) - (2c_{i-1} + c_i) + 3(c_i + c_{i-1}) = \frac{3(a_{i+1} - a_i)}{h^2}$$
$$- \frac{3(a_i - a_{i-1})}{h^2}$$

or

$$c_{i-1} + 4c_i + c_{i+1} = \frac{3}{h^2}(a_{i-1} - 2a_i + a_{i+1}), \qquad i = 1, \ldots, n-1 \quad (11)$$

Recall from (1') that $a_i = f(t_i)$. Once we solve (11) for the c_i's, then by (8) we can determine the d_i's and by (9) we can determine the b_i's. Thus, to determine the $4n$ coefficients in the n cubic polynomials of our cubic spline, we only need to solve the $(n-1)$-by-$(n-1)$ tridiagonal system (11) (recall that c_0 and c_n equal 0).

Note that it would have been possible to develop the foregoing equations without equally spaced subintervals; unequal spacing is natural for curves that are straight most of the time but change rapidly over few short intervals. If $h_i = t_{i+1} - t_i$, the tridiagonal system (11) can be shown to be

$$h_{i-1}c_i + 2(h_{i-1} + h_i)c_i + h_ic_{i+1} = \frac{3}{h_i}(a_{i+1} - a_i)$$

$$- \frac{3}{h_{i+1}}(a_i - a_{i-1})$$

(12)

Example 1. Cubic Spline Approximation of $x \cdot \sin(\pi x)$

Let us use a cubic spline to approximate the function $f(x) = x \cdot \sin(\pi x)$ over the interval $[1, 3]$. We use the values of $f(x)$ at nine mesh points: 1, 1.25, 1.5, 1.75, 2.0, 2.25, 2.5, 2.75, 3.0 ($h = .025$), yielding eight subintervals. We have the following table of values at these points.

Mesh Point t_i	Function Value $f(x) = x \cdot \sin(\pi x)$
$t_0 = 1.0$	$f(t_0) = \quad 0$
$t_1 = 1.25$	$f(t_1) = \quad -.884$
$t_2 = 1.5$	$f(t_2) = \quad -1.5$
$t_3 = 1.75$	$f(t_3) = \quad -1.237$
$t_4 = 2.0$	$f(t_4) = \quad 0$
$t_5 = 2.25$	$f(t_5) = \quad 1.591$
$t_6 = 2.5$	$f(t_6) = \quad 2.5$
$t_7 = 2.75$	$f(t_7) = \quad 1.945$
$t_8 = 3.0$	$f(t_8) = \quad 0$

(13)

By (1'), $a_i = f(t_i)$, so the second column of (13) gives the values of the a_i's. Next we write the system of seven equations in seven unknowns given by equation (11) (note that $3/h^2 = 3/.25^2 = 48$).

$$
\begin{aligned}
4c_1 + c_2 &= 48(0 + 2 \cdot .884 - 1.5) &= 12.864 \\
c_1 + 4c_2 + c_3 &= 48(-.884 + 2 \cdot 1.5 - 1.237) &= 42.192 \\
c_2 + 4c_3 + c_4 &= 48(-1.5 + 2 \cdot 1.237 + 0) &= 46.752 \\
c_3 + 4c_4 + c_5 &= 48(-1.237 - 2 \cdot 0 + 1.591) &= 16.992 \\
c_4 + 4c_5 + c_6 &= 48(0 - 2 \cdot 1.591 + 2.5) &= -32.736 \\
c_5 + 4c_6 + c_7 &= 48(1.591 - 2 \cdot 2.5 + 1.945) &= -70.272 \\
c_6 + 4c_7 &= 48(2.5 - 2 \cdot 1.945 + 0) &= -66,720
\end{aligned}
$$

(14)

Solving (14) by Gaussian elimination, we obtain

$$c_1 = 1.204, \quad c_2 = 8.048, \quad c_3 = 8.797, \quad c_4 = 3.520,$$
$$c_5 = -5.883, \quad c_6 = -12.722, \quad c_7 = -13.499$$

Using (8) to determine the d_i's and (9) to determine the b_i's, we obtain the cubic polynomials $s_i(x)$:

$$s_0(x) = 0 \qquad\quad - 3.636(x - 1) \quad + \quad 0(x - 1)^2 \qquad + \quad 1.605(x - 1)^3$$
$$s_1(x) = -.884 \quad - 3.335(x - 1.25) + \quad 1.204(x - 1.25)^2 + \quad 9.125(x - 1.25)^3$$
$$s_2(x) = -1.5 \quad\; - 1.022(x - 1.5) \; + \quad 8.048(x - 1.5)^2 \; + \quad .999(x - 1.5)^3$$
$$s_3(x) = -1.237 + 3.189(x - 1.75) + \quad 8.797(x - 1.75)^2 - \quad 7.036(x - 1.75)^3$$
$$s_4(x) = \quad 0 \quad\;\; + 6.268(x - 2.0) \; + \quad 3.520(x - 2.0)^2 \; - 12.537(x - 2.0)^3$$
$$s_5(x) = \quad 1.591 + 5.677(x - 2.25) - \quad 5.883(x - 2.25)^2 - \quad 9.119(x - 2.25)^3$$
$$s_6(x) = \quad 2.5 \quad\; + 1.025(x - 2.5) \; - 12.722(x - 2.5)^2 \; - \quad 1.036(x - 2.5)^3$$
$$s_7(x) = \quad 1.945 - 5.530(x - 2.75) - 13.499(x - 2.75)^2 + 17.999(x - 2.75)^3$$

Finally, we give a table comparing the values of the spline approximation $s(x)$ with the original function $f(x) = x \cdot \sin(\pi x)$.

x	$f(x)$	$s(x)$
1.0	0	0
1.1	$-.340$	$-.362$
1.2	$-.705$	$-.714$
1.3	-1.052	-1.046
1.4	-1.331	-1.326
1.5	-1.5	-1.5
1.6	-1.522	-1.521
1.7	-1.375	-1.375
1.8	-1.058	-1.056
2.0	0	0
2.1	.649	.649
2.2	1.293	1.294
2.3	1.861	1.859
2.4	2.282	2.279
2.5	2.5	2.5
2.6	2.473	2.474
2.7	2.184	2.188
2.8	1.646	1.637
2.9	.896	.873
3.0	0	0

(15)

A very good fit.   ∎

There are other ways to obtain a tridiagonal system for determining the coefficients in a cubic spline; see Cheney and Kincaid (mentioned in the References) for a method starting with the second derivative of $s(x)$ and integrating twice.

Section 4.7 **Appendix Exercises**

Summary of Exercises
These exercises require the construction of cubic splines to approximate various functions. The reader should mimic the steps in Example 1.

1. Repeat the calculations in Example 1 but now use the interval [3, 4] with mesh points 3, 3.25, 3.5, 3.75, 4. Compare the values of your approximation with the true values of $f(x)$ at $x = 3.1$ and 3.9.

2. (a) Use a cubic spline to approximate the function $f(x) = \sin(\pi x)$ over the interval [0, 1] with mesh points 0, .25, .5, .75, 1. Compare the values of your approximation with the true values of $f(x)$ at $x = .1$ and .65.
   (b) Use your cubic spline to approximate the integral of $\sin(\pi x)$ from 0 to 1. Note the exact answer is $2/\pi$.

3. Repeat Exercise 2 using mesh points 0, .2, .4, .6, .8, 1.

4. (a) Use a cubic spline to approximate the function $f(x) = e^{-x}$ over the interval [0, 4] with mesh points 0, 1, 2, 3, 4. Compare the values of your approximation with the true values of $f(x)$ at $x = .5$ and 2.5.
   (b) Use your cubic spline to approximate the integral of e^{-x} from 0 to 4. Integrate to compute the exact answer.

5. (a) Use a cubic spline to approximate the function $f(x) = \log_e x$ over the interval [.5, 1.5] with mesh points .5, .75, 1, 1.25, 1.5.
   (b) Use your cubic spline to approximate the integral of $\log_e x$ from .5 to 1.5. Integrate by parts to determine the exact answer.

Interlude: Abstract Linear Transformations and Vector Spaces

In this interlude, we take a step back from matrix algebra and give a more general setting to the linear problems addressed in this book. By "linear problems" we mean problems involving equations consisting of linear combinations of variables. These problems have come in two general forms: (i) solving a system of linear equations $\mathbf{Ax} = \mathbf{b}$, and (ii) describing the behavior of iterative models of the form $\mathbf{p}' = \mathbf{Ap}$.

In Section 4.1 we used the term *linear transformation* to refer to mappings $T : \mathbf{w} \rightarrow \mathbf{w}' = T(\mathbf{w})$, where $\mathbf{w}' = \mathbf{Aw}$. In that section we treated the system $\mathbf{w}' = \mathbf{Aw}$ as defining a computer graphics transformation $T(\mathbf{w})$ that might be applied to a stick-figure drawing or possibly the whole x-y plane (or x-y-z space). The key property of linear transformations is (Theorem 1 of Section 4.1)

$$T(a\mathbf{w} + b\mathbf{v}) = aT(\mathbf{w}) + bT(\mathbf{v}) \tag{1}$$

Property (1) led to the observation that linear transformations take lines into lines.

Property (1) turns out to be at the heart of all linear models. If a transformation $T(\mathbf{w})$ satisfies (1), where $\mathbf{w}$ is an n-vector, then $T(\mathbf{w})$ must actually be a matrix transformation: $T(\mathbf{w}) = \mathbf{Aw}$. That is, for vectors, property (1) defines a matrix-vector product.

We define an **abstract linear transformation** on n-vectors to be any mapping $\mathbf{w}' = T(\mathbf{w})$ of an n-vector $\mathbf{w}$ to an m-vector $\mathbf{w}'$ such that T satisfies (1).

Theorem 1. Any abstract linear transformation $\mathbf{w}' = T(\mathbf{w})$ can be represented by matrix multiplication: $\mathbf{w}' = \mathbf{Aw}$.

Proof: If $\mathbf{w}$ is an n-vector and $\mathbf{w}'$ is an m-vector, then $\mathbf{A}$ will have to be an m-by-n matrix. When transforming a set of points in Section 4.1, we used property (1) to simplify the calculations by expressing

387

all points as linear combinations of a few key points. In this proof we reason the same way. Here the key points will be unit vectors, such as $(1, 0, 0, \ldots, 0)$. Let e_j be the jth unit vector in n dimensions (with a 1 in the jth position and 0's in the other $n - 1$ positions).

We define the jth column a_j^C of A to be $T(e_j)$, the vector to which e_j is mapped by T:

$$a_j^C = T(e_j) \tag{2}$$

If $w = (w_1, w_2, \ldots, w_n)$, we can write w as

$$w = w_1 e_1 + w_2 e_2 + \cdots + w_n e_n \tag{3}$$

Then by (1) and (2)

$$T(w) = w_1 T(e_1) + w_2 T(e_2) + \cdots + w_n T(e_n) \tag{4a}$$
$$= w_1 a_1^C + w_2 a_2^C + \cdots + w_n a_n^C \tag{4b}$$
$$= Aw$$

The linear combination in (4b) of the columns of A is exactly the definition of Aw.   ∎

The preceding proof shows that any linear transformation is specified by knowing what it does with a set of coordinate vectors. In the proof we used the unit vectors e_j, but any set of vectors whose linear combinations yield all other vectors would work.

In Section 2.5 we saw that for square matrices, eigenvector coordinates made matrix multiplication very easy. If u_i is an eigenvector of T, then $T(u_i) = \lambda_i u_i$. In the proof above, if w has the representation in eigenvector coordinates of

$$w = r_1 u_1 + r_2 u_2 + \cdots + r_n u_n \tag{5}$$

then (4a) and (4b) become

$$T(w) = r_1 T(u_1) + r_2 T(u_2) + \cdots + r_n T(u_n) \tag{6}$$
$$= \lambda_1 r_1 u_1 + \lambda_2 r_2 u_2 + \cdots + \lambda_n r_n u_n$$

For example, the rabbit–fox growth model (Example 5 of Section 3.1)

$$p' = Ap, \qquad \text{where } A = \begin{bmatrix} 1.1 & -.15 \\ .1 & .85 \end{bmatrix} \tag{7}$$

has eigenvectors $u_1 = [3, 2]$ and $u_2 = [1, 1]$ with associated eigenvalues $\lambda_1 = 1$ and $\lambda_2 = .95$. So if the initial population $p = [R, F]$ were written in eigenvector coordinates as $p = [s_1, s_2]$ ($= s_1 u_1 + s_2 u_2$), then the linear transformation T given by A becomes

$$T(\mathbf{p}) = s_1 T(\mathbf{u}_1) + s_2 T(\mathbf{u}_2) \qquad \text{or}$$

$$\mathbf{T}\begin{bmatrix} s_1 \\ s_2 \end{bmatrix} = \begin{bmatrix} 1 & 0 \\ 0 & .95 \end{bmatrix}\begin{bmatrix} s_1 \\ s_2 \end{bmatrix} = \begin{bmatrix} s_1 \\ .95 s_2 \end{bmatrix} \tag{8}$$

$$= s_1(1\mathbf{u}_1) + s_2(.95\mathbf{u}_2)$$

The concept of a linear transformation helps to remind us that (7) and (8) are the same thing—that is, are the same linear transformation—but expressed in different coordinate systems.

The lesson from Section 2.5 is that, when possible, linear transformations should be expressed in eigenvector coordinates. To convert to eigenvector coordinates and afterwards to convert back to standard coordinates, we can use the matrix equation (Theorem 5 of Section 3.3)

$$\mathbf{A} = \mathbf{U}\mathbf{D}_\lambda\mathbf{U}^{-1} \tag{9}$$

The concept of a linear transformation also gives new understanding to the problem of solving a system of equations. Consider our refinery problem:

$$\mathbf{Ax} = \mathbf{b}: \quad \begin{array}{lrrrr} \text{Heating oil:} & 20x_1 + & 4x_2 + & 4x_3 = & 500 \\ \text{Diesel oil:} & 10x_1 + & 14x_2 + & 5x_3 = & 850 \\ \text{Gasoline:} & 5x_1 + & 5x_2 + & 12x_3 = & 1000 \end{array} \tag{10}$$

Viewing (10) as a linear transformation problem,

$$T(\mathbf{x}) = \mathbf{b} \tag{11}$$

we see that solving (11) for a vector $\mathbf{x}$ of production levels is asking for a vector $\mathbf{x}$ in the domain of T that is mapped by T to $\mathbf{b}$. *This is a vector-valued version of the problem: Given a function f(x) and a constant b, find an x for which f(x) = b.*

Just like the function version of this problem, if $\mathbf{b}$ is in the range of T, there will be at least one solution; if T is a one-to-one mapping, there will be at most one solution; otherwise, there may be many solutions.

Linear transformations provide a convenient way to abstract matrix problems. However, matrix problems are only the beginning. Linear transformations can be defined to act on more complicated sets than vectors, such as functions.

Functions can be thought of as an infinite dimensional extension of vectors. An **abstract vector space** is any collection $\mathbf{C}$ of elements that obey the law of linearity. That is, if A and B are elements of $\mathbf{C}$, then $rA + sB$ are in $\mathbf{C}$, for any constants r, s. The set of all continuous functions forms an abstract vector space. The same is true for the set of all functions, continuous or not.

Example 1. **Linear Transformations on Functions**

(i) *Shift transformation S_a.* Let $S_a(f(x)) = f(x + a)$. That is, S_5 shifts the values of a function $f(x)$ 5 units to the left:

$$\text{if } f(x) = x^2 - 2x, \qquad \text{then } S_5(f(x)) = (x + 5)^2 - 2(x + 5)$$

(ii) *Reflection transformation R.* Let $R(fx)) = f(-x)$. So R reflects the graph of any function about the y-axis.

(iii) *Differentiation transformation D.* Let $D(f(x)) = df(x)/dx$, the derivative of $f(x)$ (assuming that the derivative exists).

(iv) *Integration transformation I.* Let $I(f(x)) = \int f(x)\, dx$, the integral of $f(x)$ (integration actually requires a constant term; here we will assume that the constant is 0).   ■

It is left to the reader to check that the transformations in Example 1, parts (i) and (ii) are indeed linear transformations. The required property, generalizing (1), is

$$T(af(x) + bg(x)) = aT(f(x)) + bT(g(x)) \qquad \text{for any constants } a, b \qquad (12)$$

Because differentiation is so important, let us check that it is a linear transformation. Property (12) is

$$\frac{d}{dx}[af(x) + bg(x)] = a\frac{d}{dx}f(x) + b\frac{d}{dx}g(x) \qquad (13)$$

But (13) is the linearity rule of derivatives. Similarly, for integration we have

$$\int [af(x) + bg(x)]\, dx = a\int f(x)\, dx + b\int g(x)\, dx \qquad (14)$$

Virtually all of the theory for analyzing matrix equations extends to linear transformations of functions. For example, we can talk about inverse transformations and about eigenfunctions $u(x)$: $T(u(x)) = \lambda u(x)$.

Example 2. **Inverse Transformations of Linear Transformations**

(i) For the shift transformation $S_a(f(x)) = f(x + a)$, the inverse transformation S^{-1} is S_{-a} since $S_{-a}[S_a(f(x))] = f(x + a - a) = f(x)$.

(ii) For the reflection transformation $R(f(x)) = f(-x)$, the inverse R^{-1} is simply the reflection transformation itself. So $R^{-1} = R$.

(iii) For differentiation, there is no (unique) inverse D^{-1}. If two functions differ by a constant, say, $x^2 + x + 2$ and $x^2 + x + 5$, they have the same derivative, $2x + 1$. So $D^{-1}(2x + 1)$ cannot be uniquely defined.

(iv) For integration, the inverse I^{-1} is differentiation. That is,

$$\frac{d}{dx} \left[\int f(x) \, dx \right] = f(x)$$

This is the fundamental theorem of calculus!   ■

Example 3. **Eigenfunctions of Linear Transformations**

(i) For the shift transformation $S_a(f(x)) = f(x + a)$, an eigenfunction $u(x)$ must have the property that $u(x + a) = u(x)$. For example, when $a = 2\pi$, the trigonometric functions, such as $\sin x$ or $\cos x$, are eigenfunctions of $S_{2\pi}$, with eigenvalue 1.

(ii) For the reflection transformation $R(f(x)) = f(-x)$, an eigenfunction $u(x)$ associated with $\lambda = 1$ is any symmetric $u(x)$, that is, $u(x) = u(-x)$.

(iii) For differentiation, $e^{\lambda x}$ is the eigenfunction of D associated with eigenvalue $\lambda = k$, since $de^{kx}/dx = ke^{kx}$.

(iv) For integration e^x/λ is the eigenfunction of I associated with eigenvalue $\lambda = k$, since integration is the reverse operation of differentiation.   ■

There is one very important generalization of differentiation that bears special mention, namely differential equations. The differential equation

$$y''(x) - 2y'(x) = f(x) \tag{15}$$

can be considered a linear transformation DE of $y(x)$ to $f(x)$, that is, $DE(y(x)) = f(x)$. It is left to the reader to check that DE satisfies property (1). Any differential equation whose left side is a linear combination of derivatives will be a linear transformation. The advanced theory of differential equations is based heavily on eigenfunctions and inverse transformations.

For linear transformations defined in terms of matrices, we noted that the "right" coordinates for describing the transformation are eigenvector coordinates. The same applies to linear transformations of functions. Functions should be expressed as linear combinations (infinite series) of eigenfunctions of the linear transformation.

This book is not about linear transformations of functions. It is about matrices and vectors. But it is important to be aware of the powerful generalizations of matrices and matrix algebra which are the basis for much of higher mathematics. If the reader masters the matrix-based linear algebra in this book, he or she will have an excellent foundation for any future work with functional linear algebra.

The purpose of this interlude has been to implant the seed in the reader's mind that many operations on functions are linear transformations and that most of the theory of matrices extends to these linear transformations. In Chapter 5, occasional examples using linear transformations of functions are given.

5

Theory of Systems of Linear Equations and Eigenvalue/Eigenvector Problems

Null Space and Range of a Matrix

The **null space** Null(**A**) of a matrix **A** is the set of vectors **x** that are solutions to the system of equations $\mathbf{Ax} = \mathbf{0}$. The **range** Range(**A**) of **A** is the set of vectors **b** such that $\mathbf{Ax} = \mathbf{b}$ has a solution. Another name that is sometimes used for Null(**A**) is the *kernel* of **A**. In this section we examine these two important sets of vectors associated with any matrix. We look at their role in linear models and learn how to determine these sets.

Both the Null(**A**) and Range(**A**) are vector spaces.

Definition. A **vector space** is any set V of vectors such that if $\mathbf{x}_1$, $\mathbf{x}_2$ are in V, then any linear combination $r\mathbf{x}_1 + s\mathbf{x}_2$ is also in V.

If $\mathbf{Ax}_1 = 0$ and $\mathbf{Ax}_2 = 0$, then we have

$$\mathbf{A}(r\mathbf{x}_1 + s\mathbf{x}_2) = r(\mathbf{Ax}_1) + s(\mathbf{Ax}_2) = r(0) + s(0) = 0$$

Thus Null(**A**) is a vector space. A similarly simple proof, left as an exercise, shows that Range(**A**) is a vector space.

Suppose that **A** is an n-by-n matrix for which the system $\mathbf{Ax} = \mathbf{b}$ has a unique solution for every **b**. Then Range(**A**) is all possible n-vectors, and Null(**A**) is just the zero vector **0**, since **0** is always a solution to $\mathbf{Ax} = \mathbf{0}$ and by assumption there can only be one solution to $\mathbf{Ax} = \mathbf{0}$.

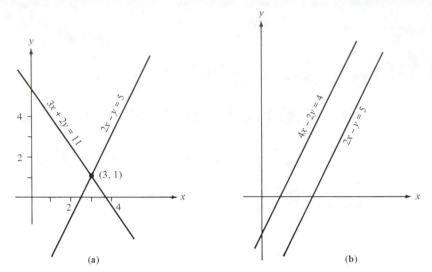

Figure 5.1 (a) Two lines intersect at a point. (b) Two parallel lines have no intersection points.

In this section we are concerned with matrices **A** for which **Ax** = **b** has nonunique solutions or no solutions. Then determining the range and null space of **A** becomes important.

Let us briefly describe the geometry of systems of equations that do not have unique solutions. Consider the pair of equations

$$\begin{aligned} 2x - y &= 5 \\ 3x + 2y &= 11 \end{aligned} \tag{1}$$

In Figure 5.1a we have plotted the graph of the two equations in (1) in $x - y$ space. The graph of a linear equation is a straight line. An (x, y) pair that solves (1) must lie on the lines of both equations. That is, the (x, y) pair must be the coordinates of the point where the two lines intersect. From Figure 5.1a we see that this intersection point has coordinates (3, 1).

Suppose the two equations produce lines that are parallel:

$$\begin{aligned} 2x - y &= 5 \\ 4x - 2y &= 4 \end{aligned} \tag{2}$$

(see Figure 5.1b). Then there is no common point—no solution to (2). Note that "parallel" means that the second equation's coefficients are multiples of the first's. We saw in the canoe-with-sail example in Section 1.1 that when two equations produce almost parallel lines, the equations can give strange results.

A system of three equations in three unknowns, such as

$$\begin{aligned} 2x + 5y + 4z &= 4 \\ x + 4y + 3z &= 1 \\ -x + 3y + 2z &= -5 \end{aligned} \tag{3}$$

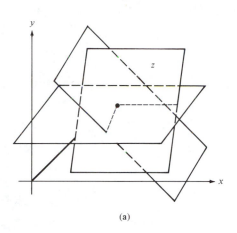

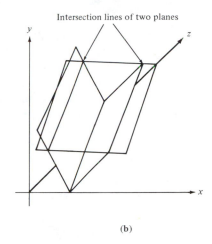

Intersection lines of two planes

(a)

(b)

Figure 5.2 (a) Three planes intersect at a point. (b) The lines formed by intersections of two planes are parallel. (c) Three planes intersect along a line.

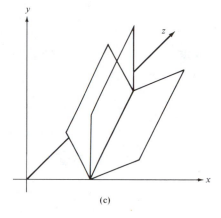

(c)

has a similar interpretation in three-dimensional space (see Figure 5.2a). Each equation now determines a plane of points. A solution of (3) will be the coordinates of a point where all three planes intersect. If two of the planes are parallel, no solution can exist.

Another possibility is that while each pair of equations intersects to form a line, there may be no point common to all three. This happens if the line formed by the intersection of two planes is parallel to the third plane. Figure 5.2b illustrates this situation; such a system of equations is

$$
\begin{aligned}
x - y + z &= 3 \\
x + y + z &= 3 \\
x + 3y + z &= 7
\end{aligned}
\tag{4}
$$

The points (x, y, z) satisfying both of the first two equations in (4) form the line $x + z = 3$, $y = 0$; the points satisfying the first and third equations form the line $x + z = 4$, $y = 1$; and the points satisfying the second and third equations form the line $x + z = 1$, $y = 2$.

Of course, when we have four or more equations in three unknowns, it is very likely that there will be no (x, y, z) that lies on all four planes (satisfies all four equations). Conversely, if we had only the first two equations in (4), all the points on the line $x + z = 3$, $y = 0$ mentioned above will be solutions.

Let us turn from systems with no solutions to 3-by-3 systems with multiple solutions. The points that lie on all three planes may happen to form a line, as in the case for the system

$$
\begin{aligned}
4x - y + 2z &= 8 \\
2x + 5y + z &= 4 \\
-2x + 3y - z &= -4
\end{aligned}
\tag{5}
$$

which has the graph shown in Figure 5.2c. The line of points (x, y, z) common to all three equations in (5) is $2x + z = 4$, $y = 0$. So (5) has an infinite number of solutions. As we go to higher dimensions, the possibilities for multiple solutions or no solution increase.

We present two theorems that show the fundamental link between multiple solutions to the system $\mathbf{Ax} = \mathbf{b}$ and the null space of $\mathbf{A}$.

Theorem 1. Let $\mathbf{A}$ be any m-by-n matrix.

  (i) If Null($\mathbf{A}$) contains one nonzero vector $\mathbf{x}^0$, then Null($\mathbf{A}$) contains an infinite number of vectors; in particular, any multiple $r\mathbf{x}^0$ is in Null($\mathbf{A}$).
 (ii) If $\mathbf{x}^0$ is in Null($\mathbf{A}$) and $\mathbf{x}^*$ is a solution to $\mathbf{Ax} = \mathbf{b}$, then $\mathbf{x}^* + \mathbf{x}^0$ is also a solution to $\mathbf{Ax} = \mathbf{b}$.
(iii) If $\mathbf{x}_1$, $\mathbf{x}_2$ are two different solutions to $\mathbf{Ax} = \mathbf{b}$, for some given $\mathbf{b}$, then their difference $\mathbf{x}_1 - \mathbf{x}_2$ is a vector in Null($\mathbf{A}$).
 (iv) Given a solution $\mathbf{x}^*$ to $\mathbf{Ax} = \mathbf{b}$, then any other solution $\mathbf{x}'$ to this matrix equation can be written as

$$
\mathbf{x}' = \mathbf{x}^* + \mathbf{x}^0 \qquad \text{for some } \mathbf{x}^0 \text{ in Null}(\mathbf{A}).
$$

  (v) If Null($\mathbf{A}$) consists of only the zero vector $\mathbf{0}$ (i.e., $\mathbf{Ax} = \mathbf{0}$ has only the solution $\mathbf{x} = \mathbf{0}$), then $\mathbf{Ax} = \mathbf{b}$ has at most one solution, for any given $\mathbf{b}$.

Proof

  (i) $\mathbf{A}(r\mathbf{x}^0) = r(\mathbf{Ax}^0) = r\mathbf{0} = \mathbf{0}$, so $r\mathbf{x}^0$ is in Null($\mathbf{A}$) for any scalar r. Thus Null($\mathbf{A}$) is infinite.
 (ii) Let $\mathbf{x}^0$ be in Null($\mathbf{A}$), so $\mathbf{Ax}^0 = \mathbf{0}$. Since $\mathbf{Ax}^* = \mathbf{b}$, then

$$
\begin{aligned}
\mathbf{A}(\mathbf{x}^* + \mathbf{x}^0) &= \mathbf{Ax}^* + \mathbf{A}(\mathbf{x}^0) \\
&= \mathbf{b} \quad + \mathbf{0} = \mathbf{b}
\end{aligned}
\tag{6}
$$

Thus $\mathbf{x}^* + \mathbf{x}^0$ is a solution to $\mathbf{Ax} = \mathbf{b}$, as claimed.
(iii) Since $\mathbf{A}(\mathbf{x}_1 - \mathbf{x}_2) = \mathbf{Ax}_1 - \mathbf{Ax}_2 = \mathbf{b} - \mathbf{b} = \mathbf{0}$, then $\mathbf{x}_1 - \mathbf{x}_2$ is in Null($\mathbf{A}$).

(iv) Given solutions $\mathbf{x}^*$ and $\mathbf{x}'$, choose $\mathbf{x}^0 = \mathbf{x}' - \mathbf{x}^*$. Then $\mathbf{x}' = \mathbf{x}^* + \mathbf{x}^0$, and by part (iii), $\mathbf{x}^0$ is in Null($\mathbf{A}$).

(v) Suppose that one solution $\mathbf{x}^*$ to $\mathbf{Ax} = \mathbf{b}$ is known. By part (iv), all solutions $\mathbf{x}'$ of $\mathbf{Ax} = \mathbf{b}$ can be expressed in the form $\mathbf{x}' = \mathbf{x}^* + \mathbf{x}^0$, where $\mathbf{x}^0$ is in Null($\mathbf{A}$). If Null($\mathbf{A}$) consists of just $\mathbf{0}$, then $\mathbf{x}' = \mathbf{x}^* + \mathbf{0}$—there is only solution, $\mathbf{x}^*$, to $\mathbf{Ax} = \mathbf{b}$. ∎

Theorem 2. Let $\mathbf{A}$ be an m-by-n matrix. If $\mathbf{Ax} = \mathbf{b}'$ has two solutions for some particular $\mathbf{b}'$:

(i) The null space of $\mathbf{A}$ has an infinite number of vectors.

(ii) For any $\mathbf{b}$, either $\mathbf{Ax} = \mathbf{b}$ has no solution or an infinite number of solutions.

Proof

(i) By Theorem 1, part (iii), the difference of two solutions is a (nonzero) vector $\mathbf{x}^0$ in Null($\mathbf{A}$), and then by Theorem 1, part (i), the multiples $r\mathbf{x}^0$ yield an infinite number of vectors in Null($\mathbf{A}$).

(ii) Suppose that $\mathbf{Ax} = \mathbf{b}$ has one solution $\mathbf{x}^*$. From Theorem 1, part (ii), $\mathbf{x}^* + \mathbf{x}^0$ is also a solution, for any $\mathbf{x}^0$ in Null($\mathbf{A}$). By part (i) of this theorem, Null($\mathbf{A}$) is infinite. ∎

Theorem 2's result that two solutions lead to an infinite number of solutions corresponds to our geometric pictures in which multiple solutions always consisted of an (infinite) set of points along a line.

A system of equations with $\mathbf{0}$ on the right side—$\mathbf{Ax} = \mathbf{0}$—is called a **homogeneous system.** Solutions to the homogeneous system $\mathbf{Ax} = \mathbf{0}$ form the null space of $\mathbf{A}$. One often speaks of the *null space of the system* $\mathbf{Ax} = \mathbf{0}$, implicitly meaning the null space of $\mathbf{A}$.

Homogeneous systems $\mathbf{Ax} = \mathbf{0}$ have arisen in several different settings in this book.

Example 1. **Multiple Solutions in an Oil Refinery Problem**

Let us suppose that in our familiar oil refinery problem, the three refineries produce only the first two products:

$$\text{Heating oil:} \quad 20x_1 + 4x_2 + 4x_3 = 500 \tag{7}$$
$$\text{Diesel oil:} \quad 10x_1 + 14x_2 + 5x_3 = 850$$

Suppose that we are given one solution, $x_1 = 15$, $x_2 = 50$, $x_3 = 0$, using just the first two refineries. We want to find another solution with $x_3 = 20$. Let us find the null space of this coefficient matrix and then, using Theorem 1, part (iv), add an appropriate null space vector to the given solution [15, 50, 0] to get a solution with $x_3 = 20$.

The null space for this system of equations is all solutions to the associated homogeneous system

$$20x_1 + 4x_2 + 4x_3 = 0 \tag{8}$$
$$10x_1 + 14x_2 + 5x_3 = 0$$

Solving with elimination by pivoting, we obtain

$$x_1 \quad + \tfrac{3}{20}x_3 = 0$$
$$x_2 + \tfrac{1}{4}x_3 = 0$$

Thus $x_1 = -\tfrac{3}{20}x_3$ and $x_2 = -\tfrac{1}{4}x_3$. So vectors in the null space have the form

$$[-\tfrac{3}{20}x_3, -\tfrac{1}{4}x_3, x_3] \quad \text{or} \quad x_3[-\tfrac{3}{20}, -\tfrac{1}{4}, 1] \tag{9}$$

Theorem 1, part (iv), says that any solution $\mathbf{x}'$ to the system (7) can be expressed in the form $\mathbf{x}' = \mathbf{x} + \mathbf{x}^0$, where $\mathbf{x}^0$ is a null-space vector, in this case $\mathbf{x}^0 = r[-\tfrac{3}{20}, -\tfrac{1}{4}, 1]$ for some constant r. We said above that we want $\mathbf{x}'$ to have $x_3' = 20$. Then

$$\mathbf{x}' = \mathbf{x} + \mathbf{x}^0 \tag{10}$$
$$[x_1', x_2', 20] = [15, 50, 0] + r[-\tfrac{3}{20}, -\tfrac{1}{4}, 1]$$

Matching the third entry on each side of (10), we have $20 = 0 + r$. So $r = 20$ and the desired solution is

$$\mathbf{x}' = \mathbf{x} + \mathbf{x}^0 = [15, 50, 0] + 20[-\tfrac{3}{20}, -\tfrac{1}{4}, 1]$$
$$= [12, 45, 20] \tag{11} \blacksquare$$

Example 2. Balancing Chemical Equations Revisited

In Example 1 of Section 4.3 we obtained a system of equations for balancing the atomic equations for the chemical reaction in which permanganate (MnO_4) and hydrogen (H) ions combine to form manganese (Mn) and water (H_2O):

$$MnO_4 + H \rightarrow Mn + H_2O \tag{12}$$

where H represents hydrogen and O oxygen. We let x_1 be the number of permanganate ions, x_2 the number of hydrogen ions, x_3 the number of manganese atoms, and x_4 the number of water molecules. To have the same number of atoms in the molecules on each side of the reaction, we obtained the system of equations

$$\text{H:} \qquad x_2 = 2x_4$$
$$\text{Mn:} \qquad x_1 = x_3 \tag{13a}$$
$$\text{0:} \qquad 4x_1 = x_4$$

or

$$x_2 \quad\quad -2x_4 = 0$$
$$x_1 \quad -x_3 \quad\quad = 0 \tag{13b}$$
$$4x_1 \quad\quad\quad -x_4 = 0$$

Notice that we have four unknowns but only three equations. When we solved system (13b) using elimination by pivoting [pivoting on entries (2, 1), (1, 2), (3, 3)], we obtained

$$x_2 \quad -2x_4 = 0 \quad\quad\quad x_2 = 2x_4$$
$$x_1 \quad -\tfrac{1}{4}x_4 = 0 \quad \text{or} \quad x_1 = \tfrac{1}{4}x_4 \tag{14}$$
$$x_3 - \tfrac{1}{4}x_4 = 0 \quad\quad\quad x_3 = \tfrac{1}{4}x_4$$

As vectors, the solutions in (14) have the form

$$[\tfrac{1}{4}x_4, 2x_4, \tfrac{1}{4}x_4, x_4] \quad\quad \text{or} \quad\quad x_4[\tfrac{1}{4}, 2, \tfrac{1}{4}, 1] \tag{15}$$

These vectors form the null space for the system (13b).

For example, if $x_4 = 4$, then $x_2 = 8$, $x_1 = x_3 = 1$, and the reaction equation becomes

$$MnO_4 + 8H \rightarrow Mn + 4H_2O$$

The solution we obtain makes the amounts of the first three types of molecules fixed ratios of the amount of the fourth type, which we are free to give any value (i.e., x_4 is a free variable). In another series of pivots, the final free variable might end up being x_1.

$$-8x_1 + x_2 \quad\quad\quad = 0 \quad\quad\quad x_2 = 8x_1$$
$$-x_1 \quad\quad + x_3 \quad\quad = 0 \quad \text{or} \quad x_3 = x_1 \tag{16}$$
$$-4x_1 \quad\quad\quad + x_4 = 0 \quad\quad\quad x_4 = 4x_1$$

yielding solution vectors

$$[x_1, 8x_1, x_1, 4x_1] \quad\quad \text{or} \quad\quad x_1[1, 8, 1, 4] \tag{17}$$

However we formulate the solution of (13b), we always have the same set of vectors, namely, the null space of the coefficient matrix in (13b). For example, the vector [1, 8, 1, 4] in (17) is a multiple of the vector $[\tfrac{1}{4}, 2, \tfrac{1}{4}, 1]$ in (15). ∎

Examples 1 and 2 show us how to determine the null space of any matrix. We simply apply elimination by pivoting to the homogeneous system $\mathbf{Ax} = \mathbf{0}$ and reduce it to the form (14), from which we obtain the null-space vectors as expressed in (15).

Example 3. **The Null Space of the
Frog Markov Chain**

We have already performed elimination by pivoting on this transition
matrix **A** in Example 6 of Section 3.3 when we were trying to invert
the matrix. We do the pivoting again, but now the right-side vector
is **0**.

$$
\begin{bmatrix}
.50 & .25 & 0 & 0 & 0 & 0 \\
.50 & .50 & .25 & 0 & 0 & 0 \\
0 & .25 & .50 & .25 & 0 & 0 \\
0 & 0 & .25 & .50 & .25 & 0 \\
0 & 0 & 0 & .25 & .50 & .25 \\
0 & 0 & 0 & 0 & .25 & .50
\end{bmatrix}
\begin{bmatrix}
p_1 \\ p_2 \\ p_3 \\ p_4 \\ p_5 \\ p_6
\end{bmatrix}
=
\begin{bmatrix}
0 \\ 0 \\ 0 \\ 0 \\ 0 \\ 0
\end{bmatrix}
$$

Pivoting on entries (1, 1), (2, 2), (3, 3), (4, 4), and (5, 5), we obtain

$$
\begin{bmatrix}
1 & 0 & 0 & 0 & 0 & 1 \\
0 & 1 & 0 & 0 & 0 & -2 \\
0 & 0 & 1 & 0 & 0 & 2 \\
0 & 0 & 0 & 1 & 0 & -2 \\
0 & 0 & 0 & 0 & 1 & 2 \\
0 & 0 & 0 & 0 & 0 & 0
\end{bmatrix}
\begin{bmatrix}
p_1 \\ p_2 \\ p_3 \\ p_4 \\ p_5 \\ p_6
\end{bmatrix}
=
\begin{bmatrix}
0 \\ 0 \\ 0 \\ 0 \\ 0 \\ 0
\end{bmatrix}
\tag{18}
$$

$$
\text{or} \quad
\begin{aligned}
p_1 + p_6 &= 0 \\
p_2 - 2p_6 &= 0 \\
p_3 + 2p_6 &= 0 \\
p_4 - 2p_6 &= 0 \\
p_5 + 2p_6 &= 0
\end{aligned}
$$

Equations (18) express the first five p_i's in terms of p_6. Rewriting (18),
we get

$$
p_1 = -p_6, \quad p_2 = 2p_6, \quad p_3 = -2p_6, \quad p_4 = 2p_6, \quad p_5 = -2p_6
$$

and thus the solutions to **Ap** = **0** have the form

$$
[-p_6, 2p_6, -2p_6, 2p_6, -2p_6, p_6] \quad \text{or} \quad p_6[-1, 2, -2, 2, -2, 1]
\tag{19}
$$

The vectors in (19) are the null space of **A**.

We can add a null-space vector $\mathbf{p}^0$ like (19) to a probability
vector **p**, and if **Ap** = **p**′, then also **A(p** + **p**^{0}) = **p**′ [this fact is
Theorem 1, part (ii)]. For example, we found earlier that **p*** =

$[.1, .2, .2, .2, .2, .1]$ is the stable distribution for the frog Markov chain, so that $\mathbf{Ap}^* = \mathbf{p}^*$. So for any null-space vector $\mathbf{p}^0$, we have $\mathbf{A}(\mathbf{p}^* + \mathbf{p}^0) = \mathbf{Ap}^* = \mathbf{p}^*$. Suppose that we select for the null-space vector $\mathbf{p}^0 = [-.1, .2, -.2, .2, -.2, .1]$ [with $p_6 = .1$ in (19)]. Then

$$
\begin{aligned}
\mathbf{p}^* + \mathbf{p}_0 &= [.1, .2, .2, .2, .2, .1] + [-.1, .2, -.2, .2, -.2, .1] \\
&= [0, .4, 0, .4, .2]
\end{aligned}
$$

Thus if we start with distribution $[0, .4, 0, .4, 0, .2]$, we will reach the stable distribution $\mathbf{p}^*$ after just one period. (The reader should verify this result numerically.)   ■

Example 4. A Two-Variable Null Space

The following system of equations formed constraints in the transportation problem presented in Section 4.6; the names of the variables have been changed for simplicity. (*Background:* The first equation represents the fact that the amount x_1 of food shipped from the first warehouse to college A plus the amount x_2 shipped from the first warehouse to college B equals 20, the amount of food in the first warehouse. The other equations represent the amounts available at the second and third warehouses and the amounts needed at colleges A and B. There are many solutions—ways to ship the food between the three warehouses and the two colleges. In Section 4.6 each x_i had an associated cost and we sought to minimize the total cost.)

$$
\begin{array}{rcl}
x_1 + x_2 & & = 20 \\
x_3 + x_4 & & = 30 \\
x_5 + x_6 & = 15 \\
x_1 \quad + x_3 \quad + x_5 & & = 25 \\
x_2 \quad + x_4 \quad + x_6 & = 40
\end{array}
\tag{20}
$$

To change from one solution $\mathbf{x}^*$ of (20) to another (cheaper) solution $\mathbf{x}^{**}$, we would add some null-space vector to $\mathbf{x}^*$ according to Theorem 1. To find the null space, we solve the associated homogeneous system

$$
\begin{array}{rcl}
x_1 + x_2 & & = 0 \\
x_3 + x_4 & & = 0 \\
x_5 + x_6 & = 0 \\
x_1 \quad + x_3 \quad + x_5 & & = 0 \\
x_2 \quad + x_4 \quad + x_6 & = 0
\end{array}
\tag{21}
$$

Pivoting on entries (1, 2), (2, 4), (3, 6), (4, 3), we obtain

$$\begin{aligned}
x_1 + x_2 && &= 0 \\
-x_1 && + x_4 - x_5 &= 0 \\
&& x_5 + x_6 &= 0 \qquad \text{or} \\
x_1 && + x_3 \quad + x_5 &= 0 \\
&& 0 &= 0
\end{aligned}
\qquad
\begin{aligned}
x_2 &= -x_1 \\
x_4 &= \quad x_1 + x_5 \\
x_6 &= \qquad - x_5 \\
x_3 &= -x_1 - x_5
\end{aligned}$$

$$(22)$$

The solutions in (22) to the homogeneous system (21) produce the set of null-space vectors

$$[x_1, \ -x_1, \ -x_1 - x_5, \ x_1 + x_5, \ x_5, \ -x_5] \qquad (23)$$

Breaking the x_1 and x_5 components apart, we can rewrite (23) as

$$x_1[1, \ -1, \ -1, \ 1, \ 0, \ 0] + x_5[0, \ 0, \ -1, \ 1, \ 1, \ -1] \qquad (24)$$

So the null space is all linear combinations of the two vectors

$$\mathbf{x}_1^* = [1, \ -1, \ -1, \ 1, \ 0, \ 0] \quad \text{and} \quad \mathbf{x}_5^* = [0, \ 0, \ -1, \ 1, \ 1, \ -1].$$

Suppose that we are given the solution to (20) $\mathbf{x}$: $x_1 = 20$, $x_3 = 5$, $x_4 = 25$, $x_6 = 15$, and $x_2 = x_5 = 0$. Thus

$$\mathbf{x} = [20, 0, 5, 25, 0, 15]$$

Also suppose that we want a solution in which $x_2 = 10$ and $x_5 = 5$. We can achieve this by adding the right linear combination of null-space vectors $\mathbf{x}_1^* = [1, \ -1, \ -1, \ 1, \ 0, \ 0]$ and $\mathbf{x}_5^* = [0, 0, -1, 1, 1, -1]$. Remember that adding any null-space vector to a solution vector yields another solution vector.

To make $x_2 = 10$ (it is now 0), we can add to $\mathbf{x}$ the vector $-10\mathbf{x}_1^* = [-10, 10, 10, -10, 0, 0]$. To make $x_5 = 5$ (also now 0), we can add to $\mathbf{x}$ the vector $5\mathbf{x}_5^* = [0, 0, -5, 5, 5, -5]$ to $\mathbf{x}$. So our desired solution is

$$\begin{aligned}
\mathbf{x} - 10\mathbf{x}_1^* + 5\mathbf{x}_5^* &= [20, 0, 5, 25, 0, 15] \\
&\quad + [-10, 10, 10, -10, 0, 0] \\
&\quad + [0, 0, -5, 5, 5, -5] \\
&= [10, 10, 10, 20, 5, 10] \qquad \blacksquare
\end{aligned}$$

The null space in Example 4 is a little more complicated. If we had performed a different pivot sequence in (22), the two vectors that generated the null space would be different from the two vectors in (24). It is not even obvious that we would end up with two vectors. Maybe this same null space could be expressed as combinations of three vectors. Could it be expressed as multiples of a single vector? Probably not.

In Section 5.2 we use the theory of vector spaces to prove that any pivot sequence yields the same number of vectors for building the null space.

Optional Example on the Null Space of the Differentiation Transformation

In the Interlude before this chapter, we noted that the operation $D(f(x))$ of taking the derivative of a function $f(x)$ was a linear transformation of functions. D acts on the abstract vector space of all differentiable functions.

Let us determine the null space of the differentiation transformation and see how the results in Theorem 1 apply to D. $\text{Null}(D)$ is the set of functions $f(x)$ for which $D(f(x)) = 0$, that is, $f'(x) = 0$. These are just the constant functions, $f(x) = c$ for some constant c.

$$\text{Null}(D) = \{\text{all constant functions}\}$$

Suppose that we want to solve the problem

$$D(f(x)) = 9x^2 + x \qquad [\text{find } f(x) \text{ such that } f'(x) = 9x^2 + x] \tag{25}$$

By Theorem 1, part (ii), if $f^*(x)$ is a solution to (25), then $f^*(x)$ plus any constant function [i.e., plus any member of $\text{Null}(D)$] is also a solution, say, $f^*(x) = 3x^3 + x^2/2$. Then $3x^3 + x^2/2 + c$, for any constant c, is also a solution.

Conversely, by Theorem 1, part (iv), every solution to (25) can be written as the sum of a specific solution $f^*(x)$ to (25) plus some constant function [a member of $\text{Null}(D)$]. In this case, every solution has the form $3x^3 + x^2/2 + c$. This result is the basic formula about the form of the integral (the antiderivative) of a function. ∎

We close this discussion of null spaces by noting a close relationship between null spaces and eigenvectors. An eigenvector $\mathbf{u}$, with associated eigenvalue λ, satisfies the n-by-n homogeneous system

$$(\mathbf{A} - \lambda\mathbf{I})\mathbf{u} = \mathbf{0} \qquad (\text{or more familiarly, } \mathbf{A}\mathbf{u} = \lambda\mathbf{u})$$

Thus an eigenvector associated with λ is a member of the null space of $(\mathbf{A} - \lambda\mathbf{I})$. For example, if $\mathbf{A}$ is the transition matrix of the frog Markov chain, then the stable distribution $\mathbf{p}^*$ satisfies $\mathbf{A}\mathbf{p}^* = \mathbf{p}^*$ or $(\mathbf{A} - \mathbf{I})\mathbf{p}^* = \mathbf{0}$, so $\mathbf{p}^*$ is in the null space of $\mathbf{A} - \mathbf{I}$.

Conversely, $\mathbf{A}\mathbf{u} = \mathbf{0}$ is equivalent to $(\mathbf{A} - 0\mathbf{I})\mathbf{u} = \mathbf{0}$. So any nonzero vector $\mathbf{u}$ in the null space of $\mathbf{A}$ is an eigenvector of $\mathbf{A}$ associated with eigenvalue $\lambda = 0$.

So far we have considered systems with multiple solutions. Next we discuss systems with no solution—where $\mathbf{A}\mathbf{x} = \mathbf{b}$ cannot be solved for some vectors $\mathbf{b}$. Recall that the set of $\mathbf{b}$'s for which $\mathbf{A}\mathbf{x} = \mathbf{b}$ can be solved is called the range of $\mathbf{A}$. The following simple growth model presents a system with no solution.

Example 5. **An Inconsistent System of Equations**

Let c_1 be the value of currency on isle 1 and c_2 the currency on isle 2. Suppose that the growth equations for currency in the next period are

$$c_1' = .4c_1 + .5c_2 \tag{26}$$
$$c_2' = .6c_1 + .5c_2$$

Consider the problem of picking c_1 and c_2 so that in the next period, isle 1 has \$100 more and isle 2 has \$200 less: $c_1' = c_1 + 100$ and $c_2' = c_2 - 200$. Then c_1, c_2 must satisfy the equations

$$c_1 + 100 = c_1' = .4c_1 + .5c_2$$
$$c_2 - 200 = c_2' = .6c_1 + .5c_2$$

or

$$.6c_1 - .5c_2 = -100 \tag{27}$$
$$-.6c_1 + .5c_2 = 200$$

When we pivot on entry $(1, 1)$ in (27), we obtain

$$c_1 - \tfrac{5}{6}c_2 = \tfrac{100}{6} \tag{28}$$
$$0 = -100$$

The second equation in (28) is an impossibility.   ∎

A system of equations is called **inconsistent** if, when reduced, it yields an impossible equation. If a system of equations is inconsistent, no solution is possible. Conversely, if no solution is possible, elimination must reveal an inconsistency—otherwise, elimination will produce a solution.

An extreme case of inconsistency arises in regression.

Example 6. **The Regression Model $\hat{y} = qx + r$ as System of Equations to Be Solved**

Suppose that we have the set of (x, y) points $(1, 3)$, $(2, 5)$, $(3, 4)$, $(3, 6)$, $(4, 7)$, and $(4, 6)$. The x-value might represent the number of years of college and the y-value the score on some graduate admissions test. We want to fit a line of the form $y = qx + r$ to these data. That is, we want the best possible estimates $\hat{y}_i$ for each y_i when $\hat{y}$ is the linear function $qx + r$. Trying to draw a line that actually passes through these six points is equivalent to trying to solve the system of equations

$$
\begin{aligned}
3 &= 1q + r \\
5 &= 2q + r \\
4 &= 3q + r \quad \text{or} \quad \mathbf{y} = \mathbf{Xq}, \quad \text{with } \mathbf{X} = \begin{bmatrix} 1 & 1 \\ 2 & 1 \\ 3 & 1 \\ 3 & 1 \\ 4 & 1 \\ 4 & 1 \end{bmatrix}, \quad \mathbf{y} = \begin{bmatrix} 3 \\ 5 \\ 4 \\ 6 \\ 7 \\ 6 \end{bmatrix} \\
6 &= 3q + r \\
7 &= 4q + r \quad\quad \mathbf{q} = [q, r] \\
6 &= 4q + r
\end{aligned}
\tag{29}
$$

Clearly, no solution is possible here in the regular sense—system (29) is inconsistent. However, an approximate solution to $\mathbf{y} = \mathbf{Xq}$ is provided by the least-squares fit of regression theory. ∎

Let us turn our attention to the task of determining the range of a matrix, that is, to finding those **b**'s for which the system $\mathbf{Ax} = \mathbf{b}$ has a solution.

Example 7. **Range of a Projection Transformation**

In Section 4.1 we observed that the linear transformation

$$
\begin{aligned}
x' &= .5x + .5y \\
y' &= .5x + .5y
\end{aligned}
\tag{30}
$$

projects all (x, y) points onto the line $x' = y'$: It maps any $[x, y]$ to the point $[(x + y)/2, (x + y)/2]$ (see Figure 4.3).

Let **A** be the coefficient matrix in (30). The range of **A** is the set of vectors $[x', y']$ such that $x' = y'$. Let us derive this defining equation for the range directly from **A**. Letting $\mathbf{w} = [x, y]$ and $\mathbf{w}' = [x', y']$, we rewrite the transformation $\mathbf{Aw} = \mathbf{w}'$ in (30) as $\mathbf{Aw} = \mathbf{Iw}'$:

$$
\begin{aligned}
.5x + .5y &= 1x' + 0y' \\
.5x + .5y &= 0x' + 1y'
\end{aligned}
\tag{31}
$$

Now we try to perform elimination by pivoting to convert $\mathbf{Aw} = \mathbf{Iw}'$ into the form $\mathbf{Iw} = \mathbf{A}^{-1}\mathbf{w}'$, as we do when computing the inverse of **A**. We subtract the first equation from the second (to eliminate x from the second equation) and divide the first equation by .5 to obtain

$$
\begin{aligned}
x + y &= 2x' \\
0 &= -x' + y'
\end{aligned}
\tag{32}
$$

The second equation in (32), which can be rewritten $x' = y'$ gives a condition that must be true for any $\mathbf{w}' = [x', y']$ satisfying $\mathbf{A}\mathbf{w} = \mathbf{I}\mathbf{w}'$. We have verified that vectors $[x', y']$ with $x' = y'$ form the range of $\mathbf{A}$. ∎

Example 8. The Range of a Refinery Production Problem

Consider the following variation on our refinery production problem. Now there are just two refineries and their collective output vector $[b_1, b_2, b_3]$ is given by

$$
\begin{array}{lrcl}
\text{Heating oil:} & 20x_1 + & 4x_2 & = b_1 \\
\text{Diesel oil:} & 10x_1 + & 14x_2 & = b_2 \\
\text{Gasoline:} & 5x_1 + & 5x_2 & = b_3
\end{array}
\tag{33}
$$

We seek an equation describing possible output vectors. That is, if $\mathbf{A}$ is the coefficient matrix of (33), we seek a defining constraint on the range vectors of $\mathbf{A}$.

We use the technique introduced in Example 7. We write the system $\mathbf{A}\mathbf{x} = \mathbf{b}$ in (33) as $\mathbf{A}\mathbf{x} = \mathbf{I}\mathbf{b}$ and perform elimination by pivoting at entry $(1, 1)$ and then entry $(2, 2)$ in the augmented matrix $[\mathbf{A} \quad \mathbf{I}]$:

$$
\begin{bmatrix}
20 & 4 & | & 1 & 0 & 0 \\
10 & 14 & 0 & 1 & 0 \\
5 & 5 & | & 0 & 0 & 1
\end{bmatrix}
\rightarrow
\begin{bmatrix}
1 & .2 & | & \frac{1}{20} & 0 & 0 \\
0 & 12 & -\frac{1}{2} & 1 & 0 \\
0 & 4 & | & -\frac{1}{4} & 0 & 1
\end{bmatrix}
$$
$$
\rightarrow
\begin{bmatrix}
1 & 0 & | & \frac{7}{120} & -\frac{1}{60} & 0 \\
0 & 1 & -\frac{1}{24} & \frac{1}{12} & 0 \\
0 & 0 & | & -\frac{1}{12} & -\frac{1}{3} & 1
\end{bmatrix}
\tag{34}
$$

The reduced augmented matrix in (34) corresponds to the system of equations

$$
\begin{array}{rcl}
x_1 & = & \frac{7}{120}b_1 + \frac{1}{60}b_2 \\
x_2 & = & -\frac{1}{24}b_1 + \frac{1}{12}b_2 \\
0 & = & -\frac{1}{12}b_1 - \frac{1}{3}b_2 + b_3
\end{array}
\tag{35}
$$

The last equation in (35) can be rewritten as

$$
b_3 = \tfrac{1}{12}b_1 + \tfrac{1}{4}b_3 = \tfrac{1}{12}(b_1 + 3b_2)
\tag{36}
$$

This is the range constraint we were looking for. In terms of refinery production, it means that we can achieve any production vector $\mathbf{b}$ in which the gasoline production (b_3) equals $\frac{1}{12}$ the sum of the heating oil production (b_1) plus three times the diesel oil production (b_2).

Suppose that heating oil (b_1) and diesel oil (b_2) are the outputs of primary interest and we want $b_1 = 300$ and $b_2 = 300$. Then we pick a b_3 using (36) to get a vector in the range. We set

$$b_3 = \tfrac{1}{12}(b_1 + 3b_2) = \tfrac{1}{12}(300 + 3 \cdot 300) = 100$$

Now we can determine the appropriate production levels x_1, x_2 from the first two equation in (35). With $\mathbf{b} = [300, 300, 100]$, then

$$x_1 = \tfrac{7}{120}b_1 - \tfrac{1}{60}b_2 = \tfrac{7}{120} \cdot 300 - \tfrac{1}{60} \cdot 300$$
$$= 17.5 - 5 = 12.5$$
$$x_2 = -\tfrac{1}{24}b_1 + \tfrac{1}{12}b_2 = -\tfrac{1}{24} \cdot 300 + \tfrac{1}{12} \cdot 300$$
$$= -12.5 + 25 = 12.5$$

(Note that some vectors in the range may be infeasible because they would make x_1 or x_2 negative.) ■

Let us try out this method for finding the range of the transition matrix **A** in our frog Markov chain.

Example 9. Range of Frog Markov Chain

As in Example 8 we try to convert $\mathbf{Ap} = \mathbf{Ip'}$ into $\mathbf{Ip} = \mathbf{A}^{-1}\mathbf{p'}$ using elimination by pivoting. We already attempted this inversion in Example 6 of Section 3.3. We started with the augmented matrix of $\mathbf{Ap} = \mathbf{Ip'}$.

$$\begin{bmatrix} .50 & .25 & 0 & 0 & 0 & 0 \\ .50 & .50 & .25 & 0 & 0 & 0 \\ 0 & .25 & .50 & .25 & 0 & 0 \\ 0 & 0 & .25 & .50 & .25 & 0 \\ 0 & 0 & 0 & .25 & .50 & .50 \\ 0 & 0 & 0 & 0 & .25 & .50 \end{bmatrix} \mathbf{p} = \begin{bmatrix} 1 & 0 & 0 & 0 & 0 & 0 \\ 0 & 1 & 0 & 0 & 0 & 0 \\ 0 & 0 & 1 & 0 & 0 & 0 \\ 0 & 0 & 0 & 1 & 0 & 0 \\ 0 & 0 & 0 & 0 & 1 & 0 \\ 0 & 0 & 0 & 0 & 0 & 1 \end{bmatrix} \mathbf{p'}$$

and ended after pivoting on entries (1, 1), (2, 2), (3, 3), (4, 4), and (5, 5) with

$$\begin{bmatrix} 1 & 0 & 0 & 0 & 0 & 1 \\ 0 & 1 & 0 & 0 & 0 & -2 \\ 0 & 0 & 1 & 0 & 0 & 2 \\ 0 & 0 & 0 & 1 & 0 & -2 \\ 0 & 0 & 0 & 0 & 1 & 2 \\ 0 & 0 & 0 & 0 & 0 & 0 \end{bmatrix} \mathbf{p} \tag{37}$$

$$= \begin{bmatrix} 10 & -8 & 6 & -4 & 2 & 0 \\ -16 & 16 & -12 & 8 & -4 & 0 \\ 12 & -12 & 12 & -8 & 4 & 0 \\ -8 & 8 & -8 & 8 & -4 & 0 \\ 4 & -4 & 4 & -4 & 4 & 0 \\ -1 & 1 & -1 & 1 & -1 & 1 \end{bmatrix} \mathbf{p}'$$

Again the last equation in (37) gives the constraint required for a vector $\mathbf{p}'$ to be in the range of $\mathbf{A}$, namely,

$$0 = -p_1' + p_2' - p_3' + p_4' - p_5' + p_6'$$

or

$$p_2' + p_4' + p_6' = p_1' + p_3' + p_5' \tag{38}$$

This is a nice simple formula that allows us to test whether a probability vector $\mathbf{p}'$ can be a next-state distribution.

It is left as an exercise for the reader to explain in terms of the frog Markov chain model why even-state probabilities must equal odd-state probabilities.

Before leaving this example, note that we can apply this technique for determining the range not only to $\mathbf{A}$ but to powers of $\mathbf{A}$. The range of $\mathbf{A}^k$ tells us possible distributions for the Markov chain after k periods. What happens as k goes to infinity? The range should contract to multiples of the stable distribution [.1, .2, .2, .2, .2, .1]. This behavior is explored in Exercise 25. ∎

To summarize what we now know about the range of an m-by-n matrix $\mathbf{A}$, elimination by pivoting when applied to $\mathbf{Ax} = \mathbf{Ib}$ results in one of three possibilities [cases 2 and 3 may both apply]:

1. *The reduced form of* $\mathbf{A}$ *is* $\mathbf{I}$. Then for each $\mathbf{b}$, $\mathbf{Ax} = \mathbf{b}$ has a unique solution, and $\text{Null}(\mathbf{A}) = \mathbf{0}$.
2. *The reduced form of* $\mathbf{A}$ *contains one or more rows of zeros.* Then the right sides of these zero rows yield defining constraints that a vector $\mathbf{b}$ in the range of $\mathbf{A}$ must satisfy.
3. *The reduced form of* $\mathbf{A}$ *contains an* m-by-m $\mathbf{I}$ *plus additional columns.* Then, for each $\mathbf{b}$, $\mathbf{Ax} = \mathbf{b}$ has an infinite number of solutions (the additional columns give rise to the null-space vectors).

Section 5.1 Exercises

Summary of Exercises
Exercises 1–4 involve plotting and graphically solving systems of equations. Exercises 5–17 require finding null spaces and related particular solutions to various matrix systems. Exercises 18–25 require finding a constraint equa-

tion for the range and finding particular range vectors. Exercises 26–28 are some simple theory questions.

1. Plot the lines of solutions to the following systems of equations.

   (a) $2x + 3y = 10$

       $6x + 9y = 30$

   (b) $x - 2y = -4$

       $-3x + 6y = 12$

2. Describe and sketch the solution sets, if any, of the following systems of equations.

   (a) $x + 2y + 3z = 0$

       $4x + 3y + 5z = 0$

       $-4x + 2y + 2z = 0$

   (b) $2x + y + z = 3$

       $x - 2y - 3z = 2$

       $3x + 4y + 5z = 3$

   (c) $x + 3y - z = 5$

       $2x + y + 2z = 8$

       $3x - y + 5z = 11$

   (d) $x + 2y + 3z = 3$

       $x - y + z = 4$

       $2x + 10y + 10z = 8$

   (e) $2x + 2y + 2z = 6$

       $x + y - z = 3$

       $-x - y + 3z = -7$

3. For each of the following systems of equations, plot each line and from your drawing determine whether there is no solution, one solution, or an infinite set of solutions.

   (a) $3x - 2y = 5$

       $-6x + 4y = 10$

   (b) $3x - 2y = 5$

       $2x - 3y = 5$

   (c) $3x - 2y = 5$

       $9x - 6y = 15$

4. For each of the systems of equations in Exercise 2, sketch as best you can the plane determined by each equation. From your sketch, guess whether there is no solution, one solution, or an infinite set of solutions. Verify your guess by solving the system.

5. Give a vector, if one exists, that generates the null space of the following systems of equations or matrices. Which of these seven systems/matrices are invertible? [Consider the coefficient matrix and ignore the particular right-side values in parts (e) and (f).]

   (a) $\begin{bmatrix} 1 & -2 \\ -2 & 4 \end{bmatrix}$

   (b) $\begin{bmatrix} 4 & -1 & 2 \\ 2 & 5 & 1 \\ -2 & 3 & -1 \end{bmatrix}$

   (c) $\begin{bmatrix} -1 & 3 & 1 \\ 5 & -1 & 3 \\ 2 & 1 & 2 \end{bmatrix}$

   (d) $\begin{bmatrix} 2 & 1 & 1 \\ 1 & -2 & -3 \\ 3 & 4 & 5 \end{bmatrix}$

   (e) $x - 2y + z = 6$

       $-x + y - 2z = 4$

   (f) $x - y + z = 3$

       $x + y + z = 3$

       $2x - 3y + z = 7$

(g) The coefficient matrix $\mathbf{X}$ in the six regression equations in Example 6.

6. For the following matrices, describe the set of vectors in the null space.

(a) $\begin{bmatrix} -3 & 4 & 1 \\ 2 & -1 & 1 \end{bmatrix}$

(b) $\begin{bmatrix} 2 & 1 & 5 & 0 \\ 1 & 2 & 4 & -3 \\ 1 & 1 & 3 & -1 \end{bmatrix}$

(c) $\begin{bmatrix} 1 & 1 & 0 & 0 & 1 \\ 1 & 0 & 1 & 0 & 1 \\ 0 & 0 & 1 & 1 & 0 \\ 0 & 1 & 0 & 1 & 0 \\ 1 & 1 & 1 & 1 & 1 \end{bmatrix}$

7. (a) For $\mathbf{A}$ in Exercise 6, part (a) and $\mathbf{b} = [10, 10]$, if $\mathbf{Ax} = \mathbf{b}$ has the given solution $\mathbf{x}' = [0, 0, 10]$, find the family of all solutions to $\mathbf{Ax} = \mathbf{b}$.
   (b) Find a solution to $\mathbf{Ax} = \mathbf{b}$ in part (a) with $x_1 = 3$.

8. (a) For $\mathbf{A}$ in Exercise 6, part (b) and $\mathbf{b} = [30, 30, 20]$, if $\mathbf{Ax} = \mathbf{b}$ has the given solution $\mathbf{x}'$ $[10, 10, 0, 0]$, find the family of all solutions to $\mathbf{Ax} = \mathbf{b}$.
   (b) Find a solution to $\mathbf{Ax} = \mathbf{b}$ in part (a) with $x_1 = 5$.

9. (a) For $\mathbf{A}$ in Exercise 6, part (c) and $\mathbf{b} = [10, 15, 5, 0, 15]$, if $\mathbf{Ax} = \mathbf{b}$ has the given solution $\mathbf{x}' = [5, 0, 5, 0, 5]$, find the family of all solutions to $\mathbf{Ax} = \mathbf{b}$.
   (b) Find a solution to $\mathbf{Ax} = \mathbf{b}$ in part (a) with $x_4 = 10$ and $x_5 = 10$.
   (c) Find a solution to $\mathbf{Ax} = \mathbf{b}$ in part (a) with $x_1 = 10$ and $x_2 = 5$.

10. Consider the modified refinery system from Example 1:

$$20x_1 + 4x_2 + 4x_3 = 700$$
$$10x_1 + 14x_2 + 5x_3 = 500$$

Given the solution $x_1 = 31$, $x_2 = 10$, $x_3 = 10$, use the appropriate null-space vector to obtain a second solution in which the following is true.
   (a) $x_3 = 22$ (b) $x_2 = 25$ (c) $x_1 = 28$

11. Consider the following refinery-type problem:

$$10x_1 + 5x_2 + 5x_3 = 300$$
$$5x_1 + 10x_2 + 8x_3 = 300$$

(a) Find the null space of the system.
(b) Given the solution $x_1 = x_2 = 20$, $x_3 = 0$, find a second solution with $x_3 = 10$.
(c) Repeat part (b) to find a second solution with $x_1 = 15$.

12. The rabbit–fox growth model from Section 1.3,

$$R' = R + .1R - .15F$$
$$F' = F + .1R - .15F$$

had stable values for which $R' = R$ and $F' = F$ when the monthly change was zero:

$$R = R' \rightarrow .1R - .15F = 0$$
$$F = F' \rightarrow .1R - .15F = 0$$

Solutions for these homogeneous equations were of the form $[R, F] = r[3, 2]$. Suppose we want a vector $[R, F]$ that remains stable when 30 rabbits and 30 foxes are killed each month by hunters.

$$R = R + .1R - .15F - 30 \qquad .1R - .15F = 30$$
$$F = F + .1R - .15F - 30 \quad \rightarrow \quad .1R - .15F = 30$$

Find the family of stable population vectors in this case by adding one particular solution to the set of homogeneous solutions. Find a stable vector with $F = 400$.

13. Write out a system of equations required to balance the following chemical reaction and solve.

$$SO_2 + NO_3 + H_2O \rightarrow H + SO_4 + NO$$

where S represents sulfur, N nitrogen, H hydrogen, and O oxygen.

14. Write out a system of equations required to balance the following chemical reaction and solve.

$$PbN_6 + CrMn_2O_8 \rightarrow Cr_2O_3 + MnO_2 + Pb_3O_4 + NO$$

where Pb represents lead, N nitrogen, Cr chromium, Mn manganese, and O oxygen.

15. Find the null space for these Markov transition matrices.

(a) $\begin{bmatrix} 1 & .5 & 0 \\ 0 & 0 & 0 \\ 0 & .5 & 1 \end{bmatrix}$
(b) $\begin{bmatrix} .4 & 0 & .2 \\ .3 & .5 & .4 \\ .3 & .5 & .4 \end{bmatrix}$
(c) $\begin{bmatrix} .5 & 0 & .5 \\ 0 & 1 & 0 \\ .5 & 0 & .5 \end{bmatrix}$

(d) $$\begin{bmatrix} .50 & .25 & 0 & 0 & 0 \\ .50 & .50 & .25 & 0 & 0 \\ 0 & .25 & .50 & .25 & 0 \\ 0 & 0 & .25 & .50 & .50 \\ 0 & 0 & 0 & .25 & .50 \end{bmatrix}$$

(e) $$\begin{bmatrix} \frac{2}{3} & \frac{1}{3} & 0 & 0 & 0 & 0 \\ \frac{1}{3} & \frac{1}{3} & \frac{1}{3} & 0 & 0 & 0 \\ 0 & \frac{1}{3} & \frac{1}{3} & \frac{1}{3} & 0 & 0 \\ 0 & 0 & \frac{1}{3} & \frac{1}{3} & \frac{1}{3} & 0 \\ 0 & 0 & 0 & \frac{1}{3} & \frac{1}{3} & \frac{1}{3} \\ 0 & 0 & 0 & 0 & \frac{1}{3} & \frac{2}{3} \end{bmatrix}$$

(f) $$\begin{bmatrix} \frac{2}{3} & \frac{2}{3} & 0 & 0 & 0 & 0 \\ \frac{1}{3} & \frac{1}{6} & \frac{2}{3} & 0 & 0 & 0 \\ 0 & \frac{1}{6} & \frac{1}{6} & \frac{2}{3} & 0 & 0 \\ 0 & 0 & \frac{1}{6} & \frac{1}{6} & \frac{2}{3} & 0 \\ 0 & 0 & 0 & \frac{1}{6} & \frac{1}{6} & \frac{2}{3} \\ 0 & 0 & 0 & 0 & \frac{1}{6} & \frac{1}{3} \end{bmatrix}$$

(g) $$\begin{bmatrix} \frac{1}{2} & \frac{1}{2} & 0 & 0 & 0 & 0 \\ \frac{1}{2} & 0 & \frac{1}{2} & 0 & 0 & 0 \\ 0 & \frac{1}{2} & 0 & \frac{1}{2} & 0 & 0 \\ 0 & 0 & \frac{1}{2} & 0 & \frac{1}{2} & 0 \\ 0 & 0 & 0 & \frac{1}{2} & 0 & \frac{1}{2} \\ 0 & 0 & 0 & 0 & \frac{1}{2} & \frac{1}{2} \end{bmatrix}$$

16. For each transition matrix A in Exercise 15, find the set of probability vectors p such that the next-period vector p' ($= Ap$) is a stable probability vector (as was done in Example 3).

17. Prove that if A is the 3-by-3 transition matrix of some Markov chain and the null space of A is infinite, then either two columns of A are equal or else one column is the weighted average of the other two (one-half the sum of the other two).

18. Give a constraint equation, if one exists, on the vectors in the range of the matrices in Exercise 5.

19. (a) For the matrix A in Exercise 5, part (a), find a range vector b in which $b_1 = 5$.
    (b) For the b in part (a), solve $Ax = b$ (give the family of solutions using Theorem 2).

20. Give a constraint equation, if one exists, on the vectors in the range of the matrices in Exercise 6.

21. (a) For matrix A in Exercise 6, part (b), find a range vector b in which $b_1 = 5$ and $b_2 = 2$.
    (b) For the b in part (a), solve $Ax = b$ [give the family of solutions using Theorem 1, part (iv)].

22. (a) For matrix A in Exercise 6, part (c), find a range vector b in which $b_1 = b_2 = b_3 = 15$.
    (b) For the b in part (a), solve $Ax = b$ [give the family of solutions using Theorem 1, part (iv)].
    (c) Repeat part (a) to find a range vector with $b_1 = 20$, $b_3 = 10$, $b_4 = 20$.

23. Give a constraint equation, if one exists, on the probability vectors in the range of the transition matrices in Exercise 15.

24. By looking at the transition matrix for the frog Markov chain, explain why the even-state probabilities in the next period must equal the odd-state probabilities in the next period.

    Hint: Show that this equality is true if we start (this period) from a specific state.

25. Compute the constraint on the range of the following powers of the frog transition matrix. You will need a matrix software package.
    (a) A^2 (b) A^3 (c) A^{10}

26. Show that the range $R(A)$ of a matrix is a vector space. That is, if **b** and **b′** are in $R(A)$ (for some **x** and **x′**, $Ax = b$ and $Ax' = b'$), show that $rb + sb'$ is in $R(A)$, for any scalars r, s.

27. Using matrix algebra, show that if x_1 and x_2 are solutions to the matrix equation $Ax = b$, then any linear combination $x' = cx_1 + dx_2$, with $c + d = 1$, is also a solution.

28. Show that the intersection $V_1 \cap V_2$ of two vector spaces V_1, V_2 is again a vector space.

Section 5.2   Theory of Vector Spaces Associated with Systems of Equations

In this section we introduce basic concepts about vector spaces and use them to obtain important information about the range and null space of a matrix. Recall that a **vector space** V is a collection of vectors such that if $u, v \varepsilon V$, then any linear combination $ru + sv$ is in V. In Section 5.1 we introduced the range and null space of a matrix **A**:

$$\text{Range}(A) = \{b : Ax = b \text{ for some } x\}$$
$$\text{Null}(A) \quad = \{x : Ax = 0\}$$

We noted that Range(**A**) and Null(**A**) are both vector spaces.

In Examples 2, 3, and 4 of Section 5.1 we used the elimination process to find a vector or pair of vectors that generated the null spaces of certain matrices. For example, multiples of $[-1, 2, -2, 2, -2, 1]$ formed the null space of the frog Markov transition matrix. In Examples 7, 8, and 9 of Section 5.1, we used elimination to find constraint equations that vectors in the range must satisfy. For the frog Markov matrix, the constraint for range vectors **p** was $p_1 + p_3 + p_5 = p_2 + p_4 + p_6$.

The number of vectors generating the null space and number of constraint equations for the range were dependent on how many pivots we made during elimination. Our goal in this section is to show that the sizes of Null(**A**) and Range(**A**) are independent of how elimination is performed.

The **vector space** V **generated** by a set $Q = \{\mathbf{q}_1, \mathbf{q}_2, \ldots, \mathbf{q}_t\}$ of vectors is the collection of all vectors that can be expressed as a linear combination of the $\mathbf{q}_i$'s. That is,

$$V = \{\mathbf{v} : \mathbf{v} = r_1\mathbf{q}_1 + r_2\mathbf{q}_2 + \cdots + r_t\mathbf{q}_t, \; r_i \text{ scalars}\}$$

For example, if Q consists of the unit n-vectors $\mathbf{e}_j$ (with all 0's except for a 1 in position j), then V is the vector space of all n-vectors, that is, euclidean n-space. Another name for a generating set is a *spanning set*.

The **column space** of **A**, denoted Col (**A**), is the vector space generated by the column vectors $\mathbf{a}_j^C$ of **A**. When we write $\mathbf{Ax} = \mathbf{b}$ as

$$\mathbf{a}_1^C x_1 + \mathbf{a}_2^C x_2 + \cdots + \mathbf{a}_n^C x_n = \mathbf{b}$$

we see that the system $\mathbf{Ax} = \mathbf{b}$ has a solution if and only if **b** can be expressed as a linear combination of the column vectors of **A**, or

Lemma 1. The system $\mathbf{Ax} = \mathbf{b}$ has a solution if and only if **b** is in Col(**A**). Equivalently, Col(**A**) = Range(**A**).

The components x_i of the solution **x** give the weights in the linear combination of columns that yield **b**. Note that Lemma 1 is *true for any m-by-n matrix* **A** and any m-vector **b**.

Example 1. **Refinery Problem as a Column Space Problem**

The refinery problem introduced in Section 1.2 involved three refineries each producing different amounts of heating oil, diesel oil, and gasoline from a barrel of crude oil. Production levels of each refinery were sought to satisfy a vector of demands. The resulting system of equations was

$$
\begin{array}{lrcr}
\text{Heating oil:} & 20x_1 + 4x_2 + 4x_3 &=& 500 \\
\text{Diesel oil:} & 10x_1 + 14x_2 + 5x_3 &=& 850 \\
\text{Gasoline:} & 5x_1 + 5x_2 + 12x_3 &=& 1000
\end{array}
$$

But this system is just seeking to express the demand vector [500, 850, 1000] as a linear combination of the production vectors of the three refineries. That is, we seek x_1, x_2, x_3 such that

$$x_1 \begin{bmatrix} 20 \\ 10 \\ 5 \end{bmatrix} + x_2 \begin{bmatrix} 4 \\ 14 \\ 5 \end{bmatrix} + x_3 \begin{bmatrix} 4 \\ 5 \\ 12 \end{bmatrix} = \begin{bmatrix} 500 \\ 850 \\ 1000 \end{bmatrix} \qquad \blacksquare$$

Up to this point, solving a system $\mathbf{Ax} = \mathbf{b}$ was viewed as a problem about the rows of $\mathbf{A}$, that is, about the equations specified by the rows. Gaussian elimination involves forming linear combinations of the equations (rows of $\mathbf{A}$) to obtain a new reduced system that can be solved by back substitution. Lemma 1 says that solving $\mathbf{Ax} = \mathbf{b}$ can equally be viewed as a problem about a linear combination of the columns of $\mathbf{A}$. This vector approach to solving $\mathbf{Ax} = \mathbf{b}$ has an associated geometric picture.

Example 2. Geometric Picture of Solution to a System of Equations

Consider the system of equations

$$
\begin{array}{l} x_1 + x_2 = 4 \\ x_1 - 2x_2 = 1 \end{array}
\quad \text{or} \quad
x_1 \begin{bmatrix} 1 \\ 1 \end{bmatrix} + x_2 \begin{bmatrix} 1 \\ -2 \end{bmatrix} = \begin{bmatrix} 4 \\ 1 \end{bmatrix}
$$

Solving by elimination, we find that $x_1 = 3$ and $x_2 = 1$. Figure 5.3 graphs this solution in vector-space terms, showing the right-side vector $[4, 1]$ as a linear combination of the column vectors $[1, 1]$ and $[1, -2]$. Note that the picture gives no insight into why $x_1 = 3$, $x_2 = 1$ is the solution. ∎

To determine the size of Range($\mathbf{A}$), we analyze the structure of Col($\mathbf{A}$), the column space of $\mathbf{A}$, which by Lemma 1 equals Range($\mathbf{A}$). The key question is: How many of the columns of $\mathbf{A}$ are actually needed to generate Col($\mathbf{A}$). Some columns in $\mathbf{A}$ may be redundant.

A set of vectors $\mathbf{a}_1, \mathbf{a}_2, \ldots, \mathbf{a}_t$ is called **linearly dependent** if one of them can be expressed as a linear combination of the others. Another way to say this is that there is a nonzero solution $\mathbf{x}$ to

$$
x_1 \mathbf{a}_1 + x_2 \mathbf{a}_2 + \cdots + x_t \mathbf{a}_t = \mathbf{0} \quad \text{or, equivalently,} \quad \mathbf{Ax} = \mathbf{0} \quad (1)
$$

Figure 5.3

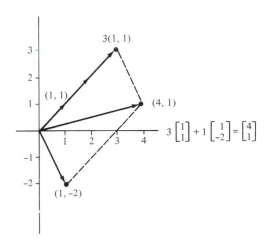

where $\mathbf{A}$ is the matrix whose columns are the vectors $\mathbf{a}_i$. Linear dependence is equivalent to (1) because if $x_i \neq 0$, we can rewrite (1) as

$$-x_i\mathbf{a}_i = x_1\mathbf{a}_1 + \cdots + x_n\mathbf{a}_n$$

or

$$\mathbf{a}_i = -\frac{x_1}{x_i}\mathbf{a}_1 - \frac{x_2}{x_i}\mathbf{a}_2 \cdots - \frac{x_n}{x_i}\mathbf{a}_n$$

So any $\mathbf{a}_i$, for which $x_i \neq 0$, can be written as a linear combination of the other $\mathbf{a}$'s.

A set of vectors are **linearly independent** if they are not linearly dependent. For example, the columns of an identity matrix

$$\mathbf{I} = \begin{bmatrix} 1 & 0 & 0 \\ 0 & 1 & 0 \\ 0 & 0 & 1 \end{bmatrix}$$

are linearly independent. If vectors $\mathbf{a}_i$ are linearly independent, then the only solution to $x_1\mathbf{a}_1 + \cdots + x_n\mathbf{a}_n = 0$ (i.e., $\mathbf{Ax} = \mathbf{0}$) can be $\mathbf{x} = \mathbf{0}$.

Example 3. **Example of a Linearly Dependent Set of Columns**

Consider the matrix $\mathbf{A} = \begin{bmatrix} 1 & 1 & 4 \\ 1 & -2 & 1 \end{bmatrix}$. By inspection we see that

$$\mathbf{a}_3 = 3\mathbf{a}_1 + \mathbf{a}_2: \qquad \begin{bmatrix} 4 \\ 1 \end{bmatrix} = 3\begin{bmatrix} 1 \\ 1 \end{bmatrix} + \begin{bmatrix} 1 \\ -2 \end{bmatrix} \qquad (2)$$

So the columns of $\mathbf{A}$ are linearly dependent.

The following method illustrates a systematic way to find this linear dependence. We perform elimination by pivoting on $\mathbf{A}$. We pivot on entry $(1, 1)$ and then on $(2, 2)$:

$$\begin{bmatrix} 1 & 1 & 4 \\ 1 & -2 & 1 \end{bmatrix} \rightarrow \begin{bmatrix} 1 & 1 & 4 \\ 0 & -3 & -3 \end{bmatrix} \rightarrow \mathbf{A}^* = \begin{bmatrix} 1 & 0 & 3 \\ 0 & 1 & 1 \end{bmatrix}$$

where $\mathbf{A}^*$ represents the reduced form of $\mathbf{A}$.

Remember that $\mathbf{x}$ is a solution to $\mathbf{Ax} = \mathbf{0}$ if and only if $\mathbf{x}$ is a solution to $\mathbf{A}^*\mathbf{x} = \mathbf{0}$. Equivalently, there is a linear dependence among the columns of $\mathbf{A}$ if and only if there are is linear dependence among the columns of $\mathbf{A}^*$. But the first two columns of $\mathbf{A}^*$ are unit vectors (they form the 2-by-2 identity matrix). So trivially, the third column of $\mathbf{A}^*$ is dependent on the first two:

$$\mathbf{a}_3^* = 3\mathbf{a}_1^* + \mathbf{a}_2^*: \qquad \begin{bmatrix} 3 \\ 1 \end{bmatrix} = 3 \begin{bmatrix} 1 \\ 0 \end{bmatrix} + \begin{bmatrix} 0 \\ 1 \end{bmatrix} \qquad (3)$$

This is exactly the relationship that we found in (2). ∎

If we were to apply the method in Example 3 to the coefficient matrix in the refinery problem of Example 1, elimination by pivoting would reduce the matrix to a 3-by-3 identity matrix. Since the columns of the identity matrix are trivially linearly independent, this means that the original columns in the refinery problem were linearly independent.

We state the method used to find linear dependence in Example 3 and its consequences as a theorem.

Theorem 1

   (i) Let **A** be any *m*-by-*n* matrix and let **A*** be the reduced matrix obtained from **A** using elimination by pivoting. Then a set of columns of **A** is linearly dependent (linearly independent) if and only if the corresponding columns in **A*** are linearly dependent (linearly independent).

  (ii) Any unpivoted column of **A*** (a column that was not reduced to a unit vector) is linearly dependent on the set of columns containing pivots.

 (iii) The columns of **A*** with pivots are linearly independent. The corresponding columns of **A** generate the column space of **A**.

The following example illustrates this method further.

Example 4. Redundant Columns in Transportation Problem Constraints

In Example 4 of Section 5.1 we examined the following system of equations (that were transportation problem constraints seen in Section 4.6).

$$
\begin{aligned}
x_1 + x_2 &= 20 \\
x_3 + x_4 &= 30 \\
x_5 + x_6 &= 15 \\
x_1 + x_3 + x_5 &= 25 \\
x_2 + x_4 + x_6 &= 40
\end{aligned}
$$

with coefficient matrix **A**

$$\mathbf{A} = \begin{bmatrix} 1 & 1 & 0 & 0 & 0 & 0 \\ 0 & 0 & 1 & 1 & 0 & 0 \\ 0 & 0 & 0 & 0 & 1 & 1 \\ 1 & 0 & 1 & 0 & 1 & 0 \\ 0 & 1 & 0 & 1 & 0 & 1 \end{bmatrix} \qquad (4)$$

In Section 5.1 we performed elimination by pivoting on **A** using entries (1, 2), (2, 4), (3, 6) and (4, 3). The reduced matrix **A*** was

$$\mathbf{A^*} = \begin{bmatrix} 1 & 1 & 0 & 0 & 0 & 0 \\ -1 & 0 & 0 & 1 & -1 & 0 \\ 0 & 0 & 0 & 0 & 1 & 1 \\ 1 & 0 & 1 & 0 & 1 & 0 \\ 0 & 0 & 0 & 0 & 0 & 0 \end{bmatrix} \tag{5}$$

First note that the last row of zeros can be ignored. Columns 2, 4, 6, and 3 of **A*** (in that order) are the unit vectors of the 4-by-4 identity matrix. Columns 1 and 5 of **A*** are each linearly dependent on the four unit-vector columns. For example,

$$\mathbf{a_1^*} = \mathbf{a_2^*} - \mathbf{a_4^*} + \mathbf{a_3^*}: \qquad \begin{bmatrix} 1 \\ -1 \\ 0 \\ 1 \\ 0 \end{bmatrix} = \begin{bmatrix} 1 \\ 0 \\ 0 \\ 0 \\ 0 \end{bmatrix} - \begin{bmatrix} 0 \\ 1 \\ 0 \\ 0 \\ 0 \end{bmatrix} + \begin{bmatrix} 0 \\ 0 \\ 0 \\ 1 \\ 0 \end{bmatrix} \tag{6}$$

The relation (6) among columns in **A*** is mirrored in **A**, where

$$\mathbf{a_1} = \mathbf{a_2} - \mathbf{a_4} + \mathbf{a_3}: \qquad \begin{bmatrix} 1 \\ 0 \\ 0 \\ 1 \\ 0 \end{bmatrix} = \begin{bmatrix} 1 \\ 0 \\ 0 \\ 0 \\ 1 \end{bmatrix} - \begin{bmatrix} 0 \\ 1 \\ 0 \\ 0 \\ 1 \end{bmatrix} + \begin{bmatrix} 0 \\ 1 \\ 0 \\ 1 \\ 0 \end{bmatrix}$$

Thus the columns where pivots were performed, columns $\mathbf{a_2}$, $\mathbf{a_3}$, $\mathbf{a_4}$, *and* $\mathbf{a_6}$, *generate the range of* **A**.   ∎

From Theorem 1, part (iii), it follows that the *number of pivots performed equals the number of linearly independent columns that generate the column space of* **A**.

A **basis** of a vector space *V* is a minimal-sized set of vectors that generate *V*. Implicit in this definition is the fact that a *basis is a set of linearly independent vectors*. As an example, the *n* coordinate vectors $\mathbf{e}_j$ form a basis for the space of all *n*-dimensional vectors. Since a basis is a minimal-sized generating set, every generating set contains a basis. For example, while the column space of a matrix **A** is defined to be generated by the columns of **A**, only the pivot columns are needed to generate the column space, as shown in Example 4.

The following result, which we prove in two ways, shows the theoretical relationship among the concepts of basis, linear independence, and unique solution.

Proposition. If $\{\mathbf{v}_1, \mathbf{v}_2, \ldots, \mathbf{v}_n\}$ is a basis for a vector space V, every vector in V has a unique representation as a linear combination of the $\mathbf{v}_i$'s.

Proof 1 (Using the Definition of Linear Independence): Suppose that $\mathbf{w}$ is a vector in V that has two representations as a linear combination of the $\mathbf{v}_i$. So

$$\mathbf{w} = a_1\mathbf{v}_1 + a_2\mathbf{v}_2 + \cdots + a_n\mathbf{v}_n \qquad \text{and}$$
$$\mathbf{w} = b_1\mathbf{v}_1 + b_2\mathbf{v}_2 + \cdots + b_n\mathbf{v}_n$$

Then

$$\mathbf{0} = \mathbf{w} - \mathbf{w} = (a_1 - b_1)\mathbf{v}_1 \qquad (7)$$
$$+ (a_2 - b_2)\mathbf{v}_2 + \cdots + (a_n - b_n)\mathbf{v}_n$$

Since the $\mathbf{v}_i$'s form a basis and hence are linearly independent, the linear combination of $\mathbf{v}_i$'s in (7) can only equal $\mathbf{0}$ if the terms $(a_i - b_i)$ are all zero. So the two representations must be the same. ∎

Proof 2 (Using Elimination). To find the representation of $\mathbf{w}$ in terms of the $\mathbf{v}_i$'s, we solve the system of equations for the x_i's:

$$x_1\mathbf{v}_1 + x_2\mathbf{v}_2 + \cdots + x_n\mathbf{v}_n = \mathbf{w} \qquad \text{or} \qquad \mathbf{Ax} = \mathbf{w} \qquad (8)$$

where $\mathbf{A}$ has the $\mathbf{v}_i$'s as its columns. Since the $\mathbf{v}_i$'s are a basis and hence linearly independent, we can pivot in every column [otherwise, by Theorem 1, part (ii), each unpivoted column is linearly dependent on the pivot columns]. Then $\mathbf{Ax} = \mathbf{w}$ has a unique solution (see the summary at the end of Section 5.1). ∎

Proof 2 shows us how to compute the unique representation of a vector $\mathbf{w}$ in terms of the $\mathbf{v}_i$'s, simply solve (8). For example, if $\mathbf{v}_1 = [1, 2, 3]$ and $\mathbf{v}_2 = [0, -1, 2]$ are a basis for vector space V and $\mathbf{w} = [3, 8, 5]$ is in V, then to determine the right linear combination of the $\mathbf{v}_i$'s to get $\mathbf{w}$, we solve the system

$$\begin{bmatrix} 1 & 0 \\ 2 & -1 \\ 3 & 2 \end{bmatrix} \begin{bmatrix} x_1 \\ x_2 \end{bmatrix} = \begin{bmatrix} 3 \\ 8 \\ 5 \end{bmatrix} \rightarrow \begin{bmatrix} 1 & 0 \\ 0 & 1 \\ 0 & 0 \end{bmatrix} \begin{bmatrix} x_1 \\ x_2 \end{bmatrix} = \begin{bmatrix} 3 \\ -2 \\ 0 \end{bmatrix} \qquad (9)$$

or $x_1 = 3$, $x_2 = -2$.

Now we prove a critical vector-space lemma that resolves the question of whether all pivots sequences have the same length.

Lemma 2. All linearly independent sets of vectors that generate a given vector space V have the same size. Any such set is a basis for V.

Proof. Let $S = \{s_i\}$ and $T = \{t_i\}$ be two sets of linearly independent vectors that generate V. Suppose that S and T have different sizes; for concreteness, let S have four vectors and T have five vectors. Then we can use linear combinations of the vectors of S to represent the vectors in T. If $t_1 = c_1 s_1 + c_2 s_2 + c_3 s_3 + c_4 s_4$, define the *S-coordinate vector* of t_1 to be $[c_1, c_2, c_3, c_4]$. Consider the equation, defined in (1), for dependence of the t_i (with the t_i represented in S-coordinates):

$$x_1 t_1 + x_2 t_2 + x_3 t_3 + x_4 t_4 + x_5 t_5 = 0 \qquad (10)$$

Since the t_i are four-dimensional vectors (in S-coordinates), (10) is a system with four equations in five variables. Solving (10) by elimination by pivoting leaves at least one unpivoted column and hence by Theorem 1, part (ii), there is linear dependence among the 5 t_i. Contradiction. ■

The **dimension** of a vector space V, written $\dim(V)$, is the number of vectors in a basis for V. For example, the set of n unit vectors e_j is a basis for "n-dimensional space"; thus this space does indeed have dimension n.

Combining Lemma 2 with Theorem 1, part (iii), we have

Theorem 2
  (i) The columns of $\mathbf{A}$ used in a pivot sequence are a basis for the range of A.
  (ii) All pivot sequences have the same size; the size is the dimension of the range of $\mathbf{A}$.

By Theorem 2, it now makes sense to talk about *the* number of pivots in a pivot sequence. The **rank** of a matrix $\mathbf{A}$, written $\text{rank}(\mathbf{A})$, is the number of pivots in any pivot sequence.

Corollary

$$\text{Rank}(\mathbf{A}) = \text{Dim}(\text{Range}(\mathbf{A}))$$

We have been concerned about which sets of columns of $\mathbf{A}$ are linearly dependent, that is, when there is a nonzero $\mathbf{x}$ so that

$$x_1 \mathbf{a}_1 + x_2 \mathbf{a}_2 + \cdots + x_n \mathbf{a}_n = \mathbf{0} \qquad \text{or} \qquad \mathbf{Ax} = \mathbf{0} \qquad (11)$$

Such an $\mathbf{x}$ in (11) is a vector in $\text{Null}(\mathbf{A})$, the null space of $\mathbf{A}$.

If $\mathbf{A}^*$ is the reduced matrix, then we know that $\mathbf{x}$ is a solution of

$\mathbf{Ax} = \mathbf{0}$ if and only if it is a solution to $\mathbf{A}^*\mathbf{x} = \mathbf{0}$. Thus *Null(A*) equals Null(A)*.

Example 5. Relation Between Null Space and Column Space

Consider the matrix

$$\mathbf{A} = \begin{bmatrix} 1 & 2 & 3 & 1 \\ 1 & -2 & -1 & -3 \\ -2 & 3 & 1 & 5 \end{bmatrix}$$

and perform elimination by pivoting.

Pivoting on entry $(1, 1)$ yields

$$\begin{bmatrix} 1 & 2 & 3 & 1 \\ 0 & -4 & -4 & -4 \\ 0 & 7 & 7 & 7 \end{bmatrix}$$

Pivoting on entry $(2, 2)$ yields

$$\mathbf{A}^* = \begin{bmatrix} 1 & 0 & 1 & -1 \\ 0 & 1 & 1 & 1 \\ 0 & 0 & 0 & 0 \end{bmatrix}$$

Clearly, the first two columns of $\mathbf{A}^*$, the pivot columns, are a basis for the column space of $\mathbf{A}^*$—they are linearly independent and generate $\text{Col}(\mathbf{A}^*)$. Then by Theorem 1, part (iii), the first two columns in $\mathbf{A}$ generate $\text{Col}(\mathbf{A})$.

Since $\text{Null}(\mathbf{A}^*) = \text{Null}(\mathbf{A})$, a basis for $\text{Null}(\mathbf{A}^*)$ will be a basis for $\text{Null}(\mathbf{A})$. Looking at $\mathbf{A}^*$, we see that

$$\mathbf{a}_3^* = \mathbf{a}_1^* + \mathbf{a}_2^* \qquad \text{and} \qquad \mathbf{a}_4^* = -\mathbf{a}_1^* + \mathbf{a}_2^* \qquad (12)$$

(where $\mathbf{a}_i^*$ denotes the *i*th column of $\mathbf{A}^*$). The vector equations in (12) can be rewritten as

$$-\mathbf{a}_1^* - \mathbf{a}_2^* + \mathbf{a}_3^* = 0 \qquad \text{or} \qquad \mathbf{A}^* \begin{bmatrix} -1 \\ -1 \\ 1 \\ 0 \end{bmatrix} = \mathbf{0}$$

$$ \tag{13}$$

$$\mathbf{a}_1^* - \mathbf{a}_2^* + \mathbf{a}_4^* = 0 \qquad \text{or} \qquad \mathbf{A}^* \begin{bmatrix} 1 \\ -1 \\ 0 \\ 1 \end{bmatrix} = \mathbf{0}$$

Let $\mathbf{x}_3^* = [-1, -1, 1, 0]$ and $\mathbf{x}_4^* = [1, -1, 0, 1]$. Since $\mathbf{x}_3^*$ and $\mathbf{x}_4^*$ are linearly independent (look at their last two entries) and generate Null($\mathbf{A}^*$), they form a basis for Null($\mathbf{A}^*$) and hence for Null($\mathbf{A}$). ∎

In general, given a reduced matrix $\mathbf{A}^*$, each unpivoted column can be expressed as a linear combination of the pivoted columns, which are unit vectors in $\mathbf{A}^*$ [see (12)]. This linear combination yields a solution to $\mathbf{A}^*\mathbf{x} = \mathbf{0}$, as shown in (13). If unpivoted column h of $\mathbf{A}^*$ has an entry a_{ih}^* in the ith row, the solution $\mathbf{x}_h^*$ we obtain is

$$\mathbf{x}_h^* = [-a_{1h}^*, \, - a_{2h}^*, \, \ldots, \, -a_{mh}^*, 0, 0, \ldots, 1, \ldots, 0] \quad (14)$$

where the entries for unpivoted columns are all 0 except for entry h, which is 1 (assuming pivots were performed in the first m rows and m columns). For example, in Example 5, the fourth column of $\mathbf{A}^*$ begins

$$\begin{bmatrix} -1 \\ 1 \\ \vdots \\ \vdots \end{bmatrix}$$

(in the two pivot rows), so $\mathbf{x}_4^* = [1, -1, 0, 1]$. The entries in (14) from the unpivoted columns form a unit vector, so the set of $\mathbf{x}_h^*$'s are linearly independent and form a basis of the null space of $\mathbf{A}$.

Observe that every column in $\mathbf{A}^*$ is now either (i) a unit vector that is in the basis of the column space; or else (ii) gives rise to a vector in the basis of the null space. That is, every column contributes to the size of the range of $\mathbf{A}$ or to the diversity of different solutions possible to $\mathbf{A}\mathbf{x} = \mathbf{b}$, for a given $\mathbf{b}$.

Theorem 3. Let $\mathbf{A}$ be a matrix with n columns. The vectors $\mathbf{x}_h^*$ in (14) corresponding to unpivoted columns form a basis for Null($\mathbf{A}$). Furthermore,

$$\dim(\text{Range}(\mathbf{A})) + \dim(\text{Null}(\mathbf{A})) = n$$

Corollary A

$$\begin{aligned} \text{Dim}(\text{Null}(\mathbf{A})) &= n - \dim(\text{Range}(A)) \\ &= n - \text{rank}(\mathbf{A}) \end{aligned}$$

Corollary B. Any solution $\mathbf{x}'$ to $\mathbf{A}\mathbf{x} = \mathbf{b}$ can be written in the form

$$\mathbf{x}' = \mathbf{x}^* + r_1\mathbf{x}_1^* + r_2\mathbf{x}_2^* + \cdots + r_k\mathbf{x}_k^* \quad (15)$$

where $\mathbf{x}^*$ is a given particular solution to $\mathbf{A}\mathbf{x} = \mathbf{b}$ and the $\mathbf{x}_h^*$'s are as given in (14).

Proof of Corollary B. By Theorem 1, part (iv) of Section 5.1, any solution $\mathbf{x}'$ can be written $\mathbf{x}' = \mathbf{x}^* + \mathbf{x}^\circ$, the sum of a particular solution $\mathbf{x}^*$ and some null space vector $\mathbf{x}^\circ$. Since the $\mathbf{x}_h^*$ generate Null($\mathbf{A}$), any such $\mathbf{x}^\circ$ is some linear combination of the $\mathbf{x}_h^*$'s. ∎

We complete our brief survey of vector spaces of $\mathbf{A}$ with Row($\mathbf{A}$), the vector space generated by the rows of $\mathbf{A}$. As noted when elimination was introduced in Section 3.2, the elimination process repeatedly replaces a row with a linear combination of rows. When rows are zeroed out in the elimination process, they are linearly dependent on the preceding rows *in which pivots were performed*. Conversely, every nonzero row in $\mathbf{A}^*$ is a pivot row (where a pivot was performed).

Because the submatrix of $\mathbf{A}^*$ formed by the pivot rows and pivot columns is an identity matrix, these pivot rows are linearly independent (see Exercises for details) and will be shown shortly to form a basis for Row($\mathbf{A}$). Hence the dimension of the row space equals rank($\mathbf{A}$) (= number of pivots).

Theorem 4. Let $\mathbf{A}$ be any *m*-by-*n* matrix. The maximum number of linearly independent rows in $\mathbf{A}$ and the maximum number of linearly independent columns in $\mathbf{A}$ are equal. Both are rank($\mathbf{A}$). That is,

$$\dim(\text{Row}(\mathbf{A})) = \text{rank}(\mathbf{A}) = \dim(\text{Col}(\mathbf{A}))$$

The results in Theorems 2, 3, and 4 yield several more equivalent conditions for when a system of equations has a unique solution.

Theorem 5. Let $\mathbf{A}$ be an *n*-by-*n* matrix. The system $\mathbf{Ax} = \mathbf{b}$ has a unique solution, for any $\mathbf{b}$, if and only if any of the following equivalent conditions are satisfied.
   (i) The dimension of Range($\mathbf{A}$) is *n*.
   (ii) The column vectors of $\mathbf{A}$ are linearly independent.
   (iii) The dimension of Row($\mathbf{A}$) is *n*.
   (iv) The row vectors of $\mathbf{A}$ are linearly independent.
   (v) The null space of $\mathbf{A}$ has dimension 0 (consists of only the $\mathbf{0}$ vector).

The following example illustrates the uses of Theorem 5.

Example 6. **Row Space Test for Unique Solution**

Let us consider the following variation of our refinery model introduced in Section 1.2. Suppose that we change the numbers in gasoline production so that the third row is the sum of the first two rows.

$$
\begin{array}{lrcl}
\text{Heating oil:} & 20x_1 + 4x_2 + 4x_3 & = & 500 \\
\text{Diesel oil:} & 10x_1 + 14x_2 + 5x_3 & = & 850 \\
\text{Gasoline:} & 30x_1 + 18x_2 + 9x_3 & = & 1000
\end{array}
$$

Once we observe that the last row in the coefficient matrix is the sum (a linear combination) of the first two rows, so that the rows are not linearly independent, then we know by Theorem 5, part (iv), that there will not be a unique solution to this production problem: either no solution or multiple solutions. If we tried to perform elimination, we would only be able to pivot twice (if we could pivot in all three rows, they would have to be linearly independent).

Theorem 4 tells us that if the rows are dependent, the columns also are. However, that column dependence is far from obvious. ∎

We now give a theoretical application of Theorems 3 and 4. Suppose that A is m-by-n, where $m < n$. All the columns cannot be linearly independent, since $\dim(\text{Col}(A)) = \dim(\text{Row}(A))$ and there are only m rows, $m < n$. Therefore, $\text{rank}(A) \leq m < n$. We conclude from Theorem 3, $\dim(\text{Null}(A)) = n - \text{rank}(A) = n - \dim(\text{Row}(A)) > 0$. Then $\text{Null}(A)$ is infinite and $Ax = b$ cannot have a unique solution:

Theorem 6. If A is an m-by-n matrix, where $m < n$, the system $Ax = b$ can never have a unique solution (either multiple solutions or no solution).

We close this section with a discussion of another way to interpret the elimination process and the rank of a matrix. To do this, we must introduce the concept of a simple matrix.

A **simple matrix** K is formed by the product $c * d$ of two vectors c and d in which c is treated as an m-by-1 matrix and d as a 1-by-n matrix. Thus entry k_{ij} of K equals $a_i b_j$. We refer to this product $c * d$ as a *matrix product of vectors*. For example, the following matrix product of vectors yields a simple matrix.

$$[3, -1] * [1, 2, 3] = \begin{bmatrix} 3 \\ -1 \end{bmatrix} [1, 2, 3]$$

$$= \begin{bmatrix} 3 \cdot 1 & 3 \cdot 2 & 3 \cdot 3 \\ -1 \cdot 1 & -1 \cdot 2 & -1 \cdot 3 \end{bmatrix} \quad (16)$$

$$= \begin{bmatrix} 3 & 6 & 9 \\ -1 & -2 & -3 \end{bmatrix}$$

All rows in a simple matrix are multiples of each other, and similarly for columns. If we pivot on an entry (i, j) in a simple matrix, the elimination computation will convert all other rows to 0's (verification is left as an exercise). This means that simple matrices have rank 1.

Simple matrices will be used extensively in Section 5.5. For now, the property of simple matrices of interest is

Theorem 7. Let C be a m-by-r matrix with columns c_j^C and D be an r-by-n matrix with rows d_i^R. Then the matrix multiplication CD can be decomposed into a sum of the simple matrices $\mathbf{c}_j^C * \mathbf{d}_i^R$ of the column vectors of $\mathbf{C}$ times the row vectors of $\mathbf{D}$.

$$\mathbf{CD} = \mathbf{c}_1^C * \mathbf{d}_1^R + \mathbf{c}_2^C * \mathbf{d}_2^R + \cdots + \mathbf{c}_r^C * \mathbf{d}_r^R \qquad (17)$$

One way to verify (17) is using the rules for partitioned matrices, that is, we partition $\mathbf{C}$ into r m-by-1 matrices (the $\mathbf{c}_i^C$) and partition $\mathbf{D}$ into r 1-by-n matrices (the $\mathbf{d}_j^R$); see Exercise 15 of Section 2.6 for details.

We illustrate this theorem with the following product of two matrices.

Example 7. **Decomposition of Matrix Multiplication into a Sum of Simple Matrices**

Let

$$\mathbf{C} = \begin{bmatrix} 1 & 2 & 3 \\ 4 & 5 & 6 \end{bmatrix} \quad \text{and} \quad \mathbf{D} = \begin{bmatrix} 11 & 12 & 13 \\ 14 & 15 & 16 \\ 17 & 18 & 19 \end{bmatrix}$$

Then

$$\mathbf{CD} = \begin{bmatrix} 1 \times 11 + 2 \times 14 + 3 \times 17 & 1 \times 12 + 2 \times 15 + 3 \times 18 & 1 \times 13 + 2 \times 16 + 3 \times 19 \\ 4 \times 11 + 5 \times 14 + 6 \times 17 & 4 \times 12 + 5 \times 15 + 6 \times 18 & 4 \times 13 + 5 \times 16 + 6 \times 19 \end{bmatrix} \quad (18a)$$

$$= \begin{bmatrix} 1 \times 11 & 1 \times 12 & 1 \times 13 \\ 4 \times 11 & 4 \times 12 & 4 \times 13 \end{bmatrix} + \begin{bmatrix} 2 \times 14 & 2 \times 15 & 2 \times 16 \\ 5 \times 14 & 5 \times 15 & 5 \times 16 \end{bmatrix} + \begin{bmatrix} 3 \times 17 & 3 \times 18 & 3 \times 19 \\ 6 \times 17 & 6 \times 18 & 6 \times 19 \end{bmatrix} \quad (18b)$$

$$= \begin{bmatrix} 1 \\ 4 \end{bmatrix} * [11 \quad 12 \quad 13] + \begin{bmatrix} 2 \\ 5 \end{bmatrix} * |14 \quad 15 \quad 16| + \begin{bmatrix} 3 \\ 6 \end{bmatrix} * [17 \quad 18 \quad 19] \quad (18c)$$

$$= \mathbf{c}_1^C * \mathbf{d}_1^R + \mathbf{c}_2^C * \mathbf{d}_2^R + \mathbf{c}_3^C * \mathbf{d}_3^R$$

The first simple matrix $\mathbf{c}_1^C * \mathbf{d}_1^R$ in (18c) is a matrix containing the first term of each scalar product in the entries of $\mathbf{CD}$ in (18a). Similarly for the second and third simple matrices. ∎

We now show how an m-by-n matrix $\mathbf{A}$ can be decomposed into a sum of k simple matrices, where $k = \text{rank}(\mathbf{A})$. Another way to say this is that we subtract a set of simple matrices from $\mathbf{A}$ to eliminate all entries in $\mathbf{A}$ (to reduce $\mathbf{A}$ to the $\mathbf{O}$ matrix).

Our strategy will be to form a simple matrix $\mathbf{K}_1 = \mathbf{l}_1 * \mathbf{u}_1$ whose first row equals $\mathbf{a}_1^R$ (the first row of $\mathbf{A}$) and whose first column equals $\mathbf{a}_1^C$ (the first column of $\mathbf{A}$). Then $\mathbf{A} - \mathbf{K}_1$ will have 0's in its first row and column. We form $\mathbf{K}_2$ to remove the second row and column of $\mathbf{A}$; possibly we zero out additional rows and columns in the process. We continue similarly with $\mathbf{K}_3$, and so on.

Let $\mathbf{u}_1 = \mathbf{a}_1^R$ and let

$$\mathbf{l}_1 = \left(\frac{1}{a_{11}}\right)\mathbf{a}_1^C = \left[\frac{a_{11}}{a_{11}}, \frac{a_{21}}{a_{11}}, \frac{a_{31}}{a_{11}}\right]$$

(actually the first entry of $\mathbf{l}_1$ is 1, $= a_{11}/a_{11}$). Then the entries of first row of $\mathbf{l}_1 * \mathbf{u}_1$ are 1 (first entry of $\mathbf{l}_1$) times $\mathbf{u}_1$, which equals $\mathbf{u}_1$ ($= \mathbf{a}_1^R$), as required. And the entries in the first column of $\mathbf{l}_1 * \mathbf{u}_1$ are $\mathbf{l}_1$ times the first entry of $\mathbf{u}_1$, a_{11}. We have

$$\begin{bmatrix} 1 \\ \dfrac{a_{21}}{a_{11}} \\ \dfrac{a_{31}}{a_{11}} \end{bmatrix} * [a_{11}, a_{12}, a_{13}] = \begin{bmatrix} a_{11} & a_{12} & a_{13} \\ a_{21} & \cdots & \cdots \\ a_{31} & \cdots & \cdots \end{bmatrix}$$

So $\mathbf{A} - \mathbf{K}_1$ ($= \mathbf{A} - \mathbf{l}_1 * \mathbf{u}_1$) has zeros in the first row and column.

Observe that outside the first row and column, the new entry (i, j) in $\mathbf{A} - \mathbf{K}_1$ equals

$$a_{ij} - \left(\frac{a_{i1}}{a_{11}}\right)a_{1j}$$

Surprise! This is our old friend the elimination operation [when we pivot on entry $(1, 1)$]. Vector $\mathbf{l}_1$ is just the first column in the matrix $\mathbf{L}$ of elimination multipliers from the $\mathbf{A} = \mathbf{LU}$ decomposition, and $\mathbf{u}_1$ is the first row of $\mathbf{U}$ (which equals the first row of $\mathbf{A}$). So we have shown that when we subtract from $\mathbf{A}$, the simple matrix $\mathbf{l}_1^C * \mathbf{u}_1^R$ formed by $\mathbf{l}_1^C$ (the first column of $\mathbf{L}$) and $\mathbf{u}_1^R$ (the first row of $\mathbf{U}$), we obtain a matrix with 0's in the first row and first column. This new matrix is just the coefficient matrix (ignoring the first row of 0's) for the remaining $n - 1$ equations in $\mathbf{Ax} = \mathbf{b}$ when we pivot on entry $(1, 1)$.

Repeating this argument, we let $\mathbf{K}_2 = \mathbf{l}_2^C * \mathbf{u}_2^R$ and subtracting $\mathbf{K}_2$ from $\mathbf{A} - \mathbf{K}_1$ will have the effect of next pivoting on entry $(2, 2)$, and zeroing out the second row and column. The other $\mathbf{K}_i$ are defined and perform similarly, so ultimately we see that

$$\mathbf{A} = \mathbf{l}_1^C * \mathbf{u}_1^R + \mathbf{l}_2^C * \mathbf{u}_2^R + \cdots + \mathbf{l}_m^C * \mathbf{u}_m^R \tag{19}$$

Example 8. Refinery Matrix Expressed as a Sum of Simple Matrices

In Section 3.2 we gave the **LU** decomposition of our refinery matrix

$$\mathbf{A} = \begin{bmatrix} 20 & 4 & 4 \\ 10 & 14 & 5 \\ 5 & 5 & 12 \end{bmatrix} = \begin{bmatrix} 1 & 0 & 0 \\ \frac{1}{2} & 1 & 0 \\ \frac{1}{4} & \frac{1}{3} & 1 \end{bmatrix}\begin{bmatrix} 20 & 4 & 4 \\ 0 & 12 & 3 \\ 0 & 0 & 10 \end{bmatrix}$$

By (19), we can write **A** as

$$\mathbf{A} = \mathbf{l}_1^C * \mathbf{u}_1^R + \mathbf{l}_2^C * \mathbf{u}_2^R + \mathbf{l}_3^C * \mathbf{u}_3^R$$

$$= \begin{bmatrix} 1 \\ \frac{1}{2} \\ \frac{1}{4} \end{bmatrix} * [20 \quad 4 \quad 4] + \begin{bmatrix} 0 \\ 1 \\ \frac{1}{3} \end{bmatrix} * [0 \quad 12 \quad 3] + \begin{bmatrix} 0 \\ 0 \\ 1 \end{bmatrix} * [0 \quad 0 \quad 10]$$

$$= \begin{bmatrix} 20 & 4 & 4 \\ 10 & 2 & 2 \\ 5 & 1 & 1 \end{bmatrix} + \begin{bmatrix} 0 & 0 & 0 \\ 0 & 12 & 3 \\ 0 & 4 & 1 \end{bmatrix} + \begin{bmatrix} 0 & 0 & 0 \\ 0 & 0 & 0 \\ 0 & 0 & 10 \end{bmatrix} \tag{20}$$

The reader should check that this set of three simple matrices adds up to **A**. ∎

Theorem 8. Gaussian elimination can be viewed as a decomposition of **A** into a sum of rank(**A**) simple matrices:

$$\mathbf{A} = \mathbf{l}_1^C * \mathbf{u}_1^R + \mathbf{l}_2^C * \mathbf{u}_2^R + \cdots + \mathbf{l}_k^C * \mathbf{u}_k^R \tag{21}$$

where $k = \text{rank}(\mathbf{A})$, $\mathbf{l}_i^C$ is the ith column of **L** (the matrix of elimination multipliers), and $\mathbf{u}_i^R$ is the ith row of **U** (the reduced matrix in Gaussian elimination).

The minimum number of simple matrices whose sum equals matrix **A** is rank(**A**).

The last sentence of Theorem 8 is proved in Exercise 32. The symmetric role of columns and rows in Theorem 8 explains why the dimensions of the row and column spaces of a matrix are equal.

It is not hard to show (see Exercise 31) that if **A** has the simple-matrix decomposition $\mathbf{A} = \mathbf{c}_1 * \mathbf{d}_1 + \cdots + \mathbf{c}_k * \mathbf{d}_k$, then the $\mathbf{c}_i$ are a basis for the column space of **A** and the $\mathbf{d}_i$ are a basis for the row space of **A**. It follows that

Corollary
   (i) The nonzero rows of **U** generate the row space of **A** and the nonzero (below main diagonal) columns of **L** generate the column space of **A**.
   (ii) Theorem 8 reproves the fact that the dimension of the column space of **A** equals the dimension of the row space of **A**, equals rank(**A**).

From Theorem 7 it follows if **A** equals the sum of simple matrices $\mathbf{l}_i^C * \mathbf{u}_i^R$, then $\mathbf{A} = \mathbf{LU}$—we have proved the **LU** decomposition.

The decomposition (19) of a matrix **A** into a sum of rank(**A**) simple matrices is of more theoretical than practical interest.

Optional (Based on Section 4.6)

Let us reinterpret the simplex algorithm of linear programming using the concept of a basis for the column space. When slack variables were added, as say to the table–chair production problem in Section 4.6, the form of the constraint equations was

$$
\begin{array}{c}
\\
\text{Objective Function} \\
\text{Wood} \\
\text{Labor} \\
\text{Braces} \\
\text{Upholstery}
\end{array}
\begin{array}{cccccc}
x_1 & x_2 & x_3 & x_4 & x_5 & x_6 \\
\end{array}
\left[
\begin{array}{cccccc|c}
40 & 200 & 0 & 0 & 0 & 0 & 0 \\
1 & 4 & 1 & 0 & 0 & 0 & 1400 \\
2 & 3 & 0 & 1 & 0 & 0 & 2000 \\
1 & 12 & 0 & 0 & 1 & 0 & 3600 \\
2 & 0 & 0 & 0 & 0 & 1 & 1800
\end{array}
\right]
\tag{22}
$$

Observe that the columns associated with the slack variables x_3, x_4, x_5, x_6 form an identity matrix and hence are the basis for the column space. For this reason, x_3, x_4, x_5, x_6 are called *basic variables* and variables x_1, x_2 *nonbasic*, for the linear program (22): Clearly, the columns of nonbasic variables x_1, x_2 in (22) are linearly dependent on the basic variables' columns. Recall that the simplex algorithm sets the nonbasic variables equal to 0 so that the basic variables then have nonnegative values equal to the corresponding right-side entry.

The pivot step in the simplex algorithm can be viewed as picking some nonbasic variable to enter the basis while a basic variable leaves in the basis. For (22), we chose x_2 (whose coefficient 200 in the objective function is largest) to enter and x_5 to leave the basis. After pivoting, we have

$$
\begin{array}{c}
\\
\text{Objective Function} \\
\text{Wood} \\
\text{Labor} \\
\text{Braces} \\
\text{Upholstery}
\end{array}
\begin{array}{cccccc}
x_1 & x_2 & x_3 & x_4 & x_5 & x_6 \\
\end{array}
\left[
\begin{array}{cccccc|c}
\frac{70}{3} & 0 & 0 & 0 & -\frac{50}{3} & 0 & -60{,}000 \\
\frac{2}{3} & 0 & 1 & 0 & -\frac{1}{3} & 0 & 200 \\
\frac{7}{4} & 0 & 0 & 1 & -\frac{1}{4} & 0 & 1{,}100 \\
\frac{1}{12} & 1 & 0 & 0 & \frac{1}{12} & 0 & 300 \\
2 & 0 & 0 & 0 & 0 & 1 & 1{,}800
\end{array}
\right]
\tag{23}
$$

Note that now the columns of x_2, x_3, x_4, x_6 form the basis for the column space.

Whereas our discussion in Section 4.6 focused on which were the independent (nonbasic) variables, the traditional approach is to concentrate on which are the basic variables.

Section 5.2 Exercises

Summary of Exercises
Exercises 1–10 involve the column space, linear dependence, and generators of the column space. Exercises 11–25 involve associated theory. Exercises 26–32 involve simple matrices and the representation of a matrix as a sum of simple matrices.

1. Three soup factories F_1, F_2, and F_3 generate production vectors, $\mathbf{f}_1 = [20, 100, 20]$, $\mathbf{f}_2 = [200, 0, 50]$, and $\mathbf{f}_3 = [0, 100, 200]$, of the amounts (in gallons) of tomato, chicken, and split-pea soup produced each hour. If the demand is $\mathbf{d} = [5000, 3000, 3000]$, write a system of equations for determining the right linear combination (how long each factory should work) of production vectors to meet the demand. Determine the weights.

2. There are three refineries producing heating oil, diesel oil, and gasoline. The production vector of refinery A (per barrel of crude oil) is $[10, 5, 10]$ and of refinery B is $[4, 11, 8]$. The production vector for refinery C is the average of the vector for refineries A and B. If the demand vector is $[380, 370, 460]$, write a system of equations for finding the right linear combination of refinery production vectors to equal the demand vector. Find the set of such linear combinations.

3. For each of the following sets of vectors, express the first vector as a linear combination of the remaining vectors if possible.
    (a) $[1, 1]$: $[2, 1]$, $[2, -1]$ (b) $[3, 2]$: $[2, -3]$, $[-3, 6]$
    (c) $[3, -1]$: $[1, 3]$, $[-2, 3]$
    (d) $[1, 1, 1]$: $[2, 1, 0]$, $[0, 1, 2]$, $[3, 2, 1]$

4. For each of the following pairs of a matrix and a vector, express the vector as a linear combination of the columns of the matrix. Plot this linear combination as was done in Figure 5.3.

    (a) $\begin{bmatrix} 4 & 0 \\ 0 & 3 \end{bmatrix}, \begin{bmatrix} 2 \\ 1 \end{bmatrix}$ (b) $\begin{bmatrix} 1 & -1 \\ 2 & 1 \end{bmatrix}, \begin{bmatrix} 3 \\ 0 \end{bmatrix}$ (c) $\begin{bmatrix} 2 & 3 \\ 5 & 8 \end{bmatrix}, \begin{bmatrix} 2 \\ 3 \end{bmatrix}$

5. The first column in the inverse $\mathbf{A}^{-1}$ of a 2-by-2 matrix $\mathbf{A}$ gives the weights in a linear combination of $\mathbf{A}$'s columns that equals $\mathbf{e}_1 = [1, 0]$. The second column in $\mathbf{A}^{-1}$ gives the weights in a linear combination of $\mathbf{A}$'s columns that equals $\mathbf{e}_2 = [0, 1]$. Find these weights for expressing $\mathbf{e}_1$ and $\mathbf{e}_2$ as linear combinations of the columns and plot the linear combinations as in Figure 5.3 for the following matrices.

    (a) $\begin{bmatrix} 4 & 0 \\ 0 & 3 \end{bmatrix}$ (b) $\begin{bmatrix} 1 & -1 \\ 2 & 1 \end{bmatrix}$ (c) $\begin{bmatrix} 2 & 3 \\ 5 & 8 \end{bmatrix}$

6. Tell which of the following sets of vectors are linearly independent. If linearly dependent, express one vector as a linear combination of the others.
    (a) $[1, 2]$, $[-2, 4]$ (b) $[1, 3]$, $[3, -1]$
    (c) $[2, 1]$, $[2, 3]$, $[2, 8]$ (d) $[1, 1, 1]$, $[-2, 0, -2]$, $[2, 1, 2]$
    (e) $[2, 1, 0]$, $[1, 1, 3]$, $[0, 2, 1]$

7. Find a set of columns that form a basis for the column space of each of the following matrices (use the reduced matrix $\mathbf{A}^*$ as in Examples 3 and 4). Give the rank of each matrix.

(a) $\begin{bmatrix} 1 & -2 \\ -2 & 4 \end{bmatrix}$

(b) $\begin{bmatrix} 4 & -1 & 2 \\ 2 & 5 & 1 \\ -2 & 3 & -1 \end{bmatrix}$

(c) $\begin{bmatrix} -1 & 3 & 1 \\ 5 & -1 & 3 \\ 2 & 1 & 2 \end{bmatrix}$

(d) $\begin{bmatrix} 2 & 1 & -1 \\ 1 & -2 & 2 \\ 3 & 4 & -4 \end{bmatrix}$

8. For each matrix in Exercise 7, find all sets of columns that form a basis for the column space.

9. Find a set of columns that form a basis for the column space of each of the following matrices. Give the rank of each matrix. Also find a basis for the null space of each matrix.

(a) $\begin{bmatrix} -3 & 5 \\ 6 & -10 \end{bmatrix}$

(b) $\begin{bmatrix} 2 & 1 & 7 \\ 1 & 2 & 5 \\ 1 & 1 & 4 \end{bmatrix}$

(c) $\begin{bmatrix} 1 & 1 & 0 & 0 & 1 \\ 1 & 0 & 1 & 0 & 1 \\ 0 & 0 & 1 & 1 & 0 \\ 0 & 1 & 0 & 1 & 0 \\ 1 & 1 & 1 & 1 & 1 \end{bmatrix}$

(d) $\begin{bmatrix} 1 & 0 & 2 & -1 & 1 \\ 1 & 1 & 1 & 1 & 2 \\ 1 & 0 & 2 & -1 & 1 \\ 0 & 1 & -1 & 2 & 1 \\ 1 & 0 & 2 & -1 & 1 \end{bmatrix}$

(e) $\begin{bmatrix} 1 & 1 & 1 & 1 & 0 & 0 \\ 0 & 1 & 0 & 1 & 0 & 1 \\ 1 & 0 & 0 & 1 & 1 & 0 \\ 0 & 1 & 1 & 0 & 1 & 0 \\ 0 & 1 & 0 & 1 & 0 & 1 \\ 1 & 0 & 1 & 0 & 0 & 1 \end{bmatrix}$

10. Let **A** be a coefficient matrix in a refinery problem, as in Example 1, with each column representing the production vector of a refinery. Explain the practical significance of having one column be a linear combination of the others. What constraints and what freedom does this permit the manager of the refineries?

11. For a *m*-by-*n* matrix **A**, the reduced matrix **A*** can be written in the partitioned form $\mathbf{A}^* = \begin{bmatrix} \mathbf{I} & \mathbf{R} \\ \mathbf{O} & \mathbf{O} \end{bmatrix}$, where **I** is an *r*-by-*r* identity matrix ($r = $ rank(**A**)) and **R** is *r*-by-($n - r$). Using the submatrix **R** and an appropriate size identity matrix **I**, give a matrix **N** in partitioned form whose columns are the basis of Null(**A**).

Hint: See expression (14).

12. Determine the rank of matrix **A**, if possible, from the given information.
    (a) **A** is an *n*-by-*n* matrix with linearly independent columns.
    (b) **A** is a 6-by-4 matrix and Null(**A**) = {**0**}.
    (c) **A** is a 5-by-6 matrix and dim(Null(**A**)) = 3.
    (d) **A** is a 3-by-3 matrix and det (**A**) = 17.
    (e) **A** is a 5-by-5 and dim(Row(**A**)) = 3.
    (f) **A** is an invertible 4-by-4 matrix.
    (g) **A** is a 4-by-3 matrix and **Ax** = **b** has either a unique solution or else no solution.
    (h) **A** is a 8-by-8 matrix and dim(Row(**A**^{T})) = 6.
    (i) **A** is a 7-by-5 matrix in which dim(Null(**A**^{T})) = 3.

13. In this exercise, the reader should try to find by inspection a linear dependence among the rows of each matrix. If dependence is found, use elimination by pivoting to find a linear dependence among the columns (as in Examples 3 and 4).

    (a) $\begin{bmatrix} 5 & 2 & 3 \\ 0 & 1 & 2 \\ 5 & 3 & 5 \end{bmatrix}$ (b) $\begin{bmatrix} 1 & 2 & 0 \\ 1 & 1 & 1 \\ 3 & 4 & 2 \end{bmatrix}$ (c) $\begin{bmatrix} 5 & 7 & 9 \\ 4 & 5 & 6 \\ 1 & 2 & 3 \end{bmatrix}$

14. Show that the number of pivots performed in Gaussian elimination will be the same as the number of pivots in elimination by (full) pivoting. (Thus the rank of a matrix can be defined in terms of either type of elimination.)

15. Let **A*** be the reduced-form matrix of **A**.
    (a) Show that nonzero rows of **A*** generate Row(**A**) (i.e., linear combinations of the rows of **A*** generate the same vectors as linear combinations of rows of **A**).
    (b) Show that the nonzero rows of **A*** must be linearly independent. *Hint:* Look at the form of **A***.
    (c) Conclude that the nonzero rows of **A*** are a basis of Row(**A**) and hence that the dim(Row(**A**)) = rank(**A**).

16. Let **U** be the upper triangular matrix produced at the end of Gaussian elimination on the matrix **A**.
    (a) Show that nonzero rows of **U** generate Row(**A**) (i.e., linear combinations of the rows of **U** generate the same vectors as linear combinations of rows of **A**).
    (b) Show that the nonzero rows of **U** must be linearly independent. *Hint:* Look at the form of **U**.
    (c) Conclude that the nonzero rows of **U** are a basis of Row(**A**) and hence that the dim(Row(**A**)) = rank(**A**).

17. (a) Suppose that the rows of **A** are linearly dependent. Show that at the end of Gaussian elimination, the resulting upper triangular matrix **U** will have at least one row of zeros.
    (b) Suppose that **A** is a square matrix with linearly dependent columns.

Show that at the end of Gaussian elimination, the resulting matrix **U** will have at least one row of zeros.

Hint: Use part (a) and Theorem 5.

18. Show that Row($\mathbf{A}^T$) ($\mathbf{A}^T$ is the transpose of **A**) equals the Col(**A**) and that Col($\mathbf{A}^T$) equals Row(**A**). Show that rank(**A**) = rank($\mathbf{A}^T$).

19. (a) Use the results of Exercise 17 and 18 to show that the nonzero rows in the reduced form $\mathbf{A}^T*$ of $\mathbf{A}^T$ are a basis for the Col(**A**).
    (b) Use part (a) to compute a basis for Col(**A**) for the matrix **A** in Example 3.
    (c) Repeat part (b) for the matrix in Example 4.

20. (a) Show that the rank of a matrix does not change when a multiple of one row is subtracted from another row.
    (b) Show that the rank of a matrix does not change when a multiple of one column is subtracted from another column.

    Hint: Use $\mathbf{A}^T$.

21. (a) Show that if **A** is an m-by-n matrix and **b** an m-vector, **b** is in Range(**A**) [= Col(**A**)] if and only if rank([**A** **b**]) = rank(**A**), where [**A** **b**] denotes the augmented m-by-$(n + 1)$ matrix with **b** added as an extra column to **A**.
    (b) If **Ax** = **b** has no solution, show that rank([**A** **b**]) must be rank(**A**) + 1.

22. Let **A** be an n-by-n matrix. Show that det(**A**) = 0 if and only if the rows of **A** are linearly dependent or if the columns of **A** are linearly dependent.

    Hint: Use Theorem 5 and Theorem 4 of Section 3.3.

23. This exercise examines the vectors in the column space of two matrices **A** and **B**, that is, vectors in Col(**A**) ∩ Col(**B**). If **d** is such a vector, then **Ax**′ = **d** and **Bx**″ = **d**, for some **x**′, **x**″. Show that if **C** = [**A** −**B**] and **x*** = [**x**′ **x**″], then **d** is in Col(**A**) ∩ Col(**B**) if and only if **x*** is in Null(**C**).

24. Show that any set H of k linearly independent n-vectors, $k < n$, can be extended to a basis for all n-vectors.

    Hint: Form an n-by-$(k + n)$ matrix **A** whose first k columns come from H and whose last n columns are the identity matrix—thus dim(Col(**A**)) = n; show that a basis for Col(**A**) using the elimination by pivoting approach in Example 5 will include the columns of H.

25. Show that λ is an eigenvalue of **A** if and only if det($\mathbf{A} - \lambda\mathbf{I}$) = **0**.

    Hint: "If" part is immediate; for the "only if" part, use Theorem 3 and Theorem 4 of Section 3.3.

26. Let $\mathbf{a} = [1, 2, 3]$, $\mathbf{b} = [2, 0]$, $\mathbf{c} = [-1, 2, 1]$. Compute the following simple matrices.
    (a) $\mathbf{a} * \mathbf{b}$ (b) $\mathbf{a} * \mathbf{c}$ (c) $\mathbf{c} * \mathbf{c}$

27. Verify that the simple matrices in Exercise 26 have rank 1.

28. (a) Show that $\mathbf{A} = \begin{bmatrix} 1 & 2 \\ 2 & 4 \end{bmatrix}$ is a simple matrix by giving the two vectors whose matrix product is $\mathbf{A}$.
    (b) Repeat part (a) for

$$\mathbf{B} = \begin{bmatrix} 12 & -6 & 9 \\ 8 & -4 & 6 \\ 4 & -2 & 3 \end{bmatrix}$$

29. Write each of the following matrices as the sum of two simple matrices.
    (a) $\begin{bmatrix} 1 & 2 \\ 2 & 3 \end{bmatrix}$ (b) $\begin{bmatrix} 1 & 2 & 3 \\ 3 & 4 & 5 \\ 7 & 8 & 9 \end{bmatrix}$

30. (a) Find the **LU** decomposition of matrix in Exercise 9, part (b) and use the decomposition to write the matrix as the sum of *two* simple matrices as in Example 8.
    (b) Repeat part (a) for the matrix in Example 5.
    (c) Repeat part (a) for the matrix in Exercise 9, part (d).

31. Describe the column space and row space of a simple matrix $\mathbf{a} * \mathbf{b}$ and give a basis for each.

32. (a) Show that if a matrix $\mathbf{A}$ of rank k is expressed as the sum of k simple matrices $\mathbf{c}_i * \mathbf{d}_i$, $i = 1, 2, \ldots, k$, then the $\mathbf{c}_i$ are a basis of Col($\mathbf{A}$) and the $\mathbf{d}_i$ are a basis of Row($\mathbf{A}$).
    (b) Prove the last sentence in Theorem 8, the minimum number of simple matrices whose sum is matrix $\mathbf{A}$ is rank($\mathbf{A}$), as follows: The *LU* decomposition yields a sum of k simple matrices equaling $\mathbf{A}$, where $k = \text{rank}(\mathbf{A})$, by the first part of Theorem 8; if fewer than k simple matrices could sum to $\mathbf{A}$, use part (a) to show that then $\dim(\text{Col}(\mathbf{A})) < \text{rank}(\mathbf{A})$—impossible.

Section 5.3 Approximate Solutions and Pseudoinverses

This section presents a method for obtaining an approximate solution that can be used to "solve" an m-by-n system $\mathbf{A}\mathbf{x} = \mathbf{b}$ that has no solution, that is, when $\mathbf{b}$ is not in the range of $\mathbf{A}$. We seek a "solution" $\mathbf{w}$ that gives a vector $\mathbf{p} = \mathbf{A}\mathbf{w}$ which is as close as possible to $\mathbf{b}$. In the following discussion, we use the euclidean norm $|\mathbf{a}| = \sqrt{a_1^2 + a_2^2 + \cdots + a_n^2}$, because we will be treating the vectors **p**, **b** as points in euclidean space and minimizing the distance of the error $|\mathbf{b} - \mathbf{p}|$.

Example 1. **Refinery Problem Revisited**

Recall the refinery model first presented in Section 1.2 with three refineries producing three petroleum-based products, heating oil, diesel oil, and gasoline.

$$
\begin{array}{lrcrcrcl}
\text{Heating oil:} & 20x_1 & + & 4x_2 & + & 4x_3 & = & 500 \\
\text{Diesel oil:} & 10x_1 & + & 14x_2 & + & 5x_3 & = & 850 \\
\text{Gasoline:} & 5x_1 & + & 5x_2 & + & 12x_3 & = & 1000
\end{array}
\tag{1}
$$

Suppose that the third refinery is out of service. We still want to attempt to produce the same amounts of these products. That is, we want to satisfy the system (as best we can)

$$
\begin{array}{lrcrcl}
\text{Heating oil:} & 20x_1 & + & 4x_2 & = & 500 \\
\text{Diesel oil:} & 10x_1 & + & 14x_2 & = & 850 \quad \text{or} \\
\text{Gasoline:} & 5x_1 & + & 5x_2 & = & 1000
\end{array}
\tag{2}
$$

$$
x_1 \begin{bmatrix} 20 \\ 10 \\ 5 \end{bmatrix} + x_2 \begin{bmatrix} 4 \\ 14 \\ 5 \end{bmatrix} = \begin{bmatrix} 500 \\ 850 \\ 1000 \end{bmatrix}
$$

In Section 3.2 we solved (1) by Gaussian elimination and obtained the solution $\mathbf{x} = [4\frac{7}{8}, 33\frac{1}{8}, 67\frac{1}{2}]$. Since this solution is unique and involves a nonzero value for x_3, we shall not be able to solve (2) exactly.

Let **A** be the matrix of coefficients in (2) and let **b** be the right-side vector. We seek to minimize $|\mathbf{b} - \mathbf{Aw}|$ (recall that we are using the euclidean distance as our norm). The approximation we want requires a vector **w** so that

$$
\mathbf{Aw} = w_1 \begin{bmatrix} 20 \\ 10 \\ 5 \end{bmatrix} + w_2 \begin{bmatrix} 4 \\ 14 \\ 5 \end{bmatrix}
$$

is as close to

$$
\begin{bmatrix} 500 \\ 850 \\ 1000 \end{bmatrix}
$$

as possible.   ■

This type of approximate solution is called a **least-squares solution,** because the euclidean distance $|\mathbf{b} - \mathbf{Aw}|$ to be minimized involves a sum of squares. For such approximate solutions to be meaningful, we require that **A** have more rows than columns; otherwise, a more sophisticated theory is needed.

Recall that we encountered least-squares solutions in regression. We review the regression problem presented in Section 4.2.

Example 2. Simple Linear Regression Reviewed

We wanted to fit the three (x, y) points $(0, 1)$, $(2, 1)$, and $(4, 4)$ to a line of the form $y = qx$ (where the x-value might be the number of college mathematics courses taken and the y-value a score on some test). The requirement $qx = y$ for these points yields a system of equations

$$
\begin{aligned}
0q &= 1 \\
2q &= 1 \\
4q &= 4
\end{aligned}
\quad \text{or} \quad q\mathbf{x} = \mathbf{y}, \quad \text{where } \mathbf{x} = \begin{bmatrix} 0 \\ 2 \\ 4 \end{bmatrix}, \quad \mathbf{y} = \begin{bmatrix} 1 \\ 1 \\ 4 \end{bmatrix} \quad (3)
$$

Figure 5.4a shows the points to be estimated by this line, and Figure 5.4b shows the vectors $\mathbf{y}$ and $q\mathbf{x}$ in 3-space. Our goal is to find q so that the estimates $\hat{y}_i = qx_i$ in (3) are as close as possible to the true y_i. A way to view $q\mathbf{x}$ in Figure 5.4b is as the *projection* of $\mathbf{y}$ onto the line through $\mathbf{x}$ from the origin.

To obtain the value for q, we minimized the sum of the squares of the errors (SSE).

$$
\begin{aligned}
\text{SSE} = \Sigma \, (qx_i - y_i)^2 &= (0q - 1)^2 + (2q - 1)^2 + (4q - 4)^2 \\
&= 20q^2 - 36q + 18 \quad (4)
\end{aligned}
$$

In Section 4.2 we found the optimal q by differentiating (4), setting the derivative equal to 0, and obtaining $q = .9$ (see line $\hat{y} = .9x$ in Figure 5.4a). We also calculated how to minimize SSE for an arbitrary number of $x - y$ pairs and obtained the formula

$$
q = \frac{\Sigma \, x_i y_i}{\Sigma \, x_i^2} = \frac{\mathbf{x} \cdot \mathbf{y}}{\mathbf{x} \cdot \mathbf{x}} \quad (5)
$$

where $\mathbf{x}$ and $\mathbf{y}$ are the vectors of x- and y-values. Note that for this example

$$
q = \frac{0 \times 1 + 2 \times 1 + 4 \times 4}{0 \times 0 + 2 \times 2 + 4 \times 4} = \frac{18}{20} = .9
$$

When the model $y = qx + r$ is used, (3) is changed to

$$
\begin{aligned}
0q + r &= 1 \\
2q + r &= 1 \\
4q + r &= 4
\end{aligned}
\quad \text{or} \quad \mathbf{Aw} = \mathbf{y}, \quad \text{where } \mathbf{A} = \begin{bmatrix} 0 & 1 \\ 2 & 1 \\ 4 & 1 \end{bmatrix}, \quad \mathbf{w} = \begin{bmatrix} q \\ r \end{bmatrix}
$$

$$
(6)
$$

Figure 5.4 (a) Regression estimates for points (0, 1), (2, 1), and (4, 4).
(b) $\hat{y} = 0.9x$ is regression solution when $\mathbf{x} = [0, 2, 4]$, $\mathbf{y} = [1, 1, 4]$. (c) $\mathbf{p}$ is closest vector to $\mathbf{b}$ in range of $\mathbf{A}$.

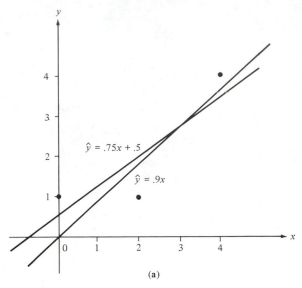

(a)

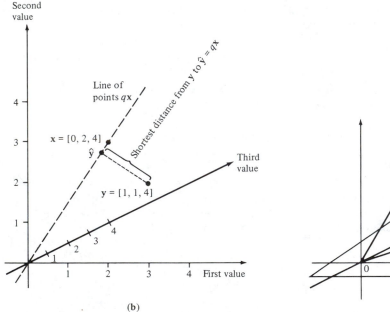

(b)

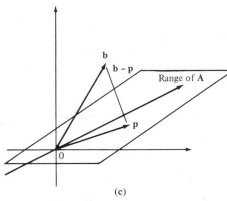

(c)

Now the least-squares solution is a pair q, r such that the vector $q[0, 2, 4] + r[1, 1, 1]$ (which is in the range of $\mathbf{A}$) is as close as possible to $\mathbf{y}$ (see Figure 5.4c, where $\mathbf{p} = \mathbf{Aw}$, $\mathbf{b} = \mathbf{y}$). Again, we can view $\mathbf{p}$ as the *projection* of $\mathbf{b}$ onto the range of $\mathbf{A}$. ∎

We have now motivated the importance of finding $\mathbf{w}$ so that the vector $\mathbf{p} = \mathbf{Aw}$ is as close as possible to a given vector $\mathbf{b}$, that is, so that $\mathbf{p}$ is the projection of $\mathbf{b}$ onto the range of $\mathbf{A}$. To determine $\mathbf{w}$ and $\mathbf{p}$, we shall use the following geometric property of the projection $\mathbf{p}$ of $\mathbf{b}$ onto the range of $\mathbf{A}$: *The error vector $\mathbf{b} - \mathbf{p}$ is at right angles to—perpendicular to—vectors in the range of $\mathbf{A}$* (see Figure 5.4b and c).

The term **orthogonal** is used in linear algebra instead of ''perpendicular'' to describe two vectors at right angles. The following theorem provides a simple numerical test for orthogonality.

Theorem 1. Two n-vectors $\mathbf{a}$, $\mathbf{b}$ are orthogonal if and only if their scalar product is zero, $\mathbf{a} \cdot \mathbf{b} = 0$.

This theorem is a consequence of a more general fact about angles between vectors that is proved later in this section.

We use Theorem 1 to obtain a matrix equation satisfied by $\mathbf{p}$ (where $\mathbf{p} = \mathbf{Aw}$) when $\mathbf{b} - \mathbf{p}$ is orthogonal to vectors in the range of $\mathbf{A}$ (implying that $\mathbf{p}$ is the closest vector to $\mathbf{b}$ in the range of $\mathbf{A}$).

First we illustrate the procedure by obtaining a formula for q in the simple regression model $\hat{y} = qx$ in Example 1, that is, to find q so that $q\mathbf{x}$ is as close as possible to $\mathbf{y}$. The error vector in this case is $\mathbf{y} - q\mathbf{x}$, and the range is simply all multiples of $\mathbf{x}$. The error vector $\mathbf{y} - q\mathbf{x}$ should be orthogonal to the range, that is, orthogonal to $\mathbf{x}$. By Theorem 1, this yields

$$\mathbf{x} \cdot (\mathbf{y} - q\mathbf{x}) = 0 \qquad \text{or} \qquad \mathbf{x} \cdot \mathbf{y} - q\mathbf{x} \cdot \mathbf{x} = 0$$

Solving for q, we obtain the same regression formula as in (5):

$$q = \frac{\mathbf{x} \cdot \mathbf{y}}{\mathbf{x} \cdot \mathbf{x}} \tag{7}$$

As noted above, $q\mathbf{x}$ is the projection of vector $\mathbf{y}$ onto vector $\mathbf{x}$. Thus

Theorem 2. The projection of $\mathbf{y}$ onto $\mathbf{x}$ (i.e., onto the line from the origin through $\mathbf{x}$) is $q\mathbf{x}$, where $q = \mathbf{x} \cdot \mathbf{y} / \mathbf{x} \cdot \mathbf{x}$.

Next consider the general case where we want an approximate solution to $\mathbf{Ax} = \mathbf{b}$ for any m-by-n matrix $\mathbf{A}$. The error vector $\mathbf{b} - \mathbf{p} = \mathbf{b} - \mathbf{Aw}$ should be orthogonal to every vector in the range of $\mathbf{A}$. Recall that the range of $\mathbf{A}$ is formed by linear combinations of the column vectors of $\mathbf{A}$, $r_1\mathbf{a}_1^C + r_2\mathbf{a}_2^C + \cdots + r_n\mathbf{a}_n^C$. If $\mathbf{b} - \mathbf{Aw}$ is orthogonal to any linear combination of the column vectors, then it certainly must be orthogonal to these column vectors $\mathbf{a}_i^C$ themselves. By Theorem 1 we have

$$\mathbf{a}_i^C \cdot (\mathbf{b} - \mathbf{Aw}) = 0 \qquad \text{for } i = 1, 2, \ldots, n \tag{8}$$

If we make a matrix $\mathbf{A}^*$ whose *rows* are the columns of $\mathbf{A}$, then (8) gives

$$\mathbf{A}^*(\mathbf{b} - \mathbf{Aw}) = 0 \tag{9}$$

But this matrix $\mathbf{A}^*$ is simply $\mathbf{A}^T$, the transpose of $\mathbf{A}$ (whose rows are the columns of $\mathbf{A}$). So (9) is

$$\mathbf{A}^T(\mathbf{b} - \mathbf{A}\mathbf{w}) = \mathbf{0} \qquad \text{or} \qquad (\mathbf{A}^T\mathbf{A})\mathbf{w} = \mathbf{A}^T\mathbf{b} \qquad (10)$$

Assuming that the matrix $\mathbf{A}^T\mathbf{A}$ is invertible, we can solve (10) for $\mathbf{w}$ to obtain

$$\mathbf{w} = (\mathbf{A}^T\mathbf{A})^{-1}\mathbf{A}^T\mathbf{b} \qquad (11)$$

The right side of (11) is a pretty messy expression. Since the rows of $\mathbf{A}^T$ are the columns of $\mathbf{A}$, entry (i, j) in the matrix product $\mathbf{A}^T\mathbf{A}$ is just the scalar product $\mathbf{a}_i^C \cdot \mathbf{a}_j^C$ of the ith column of $\mathbf{A}$ times the jth column of $\mathbf{A}$. When $\mathbf{A}$ consists of a single column $\mathbf{x}$, as in the regression model $q\mathbf{x} = \mathbf{y}$, (11) reduces to $q = (\mathbf{x} \cdot \mathbf{x})^{-1}\mathbf{x} \cdot \mathbf{y}$ or $q = \mathbf{x} \cdot \mathbf{y}/\mathbf{x} \cdot \mathbf{x}$—the formula we obtained above.

We call the product of matrices on the right in (11)

$$\mathbf{A}^+ = (\mathbf{A}^T\mathbf{A})^{-1}\mathbf{A}^T \qquad (12)$$

the **pseudoinverse** of $\mathbf{A}$ (the term *generalized inverse* is also used). If $\mathbf{A}$ is an m-by-n matrix, $\mathbf{A}^+$ will be an n-by-m matrix.

Theorem 3.   The least-squares solution $\mathbf{w}$ to the system of equations $\mathbf{A}\mathbf{x} = \mathbf{b}$ is $\mathbf{w} = \mathbf{A}^+\mathbf{b}$, where $\mathbf{A}^+ = (\mathbf{A}^T\mathbf{A})^{-1}\mathbf{A}^T$. Further, $\mathbf{A}^+$ is the left inverse of $\mathbf{A}$: $\mathbf{A}^+\mathbf{A} = \mathbf{I}$.

The second sentence of the theorem is easily verified: $\mathbf{A}^+\mathbf{A} = (\mathbf{A}^T\mathbf{A})^{-1}(\mathbf{A}^T\mathbf{A}) = \mathbf{I}$, since we are multiplying $\mathbf{A}^T\mathbf{A}$ times its inverse. *The identity,* $\mathbf{A}^+\mathbf{A} = \mathbf{I}$, *can be used to check that you have computed* $\mathbf{A}^+$ *correctly.*

If $\mathbf{A}$ is an invertible n-by-n matrix, the pseudoinverse $\mathbf{A}^+$ equals the regular inverse $\mathbf{A}^{-1}$ (see Exercise 22). If $\mathbf{b}$ happens to lie in the range of $\mathbf{A}$, then $\mathbf{A}\mathbf{w}$ will the exact solution, that is, $\mathbf{A}\mathbf{w}$ equals $\mathbf{b}$.

Although (12) is complex, the fact that such a matrix $\mathbf{A}^+$ exists at all is impressive. Applying Theorem 2 to the general regression model, we obtain

Corollary.   Consider the regression model $\hat{y} = q_1 x_1 + q_2 x_2 + \cdots + q_n x_n + r$ with associated matrix equation

$$\mathbf{y} = \mathbf{X}\mathbf{q}$$

where $\mathbf{y}$ is the set of y-value observations, $\mathbf{X}$ is the matrix whose jth column is the set of x_j-value observations and whose last column is the 1's vector, and $\mathbf{q} = [q_1, q_2, \ldots, q_n, r]$. Then the regression model parameters $\mathbf{q}$ are given by

$$\mathbf{q} = (\mathbf{X}^T\mathbf{X})^{-1}\mathbf{X}^T\mathbf{y} \qquad (13)$$

Example 3. **Least-Squares Solution to Refinery Problem**

Let us find the least-squares solution to the system of equations we had in Example 1, where the first two refineries alone had to try to satisfy the demand vector.

$$
\begin{aligned}
20x_1 + 4x_2 &= 500 \\
10x_1 + 14x_2 &= 850 \\
5x_1 + 5x_2 &= 1000
\end{aligned} \tag{14}
$$

If **A** is the coefficient matrix in (14), then we compute $\mathbf{A}^T\mathbf{A}$ and $(\mathbf{A}^T\mathbf{A})^{-1}$ to be [recall that entry (i, j) in $\mathbf{A}^T\mathbf{A}$ is the scalar product of columns i and j of **A**].

$$
\mathbf{A}^T\mathbf{A} = \begin{bmatrix} 525 & 245 \\ 245 & 237 \end{bmatrix} \quad \text{and} \quad (\mathbf{A}^T\mathbf{A})^{-1} = \begin{bmatrix} .00368 & -.00380 \\ -.00380 & .00815 \end{bmatrix}
$$

The pseudoinverse $\mathbf{A}^+$ of **A** is

$$
\begin{aligned}
\mathbf{A}^+ = (\mathbf{A}^T\mathbf{A})^{-1}\mathbf{A}^T &= \begin{bmatrix} .00368 & -.00380 \\ -.00380 & .00815 \end{bmatrix} \begin{bmatrix} 20 & 10 & 5 \\ 4 & 14 & 5 \end{bmatrix} \\
&= \begin{bmatrix} .0584 & -.0164 & -.0006 \\ -.0435 & .0761 & .0217 \end{bmatrix}
\end{aligned} \tag{15}
$$

With (15), we can now find the least-squares solution **w** to (14):

$$
\begin{aligned}
\mathbf{w} = \mathbf{A}^+\mathbf{b} &= \begin{bmatrix} .0584 & -.0164 & -.0006 \\ -.0435 & .0761 & .0217 \end{bmatrix} \begin{bmatrix} 500 \\ 850 \\ 1000 \end{bmatrix} \\
&= \begin{bmatrix} 14.6 \\ 64.7 \end{bmatrix}
\end{aligned} \tag{16}
$$

This solution produces the following approximating output vector:

$$
\mathbf{Aw} = [551, 1051, 394]
$$

with an error vector of

$$
\begin{aligned}
\mathbf{b} - \mathbf{Aw} &= [500 - 551, 850 - 1051, 1000 - 394] \\
&= [-51, -201, 606]
\end{aligned}
$$

This is a terrible approximation. We vastly underproduce the third product (gasoline). With the third refinery shut down, we shall always get much more of the second product than the third product [see (14)]. ∎

Example 4. **Solution to Regression Model**
$$\hat{y} = qx + r$$

Let us use the pseudoinverse to solve our regression problem with points $(0, 1)$, $(2, 1)$, $(4, 4)$ and the model $\hat{y} = qx + r$. The system of equations is

$$0q + r = 1$$
$$2q + r = 1 \qquad \text{or} \qquad \mathbf{Xq} = \mathbf{y},$$
$$4q + r = 4$$

$$\text{where } \mathbf{X} = \begin{bmatrix} 0 & 1 \\ 2 & 1 \\ 4 & 1 \end{bmatrix}, \quad \mathbf{y} = \begin{bmatrix} 1 \\ 1 \\ 4 \end{bmatrix}, \quad \mathbf{q} = \begin{bmatrix} q \\ r \end{bmatrix} \qquad (17)$$

Then

$$\mathbf{X}^T\mathbf{X} = \begin{bmatrix} 20 & 6 \\ 6 & 3 \end{bmatrix}, \qquad (\mathbf{X}^T\mathbf{X})^{-1} = \begin{bmatrix} \frac{1}{8} & -\frac{1}{4} \\ -\frac{1}{4} & \frac{5}{6} \end{bmatrix}$$

Using $(\mathbf{X}^T\mathbf{X})^{-1}$, we obtain the pseudoinverse

$$\mathbf{X}^+ = (\mathbf{X}^T\mathbf{X})^{-1}\mathbf{X}^T = \begin{bmatrix} \frac{1}{8} & -\frac{1}{4} \\ -\frac{1}{4} & \frac{5}{6} \end{bmatrix} \begin{bmatrix} 0 & 2 & 4 \\ 1 & 1 & 1 \end{bmatrix}$$
$$= \begin{bmatrix} -\frac{1}{4} & 0 & \frac{1}{4} \\ \frac{5}{6} & \frac{1}{3} & -\frac{1}{6} \end{bmatrix} \qquad (18)$$

Then

$$\mathbf{q} = \begin{bmatrix} q \\ r \end{bmatrix} = \mathbf{X}^+\mathbf{y} = \begin{bmatrix} -\frac{1}{4} & 0 & \frac{1}{4} \\ \frac{5}{6} & \frac{1}{3} & -\frac{1}{6} \end{bmatrix} \begin{bmatrix} 1 \\ 1 \\ 4 \end{bmatrix} = \begin{bmatrix} \frac{3}{4} \\ \frac{1}{2} \end{bmatrix}$$

So $q = .75$, $r = .5$; this is the same answer that we obtained for this problem in Section 4.2 (see Figure 5.4a). Our regression estimates for the y-values are given by $\mathbf{Xq} = [.5, 2, 3.5]$. ∎

Example 5. **Least-Squares Polynomial Fitting**

Suppose that we want to try to fit a quadratic curve through the set of points $(0, 7)$, $(1, 5)$, $(2, 4)$, $(3, 4)$, $(4, 8)$, and $(5, 12)$ using a least-squares approximation. Our model is

$$\hat{y} = ax^2 + bx + c \qquad (19)$$

We shall treat x^2 as a separate variable, say let $z = x^2$, so that a linear multivariate regression model can be used

$$\hat{y} = az + bx + c \tag{20}$$

For the given set of points our $\mathbf{X}$ matrix is

$$\mathbf{X} = \begin{bmatrix} 0 & 0 & 1 \\ 1 & 1 & 1 \\ 4 & 2 & 1 \\ 9 & 3 & 1 \\ 16 & 4 & 1 \\ 25 & 5 & 1 \end{bmatrix} \quad \text{with} \quad \mathbf{y} = \begin{bmatrix} 7 \\ 5 \\ 4 \\ 4 \\ 8 \\ 12 \end{bmatrix} \quad \text{and} \quad \mathbf{q} = \begin{bmatrix} a \\ b \\ c \end{bmatrix} \tag{21}$$

Using a computer program, we obtain

$$\mathbf{X}^+ = (\mathbf{X}^T\mathbf{X})^{-1}\mathbf{X}^T = \frac{1}{56}\begin{bmatrix} 5 & -1 & -4 & -4 & -1 & 5 \\ -33 & .2 & 18.4 & 21.6 & 9.8 & -17 \\ 46 & 18 & 0 & -8 & -6 & 6 \end{bmatrix}$$

and

$$\mathbf{q} = \mathbf{X}^+\mathbf{y} \simeq \begin{bmatrix} .893 \\ -3.493 \\ 7.214 \end{bmatrix} \quad \text{with estimates} \quad \hat{\mathbf{y}} = \mathbf{X}\mathbf{q} \simeq \begin{bmatrix} 7.2 \\ 4.6 \\ 3.8 \\ 4.8 \\ 5.3 \\ 12.1 \end{bmatrix} \tag{22}$$

Our quadratic estimate is thus $\hat{y} \simeq .893x^2 - 3.493x + 7.214$.

Although the estimated y-values work out closely to the observed y-values, a word of warning is important. This is a very poorly conditioned problem—the columns of $\mathbf{X}$ are all fairly similar. In fact, the condition number of the matrix $(\mathbf{X}^T\mathbf{X})$ is 2000 (in the sum norm)! A small change in the data could produce a large change in our answer. ∎

To compute the pseudoinverse $(\mathbf{A}^T\mathbf{A})^{-1}\mathbf{A}^T$, we need to know that the matrix $\mathbf{A}^T\mathbf{A}$ is invertible. The following result gives us the information we need.

Theorem 4. Let $\mathbf{A}$ be an m-by-n matrix. Then the n-by-n matrix $\mathbf{A}^T\mathbf{A}$ is invertible (and the pseudoinverse $\mathbf{A}^+$ exists) if the n columns of $\mathbf{A}$ are linearly independent, or equivalently, if rank$(\mathbf{A}) = n$.

Proof (Optional). We shall prove the stronger result that $\text{rank}(\mathbf{A}^T\mathbf{A}) = \text{rank}(\mathbf{A})$. Since $\text{rank}(\mathbf{A}) = n$, then $\text{rank}(\mathbf{A}^T\mathbf{A}) = n$ and any n-by-n matrix of rank n is invertible.

We work with null spaces. By Corollary A to Theorem 3 of Section 5.2, for a matrix $\mathbf{B}$ with n columns, $\dim(\text{Null}(\mathbf{B})) = n - \text{rank}(\mathbf{B})$. Then $\text{rank}(\mathbf{A}^T\mathbf{A}) = \text{rank}(\mathbf{A})$ is a consequence of showing that $\text{Null}(\mathbf{A}^T\mathbf{A}) = \text{Null}(\mathbf{A})$.

Let $\mathbf{x}$ be any vector in $\text{Null}(\mathbf{A})$, that is, $\mathbf{Ax} = \mathbf{0}$. Then

$$(\mathbf{A}^T\mathbf{A})\mathbf{x} = \mathbf{A}^T(\mathbf{Ax}) = \mathbf{A}^T(\mathbf{0}) = \mathbf{0}$$

so $\mathbf{x}$ is in $\text{Null}(\mathbf{A}^T\mathbf{A})$.

Conversely, let $\mathbf{y}$ be any vector in $\text{Null}(\mathbf{A}^T\mathbf{A})$. We want to show that $\mathbf{y}$ is in $\text{Null}(\mathbf{A})$. To do this, we show that $(\mathbf{Ay}) \cdot (\mathbf{Ay}) = 0$, which implies that $\mathbf{Ay} = \mathbf{0}$ (since $\mathbf{c} \cdot \mathbf{c} = 0$ means $\Sigma\, c_i^2 = 0$). We need the fact that $(\mathbf{Ay}) \cdot (\mathbf{Ay})$ is the same as $(\mathbf{Ay})^T(\mathbf{Ay})$ (the latter is the product of the 1-by-m matrix $(\mathbf{Ay})^T$ times the m-by-1 matrix $\mathbf{Ay}$). Then we have

$$\begin{aligned}(\mathbf{Ay}) \cdot (\mathbf{Ay}) = (\mathbf{Ay})^T(\mathbf{Ay}) &= (\mathbf{y}^T\mathbf{A}^T)(\mathbf{Ay}) \\ &= \mathbf{y}^T(\mathbf{A}^T\mathbf{Ay}) = \mathbf{y}^T \cdot \mathbf{0} = \mathbf{0} \qquad \blacksquare\end{aligned}$$

We note that in regression problems, practical considerations dictate that the matrix $\mathbf{X}$ is virtually certain to have linearly independent columns.

There is an important special case in which the computation of the pseudoinverse becomes very easy. This is when the columns of the matrix $\mathbf{A}$ are orthogonal. Now by Theorem 1, the scalar product of columns $\mathbf{a}_i^C \cdot \mathbf{a}_j^C$ equals 0. Since entry (i, j) in $\mathbf{A}^T\mathbf{A}$ is exactly this scalar product, $\mathbf{A}^T\mathbf{A}$ will be all 0's except on the main diagonal. This simple form of $\mathbf{A}^T\mathbf{A}$ leads to a simple form for $(\mathbf{A}^T\mathbf{A})^{-1}$ and for $\mathbf{A}^+$.

Example 6. **Regression with Orthogonal Columns**

We shall repeat the analysis of Example 5 with points $(0, 1)$, $(2, 1)$, and $(4, 4)$ and regression model $\hat{y} = qx + r$, but we shall shift the x-values so that the average x-value is 0 (in Section 4.2 we noted that such a shift simplified our regression formulas). The average x-value is $(0 + 2 + 4)/3 = 2$. If we subtract 2 from each x-value, obtaining points $(-2, 1)$, $(0, 1)$, and $(2, 4)$, then the new average x-value is 0 (subtracting off the average value always makes the new average be 0).

Let us repeat the pseudoinverse computations of Example 5 for these new points.

$$\begin{aligned}-2q + r &= 1 \\ 0q + r &= 1 \qquad \text{or} \qquad \mathbf{Xq} = \mathbf{y}, \\ 2q + r &= 4\end{aligned}$$

$$\text{where } \mathbf{X} = \begin{bmatrix} -2 & 1 \\ 0 & 1 \\ 2 & 1 \end{bmatrix}, \quad \mathbf{y} = \begin{bmatrix} 1 \\ 1 \\ 4 \end{bmatrix} \tag{23}$$

Observe that the two columns of $\mathbf{X}$ are now orthogonal. The scalar product of the two columns $\mathbf{x} \cdot \mathbf{1} = \Sigma\, x_i$ is 0, since the average x-value $(= \Sigma\, (x_i)/m)$ is 0. Then

$$\mathbf{X}^T\mathbf{X} = \begin{bmatrix} 8 & 0 \\ 0 & 3 \end{bmatrix} \quad \text{and} \quad (\mathbf{X}^T\mathbf{X})^{-1} = \begin{bmatrix} \frac{1}{8} & 0 \\ 0 & \frac{1}{3} \end{bmatrix} \tag{24}$$

The inverse $(\mathbf{X}^T\mathbf{X})^{-1}$ in (24) can be computed by the determinant formula, but we note that *the inverse of a diagonal matrix* $\mathbf{D}$ *(with 0's everywhere off the main diagonal) is obtained by replacing each diagonal entry with its inverse,* as in (24).

The two diagonal entries in $\mathbf{X}^T\mathbf{X}$ are, in symbolic terms, $\mathbf{x} \cdot \mathbf{x}$ and $\mathbf{1} \cdot \mathbf{1} = m$ (number of points in regression problem). Thus, when the average x-value is 0, $(\mathbf{X}^T\mathbf{X})^{-1}$ has the form

$$(\mathbf{X}^T\mathbf{X})^{-1} = \begin{bmatrix} \dfrac{1}{\mathbf{x} \cdot \mathbf{x}} & 0 \\ 0 & 1/m \end{bmatrix} \tag{25}$$

The pseudoinverse $\mathbf{X}^+$ is now

$$\mathbf{X}^+ = (\mathbf{X}^T\mathbf{X})^{-1}\mathbf{X}^T = \begin{bmatrix} \frac{1}{8} & 0 \\ 0 & \frac{1}{3} \end{bmatrix} \begin{bmatrix} -2 & 0 & 2 \\ 1 & 1 & 1 \end{bmatrix} = \begin{bmatrix} -\frac{2}{8} & 0 & \frac{2}{8} \\ \frac{1}{3} & \frac{1}{3} & \frac{1}{3} \end{bmatrix} \tag{26}$$

When we premultiply any matrix $\mathbf{B}$ *by a diagonal matrix* $\mathbf{D}$*, then* $\mathbf{D}$ *has the effect of multiplying the ith row of* $\mathbf{B}$ *by the ith diagonal entry of* $\mathbf{D}$, as in (26).

Looking at the values of the diagonal entries in $(\mathbf{X}^T\mathbf{X})^{-1}$ [see (25)], we see that $\mathbf{X}^+$ *is simply the transpose of* $\mathbf{X}$ *with the first column of* $\mathbf{X}$ *divided by its sum of squares* $(\mathbf{x} \cdot \mathbf{x})$ *and the second column divided by* m (the number of points).

Finally,

$$\mathbf{q} = \begin{bmatrix} q \\ r \end{bmatrix} = \mathbf{X}^+\mathbf{y} = \begin{bmatrix} -\frac{2}{8} & 0 & \frac{2}{8} \\ \frac{1}{3} & \frac{1}{3} & \frac{1}{3} \end{bmatrix} \begin{bmatrix} 1 \\ 1 \\ 4 \end{bmatrix} = \begin{bmatrix} \frac{3}{4} \\ 2 \end{bmatrix}$$

Observe that q is the scalar product of the first row of $\mathbf{X}^+$ with $\mathbf{y}$. But we just noted that the first row of $\mathbf{X}^+$ is simply $\mathbf{x}$ divided by the number $\mathbf{x} \cdot \mathbf{x}$. Similarly, r equals the scalar product of the second row of $\mathbf{X}^+$ times $\mathbf{y}$, and this second row is just $(1/m)\mathbf{1}$. Thus we have the simple formulas

$$q = \frac{\mathbf{x} \cdot \mathbf{y}}{\mathbf{x} \cdot \mathbf{x}}, \qquad r = \frac{\mathbf{1} \cdot \mathbf{y}}{\mathbf{1} \cdot \mathbf{1}} \; (= \text{ average } y\text{-value}) \qquad (27)$$

■

By Theorem 2, q and r are simply the projections of $\mathbf{y}$ onto $\mathbf{x}$ and $\mathbf{1}$, respectively. The nice results obtained in Example 6 will be true for the pseudoinverse of any matrix with orgthogonal columns.

Theorem 5. If the *m*-by-*n* matrix $\mathbf{A}$ ($m > n$) has orthogonal columns, the pseudoinverse $\mathbf{A}^+$ is obtained by dividing each column of $\mathbf{A}$ by the sum of the squares of the column's entries and then taking the transpose of the resulting matrix; the *i*th row of $\mathbf{A}^+$ is $\mathbf{a}_i^C/(\mathbf{a}_i^C \cdot \mathbf{a}_i^C)$. Further, the least-squares solution $\mathbf{w} = \mathbf{A}^+\mathbf{b}$ is just the projection of $\mathbf{b}$ onto the columns of $\mathbf{A}$: $w_i = \mathbf{a}_i^C \cdot \mathbf{b}/\mathbf{a}_i^C \cdot \mathbf{a}_i^C$.

Suppose that we have a regression model with several input variables, such as

$$\hat{y} = q_1 u + q_2 v + q_3 x + r \qquad (28)$$

and suppose that the vectors $\mathbf{u}$, $\mathbf{v}$, $\mathbf{x}$, and $\mathbf{1}$ (of the *u*-values, *v*-values, *x*-values, and 1's vector) are orthogonal. Then Theorem 5 tells us, generalizing (27), that the regression parameters are the projections of $\mathbf{y}$ onto $\mathbf{u}$, $\mathbf{v}$, $\mathbf{x}$, and $\mathbf{1}$:

$$q_1 = \frac{\mathbf{u} \cdot \mathbf{y}}{\mathbf{u} \cdot \mathbf{u}}, \qquad q_2 = \frac{\mathbf{v} \cdot \mathbf{y}}{\mathbf{v} \cdot \mathbf{v}}, \qquad q_3 = \frac{\mathbf{x} \cdot \mathbf{y}}{\mathbf{x} \cdot \mathbf{x}}, \qquad r = \frac{1}{m} \Sigma y_i \quad (29)$$

But what chance is there that the $\mathbf{u}$, $\mathbf{v}$, $\mathbf{x}$, and $\mathbf{1}$ vectors will be orthogonal? The answer is often up to the person who collects the data. If the *u*-, *v*-, and *x*-values measure settings of control knobs on a complex machine and the *y*-value measures the task performed by the machine, then a researcher who knows about Theorem 5 could pick settings to make the vectors orthogonal. This is a problem in a statistical subject called design of experiments.

We asserted in Theorem 1 that if $\mathbf{a}$ and $\mathbf{b}$ are orthogonal, that is, they form an angle of 90°, then $\mathbf{a} \cdot \mathbf{b} = 0$. This result was central to the derivation of the pseudoinverse. Now we prove the following theorem about the angle between two vectors, and Theorem 1 follows directly from this result. We measure the angle between two vectors $\mathbf{a}$, $\mathbf{b}$ by treating the vectors as line segments from the origin to the points with coordinates given by $\mathbf{a}$ and $\mathbf{b}$ (see Figure 5.5).

Theorem 6. The cosine of the angle θ between any vectors $\mathbf{a}$, $\mathbf{b}$ is

$$\cos \theta = \frac{\mathbf{a} \cdot \mathbf{b}}{|\mathbf{a}| \, |\mathbf{b}|} \qquad (30)$$

If $\mathbf{a}$ and $\mathbf{b}$ are unit-length vectors, (30) becomes $\cos \theta = \mathbf{a} \cdot \mathbf{b}$.

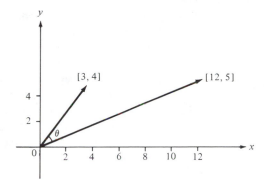

Figure 5.5 θ is the angle between **a** = [3, 4] and **b** = [12, 5].

Example 7. **Examples of Angles Between Vectors**

(i) If **a** = [1, 0, 0] and **b** = [0, 1, 0], then cos θ = **a** · **b** = 0 and we conclude that **a**, **b** form a 90° angle, that is, they are orthogonal.

(ii) If **a** = [3, 4] and **b** = [12, 5], then $|\mathbf{a}| = 5 \,(= \sqrt{9 + 16})$ and $|\mathbf{b}| = 13 \,(= \sqrt{144 + 25})$. So cos θ = **a** · **b**/|**a**| |**b**| = $(3\cdot12 + 4\cdot5)/5 \cdot 13 = \frac{56}{65} \simeq .86$. The angle with a cosine of .86 is 36° (see Figure 5.5).

(iii) If **a** = [.6, .8], with |**a**| = 1 and **b** = [1, 0], then cos θ = **a** · **b** = .6 · 1 + .8 · 0 = .6—just the first coordinate. ■

The proof of Theorem 6 uses the law of cosines:

$$|\mathbf{a} - \mathbf{b}|^2 = |\mathbf{a}|^2 + |\mathbf{b}|^2 - 2|\mathbf{a}| \, |\mathbf{b}| \cos \theta \tag{31}$$

The square of the euclidean norm $|\mathbf{c}|^2$ is simply **c** · **c**, and we can write $|\mathbf{a} - \mathbf{b}|^2$ as (**a** − **b**) · (**a** − **b**). Expanding with matrix algebra, we have

$$|\mathbf{a} - \mathbf{b}|^2 = (\mathbf{a} - \mathbf{b}) \cdot (\mathbf{a} - \mathbf{b}) = \mathbf{a} \cdot \mathbf{a} + \mathbf{b} \cdot \mathbf{b} - 2\mathbf{a} \cdot \mathbf{b} \tag{32}$$
$$= |\mathbf{a}|^2 + |\mathbf{b}|^2 - 2\mathbf{a} \cdot \mathbf{b}$$

The right side of (32) is the same as the right side of (31) except for the last terms. So these last terms must be equal:

$$-2|\mathbf{a}| \, |\mathbf{b}| \cos \theta = -2\mathbf{a} \cdot \mathbf{b}$$

Solving for cos θ yields Theorem 6.

Theorem 6 has a very important application in statistics. The cosine of the angle θ(**x**, **y**) between two vectors **x** and **y** tells us if the vectors are close together [when cos θ(**x**, **y**) is near 1] or opposites of one another [when cos θ(**x**, **y**) is near −1], or are unrelated, that is, close to orthogonal [when cos θ(**x**, **y**) is near 0].

Suppose that **x** and **y** are vectors of data from an experiment, say **x** is the scores of 10 students on a math test and **y** is the scores of the 10 students on a language test. Then cos θ(**x**, **y**) tells us how closely related these two sets of data are and helps us predict future relations between math and language scores. If cos $\theta(\mathbf{x}, \mathbf{y})$ is .8, performance on these two tests is closely related and we can view a student's score on one test as a reasonably good predictor of how he or she will do on the other test. If cos $\theta(\mathbf{x}, \mathbf{y})$ is $-.7$, the score vectors point in almost opposite directions and a high score on one test is very likely to produce a below-average score on the other test. If cos $\theta(\mathbf{x}, \mathbf{y}) = 0$ (the vectors $\mathbf{x}, \mathbf{y}$ are orthogonal), then performance on one test tells us nothing about the likely performance on the other test (in statistics, one says that the two sets of data are *independent*).

Definition. Let $\mathbf{x} = [x_1, x_2, \ldots, x_n]$ and $\mathbf{y} = [y_1, y_2, \ldots, y_n]$ be two sets of observations with the property that *the average x-value and the average y-value are each 0*. Then the **correlation coefficient** Cor$(\mathbf{x}, \mathbf{y})$ of $\mathbf{x}$ and $\mathbf{y}$ is defined to be cos $\theta(\mathbf{x}, \mathbf{y})$.

$$\text{Cor}(\mathbf{x}, \mathbf{y}) = \frac{\mathbf{x} \cdot \mathbf{y}}{|\mathbf{x}|\,|\mathbf{y}|} = \frac{\Sigma\, x_i y_i}{\sqrt{\Sigma\, x_i^2}\,\sqrt{\Sigma\, y_i^2}} \tag{33}$$

Recall that the average x-value is $\bar{x} = (1/n)\,\Sigma\, x_i$. If $\bar{x} \neq 0$, we can subtract $\bar{x}$ from each x_i to get a revised vector that does have an average value of 0. Similarly for y-values. We need an average value of 0 so that the opposite of a high score (a positive value) will be a low score (a negative value). This way the terms $x_i y_i$ in (33) for pairs of oppositely correlated entries x_i, y_i will be negative (when $x_i y_i$ is the product of a positive and a negative number), leading to a negative correlation.

Example 8. Correlation Coefficient

Suppose that we ask the eight faculty members of the Podunk University Alchemy Department to rate the quality of their graduate students and we poll the students to get a rating of the quality of each of the eight. The results of our experiment are presented in Table 5.1 (where we have processed the data to make the average value 0 in each category).

Table 5.1

Faculty	Quality of Students (x_i)	Student Rating (y_i)
1. Aristotle	$+5$	$+2$
2. Galileo	-5	-7
3. Goldbrick	-2	0
4. Hasbeen	$+3$	-1
5. Leadbottom	-4	-5
6. Merlin	$+5$	$+3$
7. Midas	$+5$	-0
8. Santa Claus	-7	$+8$

Applying formula (33) to these data, we obtain

$$\text{Cor}(\mathbf{x}, \mathbf{y}) = \frac{\Sigma \, x_i y_i}{\sqrt{\Sigma \, x_i^2} \, \sqrt{\Sigma \, y_i^2}} = \frac{5 \cdot 2 + (-5)(-7) + \cdots + (-7)8}{\sqrt{178} \cdot \sqrt{152}}$$

$$= \frac{21}{13.3 \cdot 12.3} = .1$$

Looking back at the data in Table 5.1, we are a little surprised to see such a low correlation, since the numbers in the two columns correspond fairly well for most faculty with the glaring exception of Santa Claus. Statisticians would call Santa Claus's data pair $(-7, 8)$ an **outlier,** an observation that fits poorly with the rest of the data. We warned in the regression section (Section 4.2) that one or two outliers can distort a statistical analysis. (A little investigating reveals that Santa Claus is a terrible teacher but is still well liked because he gives the students lots of candy every December.)

Let us throw out Santa Claus's numbers and recompute the correlation coefficient. This requires us to adjust the data so that the averages in each column are again 0. The new numbers are shown in Table 5.2.

Table 5.2

Faculty	Quality of Students (x_i)	Students' Rating (y_i)
1. Aristotle	$+4$	$+3$
2. Galileo	-6	-6
3. Goldbrick	-3	1
4. Hasbeen	$+2$	0
5. Leadbottom	-5	-4
6. Merlin	$+4$	$+4$
7. Midas	$+4$	$+1$

$$\text{Cor}(\mathbf{x}, \mathbf{y}) = \frac{85}{\sqrt{122} \cdot \sqrt{79}} \simeq \frac{85}{11 \cdot 8.9} \simeq .9$$

a high degree of correlation. ∎

We finish this section with a discussion of orthogonal vector spaces associated with least-squares solutions. A least-squares solution $\mathbf{w}$ to $\mathbf{Ax} = \mathbf{b}$ involves breaking $\mathbf{b}$ into two parts,

$$\mathbf{b} = \mathbf{p} + \mathbf{b} - \mathbf{p} \tag{34}$$

where $\mathbf{p} = \mathbf{Aw}$ is the least-squares solution and $\mathbf{b} - \mathbf{p}$ is the error vector.

Recall that $\mathbf{p}$ and $\mathbf{b} - \mathbf{p}$ must be orthogonal. The decomposition of $\mathbf{b}$ in (34) into a range and an error vector is unique ($\mathbf{p}$ is the unique vector $\mathbf{A}\mathbf{w}$). We assume here that $\mathbf{A}$ has linearly independent columns.

The vector $\mathbf{p}$ is in the range of $\mathbf{A}$ ($=$ column space of $\mathbf{A}$). Now we shall identify the vector space V containing $\mathbf{b} - \mathbf{p}$. V is the *error space* of $\mathbf{A}$ consisting of all error vectors; these are vectors $\mathbf{e}$ orthogonal to the columns of $\mathbf{A}$. This means that $\mathbf{a}_j^C \cdot \mathbf{e} = 0$ for all columns $\mathbf{a}_j^C$ of $\mathbf{A}$.

Earlier in this section we expressed the fact that the error vector $\mathbf{b} - \mathbf{A}\mathbf{w}$ was orthogonal to all the columns of $\mathbf{A}$ as $\mathbf{A}^T(\mathbf{b} - \mathbf{A}\mathbf{w}) = \mathbf{0}$. That is, $\mathbf{A}^T$ has rows that are the columns of $\mathbf{A}$. Then the error space V can be defined:

$$V = \{\mathbf{v} : \mathbf{A}^T\mathbf{v} = \mathbf{0}\} \tag{35}$$

But from (35), we see that V is simply the null space Null($\mathbf{A}^T$) of $\mathbf{A}^T$. Thus *Range(A) is orthogonal to Null($\mathbf{A}^T$)*. Here we are calling two vector spaces orthogonal if all pairs of vectors, one from each, are orthogonal.

Let us next determine the dimension of the error space [$=$ Null($\mathbf{A}^T$)]. By Theorem 3 of Section 5.2, the dimension of the Null($\mathbf{A}^T$) is $m -$ rank($\mathbf{A}^T$) (where m is the number of columns in $\mathbf{A}^T$). So

$$\dim(\text{Error space}(\mathbf{A})) = m - \text{rank}(\mathbf{A}^T) \tag{36}$$

But rank($\mathbf{A}^T$) $=$ rank($\mathbf{A}$) (this simple consequence of Theorem 4 of Section 5.2 is proved in Exercise 18 of that section). So (36) is the same as

$$\dim(\text{Error space}(\mathbf{A})) = m - \text{rank}(\mathbf{A}) \tag{37}$$

Using the fact that the dimension of Range($\mathbf{A}$) is rank($\mathbf{A}$), we have the expected result [in light of (34)]:

$$\dim(\text{Range}(\mathbf{A})) + \dim(\text{Error space}(\mathbf{A})) = m \tag{38}$$

Summarizing, we have

Theorem 7

(i) Let $\mathbf{A}$ be a m-by-n matrix and $\mathbf{b}$ be any m-vector. Then $\mathbf{b}$ can be written as a unique sum

$$\mathbf{b} = \mathbf{b}_1 + \mathbf{b}_2 \tag{39}$$

where $\mathbf{b}_1$ is in Range($\mathbf{A}$), $\mathbf{b}_2$ is in Error space($\mathbf{A}$), and $\mathbf{b}_1$, $\mathbf{b}_2$ are orthogonal. Further,

$$\dim(\text{Range}(\mathbf{A})) + \dim(\text{Error space}(\mathbf{A})) = m$$

The vector $\mathbf{b}_1$ in (39) equals $\mathbf{A}\mathbf{w}$ and is the projection of $\mathbf{b}$ onto the column space of $\mathbf{A}$; $\mathbf{w}$ is the least-squares solution $\mathbf{w} = \mathbf{A}^+\mathbf{b}$.

(ii) The error space of A equals $\text{Null}(A^T)$, so $\text{Null}(A^T)$ is orthogonal to $\text{Range}(A)$.

Theorem 7 is valid even if A does not have linearly independent columns.

Recall that the $\text{Range}(A)$ equals $\text{Col}(A)$, the column space of A, which equals $\text{Row}(A^T)$. So Theorem 7, part (ii) implies that $\text{Null}(A^T)$ is orthogonal to $\text{Row}(A^T)$, or, interchanging the names of A and A^T, $\text{Null}(A)$ is orthogonal to $\text{Row}(A)$. Reinterpreting (39) in terms of these vector spaces, we have

Theorem 8. Let A be an m-by-n matrix. The row space of A and the null space of A are orthogonal. Further, any n-vector x can be written as a unique sum

$$x = x_1 + x_2 \tag{40}$$

where x_1 is in the $\text{Row}(A)$ and x_2 is in $\text{Null}(A)$.

Note that $\text{Null}(A)$ being orthogonal to $\text{Row}(A)$ follows directly from the definition of $\text{Null}(A)$: x is in $\text{Null}(A)$ when $ax = 0$, but $Ax = 0$ just says that x is orthogonal to the rows of A. However, the unique sum (40) is not so obvious.

The reader should recall Theorem 1 of Section 5.1, which asserted that any solution x' to $Ax = b$ could be expressed as $x' = x^* + x^0$, where x^* is some particular solution to $Ax = b$ and x^0 is some solution to the homogeneous system $Ax = 0$ [i.e., x^0 is in $\text{Null}(A)$]. Theorem 8 tells us that the decomposition of x' can be chosen so that x^* is in the row space of A.

Corollary A. Let x' be a solution to the system $Ax = b$. Then x' can be uniquely decomposed $x' = x^* + x^0$, where x^* is in the row space of A and x^0 is in the null space of A. Further, x^* and x^0 are orthogonal.

Note in Corollary A that if $Ax' = b$, then $Ax^* + Ax^0 = b$ (since $x' = x^* + x^0$). But $Ax^0 = 0$ since x^0 is in $\text{Null}(A)$, so $Ax^* = b$. We have thus proved the surprising result:

Corollary B. If the system $Ax = b$ has a solution, it has a solution x^* that lies in the row space of A.

Section 5.3 Exercises

Summary of Exercises
Exercises 1–16 involve regression and least-squares solutions; when asked if a 2-by-2 matrix is poorly conditioned, say yes if the condition number is ≥ 10. Exercises 17–21 involve angles between vectors and the correlation coefficient. Exercises 22–30 involve examples and extensions of vector-space theory.

1. Determine the condition number of the matrix $(A^T A)$ in Example 3. Is this matrix poorly conditioned?

2. Seven students earned the following scores on a test after studying the subject matter different numbers of weeks.

Student	A	B	C	D	E	F	G
Length of Study (x_i)	0	1	2	3	4	5	6
Test Score (y_i)	3	4	7	6	10	6	10

   Fit these data with a regression model of the form $\hat{y} = qx + r$. Determine q and r by computing the pseudoinverse of X, the matrix whose first column is the vector of x_i's and whose second column is a 1's vector. Plot the observed scores and the predicted scores.

3. The following data indicate the numbers of accidents bus drivers had in one year as a function of the numbers of years on the job.

Years on Job (x_i)	2	4	6	8	10	12
Accidents (y_i)	10	8	3	8	4	5

   (a) Fit these data with a regression model of the form $\hat{y} = qx + r$. Determine q and r by computing the pseudoinverse of X, the matrix whose first column is the vector of x_i's and whose second column is a 1's vector.
   (b) What is the condition number of the matrix $(X^T X)$? Is the problem poorly conditioned?
   (c) Repeat the calculations in part (a) by first shifting the x-values to make the average x-value be 0 (see Example 6).

4. The following data shows the GPA and the job salary (five years after graduation) of six mathematics majors from Podunk U.

GPA	2.3	3.1	2.7	3.4	3.7	2.8
Salary	25,000	38,000	28,000	35,000	30,000	32,000

   (a) Fit these data with a regression model of the form $\hat{y} = qx + r$ using pseudoinverses.
   (b) What is the condition number of the matrix $(X^T X)$? Is the problem poorly conditioned?
   (c) Repeat the calculations in part (a) by first shifting the x-values to make the average x-value be 0 (see Example 6).

5. Compute the pseudoinverse, and then solve the refinery problem in Example 1 when refinery 1 is shut down (the other two refineries operate).

6. (a) Compute the pseudoinverse, and then solve, the refinery problem in Example 1 when refinery 2 is shut down (the other two refineries operate).

   (b) Compute the error vector $\mathbf{e} = \mathbf{b} - \mathbf{Aw}$ for part (a). Compute the angle between the error vector $\mathbf{e}$ in part (a) and the solution vector $\mathbf{Aw}$. It should be about 90°. Is it?

   (c) Which refinery closing, of the three refineries, has the smallest error vector (in the sum norm)—this assumes that you have done Exercise 5.

7. Compute the pseudoinverse of the following matrices.

   (a) $\begin{bmatrix} 1 \\ 2 \end{bmatrix}$ (b) $\begin{bmatrix} 1 \\ 2 \\ 3 \end{bmatrix}$ (c) $\begin{bmatrix} 1 & 0 \\ 2 & -1 \\ 1 & 1 \end{bmatrix}$

   (d) $\begin{bmatrix} 2 & 0 & 1 \\ 1 & 1 & 0 \\ 0 & -2 & 1 \\ -1 & 1 & 2 \end{bmatrix}$ (e) $\begin{bmatrix} 0 & 2 & -1 \\ 1 & -1 & 3 \\ 2 & 1 & 0 \\ -1 & 4 & 1 \end{bmatrix}$

8. In each case, find the linear combination of the first two vectors that is as close as possible to the third vector.

   (a) $[1, 2, 1], [2, 0, -1]; [3, -1, 0]$
   (b) $[1, 0, 1], [0, 1, 1]; [0, 0, 5]$
   (c) $[0, -2, 3], [1, 1, 1]; [1, -5, 10]$
   (d) $[2, 0, 1], [-1, 0, 1]; [4, 3, 2]$
   (e) $[0, 1, 1, 0], [1, -1, -1, 1]; [2, 0, 2, 0]$

9. (a) Factory A produces 30 cars, 40 light trucks, and 20 heavy trucks per day, while factory B produces 60 cars, 20 light trucks, and 20 heavy trucks a day. If the monthly demand is 1000 cars, 500 light trucks, and 400 heavy trucks, what is the least-squares solution (days of production for each factory)?

   (b) If the monthly demand increased by 10 cars, how much longer would factory A have to work each month?
   Hint: See Example 4 of Section 3.3.

   (c) If the monthly demand increased by 10 light trucks and 5 heavy trucks, how much longer would factory B have to work each month?

10. (a) Bureaucratic office A produces 40 new regulations, inspects 90 defective appliances, and approves 300 applications a week. Bureaucratic office B produces 80 new regulations, inspects 40 defec-

tive appliances, and approves 200 applications a week. How many weeks would each office have to work in order to best approximate (in the least-squares sense) a demand of producing 1000 new regulations, inspecting 700 defective appliances, and approving 2000 applications.

(b) What is the condition number of the matrix $(\mathbf{A}^T\mathbf{A})$ in the pseudoinverse computations? Is this problem poorly conditioned?

(c) If the demand for new regulations increased by 10, how much longer would office A have to work?

11. Consider the regression problem in which high school GPA and total SAT score (verbal plus math) are used to predict a person's college GPA:

$$\text{GPA college} = q_1(\text{GPA hi sch}) + q_2\left(\frac{\text{total SAT}}{1000}\right) + r$$

Suppose that our data for five people are as follows:

	GPA_{col}	GPA_{hi}	SAT
A	2.8	3.0	1.05
B	3.0	2.8	1.15
C	3.6	3.8	1.30
D	3.2	3.6	1.00
E	3.8	3.4	1.35

(a) Compute the pseudoinverse $(\mathbf{X}^T\mathbf{X})^{-1}\mathbf{X}^T$. In the process, determine the condition number of $(\mathbf{X}^T\mathbf{X})$. Is this problem poorly conditioned?

(b) Determine q_1, q_2, and r.

(c) Determine the error vector $\mathbf{e}$ (differences between true GPA-college and estimated GPA-college). Is it orthogonal to the estimated GPA-college vector?

12. In Example 5, re-solve the quadratic least-squares approximation problem for the following data points. Note that for parts (a) and (b) the x-values are the same, so the pseudoinverse will be the same (just use the $\mathbf{X}^+$ in the text).

(a) Same as in Example 5 but the fourth point is $(3, 5)$.

(b) Same as in Example 5 but the first point is $(0, 9)$.

(c) $(0, 7)$, $(1, 5)$, $(2, 7)$, $(3, 9)$, $(4, 13)$

(d) $(-2, 7)$, $(-1, 5)$, $(0, 4)$, $(1, 4)$, $(2, 8)$, $(3, 12)$; how is this problem related to the original problem in Example 5?

(e) $(0, 2)$, $(1, 4)$, $(2, 10)$

13. Fit a cubic polynomial to the following data points using the same idea as in the quadratic fit in Example 5: $(-1, -2)$, $(0, 3)$, $(1, 2)$, $(2, 8)$, $(3, 12)$, $(4, 100)$. What is the condition number of $(\mathbf{X}^T\mathbf{X})$?

14. Consider the regression model $\hat{z} = qx + ry + s$ for the following data, where the x-value is a scaled score (to have average value of 0) of high school grades, the y-value is a scaled score of SAT scores, and the z-value is a scaled score of college grades.

x	-4	-2	0	2	4
y	2	-1	-2	-1	2
z	3	6	7	7	6

Determine q, r, and s. Note that the **x**, **y**, and **1** vectors (in the regression matrix equation $\mathbf{z} = q\mathbf{x} + r\mathbf{y} + s\mathbf{1}$) are orthogonal.

15. Consider the regression model $y_i = qx_i$, $i = 1, 2, \ldots, n$. Compute the pseudoinverse for this regression problem and solve for q (in terms of the x_i, y_i values). As a matrix system $\mathbf{X}q = \mathbf{y}$, the matrix $\mathbf{X}$ is the n-by-1 column vector of x-values and $\mathbf{y}$ is the vector of y-values. Your answer should agree with the formula for q in equation (7).

16. Use a geometric picture to explain why if $\mathbf{Aw}$ is very close to $\mathbf{b}$, then the error vector $\mathbf{b} - \mathbf{Aw}$ may not be exactly orthogonal to the projection vector $\mathbf{Aw}$ in a least-squares solution to $\mathbf{Ax} = \mathbf{b}$.

17. Compute the cosine of angle, and determine the angle, made by the following pairs of vectors.
(a) $[1, 0], [1, 1]$ **(b)** $[3, 4], [-3, 4]$ **(c)** $[1, 2], [3, 1]$
(d) $[1, 0, 1], [0, 1, 0]$ **(e)** $[1, 1, 1], [1, -1, 2]$
(f) $[1, 1, 1], [2, -1, 3]$

18. Compute the correlation coefficient between the vectors of x- and y-values in Exercise 14, and between the vectors of x- and z-values in Exercise 14.

19. The following data show scores that three students received on a battery of six different tests.

	Gerry	Jimmie	Ronnie
General IQ	12	20	10
Mathematics	8	22	4
Reading	16	14	10
Running	24	16	12
Speaking	12	10	30
Watching	12	14	8

Compute the correlation coefficient between
(a) Gerry and Jimmie   (b) Gerry and Ronnie
(c) Jimmie and Ronnie

Hint: Remember first to subtract the average value from each number.

20. (a) Compute the correlation coefficient of the following readings from eight students of their IQs and scores at Zaxxon.

Student	A	B	C	D	E	F	G	H
IQ	120	130	105	90	125	120	110	160
Zaxxon	11,000	7,000	10,000	12,000	8,000	100,000	8,000	6,000

(b) Delete student *F* and recompute the correlation coefficient.

Hint: Remember first to subtract the average value from each number.

21. Suppose that there are two dials *A* and *B* on a machine that produces steel. We want to find out how settings a_i, b_i of the two dials affect the quality c_i of the steel. We use a regression model $\hat{c} = pa + qb + r$. For each of the following vectors **a** of settings for dial *A*, find a vector **b** of settings of dial *B* that is orthogonal to the dial *A* vector and also orthogonal to the 1's vector.
(a) $\mathbf{a} = [2, 1, 0, -1, -2]$ (b) $\mathbf{a} = [-4, -1, 0, 2, 3]$
(c) $\mathbf{a} = [2, 6, 1, -4, -3, -2]$

22. Prove that the pseudoinverse $\mathbf{A}^+$ equals the true inverse $\mathbf{A}^{-1}$ if the *n*-by-*n* matrix **A** is invertible (and hence has rank *n*).

23. (a) Show that in the euclidean norm $|\mathbf{a} - \mathbf{b}| \le |\mathbf{a}| + |\mathbf{b}|$ by squaring both sides of this inequality and using the law of cosines (on the left side).
(b) Show that in the euclidean norm $|\mathbf{a} + \mathbf{c}| \le |\mathbf{a}| + |\mathbf{c}|$ by letting $\mathbf{b} = -\mathbf{c}$ (and hence $-\mathbf{b} = \mathbf{c}$) and using part (a).

24. Show that if **a** is orthogonal to **b** and **c**, then **a** cannot be linearly dependent on **b** and **c**.

25. If the vectors in a basis of vector space *V* are mutually orthogonal to the vectors in a basis of vector space *W*, show that every vector in *V* is orthogonal to every vector in *W*.

26. For each of the following matrices **A**, express the 1's vector **1** (of the appropriate length) as the unique sum of two vectors, $\mathbf{1} = \mathbf{b}_1 + \mathbf{b}_2$, such that $\mathbf{b}_1$ is in Range(**A**) and $\mathbf{b}_2$ is in the error space of **A**. This unique sum exists by Theorem 7, part (i).

(a) $\begin{bmatrix} 1 \\ 2 \end{bmatrix}$ (b) $\begin{bmatrix} 1 \\ 2 \\ 3 \end{bmatrix}$ (c) $\begin{bmatrix} 1 & 0 \\ 2 & -1 \\ 1 & 1 \end{bmatrix}$

(d) $\begin{bmatrix} 2 & 0 & 1 \\ 1 & 1 & 0 \\ 0 & -2 & 1 \\ -1 & 1 & 2 \end{bmatrix}$ (e) $\begin{bmatrix} 0 & 2 & -1 \\ 1 & -1 & 3 \\ 2 & 1 & 0 \\ -1 & 4 & 1 \end{bmatrix}$

27. For each matrix **A** in Exercise 26, find a basis $\{\mathbf{v}_i\}$ for the Range(**A**) [= Col(**A**)] and a basis $\{\mathbf{w}_i\}$ for Null($\mathbf{A}^T$). Then verify that the $\mathbf{v}_i$ are orthogonal to the $\mathbf{w}_i$, as required by Theorem 7, part (ii).

28. For each of the following matrices **A**, express the 1's vector **1** as a unique sum, $\mathbf{1} = \mathbf{x}_1 + \mathbf{x}_2$ of a vector $\mathbf{x}_1$ in Row(**A**) and a vector $\mathbf{x}_2$ in Null(**A**).

(a) $\begin{bmatrix} 1 & -2 \\ -2 & 4 \end{bmatrix}$ (b) $\begin{bmatrix} 2 & 1 \\ 1 & 2 \end{bmatrix}$ (c) $\begin{bmatrix} 1 & 2 & 0 \\ 0 & 1 & 2 \end{bmatrix}$ (d) $\begin{bmatrix} 1 & 1 & 0 \\ 2 & 1 & 2 \\ 0 & 1 & 1 \end{bmatrix}$

29. Find a solution to $\mathbf{Ax} = \mathbf{1}$, for **A** the matrix in Exercise 28, part (c), in which **x** is in Row(**A**).

30. Use Theorem 8 to prove that if $\mathbf{v}_1, \mathbf{v}_2, \ldots, \mathbf{v}_k$ are a linearly independent set of vectors in the row space of a matrix **A**, then $\mathbf{w}_i = \mathbf{A}\mathbf{v}_i$ are a linearly independent set of vectors in the range of **A**. Thus, if $\{\mathbf{v}_i\}$ are a basis for Row(**A**), then $\{\mathbf{A}\mathbf{v}_i\}$ are a basis for Col(**A**).

Section 5.4 **Orthogonal Systems**

In Section 5.3 we saw that the calculation of the pseudoinverse of a matrix **A** simplified greatly if the columns of **A** were orthogonal. In this section we examine sets of orthogonal vectors further. If a set of vectors, such as the columns of a matrix, are not orthogonal, we give a procedure to transform them into an equivalent set of orthogonal vectors. Finally, we generalize the idea of an orthogonal set of vectors to build vector spaces for continuous functions generated by an orthogonal set of functions. *The idea of projecting one vector onto another vector is used over and over again in this section.* Such projections provide simple solutions to systems of equations $\mathbf{Ax} = \mathbf{b}$ for which the columns of **A** are orthogonal.

The underlying computational property that makes it easy to work with orthogonal columns is, if **a**, **b** are orthogonal, their scalar product $\mathbf{a} \cdot \mathbf{b} = 0$. Scalar products are the building blocks for much of matrix algebra (e.g., each entry in the product of two matrices is a scalar product). Thus computations with orthogonal vectors create a lot of 0's and hence yield simple results.

The inverse $\mathbf{A}^{-1}$ of a matrix $\mathbf{A}$ with orthogonal columns $\mathbf{a}_i^C$ is easy to describe. It is essentially the same as the pseudoinverse: $\mathbf{A}^{-1}$ is formed by dividing each column $\mathbf{a}_i^C$ by $\mathbf{a}_i^C \cdot \mathbf{a}_i^C$, the sum of the squares of its entries, and forming the transpose of the resulting matrix. Thus, if $s_i = \mathbf{a}_i^C \cdot \mathbf{a}_i^C$, then

$$
\mathbf{A}^{-1} = \begin{bmatrix} \dfrac{1}{s_1}\,\mathbf{a}_1^C \\[2ex] \dfrac{1}{s_2}\,\mathbf{a}_2^C \\[2ex] \vdots \\[1ex] \dfrac{1}{s_n}\,\mathbf{a}_n^C \end{bmatrix}
\tag{1}
$$

We verify (1) by noting that entry (i, j) in $\mathbf{A}^{-1}\mathbf{A}$ will be 0 if $i \neq j$ because $\mathbf{a}_i^C \cdot \mathbf{a}_j^C = 0$ (the columns are orthogonal). Entry (i, i) equals $(\mathbf{a}_i^C/s_i) \cdot \mathbf{a}_i^C = \mathbf{a}_i^C \cdot \mathbf{a}_i^C/(\mathbf{a}_i^C \cdot \mathbf{a}_i^C) = 1$.

Example 1. **Inverse of Matrix with Orthogonal Columns**

(i) Consider the matrix $\mathbf{A} = \begin{bmatrix} 3 & -4 \\ 4 & 3 \end{bmatrix}$, whose columns are othogonal. The sum of the squares of the entries in each column of $\mathbf{A}$ is $3^2 + 4^2 = 25$. If we divide each column by 25 and take the transpose, we obtain

$$
\mathbf{A}^{-1} = \begin{bmatrix} \frac{3}{25} & \frac{4}{25} \\[1ex] -\frac{4}{25} & \frac{3}{25} \end{bmatrix}
$$

The reader should check that this matrix is exactly what one would get by computing this 2-by-2 inverse using elimination.

(ii) Consider the orthogonal-column matrix

$$
\mathbf{A} = \begin{bmatrix} 2 & 1 & 0 \\ 1 & -1 & 1 \\ 1 & -1 & -1 \end{bmatrix}
$$

Its inverse, by (1), is

$$
\mathbf{A}^{-1} = \begin{bmatrix} \frac{2}{6} & \frac{1}{6} & \frac{1}{6} \\[1ex] \frac{1}{3} & -\frac{1}{3} & -\frac{1}{3} \\[1ex] 0 & \frac{1}{2} & -\frac{1}{2} \end{bmatrix}
$$

Again the reader should check that $\mathbf{A}^{-1}\mathbf{A} = \mathbf{I}$.   ■

Let use use (1) to obtain a formula for the ith component x_i in the solution $\mathbf{x}$ to $\mathbf{Ax} = \mathbf{b}$. Given the inverse $\mathbf{A}^{-1}$, we can find $\mathbf{x}$ as $\mathbf{x} = \mathbf{A}^{-1}\mathbf{b}$. The ith component in $\mathbf{A}^{-1}\mathbf{b}$ is the scalar product of ith row of $\mathbf{A}^{-1}$ with $\mathbf{b}$. By (1), the ith row of $\mathbf{A}^{-1}$ is $\mathbf{a}_i^C/(\mathbf{a}_i^C \cdot \mathbf{a}_i^C)$ and thus

$$x_i = \frac{\mathbf{a}_i^C \cdot \mathbf{b}}{\mathbf{a}_i^C \cdot \mathbf{a}_i^C} \tag{2}$$

Our old friend, the length of the projection of $\mathbf{b}$ onto column $\mathbf{a}_i^C$ (see Theorem 2 of Section 5.3).

A set of orthogonal vectors of unit length (whose norm is 1) are called **orthonormal**. The preceding formulas for x_i and $\mathbf{A}^{-1}$ become even nicer if the columns of $\mathbf{A}$ are orthonormal. In this case, $\mathbf{a}_i^C \cdot \mathbf{a}_i^C = 1$. Then the denominator in (2) is 1, so now the projection formula is $x_i = \mathbf{a}_i^C \cdot \mathbf{b}$. To obtain $\mathbf{A}^{-1}$, we divide each column of $\mathbf{A}$ by 1 and form the transpose: that is, $\mathbf{A}^{-1} = \mathbf{A}^T$. Summarizing this discussion, we have

Theorem 1

  (i) If $\mathbf{A}$ is an n-by-n matrix whose columns are orthogonal, then $\mathbf{A}^{-1}$ is obtained by dividing the ith column of $\mathbf{A}$ by the sum of the squares of its entries and transposing the resulting matrix [see (1)]. The ith component x_i in the solution of $\mathbf{Ax} = \mathbf{b}$ is the length of the projection of $\mathbf{b}$ on $\mathbf{a}_i^C$: $x_i = \mathbf{a}_i^C \cdot \mathbf{b}/\mathbf{a}_i^C \cdot \mathbf{a}_i^C$.
  (ii) If the columns of $\mathbf{A}$ are orthonormal, then the inverse $\mathbf{A}^{-1}$ is $\mathbf{A}^T$ and the length of the projection is just $x_i = \mathbf{a}_i^C \cdot \mathbf{b}$.

Suppose that we have a basis of n orthogonal vectors $\mathbf{q}_i$ for n-space. If $\mathbf{Q}$ has the $\mathbf{q}_i$ as its columns, the solution $\mathbf{x} = \mathbf{b}^*$ of $\mathbf{Qx} = \mathbf{b}$ will be a vector $\mathbf{b}^*$ of lengths of the projections of $\mathbf{b}$ onto each $\mathbf{q}_i$:

$$\mathbf{Qb}^* = b_1^*\mathbf{q}_1 + b_2^*\mathbf{q}_2 + \cdots + b_n^*\mathbf{q}_n = \mathbf{b} \tag{3}$$

Here the term $b_1^*\mathbf{q}_1$ is just the projection of $\mathbf{b}$ onto $\mathbf{q}_1$. So (3) simply says

Corollary. Any n-vector $\mathbf{b}$ can be expressed as the sum of the projections of $\mathbf{b}$ onto a set of n orthogonal vectors $\mathbf{q}_i$.

Example 2. Conversion of Coordinates from One Basis to Another

Consider the orthonormal basis $\mathbf{q}_1 = [.8, .6]$, $\mathbf{q}_2 = [-.6, .8]$ for 2-space. To express the vector $\mathbf{b} = [1, 2]$ in terms of $\mathbf{q}_1, \mathbf{q}_2$ coordinates, we need to solve the system

$$b_1^* \begin{bmatrix} .8 \\ .6 \end{bmatrix} + b_2^* \begin{bmatrix} -.6 \\ .8 \end{bmatrix} = \begin{bmatrix} 1 \\ 2 \end{bmatrix}$$

Figure 5.6

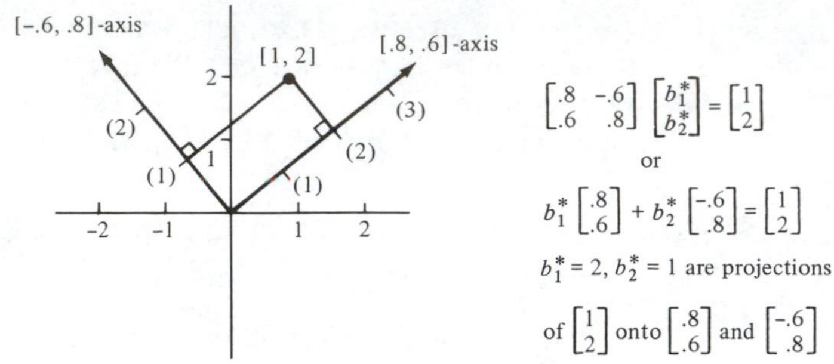

$$\begin{bmatrix} .8 & -.6 \\ .6 & .8 \end{bmatrix} \begin{bmatrix} b_1^* \\ b_2^* \end{bmatrix} = \begin{bmatrix} 1 \\ 2 \end{bmatrix}$$

or

$$b_1^* \begin{bmatrix} .8 \\ .6 \end{bmatrix} + b_2^* \begin{bmatrix} -.6 \\ .8 \end{bmatrix} = \begin{bmatrix} 1 \\ 2 \end{bmatrix}$$

$b_1^* = 2, b_2^* = 1$ are projections

of $\begin{bmatrix} 1 \\ 2 \end{bmatrix}$ onto $\begin{bmatrix} .8 \\ .6 \end{bmatrix}$ and $\begin{bmatrix} -.6 \\ .8 \end{bmatrix}$

or

$$\mathbf{Qb}^* = \mathbf{b}, \qquad \text{where } \mathbf{Q} = [\mathbf{q}_1 \quad \mathbf{q}_2]$$

By Theorem 1,

$$b_1^* = \mathbf{q}_1 \cdot \mathbf{b} = .8 \times 1 + .6 \times 2 = 2,$$
$$b_2^* = \mathbf{q}_2 \cdot \mathbf{b} = -.6 \times 1 + .8 \times 2 = 1$$

where $b_1^* \mathbf{q}_1 = 2[.8, .6]$ is the projection of $\mathbf{b}$ on $\mathbf{q}_1$, and $b_2^* \mathbf{q}_2 = [-.6, .8]$ is the projection of $\mathbf{b}$ on $\mathbf{q}_2$. Thus $\mathbf{b} = [1, 2]$ is expressed as an $\mathbf{e}_1 - \mathbf{e}_2$ coordinate vector, while $\mathbf{b}^* = [2, 1]$ is the same vector expressed in $\mathbf{q}_1 - \mathbf{q}_2$ coordinates. A geometric picture of this conversion is given in Figure 5.6, where the vector $[2, 1]$ is depicted as the sum of its projection onto $\mathbf{q}_1$ and onto $\mathbf{q}_2$. ∎

Theorem 1 is a carbon copy of Theorem 5 of Section 5.3 about pseudoinverses when columns are orthogonal. As with the inverse, if $\mathbf{A}$'s columns are orthonormal, the pseudoinverse $\mathbf{A}^+$ of $\mathbf{A}$ will simply be $\mathbf{A}^T$. The following example gives a familiar illustration of this result and shows why orthogonal columns make inverses and pseudoinverse so similar.

Example 3. **Pseudoinverse of Matrix with Orthonormal Columns**

Let $\mathbf{I}_2$ be the first two columns of the 3-by-3 identity matrix.

$$\mathbf{I}_2 = \begin{bmatrix} 1 & 0 \\ 0 & 1 \\ 0 & 0 \end{bmatrix}$$

Then

$$\mathbf{I}_2^+ = \mathbf{I}_2^T = \begin{bmatrix} 1 & 0 & 0 \\ 0 & 1 & 0 \end{bmatrix}$$

For any vector $\mathbf{b} = [b_1, b_2, b_3]$, the least-squares solution $\mathbf{x} = \mathbf{b}^*$ to $\mathbf{I}_2\mathbf{x} = \mathbf{b}$ is

$$\mathbf{b^*} = \mathbf{I_2^+b} = \begin{bmatrix} 1 & 0 & 0 \\ 0 & 1 & 0 \end{bmatrix} \begin{bmatrix} b_1 \\ b_2 \\ b_3 \end{bmatrix} = \begin{bmatrix} b_1 \\ b_2 \end{bmatrix}$$

This result confirms our intuitive notion that $[b_1, b_2, 0]$ is the closest point in the x-y plane to the point $[b_1, b_2, b_3]$. ∎

Optional

There is another interesting geometric fact about orthonormal columns (see the Exercises for the two-dimensional case).

Theorem 2. When $\mathbf{Q}$ has orthonormal columns, then solving $\mathbf{Qx} = \mathbf{b}$ for $\mathbf{b^*} = \mathbf{Q^Tb}$ is equivalent to performing the orthonormal change of basis $\mathbf{b} \rightarrow \mathbf{b^*} = \mathbf{Q^Tb}$. Such a basis change is simply a rotation of the coordinate axes, a reflection through a plane, or a combination of both. The entries in $\mathbf{Q}$ can be expressed in terms of the sines and cosines of the angles of this rotation.

For example, the rotation of axis in the plane by $\theta°$ is a linear transformation R of 2-space:

$$R: \quad \begin{aligned} x' &= x \cos \theta° + y \sin \theta° \\ y' &= -x \sin \theta° + y \cos \theta° \end{aligned} \quad \text{or} \quad \mathbf{u'} = \mathbf{Au}$$

where

$$\mathbf{A} = \begin{bmatrix} \cos \theta° & \sin \theta° \\ -\sin \theta° & \cos \theta° \end{bmatrix}$$

It is easy to check that $\mathbf{A}$ has orthonormal columns.

It follows that the distance between a pair of vectors and the angle that they form do not change with an orthonormal change of basis.

(**Note:** End of optional material.)

Orthogonal columns have another important advantage besides easy formulas. A highly nonorthogonal set of columns—that is, columns that are almost parallel—can result in unstable computations.

Example 4. **Nonorthogonal Columns**

Consider the following system of equations:

$$\begin{aligned} 1x_1 + .75x_2 &= 5 \\ 1x_1 + 1x_2 &= 7 \end{aligned} \tag{4}$$

Let us call the two column vectors in the coefficient matrix of (4): $\mathbf{u} = [1, 1]$ and $\mathbf{v} = [.75, 1]$. The cosine of their angle is, by Theorem 6 of Section 5.3,

$$\cos \theta(\mathbf{u}, \mathbf{v}) = \frac{\mathbf{u} \cdot \mathbf{v}}{|\mathbf{u}||\mathbf{v}|} = \frac{1.75}{\sqrt{2} \cdot 1.25} = .99 \tag{5}$$

The angle with cosine of .99 is 8°. Thus $\mathbf{u}$ and $\mathbf{v}$ are almost parallel (almost the same vector). Representing *any* 2-vector $\mathbf{b}$ as a linear combination of two vectors that are almost the same is tricky, that is, unstable. For example, to solve (4) we must we find weights x_1, x_2 such that

$$x_1 \begin{bmatrix} 1 \\ 1 \end{bmatrix} + x_2 \begin{bmatrix} .75 \\ 1 \end{bmatrix} = \begin{bmatrix} 5 \\ 7 \end{bmatrix} \tag{6}$$

The system (4) is the canoe-with-sail system from Section 1.1. We already know that calculations with $\mathbf{A}$, the coefficient matrix in (4), are very unstable. In Section 3.5 we computed the condition number of $\mathbf{A}$ to be $c(\mathbf{A}) \simeq 16$. Recall that the condition number $c(\mathbf{A}) = \|\mathbf{A}\| \cdot \|\mathbf{A}^{-1}\|$ measures how much a relative error in the entries of $\mathbf{A}$ (or in $\mathbf{b}$) could affect the relative error in $\mathbf{x} = [x_1, x_2]$; in this case, a 5% error in $\mathbf{b}$ could cause an error 16 $[= c(\mathbf{A})]$ times greater in $\mathbf{x}$, a $16 \times 5\% = 80\%$ error.

We solved (4) in Section 1.1 and obtained $x_1 = -1$, $x_2 = 7$. If we had solved for $\mathbf{b}' = [7, 5]$, we would have obtained the answer $x_1 = 13$, $x_2 = -8$ (see Figure 5.7 for a picture of this result). Or for $\mathbf{b}'' = [6, 6]$, $x_1 = 6$, $x_2 = 0$. ■

Figure 5.7

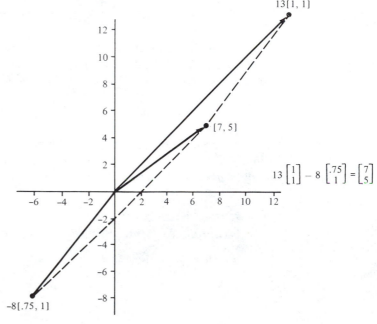

Reading the results of Example 4 in reverse, we see that when errors arise in solving an ill-conditioned system of equations $\mathbf{Ax} = \mathbf{b}$ (in which $\mathbf{A}$ has a large condition number), the problem should be that some column vector (or a linear combination of them) forms a small angle with another column vector—this means that the columns are almost linearly dependent. If the columns were close to mutually orthogonal, the system $\mathbf{Ax} = \mathbf{b}$ would be well-conditioned.

Principle. Let $\mathbf{A}$ be an *n*-by-*n* matrix with rank($\mathbf{A}$)= n so that the system of equations $\mathbf{Ax} = \mathbf{b}$ has a unique solution. The solution to $\mathbf{Ax} = \mathbf{b}$ will be more or less stable according to how close or far from orthogonal the column vectors of $\mathbf{A}$ are.

Suppose that the columns of the *n*-by-*n* matrix $\mathbf{A}$ are linearly independent but not orthogonal. We shall show how to find a new *n*-by-*n* matrix $\mathbf{A}^*$ of orthonormal columns (orthogonal and unit length) that are linear combinations of the columns of $\mathbf{A}$.

Our procedure can be applied to any basis $\mathbf{a}_1, \mathbf{a}_2, \ldots, \mathbf{a}_m$ of an *m*-dimensional space V and will yield a new basis of m orthonormal vectors $\mathbf{q}_i$ for V (unit-length vectors make calculations especially simple). The procedure is inductive in the sense that the first k $\mathbf{q}_i$ will be an orthonormal basis for the space V_k generated by the first k $\mathbf{a}_i$. The method is called **Gram–Schmidt orthogonalization**.

For $k = 1$, $\mathbf{q}_1$ should be a multiple of $\mathbf{a}_1$. To make $\mathbf{q}_1$ have norm 1, we set $\mathbf{q}_1 = \mathbf{a}_1/|\mathbf{a}_1|$. Next we must construct from $\mathbf{a}_2$ a second unit vector $\mathbf{q}_2$ orthogonal to $\mathbf{q}_1$. We divide $\mathbf{a}_2$ into two "parts": the part of $\mathbf{a}_2$ parallel to $\mathbf{q}_1$ and the part of $\mathbf{a}_2$ orthogonal (perpendicular) to $\mathbf{q}_1$ (see Figure 5.8). The component of $\mathbf{a}_2$ in $\mathbf{q}_1$'s direction is simply the projection of $\mathbf{a}_2$ onto $\mathbf{q}_1$. This projection is $s\mathbf{q}_1$, where the length s of the projection is

$$s = \frac{\mathbf{a}_2 \cdot \mathbf{q}_1}{\mathbf{q}_1 \cdot \mathbf{q}_1} = \mathbf{a}_2 \cdot \mathbf{q}_1 \tag{7}$$

since $\mathbf{q}_1 \cdot \mathbf{q}_1 = 1$. The rest of $\mathbf{a}_2$, the vector $\mathbf{a}_2 - s\mathbf{q}_1$, is orthogonal to the projection $s\mathbf{q}_1$, and hence orthogonal to $\mathbf{q}_1$. So $\mathbf{a}_2 - s\mathbf{q}_1$ is the orthogonal vector we want for $\mathbf{q}_2$. To have unit norm, we set $\mathbf{q}_2 = (\mathbf{a}_2 - s\mathbf{q}_1)/|\mathbf{a}_2 - s\mathbf{q}_1|$.

Let us show how the procedure works thus far.

Figure 5.8 Gram–Schmidt orthogonalization.

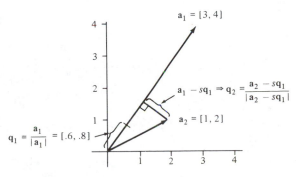

Example 5. Gram–Schmidt Orthogonalization in Two Dimensions

Suppose that $\mathbf{a}_1 = [3, 4]$ and $\mathbf{a}_2 = [2, 1]$ (see Figure 5.8). We set

$$\mathbf{q}_1 = \frac{\mathbf{a}_1}{|\mathbf{a}_1|} = \frac{[3, 4]}{5} = \left[\frac{3}{5}, \frac{4}{5}\right]$$

We project $\mathbf{a}_2$ onto $\mathbf{q}_1$ to get the part of $\mathbf{a}_2$ parallel to $\mathbf{q}_1$. From (7), the length of the projection is

$$s = \mathbf{a}_2 \cdot \mathbf{q}_1 = [2, 1] \cdot \left[\tfrac{3}{5}, \tfrac{4}{5}\right] = \tfrac{10}{5} = 2$$

and the projection is $s\mathbf{q}_1 = 2\left[\tfrac{3}{5}, \tfrac{4}{5}\right] = \left[\tfrac{6}{5}, \tfrac{8}{5}\right]$. Next we determine the other part of $\mathbf{a}_2$, the part orthogonal to $s\mathbf{q}_1$:

$$\mathbf{a}_2 - s\mathbf{q}_1 = [2, 1] - \left[\tfrac{6}{5}, \tfrac{8}{5}\right] = \left[\tfrac{4}{5}, -\tfrac{3}{5}\right]$$

Since $\left|\left[\tfrac{4}{5}, -\tfrac{3}{5}\right]\right| = 1$, then

$$\mathbf{q}_2 = \frac{\mathbf{a}_2 - s\mathbf{q}_1}{|\mathbf{a}_2 - s\mathbf{q}_1|} = \frac{\left[\tfrac{4}{5}, -\tfrac{3}{5}\right]}{1} = \left[\frac{4}{5}, -\frac{3}{5}\right]$$
∎

We extend the previous construction by finding the projections of $\mathbf{a}_3$ onto $\mathbf{q}_1$ and $\mathbf{q}_2$. Then the vector $\mathbf{a}_3 - s_1\mathbf{q}_1 - s_2\mathbf{q}_2$, which is orthogonal to $\mathbf{q}_1$ and $\mathbf{q}_2$ should be $\mathbf{q}_3$; as before, we divide $\mathbf{a}_3 - s_1\mathbf{q}_1 - s_2\mathbf{q}_2$ by its norm to make $\mathbf{q}_3$ unit length. We continue this process to find $\mathbf{q}_4$, $\mathbf{q}_5$, and so on.

Example 6. Gram–Schmidt Orthogonalization of 3-by-3 Matrix

Let us perform orthogonalization on the matrix $\mathbf{A}$ whose ith column we denote by $\mathbf{a}_i$.

$$\mathbf{A} = \begin{bmatrix} 0 & 3 & 2 \\ 3 & 5 & 5 \\ 4 & 0 & 5 \end{bmatrix} \tag{8}$$

First $\mathbf{q}_1 = \mathbf{a}_1/|\mathbf{a}_1| = [0, 3, 4]/5 = \left[0, \tfrac{3}{5}, \tfrac{4}{5}\right]$.
The length of the projection $\mathbf{a}_2$ onto $\mathbf{q}_1$ is

$$s = \mathbf{a}_2 \cdot \mathbf{q}_1 = 3 \cdot 0 + 5 \cdot \tfrac{3}{5} + 0 \cdot \tfrac{4}{5} = 3 \tag{9a}$$

So the projection of $\mathbf{a}_2$ onto $\mathbf{q}_1$ is

$$s\mathbf{q}_1 = 3\left[0, \tfrac{3}{5}, \tfrac{4}{5}\right] = \left[0, \tfrac{9}{5}, \tfrac{12}{5}\right]$$

Next we compute

$$\mathbf{a}_2 - s\mathbf{q}_1 = [3, 5, 0] - \left[0, \tfrac{9}{5}, \tfrac{12}{5}\right] = \left[3, \tfrac{16}{5}, -\tfrac{12}{5}\right]$$

where $|\mathbf{a}_2 - s\mathbf{q}_1| = \sqrt{9 + 256/25 + 144/25} = 5$. Then

$$\mathbf{q}_2 = \frac{\mathbf{a}_2 - s\mathbf{q}_1}{|\mathbf{a}_2 - s\mathbf{q}_1|} = \left[\frac{3}{5}, \frac{16}{25}, -\frac{12}{25}\right]$$

We compute the length of the projections of $\mathbf{a}_3$ onto $\mathbf{q}_1$ and $\mathbf{q}_2$:

$$s_1 = \mathbf{a}_3 \cdot \mathbf{q}_1 = 2 \cdot 0 + \frac{5 \cdot 3}{5} + \frac{5 \cdot 4}{5} = 3 + 4 = 7$$

$$s_2 = \mathbf{a}_3 \cdot \mathbf{q}_2 = \frac{2 \cdot 3}{5} + \frac{5 \cdot 16}{25} + 5 \cdot \left(\frac{-12}{25}\right) \tag{9b}$$

$$= \frac{6}{5} + \frac{16}{5} - \frac{12}{5} = 2$$

Then

$$s_1\mathbf{q}_1 = 7\left[0, \tfrac{3}{5}, \tfrac{4}{5}\right] = \left[0, \tfrac{21}{5}, \tfrac{28}{5}\right]$$

$$s_2\mathbf{q}_2 = 2\left[\tfrac{3}{5}, \tfrac{16}{25}, -\tfrac{12}{25}\right] = \left[\tfrac{6}{5}, \tfrac{32}{25}, -\tfrac{24}{25}\right]$$

and

$$\mathbf{a}_3 - s_1\mathbf{q}_1 - s_2\mathbf{q}_2 = [2, 5, 5] - \left[0, \tfrac{21}{5}, \tfrac{28}{5}\right] - \left[\tfrac{6}{5}, \tfrac{32}{25}, -\tfrac{24}{25}\right]$$

$$= \left[\tfrac{4}{5}, -\tfrac{12}{25}, \tfrac{9}{25}\right]$$

Since computation reveals that $|\mathbf{a}_3 - s_1\mathbf{q}_1 - s_2\mathbf{q}_2| = 1$, then

$$\mathbf{q}_3 = (\mathbf{a}_3 - s_1\mathbf{q}_1 - s_2\mathbf{q}_2) = \left[\tfrac{4}{5}, -\tfrac{12}{25}, \tfrac{9}{25}\right]$$

The matrix of these new orthogonal column vectors is

$$\mathbf{Q} = \begin{bmatrix} 0 & \frac{3}{5} & \frac{4}{5} \\ \frac{3}{5} & \frac{16}{25} & -\frac{12}{25} \\ \frac{4}{5} & -\frac{12}{25} & \frac{9}{25} \end{bmatrix} \tag{10}$$

■

In keeping with the principle above, the accuracy of this procedure depends on how close to and far from orthogonality the columns $\mathbf{a}_i$ are. If a linear combination of some $\mathbf{a}_i$ forms a small angle with another vector $\mathbf{a}_k$ (this means the matrix $\mathbf{A}$ has a large condition number), then the resulting $\mathbf{q}_i$ will have errors, making them not exactly orthogonal. However, more stable methods are available using advanced techniques, such as Householder transformations.

Suppose that the columns of $\mathbf{A}$ are not linearly independent. If, say, $\mathbf{a}_3$ is a linear combination of $\mathbf{a}_1$ and $\mathbf{a}_2$, then in the Gram–Schmidt procedure the error vector $\mathbf{a}_3 - s_1\mathbf{q}_1 - s_2\mathbf{q}_2$ with respect to $\mathbf{q}_1$ and $\mathbf{q}_2$ will be $\mathbf{0}$. In this case we skip $\mathbf{a}_3$ and use $\mathbf{a}_4 - s_1\mathbf{q}_1 - s_2\mathbf{q}_2$ to define $\mathbf{q}_3$. The number of vectors $\mathbf{q}_i$ formed will be the dimension of the column space of $\mathbf{A}$, that is, rank($\mathbf{A}$).

The effect of the orthogonalization process can be represented by an upper triangular matrix $\mathbf{R}$ so that one obtains the matrix factorization

Theorem 3. Any *m*-by-*n* matrix $\mathbf{A}$ can be factored in the form

$$\mathbf{A} = \mathbf{QR} \tag{11}$$

where $\mathbf{Q}$ is the *m*-by-rank($\mathbf{A}$) matrix with orthonormal columns $\mathbf{q}_i$ obtained by Gram–Schmidt orthogonalization, and $\mathbf{R}$ is an upper triangular matrix of size rank($\mathbf{A}$)-by-*n* (described below).

For $i < j$, entry r_{ij} of $\mathbf{R}$ is $\mathbf{a}_j \cdot \mathbf{q}_i$, the projection of $\mathbf{a}_j$ onto $\mathbf{q}_i$. The diagonal entries in $\mathbf{R}$ are the sizes, before normalization, of the new columns: $r_{11} = |\mathbf{a}_1|$, $r_{22} = |\mathbf{a}_2 - s\mathbf{q}_1|$, $r_{33} = |\mathbf{a}_3 - s_1\mathbf{q}_1 - s_2\mathbf{q}_2|$, and so on.

Example 7. *QR* Decomposition

Give the $\mathbf{QR}$ decomposition for the matrix $\mathbf{A}$ in Example 6.

$$\mathbf{A} = \begin{bmatrix} 0 & 3 & 2 \\ 3 & 5 & 5 \\ 4 & 0 & 5 \end{bmatrix}$$

The orthonormal matrix $\mathbf{Q}$ is given in (10). We form $\mathbf{R}$ from the information about the sizes of new columns and the projections as described in the preceding paragraph. Here $r_{12} = s = 3$ in (9a), and $r_{13} = s_1 = 7$, $r_{23} = s_2 = 2$ in (9b). Then

$$\mathbf{QR} = \begin{bmatrix} 0 & \frac{3}{5} & \frac{4}{5} \\ \frac{3}{5} & \frac{16}{25} & -\frac{12}{25} \\ \frac{4}{5} & -\frac{12}{25} & \frac{9}{25} \end{bmatrix} \begin{bmatrix} 5 & 3 & 7 \\ 0 & 5 & 2 \\ 0 & 0 & 1 \end{bmatrix}$$

Let us compute the second column of $\mathbf{QR}$—multiplying $\mathbf{Q}$ by $\mathbf{r}_2^C$, the second column of $\mathbf{R}$—and show that the result is $\mathbf{a}_2$, the second column of $\mathbf{A}$.

$$\mathbf{Qr}_2^C = \begin{bmatrix} 0 & \frac{3}{5} & \frac{4}{5} \\ \frac{3}{5} & \frac{16}{25} & -\frac{12}{25} \\ \frac{4}{5} & -\frac{12}{25} & \frac{9}{25} \end{bmatrix} \begin{bmatrix} 3 \\ 5 \\ 0 \end{bmatrix}$$

$$= 3 \begin{bmatrix} 0 \\ \frac{3}{5} \\ \frac{4}{5} \end{bmatrix} + 5 \begin{bmatrix} \frac{3}{5} \\ \frac{16}{25} \\ -\frac{12}{25} \end{bmatrix} + 0 \begin{bmatrix} \frac{4}{5} \\ -\frac{12}{25} \\ \frac{9}{25} \end{bmatrix} = \begin{bmatrix} 3 \\ 5 \\ 0 \end{bmatrix} = \mathbf{a}_2 \tag{12}$$

■

Columns of $\mathbf{Q}$ are obtained from linear combinations of the columns of $\mathbf{A}$. Reversing this procedure yields the columns of $\mathbf{A}$ as linear combinations of the columns of $\mathbf{Q}$. This reversal is what is accomplished by the matrix product $\mathbf{QR}$. Consider the computation in (12). In terms of the columns $\mathbf{q}_i$ of $\mathbf{Q}$, (12) is

$$3\mathbf{q}_1 + 5\mathbf{q}_2 + 0\mathbf{q}_3 = \mathbf{a}_2$$

or, in terms of $\mathbf{R}$,

$$r_{12}\mathbf{q}_1 + r_{22}\mathbf{q}_2 = \mathbf{a}_2 \tag{13}$$

($\mathbf{a}_2$ equals its projection onto $\mathbf{q}_1$ plus its projection onto $\mathbf{q}_2$).

Next consider the formula for $\mathbf{q}_2$:

$$\mathbf{q}_2 = \frac{\mathbf{a}_2 - s\mathbf{q}_1}{|\mathbf{a}_2 - s\mathbf{q}_1|} = \frac{\mathbf{a}_2 - r_{12}\mathbf{q}_1}{r_{22}} \tag{14}$$

since $r_{22} = |\mathbf{a}_2 - s\mathbf{q}_1|$ and $r_{12} = s$. Solving for $\mathbf{a}_2$ in (14), we obtain (13)

$$\mathbf{q}_2 = \frac{\mathbf{a}_2 - r_{12}\mathbf{q}_1}{r_{22}} \quad \rightarrow \quad r_{12}\mathbf{q}_1 + r_{22}\mathbf{q}_2 = \mathbf{a}_2$$

The same analysis shows that the jth column in the product $\mathbf{QR}$ is just a reversal of the orthogonalization steps for finding $\mathbf{q}_j$.

The matrix $\mathbf{R}$ is upper triangular because column $\mathbf{a}_i$ is only involved in building columns $\mathbf{q}_i, \mathbf{q}_{i+1}, \ldots, \mathbf{q}_n$ of $\mathbf{Q}$. *The* $\mathbf{QR}$ *decomposition is the column counterpart to the* $\mathbf{LU}$ *decomposition*, given in Section 3.2, in which the row combinations of Gaussian elimination are reversed to obtain the matrix $\mathbf{A}$ from its row-reduced matrix $\mathbf{U}$.

The $\mathbf{QR}$ decomposition is used frequently in numerical procedures. We use it to find eigenvalues in the appendix to Section 5.5.

We will sketch one of its most frequent uses, finding the inverse or pseudoinverse of an ill-conditioned matrix. If $\mathbf{A}$ is an n-by-n matrix with linearly independent columns, the decomposition $\mathbf{A} = \mathbf{QR}$ yields

$$\mathbf{A}^{-1} = (\mathbf{QR})^{-1} = \mathbf{R}^{-1}\mathbf{Q}^{-1} = \mathbf{R}^{-1}\mathbf{Q}^T \tag{15}$$

The fact that $\mathbf{Q}^{-1} = \mathbf{Q}^T$ when $\mathbf{Q}$ has orthonormal columns was part of Theorem 1. Given the QR decomposition of $\mathbf{A}$, (15) says that to get $\mathbf{A}^{-1}$, we only need to determine $\mathbf{R}^{-1}$. Since $\mathbf{R}$ is an upper triangular matrix, its inverse is obtained quickly by back substitution (see Exercise 12 of Section 3.5). When $\mathbf{A}$ is very ill-conditioned, one should compute $\mathbf{A}^{-1}$ via (15): first, determining the QR decomposition of $\mathbf{A}$, using advanced (more stable) variations of the Gram–Schmidt procedure; then determining $\mathbf{R}^{-1}$; and thus obtaining $\mathbf{A}^{-1} = \mathbf{R}^{-1}\mathbf{Q}^T$.

Equation (15) extends to pseudoinverses. That is, if $\mathbf{A}$ is an m-by-n

matrix with linearly independent columns and $m > n$, then its pseudoinverse $\mathbf{A}^+$ can be computed as

$$\mathbf{A}^+ = \mathbf{R}^{-1}\mathbf{Q}^T \tag{16}$$

See the Exercises for instructions on how to verify (16) and examples of its use. *This formula for the pseudoinverse is the standard way pseudoinverses are computed in practice.* Even if one determines $\mathbf{Q}$ and $\mathbf{R}$ using the basic Gram–Schmidt procedure given above, the resulting $\mathbf{A}^+$ from (16) will be substantially more accurate than computing $\mathbf{A}^+$ using the standard formula $\mathbf{A}^+ = (\mathbf{A}^T\mathbf{A})^{-1}\mathbf{A}^T$, because the matrix $\mathbf{A}^T\mathbf{A}$ tends to be ill-conditioned. For example, in the least-squares polynomial-fitting problem in Example 5 of Section 5.4, the condition number of the 3-by-3 matrix $\mathbf{X}^T\mathbf{X}$ was around 2000!

Principle. Because of conditioning problems, the pseudoinverse $\mathbf{A}^+$ of a matrix $\mathbf{A}$ should be computed by the formula $\mathbf{A}^+ = \mathbf{R}^{-1}\mathbf{Q}^T$, where $\mathbf{Q}$ and $\mathbf{R}$ are the matrices in the $\mathbf{QR}$ decomposition of $\mathbf{A}$.

We now introduce a very different use of orthogonality. Our goal is to make a vector space for the set of all continuous functions. To make matters a little easier, let us focus on functions that can be expressed as a polynomial or infinite series in powers of x, such as $x^3 + 3x^2 - 4x + 1$ or e^x or $\sin x$.

Recall that the defining property of a vector space V is that if $\mathbf{u}$ and $\mathbf{v}$ are in V, then $r\mathbf{u} + s\mathbf{v}$ is also in V, for any scalars r, s. Clearly, linear combinations of polynomials (or infinite series) are again polynomials (or infinite series), so these functions form a vector space.

For a vector space of functions to be useful, we need a coordinate system, that is, a basis of independent functions $u_i(x)$ (functions that are not linearly dependent on each other) so that any function $f(x)$ can be expressed as a linear combination of these basis functions.

$$f(x) = f_1 u_1(x) + f_2 u_2(x) + \cdots \tag{17}$$

This basis will need to be infinite and the linear combinations of basis functions may also be infinite. The best basis would use orthogonal, or even better, orthonormal functions.

To make an orthogonal basis, we first need to extend the definition of a scalar, or inner, product $\mathbf{c} \cdot \mathbf{d}$ of vectors to an inner product of functions. The **inner product of two functions** $f(x)$ and $g(x)$ on the interval $[a, b]$ is defined as

$$f(x) \cdot g(x) = \int_a^b f(x)g(x)\, dx \tag{18}$$

This definition is a natural generalization of the standard inner product $\mathbf{c} \cdot \mathbf{d}$ in that both $\mathbf{c} \cdot \mathbf{d}$ and $f(x) \cdot g(x)$ form sums of term-by-term products of the respective entities, but in (18) we have a continuous sum, an integral.

With an inner product defined, most of the theory and formulas defined for vector spaces can be applied to our space of functions. The inner product tells us when two vectors $\mathbf{c}$, $\mathbf{d}$ are orthogonal (if $\mathbf{c} \cdot \mathbf{d} = 0$), and allows us to compute coordinates c_i^* of $\mathbf{c}$ in an orthonormal basis $\mathbf{u}_i$: $c_i^* = \mathbf{c} \cdot \mathbf{u}_i$ (these coordinates are just the projections of $\mathbf{c}$ onto the $\mathbf{u}_i$). We can now do the same calculations for functions with (18).

The functional equivalent of the euclidean norm is defined by

$$|f(x)|^2 = f(x) \cdot f(x) = \int_a^b f(x)^2 \, dx \tag{19}$$

The counterpart of the sum norm $|\mathbf{c}|_s = \Sigma \, |c_i|$ for vectors is $|f(x)|_s = \int |f(x)| \, dx$.

An orthonormal basis for our functions on the interval $[a, b]$ will be a set of functions $\{u_i(x)\}$ which are orthogonal—by (18), $\int u_i(x)u_j(x) \, dx = 0$, for all $i \neq j$—and whose norms are 1—by (19), $\int u_i(x)^2 \, dx = 1$. Given such an orthonormal basis $\{u_i(x)\}$, the coordinates f_i of a function $f(x)$ in terms of the $u_i(x)$ are computed by the projection formula $f_i = f(x) \cdot u_i(x)$ used for n-dimensional orthonormal bases:

$$f(x) = [f(x) \cdot u_1(x)]u_1(x) + [f(x) \cdot u_2(x)]u_2(x) + \cdots \tag{20}$$

How do we find such an orthonormal basis? The first obvious choice is the set of powers of x: 1, x, x^2, x_3, ... These are linearly independent; that is, x^k cannot be expressed as a linear combination of smaller powers of x. Unfortunately, *there is no interval on which* 1, x, *and* x^2 *are mutually orthogonal.* On $[-1, 1]$, $1 \cdot x = \int x \, dx = 0$ and $x \cdot x^2 = \int x^3 \, dx = 0$, but $1 \cdot x^2 = \int x^2 \, dx = \frac{2}{3}$.

There are many sets of orthogonal functions that have been developed over the years. We shall mention two, Legendre polynomials and Fourier trigonometric functions.

The Gram–Schmidt orthogonalization procedure provides a way to build an orthonormal basis out of a basis of linearly independent vectors. The calculations in this procedure use inner products, and hence this procedure can be applied to the powers of x (which are linearly independent but, as we just said, far from orthogonal) to find an orthonormal set of polynomials.

When the interval is $[-1, 1]$, the polynomials obtained by orthogonalization are called **Legendre polynomials** $L_k(x)$. Actually, we shall not worry about making their norms equal to 1. As noted above, the functions $x^0 = 1$ and x are orthogonal on $[-1, 1]$. So $L_0(x) = 1$ and $L_1(x) = x$. Also, x^2 is orthogonal to x but not to 1 on $[-1, 1]$. We must subtract off the projection of x^2 onto 1:

$$L_2(x) = x^2 - \left(\frac{1 \cdot x^2}{1 \cdot 1}\right) 1 = x^2 - \frac{\int x^2 \, dx}{\int 1 \, dx} = x^2 - \frac{\frac{2}{3}}{2} = x^2 - \frac{1}{3} \tag{21}$$

A similar orthogonalization computation shows that $L_3(x) = x^3 - \frac{3}{5}x$.

___

Example 8. **Approximating e^x by**
Legendre Polynomials

Let us use the first four Legendre polynomials $L_0(x) = 1$, $L_1(x) = x$, $L_2(x) = x^2 - \frac{1}{3}$, $L_3(x) = x^3 - 3x/5$ to approximate e^x on the interval $[-1, 1]$. We want the first four terms in (20):

$$e^x \simeq w_0 L_0 + w_1 L_1(x) + w_2 L_2(x) + w_3 L_3(x)$$

$$\simeq w_0 + w_1 x + w_2 \left(x^2 - \frac{1}{3} \right) + w_3 \left(x^3 - \frac{3x}{5} \right) \tag{22}$$

where $w_i = e^x \cdot L_i(x)/L_i(x) \cdot L_i(x) = \int e^x L_i(x)\, dx / \int L_i(x)^2\, dx$. For example,

$$w_2 = \frac{\displaystyle\int_{-1}^{1} e^x (x^2 - \frac{1}{3})\, dx}{\displaystyle\int_{-1}^{1} (x^2 - \frac{1}{3})^2\, dx}$$

With a little calculus, we compute the w_i to be (approximately)

$$w_0 = \frac{2.35}{2} = 1.18, \qquad w_1 = \frac{.736}{.667} = 1.10,$$

$$w_2 = \frac{.096}{.178} = .53, \qquad w_3 = \frac{.008}{.046} = .18$$

Then (22) becomes

$$e^x \simeq 1.18 + 1.10x + .53 \left(x^2 - \frac{1}{3} \right) + .18 \left(x^3 - \frac{3x}{5} \right) \tag{23}$$

If we collect like powers of x together on the right side, (23) simplifies to

$$e^x \simeq 1 + x + .53x^2 + .18x^3 \tag{24}$$

Comparing our approximation against the real values of e^x at the points $-1, -.5, 0, .5, 1$, we find

x	-1	$-.5$	0	$.5$	1
e^x	.37	.61	1	1.64	2.72
Legendre approximation	.37	.61	1	1.65	2.71

A pretty good fit. In particular, it is a better fit on $[-1, 1]$ than simply using the first terms of the power series for e^x, namely, $1 + x + x^2/2 + x^3/6$. The approximation gets more accurate as more Legendre polynomials are used. ∎

Over the interval $[0, 2\pi]$ the trigonometric functions $(1/\sqrt{\pi}) \sin kx$ and $(1/\sqrt{\pi}) \cos kx$, for $k = 1, 2, \ldots$, plus the constant function $1/\sqrt{2\pi}$ are an orthonormal basis. To verify that they are orthogonal requires showing that

$$\frac{1}{\sqrt{\pi}} \sin jx \cdot \frac{1}{\sqrt{\pi}} \cos kx = \frac{1}{\pi} \int_0^{2\pi} \sin jx \cos kx \, dx = 0$$

for all j, k

$$\frac{1}{\sqrt{\pi}} \sin jx \cdot \frac{1}{\sqrt{\pi}} \sin kx = \frac{1}{\pi} \int_0^{2\pi} \sin jx \sin kx \, dx = 0$$

for all $j \neq k$

$$\frac{1}{\sqrt{\pi}} \cos jx \cdot \frac{1}{\sqrt{\pi}} \cos kx = \frac{1}{\pi} \int_0^{2\pi} \cos jx \cos kx \, dx = 0$$

for all $j \neq k$

plus showing these trigonometric functions are orthogonal to a constant function. To verify that these trigonometric functions have unit length requires showing

$$\frac{1}{\sqrt{\pi}} \sin kx \cdot \frac{1}{\sqrt{\pi}} \sin kx = \frac{1}{\pi} \int_0^{2\pi} \sin^2 kx \, dx = 1 \qquad \text{for all } k$$

$$\frac{1}{\sqrt{\pi}} \cos kx \cdot \frac{1}{\sqrt{\pi}} \cos kx = \frac{1}{\pi} \int_0^{2\pi} \cos^2 kx \, dx = 1 \qquad \text{for all } k$$

When $u_{2k-1}(x) = (1/\sqrt{\pi}) \sin kx$ and $u_{2k}(x) = (1/\sqrt{\pi}) \cos kx$, $k = 1, 2, \ldots$ and $u_0(x) = 1/\sqrt{2\pi}$ in (20), this representation of $f(x)$ is called a **Fourier series**, and the coefficients $f(x) \cdot u_i(x)$ in (20) are called **Fourier coefficients**. Using Fourier series, we see that any piecewise continuous function can be expressed as a linear combination of sine and cosine waves. One important physical interpretation of this fact is that any complex electrical signal can be expressed as a sum of simple sinusoidal signals.

Example 9. **Fourier Series Representation of a Jump Function**

Let us determine the Fourier series representation of the discontinuous function: $f(x) = 1$ for $0 < x \leq \pi$ and $= 0$ for $\pi < x \leq 2\pi$. The Fourier coefficients $f(x) \cdot u_i(x)$ in (20) are

$$f(x) \cdot u_{2k-1}(x) = f(x) \cdot \frac{1}{\sqrt{\pi}} \sin kx = \frac{1}{\sqrt{\pi}} \int_0^\pi \sin kx \, dx$$

$$= \frac{1}{k\sqrt{\pi}} [-\cos kx]_0^\pi = \begin{cases} \dfrac{2}{k\sqrt{\pi}} & k \text{ odd} \\ 0 & k \text{ even} \end{cases}$$

$$f(x) \cdot u_{2k}(x) = f(x) \cdot \frac{1}{\sqrt{\pi}} \cos kx = \frac{1}{\sqrt{\pi}} \int_0^\pi \cos kx \, dx \qquad (25)$$

$$= \frac{1}{k\sqrt{\pi}} [\sin kx]_0^\pi = 0$$

Further, we calculate $f(x) \cdot 1/\sqrt{2\pi} = \sqrt{\pi/2}$, so the constant term of the Fourier series for this $f(x)$ is $(f(x) \cdot u_0(x))u_0(x) = \frac{1}{2}$.

By (25), only the odd sine terms occur. Letting an odd k be written as $2n - 1$, we obtain the Fourier series.

$$f(x) = \frac{1}{2} + \sum_{n=1}^\infty \frac{2}{(2n-1)\sqrt{\pi}} \sin[(2n-1)x] \qquad (26)$$

Figure 5.9 shows the approximation to $f(x)$ obtained when the first three sine terms in (26) are used (dashed line) and when the first eight sine terms are used. The fit is impressive. ∎

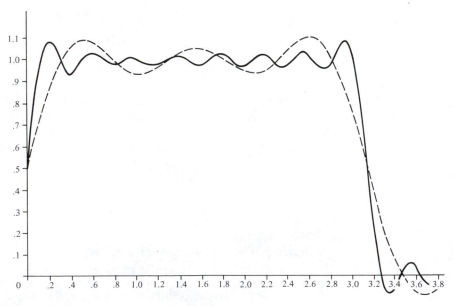

Figure 5.9 Dashed lines use first three trigonometric terms in Fourier series for $f(x)$. Solid lines use first eight terms.

Representing a function in terms of an orthonormal set of functions as in (20) has a virtually unlimited number of applications in the physical sciences and elsewhere. If one can solve a physical problem for the ortho-normal basis functions, then one can typically obtain a solution for any function as a linear combination of the solutions for the basis functions. This is true for most differential equations associated with electrical circuits, vibrating bodies, and so on. Statisticians use Fourier series to analyze time-series patterns (see Example 3 of Section 1.5). The study of Fourier series is one of the major fields of mathematics.

We complete our discussion of vector spaces of functions by showing how badly conditioned the powers of x are as a basis for representing func-tions. Remember that the powers of x, x^i, $i = 0, 1, \ldots$, are linearly independent. The problem is that they are far from orthogonal.

Let us consider how we might approximate an arbitrary function $f(x)$ as a linear combination of, say, the powers of x up to x^5:

$$f(x) \simeq w_0 + w_1 x + w_2 x^2 + w_3 x^3 + w_4 x^4 + w_5 x^5 \qquad (27)$$

using the continuous version of least-squares theory. If $f(x)$ and the powers of x were vectors, not functions, then (27) would have the familiar matrix form $\mathbf{f} = \mathbf{A}\mathbf{w}$ and the approximate solution $\mathbf{w}$ would be given by $\mathbf{w} = \mathbf{A}^+\mathbf{f}$, where $\mathbf{A}^+ = (\mathbf{A}^T\mathbf{A})^{-1}\mathbf{A}^T$.

Let us generalize $\mathbf{f} = \mathbf{A}\mathbf{w}$ to functions by letting the columns of a matrix be functions. We define the functional "matrix" $\mathbf{A}(x)$:

$$\mathbf{A}(x) = [1, x, x^2, x^3, x^4, x^5]$$

Now (27) becomes

$$f(x) = \mathbf{A}(x)\mathbf{w} \qquad (28)$$

To find the approximate solution to (28), we need to compute the functional version of the pseudoinverse $\mathbf{A}(x)^+$: $\mathbf{A}(x)^+ = (\mathbf{A}(x)^T\mathbf{A}(x))^{-1}\mathbf{A}(x)^T$ and then find the vector $\mathbf{w}$ of coefficients in (27):

$$\mathbf{w} = \mathbf{A}(x)^+ f(x) = (\mathbf{A}(x)^T\mathbf{A}(x))^{-1}(\mathbf{A}(x)^T\mathbf{f}(x)) \qquad (29)$$

The matrix $\mathbf{A}(x)^T$ has x^i as its ith "row", so the matrix product $\mathbf{A}(x)^T\mathbf{A}(x)$ involves computing the inner product of each 'row" of $\mathbf{A}(x)^T$ with each "column" of $\mathbf{A}(x)$:

entry (i, j) in $\mathbf{A}(x)^T\mathbf{A}(x)$ is $x^i \cdot x^j$ $(= \int x^i x^j \, dx)$

Similarly, the matrix-"vector" product $\mathbf{A}(x)^T f(x)$ is the vector of inner products $x^i \cdot f(x)$. The computations are simplest if we use the interval $[0, 1]$. Then entry (i, j) of $\mathbf{A}(x)^T\mathbf{A}(x)$ is

$$x^i \cdot x^j = \int_0^1 x^{i+j} \, dx = \left[\frac{x^{i+j+1}}{i+j+1}\right]_0^1 = \frac{1}{i+j+1} \tag{30}$$

For example, entry $(1, 2)$ is $\int xx^2 \, dx = \int x^3 \, dx = \frac{1}{4}$. Note that we consider the constant function $1 \ (= x^0)$ to be the zeroth row of $\mathbf{A}(x)^T$.

Computing all the inner products for $\mathbf{A}(x)^T\mathbf{A}(x)$ yields

$$\mathbf{A}(x)^T\mathbf{A}(x) = \begin{bmatrix} 1 & \frac{1}{2} & \frac{1}{3} & \frac{1}{4} & \frac{1}{5} & \frac{1}{6} \\ \frac{1}{2} & \frac{1}{3} & \frac{1}{4} & \frac{1}{5} & \frac{1}{6} & \frac{1}{7} \\ \frac{1}{3} & \frac{1}{4} & \frac{1}{5} & \frac{1}{6} & \frac{1}{7} & \frac{1}{4} \\ \frac{1}{4} & \frac{1}{5} & \frac{1}{6} & \frac{1}{7} & \frac{1}{8} & \frac{1}{9} \\ \frac{1}{5} & \frac{1}{6} & \frac{1}{7} & \frac{1}{8} & \frac{1}{9} & \frac{1}{10} \\ \frac{1}{6} & \frac{1}{7} & \frac{1}{8} & \frac{1}{9} & \frac{1}{10} & \frac{1}{11} \end{bmatrix} \tag{31}$$

This matrix is very ill-conditioned since the columns are all similar to each other. When the fractions in (31) are expressed to six decimal places, such as $\frac{1}{3} = .333333$, the inverse given by the author's microcomputer was (with entries rounded to integer values)

Fractions expressed to six decimal places

$$(\mathbf{A}(x)^T\mathbf{A}(x))^{-1} =$$

$$\begin{bmatrix} 17 & -116 & -47 & 1,180 & -1,986 & 958 \\ -116 & 342 & 7,584 & -34,881 & 49,482 & -22,548 \\ -47 & 7,584 & -76,499 & 242,494 & -301,846 & 129,004 \\ 1,180 & -34,881 & 242,494 & 644,439 & 723,636 & -289,134 \\ -1,986 & 49,482 & -301,846 & 723,636 & -747,725 & 278,975 \\ 958 & -22,548 & 129,004 & -289,134 & 278,975 & -97,180 \end{bmatrix}$$
$$\tag{32}$$

The (absolute) sum of the fifth column in (32) is about 2,000,000. The first column in (31) sums to about 2.5. So the condition number of $\mathbf{A}(x)^T\mathbf{A}(x)$, in the sum norm, is about $2,000,000 \times 2.5 = 5,000,000$. Now that is an ill-conditioned matrix!

We rounded fractions to six significant digits, but our condition number tells us that without a seventh significant digit, our numbers in (32) could be off by 500% error [a relative error of .000001 in $\mathbf{A}(x)^T\mathbf{A}(x)$ could yield answers off by a factor of 5 in pseudoinverse calculations]. Thus the numbers in (32) are worthless.

Suppose that we enter the matrix in (31) again, now expressing fractions to seven decimal places. The new inverse computation yields

Fractions expressed to seven decimal places

$$(\mathbf{A}(x)^T\mathbf{A}(x))^{-1} =$$

$$
\begin{bmatrix}
51 & -1{,}051 & 6{,}160 & -1{,}475 & 15{,}419 & -5{,}845 \\
-1{,}051 & 26{,}385 & -165{,}765 & 410{,}749 & -438{,}029 & 168{,}208 \\
6{,}160 & -165{,}765 & 1{,}079{,}198 & -2{,}731{,}939 & 2{,}955{,}103 & -1{,}146{,}281 \\
-1{,}475 & 410{,}749 & -2{,}731{,}939 & 7{,}017{,}359 & -7{,}671{,}190 & 2{,}999{,}546 \\
15{,}419 & -438{,}029 & 2{,}955{,}103 & -7{,}671{,}190 & 8{,}454{,}598 & -3{,}327{,}362 \\
-5{,}845 & 168{,}208 & -1{,}146{,}281 & 2{,}999{,}546 & -3{,}327{,}362 & 1{,}316{,}523
\end{bmatrix}
$$

$$(33)$$

We have a totally different matrix. Most of the entries in (33) are about 10 times larger than corresponding entries in (32). The sum of the fifth column in (33) is about 23,000,000. If we use (33), the condition number of $\mathbf{A}(x)^T\mathbf{A}(x)$ is around 56,000,000. Our entries in (33) were rounded to seven significant digits, but the condition number says eight significant digits were needed. Again our numbers are worthless. To compute the inverse accurately would require double-precision computation.

It is only fair to note that the ill-conditioned matrix (31) is famously bad. It is called a 6-by-6 *Hilbert matrix* [a Hilbert matrix has $1/(i + j + 1)$ in entry (i, j)].

Suppose that we used the numbers in (32) for $(\mathbf{A}(x)^T\mathbf{A}(x))^{-1}$ in computing the pseudoinverse. Let us proceed to calculate $\mathbf{A}(x)^+$ and then compute the coefficients in an approximation for a function by a fifth-degree polynomial. Let us choose $f(x) = e^x$. Then $(\mathbf{A}(x)^Te^x)$ is the vector of inner products $x^i \cdot e^i = \int x^i e^x \, dx$, $i = 0, 1, \ldots, 5$. Some calculus yields $\mathbf{A}(x)^Te^x = [2.718, 1, .718, .563, .465, .396]$ (expressed to three significant digits).

Now inserting our values for $(\mathbf{A}(x)^T\mathbf{A}(x))^{-1}$ and $\mathbf{A}^Te^x$ into (27), we obtain

$$\mathbf{w} = (\mathbf{A}(x)^T\mathbf{A}(x))^{-1}(\mathbf{A}(x)^Te^x) =$$

$$
\begin{bmatrix}
17 & -116 & -47 & 1{,}180 & -1{,}986 & 958 \\
-116 & 342 & 7{,}584 & -34{,}881 & 49{,}482 & -22{,}548 \\
-47 & 7{,}584 & -76{,}499 & 242{,}494 & -301{,}846 & 129{,}004 \\
1{,}180 & -34{,}881 & 242{,}494 & 644{,}439 & 723{,}636 & -289{,}134 \\
-1{,}986 & 49{,}482 & -301{,}846 & 723{,}636 & -747{,}725 & 278{,}975 \\
958 & -22{,}548 & 129{,}004 & -289{,}134 & 278{,}975 & -97{,}180
\end{bmatrix}
\cdot
\begin{bmatrix}
2.718 \\ 1 \\ .718 \\ .563 \\ .465 \\ .396
\end{bmatrix}
$$

$$
=
\begin{bmatrix}
17 \\ -87 \\ -219 \\ 1{,}611 \\ 2{,}449 \\ 1{,}135
\end{bmatrix}
$$

$$(34)$$

Thus our fifth-degree polynomial approximation of e^x on the interval $[0, 1]$ is

$$e^x \simeq 17 - 87x - 219x^2 + 1611x^3 + 2449x^4 + 1135x^5 \qquad (35)$$

Setting $x = 1$ in (35), we have $e^1 = 17 - 86 - 219 + 1611 + 2449 + 1135 = 4907$, pretty bad. Since our computed values in $(A(x)^T A(x))^{-1}$ are meaningless, such a bad approximation of e^x was to be expected.

Compare (35) with the Legendre polynomial approximation in Example 8.

Section 5.4 Exercises

Summary of Exercises
Exercises 1–11 involve inverses, pseudoinverses, and projections for matrices with orthogonal columns. Exercises 12–21 involve Gram–Schmidt orthogonalization and the **QR** decomposition. Exercises 22–30 present problems about functional inner products and functional approximation.

1. Compute the inverses of these matrices with orthogonal columns. Solve

$$\mathbf{Ax} = \begin{bmatrix} 1 \\ 2 \\ 3 \end{bmatrix}$$

where **A** is the matrix in part (b).

(a) $\begin{bmatrix} .6 & .8 \\ -.8 & .6 \end{bmatrix}$
 (b) $\begin{bmatrix} 2 & 2 & 1 \\ -2 & 1 & 2 \\ 1 & -2 & 2 \end{bmatrix}$

2. Compute the inverses of these matrices with orthogonal columns.

(a) $\begin{bmatrix} -1 & 4 & -1 \\ 2 & 1 & -2 \\ 1 & 2 & 3 \end{bmatrix}$
 (b) $\begin{bmatrix} 2 & -3 & 6 \\ -6 & 2 & 3 \\ 3 & 6 & 2 \end{bmatrix}$

(c) $\begin{bmatrix} .5 & -.5 & 1 \\ -.5 & .5 & 1 \\ 1 & .5 & 0 \end{bmatrix}$

Solve $\mathbf{Ax} = \mathbf{1}$, where **A** is the matrix in part (a).

3. Show that if **A** is an n-by-n upper triangular matrix with orthonormal columns, **A** is the identity matrix **I**.

4. Compute the length k of the projection of **b** onto **a** and give the projection vector **ka**.

(a) $\mathbf{a} = [0, 1, 0]$, $\mathbf{b} = [3, 2, 4]$
(b) $\mathbf{a} = [1, -1, 2]$, $\mathbf{b} = [2, 3, 1]$
(c) $\mathbf{a} = [\frac{1}{3}, \frac{2}{3}, \frac{2}{3}]$, $\mathbf{b} = [4, 1, 3]$
(d) $\mathbf{a} = [2, -1, 3]$, $\mathbf{b} = [-2, 5, 3]$

5. Express the vector $[2, 1, 2]$ as a linear combination of the following orthogonal bases for three-dimensional space.
   (a) $[1, -1, 2]$, $[2, 2, 0]$, $[-1, 1, 1]$
   (b) $[\frac{2}{3}, \frac{1}{3}, -\frac{2}{3}]$, $[\frac{1}{3}, \frac{2}{3}, \frac{2}{3}]$, $[\frac{2}{3}, -\frac{2}{3}, \frac{1}{3}]$
   (c) $[3, 1.5, 1]$, $[1, -3, 1.5]$, $[-1.5, 1, 3]$

6. Compute the pseudoinverse of

$$\mathbf{A} = \begin{bmatrix} \frac{1}{3} & -\frac{2}{3} \\ -\frac{2}{3} & \frac{1}{3} \\ \frac{2}{3} & \frac{2}{3} \end{bmatrix}$$

Find the least-squares solution to $\mathbf{Ax} = \mathbf{1}$.

7. Compute the pseudoinverse of

$$\mathbf{A} = \begin{bmatrix} 3 & 4 \\ 1 & -2 \\ 2 & -5 \end{bmatrix}$$

Find the least-squares solution to $\mathbf{Ax} = \mathbf{1}$.

8. Consider the regression model $\hat{z} = qx + ry + s$ for the following data, where the x-value is a scaled score (to have average value of 0) of high school grades, the y-value is a scaled score of SAT scores, and the z-value is a score of college grades.

x	-4	-2	0	2	4
y	2	-1	-2	-1	2
z	3	6	7	7	6

Determine q, r, and s. Note that the $\mathbf{x}$, $\mathbf{y}$, and $\mathbf{1}$ vectors are mutually orthogonal.

9. Verify that Theorem 2 is true in two dimensions, namely, that a change from the standard $\{\mathbf{e}_1, \mathbf{e}_2\}$ basis to some other orthonormal basis $\{\mathbf{q}_1, \mathbf{q}_2\}$ corresponds to a rotation (around the origin) and possibly a reflection. Note that since $\mathbf{q}_1, \mathbf{q}_2$ have unit length, they are completely determined by knowing the (counterclockwise) angles θ_1, θ_2 they make with the positive $\mathbf{e}_1$ axis; also since $\mathbf{q}_1, \mathbf{q}_2$ are orthogonal, $|\theta_1 - \theta_2| = 90°$.

10. (a) Show that an orthonormal change of basis preserves lengths (in euclidean norm).
    Hint: Verify that $(\mathbf{Qv}) \cdot (\mathbf{Qv}) = \mathbf{v} \cdot \mathbf{v}$ (where $\mathbf{Q}$ has orthonormal columns) by using the identity $(\mathbf{Ab}) \cdot (\mathbf{Cd}) = \mathbf{b}^T(\mathbf{A}^T\mathbf{C})\mathbf{d}$.
    (b) Show that an orthonormal change of basis preserves angles.
    Hint: Show that the cosine formula for the angle is unchanged by the method in part (a).

11. Compute the angle between the following pairs of nonorthogonal vectors. Which are close to orthogonal?
    (a) $[3, 2]$, $[-3, 4]$ (b) $[1, 2, 5]$, $[2, 5, 3]$
    (c) $[1, -3, 2]$, $[-2, 4, -3]$

12. Find the **QR** decomposition of the following matrices.

    (a) $\begin{bmatrix} 3 & -1 \\ 4 & 1 \end{bmatrix}$ (b) $\begin{bmatrix} 2 & 1 \\ 1 & 1 \\ 2 & 3 \end{bmatrix}$ (c) $\begin{bmatrix} 1 & -1 & 2 \\ 2 & -1 & 1 \\ 2 & -2 & 2 \end{bmatrix}$ (d) $\begin{bmatrix} 2 & 3 & 1 \\ 1 & 1 & 1 \\ 2 & 1 & 2 \end{bmatrix}$

13. Use the Gram–Schmidt orthogonalization to find an orthonormal basis that generates the same vector space as the following bases:
    (a) $[1, 1]$, $[2, -1]$ (b) $[2, 1, 2]$, $[4, 1, 1]$,
    (c) $[3, 1, 1]$, $[1, 2, 1]$, $[1, 1, 2]$

14. (a) Compute the inverse of the matrix in Exercise 12, part (c) by first finding the **QR** decomposition of the matrix and then using (15) to get the inverse. (See Exercise 12 of Section 3.5 for instructions on computing $\mathbf{R}^{-1}$.) What is its condition number?
    (b) Check your answer by computing the inverse by the regular elimination by pivoting method.

15. (a) Find the pseudoinverse $\mathbf{A}^+$ of the matrix $\mathbf{A}$ in Exercise 12, part (b) by using the **QR** decomposition of $\mathbf{A}$ and computing $\mathbf{A}^+$ as $\mathbf{A}^+ = \mathbf{R}^{-1}\mathbf{Q}^T$.
    (b) Check your answer by finding the pseudoinverse from the formula $\mathbf{A}^+ = (\mathbf{A}^T\mathbf{A})^{-1}\mathbf{A}^T$. Note that this is a very poorly conditioned matrix; compute the condition number of $(\mathbf{A}^T\mathbf{A})$.

16. Use (16) to find the pseudoinverse in solving the refinery problem in Example 3 of Section 5.3.

17. Use (16) to find the pseudoinverse in the following regression problems using the model $\hat{y} = qx + r$.
    (a) (x, y) points: $(0, 1)$, $(2, 1)$, $(4, 4)$
    (b) (x, y) points: $(3, 2)$, $(4, 5)$, $(5, 5)$, $(6, 5)$
    (c) (x, y) points: $(-2, 1)$, $(0, 1)$, $(2, 4)$

18. Use (16) to find the pseudoinverse in the least-squares polynomial-fitting problem in Example 5 of Section 5.3.

19. Verify (16): $A^+ = R^{-1}Q^T$, by substituting QR for A (and R^TQ^T for A^T) in the pseudoinverse formula $A^+ = (A^TA)^{-1}A^T$ and simplifying (remember that R is invertible; we assume that the columns of A are linearly independent).

20. Show that if the columns of the m-by-n matrix A are linearly independent, the m-by-m matrix R of the QR decomposition must be invertible.

Hint: Show main diagonal entries of R are nonzero and then see Exercise 12 of Section 3.5 for instructions on computing inverse of R.

21. Show that any set H of k orthonormal n-vectors can be extended to an orthonormal basis for n-dimensional space.

Hint: Form an n-by-$(k + n)$ matrix whose first k columns come from H and whose remaining n columns form the identity matrix; now apply the Gram–Schmidt orthogonalization to this matrix.

22. Over the interval $[0, 1]$, compute the following inner products: $x \cdot x$, $x \cdot x^3$, $x^3 \cdot x^3$.

23. Verify that the fourth Legendre polynomial is $x^3 - \frac{3}{5}x$.

24. Verify the values found for the weights w_1, w_2, w_3, and w_4 in Example 8.

Note: You must use integration by parts—or a table of integrals.

25. Approximate the following functions $f(x)$ as a linear combination of the first four Legendre polynomials over the interval $[-1, 1]$: $L_0(x) = 1$, $L_1(x) = x$, $L_2(x) = x^2 - \frac{1}{3}$, $L_3(x) = x^3 - 3x/5$.
(a) $f(x) = x^4$ **(b)** $f(x) = |x|$
(c) $f(x) = -1: x < 0$, $= 1: x \geq 0$

26. Approximate $x^3 + 2x - 1$ as a linear combination of the first four Legendre polynomials over the interval $[-1, 1]$: $L_0(x) = 1$, $L_1(x) = x$, $L_2(x) = x^2 - \frac{1}{2}$, $L_3(x) = x^3 - 3x/5$. Your "approximation" should equal $x^3 + 2x - 1$, since this polynomial is a linear combination of the functions 1, x, x^2, and x^3, from which the Legendre polynomials were derived by orthogonalization.

27. (a) Find the Legendre polynomial of degree 4.
(b) Find the Legendre polynomial of degree 5.

28. (a) Using the interval $[0, 1]$, instead of $[-1, 1]$, find three orthogonal polynomials of the form $K_0(x) = a$, $K_1(x) = bx + c$, and $K_2(x) = dx^2 + ex + f$.
(b) Find a least-squares approximation of x^4 on the interval $[0, 1]$ using your three polynomials in part (a).

(c) Find a least-squares approximation of $x^{1/2}$ on the interval $[0, 1]$ using your three polynomials in part (a).

(d) Find a least-squares approximation of $x^2 - 2x + 1$ on the interval $[0, 1]$ using your three polynomials in part (a). Hopefully, your approximation will equal $x^2 - 2x + 1$, since this polynomial is a linear combination of 1, x, and x^2, the functions used to build your set of orthogonal polynomials.

29. (a) Find a fourth polynomial $K_3(x)$ of order 3 orthogonal on $[0, 1]$ to the three polynomials in Exercise 28, part (a).

(b) Find a least-squares approximation to x^4 on the interval $[0, 1]$ using your four orthogonal polynomials.

30. Compute the inverse and find the condition number (in sum norm) of the following Hilbert-like matrices.

(a) $\begin{bmatrix} \frac{1}{8} & \frac{1}{9} \\ \frac{1}{9} & \frac{1}{10} \end{bmatrix}$
  (b) $\begin{bmatrix} \frac{1}{4} & \frac{1}{5} & \frac{1}{6} \\ \frac{1}{5} & \frac{1}{6} & \frac{1}{7} \\ \frac{1}{6} & \frac{1}{7} & \frac{1}{8} \end{bmatrix}$
  (c) $\begin{bmatrix} \frac{1}{8} & \frac{1}{9} & \frac{1}{10} \\ \frac{1}{9} & \frac{1}{10} & \frac{1}{11} \\ \frac{1}{10} & \frac{1}{11} & \frac{1}{12} \end{bmatrix}$

Section 5.5   Eigenvector Bases and the Eigenvalue Decomposition

In this section we use eigenvectors to gain insight into the structure of a matrix. We review how an eigenvector basis simplifies the computation of powers of $\mathbf{A}$. Then we present a way to decompose a matrix into simple matrices formed by the eigenvectors. This decomposition yields a way to compute all eigenvectors and eigenvalues of a symmetric matrix.

Recall that a vector $\mathbf{u}$ is an eigenvector of the n-by-n matrix $\mathbf{A}$ if for some λ (an eigenvalue), $\mathbf{Au} = \lambda\mathbf{u}$. In words, multiplying $\mathbf{u}$ by a matrix $\mathbf{A}$ has the same effect as multiplying $\mathbf{u}$ by the scalar λ. It follows that $\mathbf{A}^k\mathbf{u} = \lambda^k\mathbf{u}$. A stable distribution $\mathbf{p}$ of a Markov chain is an eigenvector of the transition matrix $\mathbf{A}$ (associated with eigenvalue 1: $\mathbf{Ap} = \mathbf{p}$). The dominant eigenvector (associated with the largest eigenvalue) gives the long-term distribution of a growth model (see Sections 2.5 and 4.5).

As noted in Section 2.5, if a vector $\mathbf{x}$ can be expressed as a linear combination of the eigenvectors $\mathbf{u}_i$ of $\mathbf{A}$;

$$\mathbf{x} = a_1\mathbf{u}_1 + a_2\mathbf{u}_2 + \cdots + a_n\mathbf{u}_n \tag{1}$$

then

$$\begin{aligned} \mathbf{Ax} &= a_1\mathbf{Au}_1 + a_2\mathbf{Au}_2 + \cdots + a_n\mathbf{Au}_n \\ &= a_1\lambda_1\mathbf{u}_1 + a_2\lambda_2\mathbf{u}_2 + \cdots + a_n\lambda_n\mathbf{u}_n \end{aligned} \tag{2}$$

In words, matrix multiplication becomes scalar multiplication of eigenvector coordinates.

Further,

$$\mathbf{A}^k\mathbf{x} = a_1\mathbf{A}^k\mathbf{u}_1 + a_2\mathbf{A}^k\mathbf{u}_2 + \cdots + a_n\mathbf{A}^k\mathbf{u}_n \qquad (3)$$
$$= a_1\lambda_1^k\mathbf{u}_1 + a_2\lambda_2^k\mathbf{u}_2 + \cdots + a_n\lambda_n^k\mathbf{u}_n$$

When we express $\mathbf{x}$ in terms of a basis of $\mathbf{A}$'s eigenvectors, $\mathbf{A}^k\mathbf{x}$ can quickly be computed by (3).

Assume that we index the eigenvalues so that $\lambda_1 > \lambda_2 \geq \lambda_3 \geq \cdots \geq \lambda_n$, λ_1^k is going to be much lárger than the other λ's. Thus the first term on the right in (3) dominates the other terms, and we have

$$\mathbf{A}^k\mathbf{x} \simeq a_1\lambda_1^k\mathbf{u}_1 \qquad (4)$$

We review the example from Section 2.5 that illustrated these results.

Example 1. Computing Powers of a Matrix with Eigenvectors

The computer (C) and dog (D) growth model from Section 2.5 is

$$\mathbf{x}' = \mathbf{A}\mathbf{x}, \qquad \text{where } \mathbf{A} = \begin{bmatrix} 3 & 1 \\ 2 & 2 \end{bmatrix}, \qquad \begin{array}{l} \mathbf{x} = [C, D] \\ \mathbf{x}' = [C', D'] \end{array}$$

The two eigenvalues and associated eigenvectors of $\mathbf{A}$ are $\lambda_1 = 4$ with $\mathbf{u}_1 = [1, 1]$ and $\lambda_2 = 1$ with $\mathbf{u}_2 = [1, -2]$. Note that since $\mathbf{u}_1$ and $\mathbf{u}_2$ are linearly independent, they form a basis for 2-space.

Suppose that we want to determine the effects of this growth model over 20 periods with the starting vector $\mathbf{x} = [1, 7]$. We want to express $\mathbf{x}$ as a linear combination of $\mathbf{u}_1$ and $\mathbf{u}_2$: $\mathbf{x} = w_1\mathbf{u}_1 + w_2\mathbf{u}_2$. Determining the set of weights $\mathbf{w} = [w_1, w_2]$ requires solving

$$\mathbf{x} = w_1\mathbf{u}_1 + w_2\mathbf{u}_2 \qquad \text{or} \qquad \mathbf{x} = \mathbf{U}\mathbf{w},$$
$$\text{where } \mathbf{U} = [\mathbf{u}_1 \quad \mathbf{u}_2] = \begin{bmatrix} 1 & 1 \\ 1 & -2 \end{bmatrix} \qquad (5)$$

The solution to (5) is $\mathbf{w} = \mathbf{U}^{-1}\mathbf{x}$, which yields $w_1 = 3$, $w_2 = -2$, so that $\mathbf{x} = 3\mathbf{u}_1 - 2\mathbf{u}_2$. (Here w_1, w_2 are simply the coordinates of $\mathbf{x}$ in the eigenvector basis for 2-space.) Then

$$\mathbf{A}\mathbf{x} = \mathbf{A}(3\mathbf{u}_1 - 2\mathbf{u}_2) = 3\mathbf{A}\mathbf{u}_1 - 2\mathbf{A}\mathbf{u}_2$$
$$= 3(4\mathbf{u}_1) - 2(1\mathbf{u}_2) \qquad (6)$$
$$= 12\mathbf{u}_1 - 2\mathbf{u}_2$$

For 20 periods, we have

$$\mathbf{A}^{20}\mathbf{x} = \mathbf{A}^{20}(3\mathbf{u}_1 - 2\mathbf{u}_2) = 3\mathbf{A}^{20}\mathbf{u}_1 - 2\mathbf{A}^{20}\mathbf{u}_2$$
$$= 3(4^{20}\mathbf{u}_1) - 2(1^{20}\mathbf{u}_2) \tag{7}$$
$$= 3 \cdot 4^{20}[1, 1] - 2[1, -2]$$

In Example 7 of Section 3.3 we observed that these steps were represented by a matrix equation

$$\mathbf{A}\mathbf{x} = \mathbf{U}\mathbf{D}_\lambda\mathbf{U}^{-1}\mathbf{x}, \quad \text{where } \mathbf{D}_\lambda = \begin{bmatrix} 4 & 0 \\ 0 & 1 \end{bmatrix} \tag{8a}$$

and

$$\mathbf{A} = \mathbf{U}\mathbf{D}_\lambda\mathbf{U}^{-1} \tag{8b}$$

where $\mathbf{U}$, as in (5), has the eigenvectors as its columns.

In words, we explain (8a) and (8b) as follows. The vector $\mathbf{w}$ in (5) can be viewed as the vector $\mathbf{x}$ expressed in terms of eigenvector coordinates, and as noted above, $\mathbf{w} = \mathbf{U}^{-1}\mathbf{x}$. Looking at $\mathbf{U}\mathbf{D}_\lambda\mathbf{U}^{-1}$ times $\mathbf{x}$ from right to left, the product $\mathbf{U}^{-1}\mathbf{x}$ converts $\mathbf{x}$ to the eigen-vector-coordinate vector $\mathbf{w}$. Next the matrix $\mathbf{D}_\lambda$ multiplies each eigen-vector coordinate by the appropriate eigenvalue ($\mathbf{D}_\lambda\mathbf{w} = [4w_1, w_2]$) to get the eigenvector-based coordinates after the matrix multiplication. Finally, multiplying $\mathbf{U}$ converts back to the original coordinate system.

The matrix-vector product $\mathbf{x}' = \mathbf{A}\mathbf{x}$ is transformed, in eigenvec-tor coordinates into $\mathbf{w}' = \mathbf{D}_\lambda\mathbf{w}$. Further, $\mathbf{A}^k = \mathbf{U}\mathbf{D}_\lambda^k\mathbf{U}^{-1}$, so

$$\mathbf{x}^{(20)} = \mathbf{A}^{20}\mathbf{x} \quad \text{becomes} \quad \mathbf{w}^{(20)} = \mathbf{D}_\lambda^{20}\mathbf{w} \tag{9}$$

For our particular $\mathbf{A}$,

$$\mathbf{w}^{(20)} = \mathbf{D}_\lambda^{20}\mathbf{w} = \begin{bmatrix} 4^{20} & 0 \\ 0 & 1 \end{bmatrix}\begin{bmatrix} w_1 \\ w_2 \end{bmatrix} = \begin{bmatrix} 4^{20}w_1 \\ w_2 \end{bmatrix} \tag{10}$$

∎

Summarizing (part of this is simply Theorem 5 of Section 3.3), we obtain

Theorem 1. Let $\mathbf{A}$ be an n-by-n matrix and let $\mathbf{U}$ be an n-by-n matrix whose columns $\mathbf{u}_i$ are n linearly independent eigenvectors of $\mathbf{A}$. If $\mathbf{c} = \mathbf{A}\mathbf{b}$, then in $\mathbf{u}_i$-coordinates $\mathbf{c}^*$, $\mathbf{b}^*$, we have

$$\mathbf{c}^* = \mathbf{D}_\lambda\mathbf{b}^* \quad \text{or} \quad c_i^* = \lambda_i b_i^* \tag{11}$$

where $\mathbf{b}^* = \mathbf{U}^{-1}\mathbf{b}$ and $\mathbf{c}^* = \mathbf{U}^{-1}\mathbf{c}$ are the $\mathbf{u}_i$-coordinate vectors for $\mathbf{b}$ and $\mathbf{c}$ and $\mathbf{D}_\lambda$ is the diagonal matrix whose diagonal entries are the eigenvalues of $\mathbf{A}$, $\lambda_1, \lambda_2, \ldots, \lambda_n$. Similarly, if $\mathbf{c} = \mathbf{A}^k\mathbf{b}$, then $c_i^* = \lambda_i^k b_i^*$.

Theorem 2. For **A**, $\mathbf{D}_\lambda$, and **U** as in Theorem 1, **A** can be written

$$\mathbf{A} = \mathbf{U}\mathbf{D}_\lambda\mathbf{U}^{-1} \tag{12a}$$

and

$$\mathbf{A}^k = \mathbf{U}\mathbf{D}_\lambda^k\mathbf{U}^{-1} \tag{12b}$$

where $\mathbf{D}_\lambda^k$ has the eigenvalues raised to the *k*th power. Further,

$$\mathbf{D}_\lambda = \mathbf{U}^{-1}\mathbf{A}\mathbf{U} \tag{13}$$

![bar]

Example 2. **Conversion of A to $\mathbf{D}_\lambda$**

Let us use the matrix **A** in Example 1, $\mathbf{A} = \begin{bmatrix} 3 & 1 \\ 2 & 2 \end{bmatrix}$. We had found

U to be: $\mathbf{U} = \begin{bmatrix} 1 & 1 \\ 1 & -2 \end{bmatrix}$. By the determinant formula for 2-by-2

inverses, $\mathbf{U}^{-1} = \begin{bmatrix} \frac{2}{3} & \frac{1}{3} \\ \frac{1}{3} & -\frac{1}{3} \end{bmatrix}$. Then, by (13),

$$
\begin{aligned}
\mathbf{D}_\lambda = \mathbf{U}^{-1}\mathbf{A}\mathbf{U} &= \left(\begin{bmatrix} \frac{2}{3} & \frac{1}{3} \\ \frac{1}{3} & -\frac{1}{3} \end{bmatrix} \begin{bmatrix} 3 & 1 \\ 2 & 2 \end{bmatrix} \right) \begin{bmatrix} 1 & 1 \\ 1 & -2 \end{bmatrix} \\
&= \left(\begin{bmatrix} \frac{8}{3} & \frac{4}{3} \\ \frac{1}{3} & -\frac{1}{3} \end{bmatrix} \right) \begin{bmatrix} 1 & 1 \\ 1 & -2 \end{bmatrix} \\
&= \begin{bmatrix} 4 & 0 \\ 0 & 1 \end{bmatrix}
\end{aligned} \tag{14}
$$

Let us next use (12b) to compute $\mathbf{A}^k$. This formula is most useful for large *k*, but for illustrative purpose we use $k = 2$. First we compute $\mathbf{A}^2$ directly:

$$\mathbf{A}^2 = \mathbf{A}\mathbf{A} = \begin{bmatrix} 3 & 1 \\ 2 & 2 \end{bmatrix}\begin{bmatrix} 3 & 1 \\ 2 & 2 \end{bmatrix} = \begin{bmatrix} 11 & 5 \\ 10 & 6 \end{bmatrix}$$

Next we compute $\mathbf{A}^2$ as

$$
\begin{aligned}
\mathbf{A}^2 = \mathbf{U}\mathbf{D}_\lambda^2\mathbf{U}^{-1} &= \left(\begin{bmatrix} 1 & 1 \\ 1 & -2 \end{bmatrix}\begin{bmatrix} 16 & 0 \\ 0 & 1 \end{bmatrix} \right)\begin{bmatrix} \frac{2}{3} & \frac{1}{3} \\ \frac{1}{3} & -\frac{1}{3} \end{bmatrix} \\
&= \left(\begin{bmatrix} 16 & 1 \\ 16 & -2 \end{bmatrix} \right)\begin{bmatrix} \frac{2}{3} & \frac{1}{3} \\ \frac{1}{3} & -\frac{1}{3} \end{bmatrix} \\
&= \begin{bmatrix} 11 & 5 \\ 10 & 6 \end{bmatrix}
\end{aligned}
$$

■

The formula $\mathbf{A} = \mathbf{U}\mathbf{D}_\lambda\mathbf{U}^{-1}$ leads to an important way to decompose a matrix into simple matrices. Just as Gaussian elimination yields the **LU** decomposition and Gram–Schmidt orthogonalization yields the **QR** decomposition, so our eigenvector coordinates formula also yields a matrix decomposition.

Recall that if $\mathbf{c} = [1, 2]$ and $\mathbf{d} = [3, 4, -1]$, the simple matrix $\mathbf{c} * \mathbf{d}$ equals

$$\mathbf{c} * \mathbf{d} = \begin{bmatrix} 1 \\ 2 \end{bmatrix} * [3 \quad 4 \quad -1] = \begin{bmatrix} 3 & 4 & -1 \\ 6 & 8 & -2 \end{bmatrix}$$

More generally, entry (i, j) in $\mathbf{c} * \mathbf{d}$ equals $c_i d_j$.

Theorem 7 of Section 5.2 says that the matrix product $\mathbf{CD}$ can be decomposed into a sum of simple matrices formed by columns $\mathbf{c}_i^C$ of $\mathbf{C}$ and the rows $\mathbf{d}_i^R$ of $\mathbf{D}$:

$$\mathbf{CD} = \mathbf{c}_1^C * \mathbf{d}_1^R + \mathbf{c}_2^C * \mathbf{d}_2^R + \cdots + \mathbf{c}_r^C * \mathbf{d}_r^R \tag{15}$$

Letting $\mathbf{C} = \mathbf{U}$ and $\mathbf{D} = \mathbf{D}_\lambda\mathbf{U}^{-1}$ (the ith row of $\mathbf{D}$ is the ith row of $\mathbf{U}^{-1}$ multiplied by λ_i), we obtain the following decomposition.

Theorem 3. Eigenvalue Decomposition. Let $\mathbf{A}$ be an n-by-n matrix with n linearly independent eigenvectors $\mathbf{u}_i$ associated with eigenvalues $|\lambda_1| \geq |\lambda_2| \geq \cdots \geq |\lambda_n|$. Let $\mathbf{u}_j'$ denote the ith row of $\mathbf{U}^{-1}$. Then $\mathbf{A}$ is the weighted sum of simple matrices:

$$\mathbf{A} = \mathbf{U}\mathbf{D}_\lambda\mathbf{U}^{-1} = \lambda_1\mathbf{u}_1 * \mathbf{u}_1' + \lambda_2\mathbf{u}_2 * \mathbf{u}_2' + \cdots + \lambda_n\mathbf{u}_n * \mathbf{u}_n' \tag{16}$$

In (16) the λ_i's are factored out in front of the simple matrices.

For typical large matrices, the eigenvalues tend to decline in size quickly. For example, if $n = 20$, perhaps $\lambda_1 = 5$, $\lambda_2 = 2$, $\lambda_3 = .6$, $\lambda_4 = .02$; so the sum of the first three simple matrices in (16) would yield a very good approximation of the matrix.

Example 3. **Eigenvalue Decomposition**
of 2-by-2 Matrix

We illustrate Theorem 3 with the 2-by-2 matrix of Examples 1 and 2. From those examples we have $\lambda_1 = 4$, $\lambda_2 = 1$ and

$$\mathbf{A} = \begin{bmatrix} 3 & 1 \\ 2 & 2 \end{bmatrix}, \quad \mathbf{U} = \begin{bmatrix} 1 & 1 \\ 1 & -2 \end{bmatrix}, \quad \mathbf{U}^{-1} = \begin{bmatrix} \frac{2}{3} & \frac{1}{3} \\ \frac{1}{3} & -\frac{1}{3} \end{bmatrix}$$

Then (16) says

$$\mathbf{A} = 4 \begin{bmatrix} 1 \\ 1 \end{bmatrix} * [\tfrac{2}{3} \quad \tfrac{1}{3}] + 1 \begin{bmatrix} 1 \\ -2 \end{bmatrix} * [\tfrac{1}{3} \quad -\tfrac{1}{3}]$$

$$= 4 \begin{bmatrix} \tfrac{2}{3} & \tfrac{1}{3} \\ \tfrac{2}{3} & \tfrac{1}{3} \end{bmatrix} + \begin{bmatrix} \tfrac{1}{3} & -\tfrac{1}{3} \\ -\tfrac{2}{3} & \tfrac{2}{3} \end{bmatrix}$$

$$= \begin{bmatrix} \tfrac{8}{3} & \tfrac{4}{3} \\ \tfrac{8}{3} & \tfrac{4}{3} \end{bmatrix} + \begin{bmatrix} \tfrac{1}{3} & -\tfrac{1}{3} \\ -\tfrac{2}{3} & \tfrac{2}{3} \end{bmatrix} = \begin{bmatrix} 3 & 1 \\ 2 & 2 \end{bmatrix} \qquad \blacksquare$$

We next state a very useful fact about symmetric matrices.

Theorem 4

(i) Any symmetric *n*-by-*n* matrix **A** has a set of *n* orthogonal eigenvectors $\mathbf{u}_i$.

(ii) Two eigenvectors associated with distinct eigenvalues are always orthogonal.

Corollary. When **A** is symmetric, the matrix **U** in Theorem 1 has orthogonal columns, and $\mathbf{U}^{-1}$ is obtained from $\mathbf{U}^T$ by dividing each row by $|\mathbf{u}_i|^2 = \mathbf{u}_i \cdot \mathbf{u}_i$. When additionally the columns have length 1 (orthonormal), the eigenvalue decomposition in Theorem 3 becomes

$$\mathbf{A} = \lambda_1 \mathbf{u}_1 * \mathbf{u}_1 + \lambda_2 \mathbf{u}_2 * \mathbf{u}_2 + \cdots + \lambda_n \mathbf{u}_n * \mathbf{u}_n \qquad (17)$$

The corollary's claim about how to obtain $\mathbf{U}^{-1}$ comes from Theorem 1 of Section 5.4.

The eigenvalue decomposition (17) sheds new light on what happens when we multiply a symmetric matrix **A** times some vector **x**. If in $\mathbf{u}_i$-coordinates ($\mathbf{u}_i$ are orthonormal), **x** is

$$\mathbf{x} = a_1 \mathbf{u}_1 + a_2 \mathbf{u}_2 + \cdots + a_n \mathbf{u}_n \qquad (18)$$

and we compute **Ax** using (17), we have

$$\mathbf{Ax} = (\lambda_1 \mathbf{u}_1 * \mathbf{u}_1 + \lambda_2 \mathbf{u}_2 * \mathbf{u}_2 + \cdots + \lambda_n \mathbf{u}_n * \mathbf{u}_n)\mathbf{x} \qquad (19)$$
$$= \lambda_1 (\mathbf{u}_1 * \mathbf{u}_1)\mathbf{x} + \lambda_2 (\mathbf{u}_2 * \mathbf{u}_2)\mathbf{x} + \cdots + \lambda_n (\mathbf{u}_n * \mathbf{u}_n)\mathbf{x}$$

and (19) is equivalent to

$$\mathbf{Ax} = \lambda_1 (\mathbf{u}_1 \cdot \mathbf{x})\mathbf{u}_1 + \lambda_2 (\mathbf{u}_2 \cdot \mathbf{x})\mathbf{u}_2 + \cdots + \lambda_n (\mathbf{u}_n \cdot \mathbf{x})\mathbf{u}_n \qquad (20)$$

since

$$(\mathbf{u}_1 * \mathbf{u}_1)\mathbf{x} = \mathbf{u}_1 (\mathbf{u}_1 \cdot \mathbf{x}) \qquad \text{or} \qquad = (\mathbf{u}_1 \cdot \mathbf{x})\mathbf{u}_1 \qquad (21)$$

and similarly for the other $\mathbf{u}_i$.

We verify (21) as follows. The first $\mathbf{u}_1$ in $\mathbf{u}_1 * \mathbf{u}_1$ is to be treated as an n-by-1 matrix, the second $\mathbf{u}_1$ as a 1-by-n matrix, and $\mathbf{x}$ as a column vector (an n-by-1 matrix). Now use the associative law of matrix multiplication in (21).

So multiplying $\mathbf{A}$ by $\mathbf{x}$, when $\mathbf{A}$ is represented as a sum of simple matrices, has the effect of first projecting $\mathbf{x}$ onto the eigenvectors $\mathbf{u}_i$ [$(\mathbf{u}_i \cdot \mathbf{x})\mathbf{u}_i$ is this projection] and then multiplying each such projection by the eigenvalue λ_i.

In Section 3.4 we used equation (4)—that $\mathbf{A}^k \mathbf{x}$ is approximately a multiple of $\mathbf{u}_1$ to determine $\mathbf{u}_1$ and λ_1. But we had no way to compute other eigenvalues or eigenvectors. The eigenvalue decomposition (17) gives us a way.

Suppose that $\mathbf{A}$ is a symmetric matrix and we have determined the dominant (largest) eigenvalue λ_1 and an associated eigenvector $\mathbf{u}_1$ (of length 1) by an iterative method. Consider then the matrix

$$\mathbf{A}_2 = \mathbf{A} - \lambda_1 \mathbf{u}_1 * \mathbf{u}_1 \tag{22}$$

The matrix $\mathbf{A}_2$ is $\mathbf{A}$ minus the first simple matrix in the eigenvalue decomposition in (17). It follows that

$$\mathbf{A}_2 = \lambda_2 \mathbf{u}_2 * \mathbf{u}_2 + \lambda_3 \mathbf{u}_3 * \mathbf{u}_3 + \cdots + \lambda_n \mathbf{u}_n * \mathbf{u}_n \tag{23}$$

Since the eigenvalue decomposition of a square matrix is unique, it follows that the (nonzero) eivenvalues of $\mathbf{A}_2$ are $\lambda_2, \lambda_3, \ldots, \lambda_n$ with associated eigenvectors $\mathbf{u}_2, \mathbf{u}_3, \ldots, \mathbf{u}_n$. In particular, λ_2 is now the dominant eigenvalue of $\mathbf{A}_2$ and applying an iterative method to $\mathbf{A}_2$ will yield λ_2 and $\mathbf{u}_2$.

If we subtract the second simple matrix in (17) from $\mathbf{A}_2$, we will get a matrix $\mathbf{A}_3$ whose dominant eigenvalue is λ_3, and so on. This method of getting the eigenvalues and eigenvectors of $\mathbf{A}$ is called **deflation.** We note that if we have a small error in λ_1 or $\mathbf{u}_1$, the resulting $\mathbf{A}_2$ still has the same dominant eigenvalue and eigenvector (the consequences of such errors are explored in the Exercises).

Deflation Method to Compute Eigenvalues and Eigenvectors of a Symmetric Matrix $\mathbf{A}$

Step 0. Set $\mathbf{A}_1 = \mathbf{A}$; set $i = 1$.

Step 1. Use the iterative method to determine $\mathbf{A}_i$'s dominant eigenvalue λ_i and an associated eigenvector $\mathbf{u}_i$ (of length 1).

Step 2. Set $\mathbf{A}_{i+1} = \mathbf{A}_i - \lambda_i \mathbf{u}_i * \mathbf{u}_i$. Increase i by 1. If $i \leq n$, go to step 1.

A faster method for determining all eigenvalues and eigenvectors of any n-by-n matrix, symmetric or not, is presented in the appendix to this section.

███████

Example 4. **Finding Eigenvalues and Eigenvectors of a 3-by-3 Symmetric Matrix**

Let us use the deflation method to find all eigenvalues and eigenvectors of the symmetric matrix

$$\mathbf{A} = \begin{bmatrix} 3 & 2 & 1 \\ 2 & 3 & 1 \\ 1 & 1 & 0 \end{bmatrix}$$

Setting $\mathbf{x} = [1, 0, 0]$ and computing $\mathbf{A}^{25}\mathbf{x}$, we get a multiple of the unit vector

$$\mathbf{u}_1 \simeq [.684, .684, .254] \quad \text{with } \lambda_1 \simeq 5.37$$

Next we compute the deflated matrix $\mathbf{A}_2$:

$$\mathbf{A}_2 = \mathbf{A} - \lambda_1 \mathbf{u}_1 * \mathbf{u}_1$$

$$= \begin{bmatrix} 3 & 2 & 1 \\ 2 & 3 & 1 \\ 1 & 1 & 0 \end{bmatrix} - \begin{bmatrix} 2.51 & 2.51 & .93 \\ 2.51 & 2.51 & .93 \\ .93 & .93 & .35 \end{bmatrix}$$

$$= \begin{bmatrix} .49 & -.51 & .07 \\ -.51 & .49 & .07 \\ .07 & .07 & -.35 \end{bmatrix}$$

Again with $\mathbf{x} = [1, 0, 0]$, we compute $A_1^{25}\mathbf{x}$ and get $[.5, -.5, 0]$. So

$$\mathbf{u}_2 = [.707, -.707, 0] \quad \text{with } \lambda_2 = 1$$

(The reader should verify that $[.5, -.5, 0]$ was an eigenvector of the original matrix $\mathbf{A}$.) Deflating again, we have

$$\mathbf{A}_3 = \mathbf{A}_2 - \lambda_2 \mathbf{u}_2 * \mathbf{u}_2$$

$$= \begin{bmatrix} .49 & -.51 & .07 \\ -.51 & .49 & .07 \\ .07 & .07 & -.35 \end{bmatrix} - \begin{bmatrix} .5 & -.5 & 0 \\ -.5 & .5 & 0 \\ 0 & 0 & 0 \end{bmatrix}$$

$$= \begin{bmatrix} -.01 & -.01 & .07 \\ -.01 & -.01 & .07 \\ .07 & .07 & -.35 \end{bmatrix} \tag{24}$$

Next we find that $\mathbf{u}_3 \simeq [.181, .181, .967]$ and $\lambda_3 \simeq -.37$. Computing the last simple matrix, we obtain

$$\lambda_3 u_3 * u_3 = \begin{bmatrix} -.01 & -.01 & .07 \\ -.01 & -.01 & .07 \\ .07 & .07 & -.35 \end{bmatrix}$$

This simple matrix equals A_3, as required. Thus we have confirmed that A is the sum of three simple matrices. ■

Substantial savings in computer storage can be realized by representing a large matrix as a sum of simple matrices. If a symmetric 20-by-20 matrix is well approximated as the sum of two simple matrices, then only 10% as much storage is needed: 2 columns instead of all 20 columns.

Example 5. Approximating a Digital Picture with Simple Matrices

Approximate the 8-by-8 digital "picture" A whose entries represent varying levels of darkness between 0 and 1.

$$A = \begin{bmatrix} 0 & 0 & 0 & .2 & .2 & 0 & 0 & 0 \\ 0 & .1 & .2 & .3 & .3 & .2 & .1 & 0 \\ 0 & .2 & .5 & .6 & .6 & .5 & .2 & 0 \\ .2 & .3 & .6 & .8 & .8 & .6 & .3 & .2 \\ .2 & .3 & .6 & .8 & .8 & .6 & .3 & .2 \\ 0 & .2 & .5 & .6 & .6 & .5 & .2 & 0 \\ 0 & .1 & .2 & .3 & .3 & .2 & .1 & 0 \\ 0 & 0 & 0 & .2 & .2 & 0 & 0 & 0 \end{bmatrix} \tag{25}$$

We first approximate A with the simple matrix $c * c$, where $c = [.1, .3, .7, .9, .9, .7, .3, .1]$ has c_i equal to the (approximate) square root of the diagonal entry (i, i) of A.

$$c * c = \begin{bmatrix} .01 & .03 & .07 & .09 & .09 & .07 & .03 & .01 \\ .03 & .09 & .14 & .27 & .27 & .14 & .09 & .03 \\ .07 & .14 & .49 & .63 & .63 & .49 & .14 & .07 \\ .09 & .27 & .63 & .81 & .81 & .63 & .27 & .09 \\ .09 & .27 & .63 & .81 & .81 & .63 & .27 & .09 \\ .07 & .14 & .49 & .63 & .63 & .49 & .14 & .07 \\ .03 & .09 & .14 & .27 & .27 & .14 & .09 & .03 \\ .01 & .03 & .07 & .09 & .09 & .07 & .03 & .01 \end{bmatrix} \tag{26}$$

Now we shall use the eigenvalue decomposition of A into simple matrices to approximate A more accurately. We use the first three terms involving the three largest eigenvalues,

$$\mathbf{A} \simeq \lambda_1 \mathbf{u}_1 * \mathbf{u}_1 + \lambda_2 \mathbf{u}_2 * \mathbf{u}_2 + \lambda_2 \mathbf{u}_3 * \mathbf{u}_3 \tag{27}$$

Actually, for this **A**, the first simple matrix $\lambda_1 \mathbf{u}_1 * \mathbf{u}_1$ alone will be a good approximation.

For large matrices, the iterative method of computing $\mathbf{A}^k \mathbf{x}$ for a large k to approximate $\mathbf{u}_1$ is unnecessarily time consuming. One of the faster methods mentioned in the appendix to this section should be used. For $\mathbf{u}_1$, we shall give the results using simple iteration because it converges quickly. Computing $\mathbf{A}^{10} \mathbf{x}$ with $\mathbf{x} = [1, 1, 1, 1, 1, 1, 1, 1]$ yields a multiple of

$$\mathbf{u}_1 = [.078, .189, .408, .540, .540, .408, .189, .078]$$
$$\text{with } \lambda_1 = 2.774$$

so

$$\lambda_1 \mathbf{u}_1 * \mathbf{u}_1 = \begin{bmatrix} .017 & .041 & .088 & .117 & .117 & .088 & .041 & .017 \\ .041 & .099 & .214 & .283 & .283 & .214 & .099 & .041 \\ .088 & .214 & .461 & .610 & .610 & .461 & .214 & .088 \\ .117 & .283 & .610 & .808 & .808 & .610 & .283 & .117 \\ .117 & .283 & .610 & .808 & .808 & .610 & .283 & .117 \\ .088 & .214 & .461 & .610 & .610 & .461 & .214 & .088 \\ .041 & .099 & .214 & .283 & .283 & .214 & .099 & .041 \\ .017 & .041 & .088 & .117 & .117 & .088 & .041 & .017 \end{bmatrix} \tag{28}$$

This first simple matrix is very close to **A**. Except for entries $(1, 3)$ and $(1, 4)$ (and symmetrically equivalent entries), every entry in (28) is within about .04 of the corresponding entry in **A**. Since the numbers in **A** were probably rounded off to one decimal digit, one could argue that (28) is as good an approximation to **A** as we should seek.

Next we compute the deflated matrix $\mathbf{A}_2$:

$$\mathbf{A}_2 = \mathbf{A} - \lambda_1 \mathbf{u}_1 * \mathbf{u}_1$$

$$= \begin{bmatrix} -.017 & -.041 & -.088 & .083 & .083 & -.088 & -.041 & -.017 \\ -.041 & .001 & -.014 & .017 & .017 & -.014 & .001 & -.041 \\ -.088 & -.014 & .039 & -.010 & -.010 & .039 & -.014 & -.088 \\ .083 & .017 & -.010 & -.008 & -.008 & -.010 & .017 & .083 \\ .083 & .017 & -.010 & -.008 & -.008 & -.010 & .017 & .083 \\ -.088 & -.014 & .039 & -.010 & -.010 & .039 & -.014 & -.088 \\ -.041 & .001 & -.014 & .017 & .017 & -.014 & .001 & -.041 \\ -.017 & -.041 & -.088 & .083 & .083 & -.088 & -.041 & -.017 \end{bmatrix} \tag{29}$$

Using the iterative method on A_2 takes a long time to converge but finally gives us a multiple of

$$\mathbf{u}_2 = [.514, .224, .258, -.345, -.345, .258, .224, .514]$$

with $\lambda_2 = -.27$

Instead of computing the second simple matrix, we give the sum of the first two simple matrices,

$$\lambda_1\mathbf{u}_1 * \mathbf{u}_1 + \lambda_2\mathbf{u}_2 * \mathbf{u}_2$$

$$= \begin{bmatrix}
-.054 & .010 & .052 & .165 & .165 & .052 & .010 & -.054 \\
.010 & .084 & .198 & .304 & .304 & .198 & .084 & .010 \\
.052 & .198 & .443 & .634 & .634 & .443 & .198 & .052 \\
.165 & .304 & .634 & .776 & .776 & .634 & .304 & .165 \\
.165 & .304 & .634 & .776 & .776 & .634 & .304 & .165 \\
.052 & .198 & .443 & .634 & .634 & .443 & .198 & .052 \\
.010 & .084 & .198 & .304 & .304 & .198 & .084 & .010 \\
-.054 & .010 & .052 & .165 & .165 & .052 & .010 & -.054
\end{bmatrix}$$

$$(30)$$

Next we form A_3 and find that

$$\mathbf{u}_3 = [.451, -.054, -.454, .295, .295, -.454, -.054, .451]$$

with $\lambda_3 = .263$

Note how close in absolute value λ_2 and λ_3 are; this is why the iterative method converged so slowly for A_2.

Now we can give the desired approximation of A by the first three simple matrices associated with the eigenvalue decomposition of A.

$$A \simeq \lambda_1\mathbf{u}_1 * \mathbf{u}_1 + \lambda_2\mathbf{u}_2 * \mathbf{u}_2 + \lambda_1\mathbf{u}_3 * \mathbf{u}_3$$

$$= \begin{bmatrix}
-.001 & .004 & -.002 & .200 & .200 & -.002 & .004 & -.001 \\
.004 & .089 & .204 & .300 & .300 & .204 & .089 & .004 \\
-.002 & .204 & .497 & .599 & .599 & .497 & .204 & -.002 \\
.200 & .300 & .599 & .799 & .799 & .599 & .300 & .200 \\
.200 & .300 & .599 & .799 & .799 & .599 & .300 & .200 \\
-.002 & .204 & .497 & .599 & .599 & .497 & .204 & -.002 \\
.004 & .089 & .204 & .300 & .300 & .204 & .089 & .004 \\
-.001 & .004 & -.002 & .200 & .200 & -.002 & .004 & -.001
\end{bmatrix}$$

$$(31)$$

The average deviation of an entry of (31) from A is .002, and only one entry has an error exceeding .004. ∎

We next give a statistical application of the eigenvalue decomposition of a symmetric matrix.

Example 6. **Principal Components in Statistical Analysis**

Suppose that an anthropologist has collected data on 25 physical characteristics, variables $x_1, x_2, \ldots, x_{25}$, for 100 prehuman fossil remains. The researcher computes a measure of the variability V_i of each variable x_i called the **variance** of x_i. A large variance means that the x_i-variable varies substantially from fossil to fossil. The anthropologist also computes a measure of the joint variability Cov_{ij} of each pair of variables x_i, x_j called the **covariance.** The covariance is proportional to the correlation coefficient (which was discussed in Section 5.3). A positive Cov_{ij} means variables x_i and x_j have similar values; a negative Cov_{ij} means the variables are opposites [if the kth fossil has a large x_i-value, the kth fossil probably has a small or negative x_j-value; and Cov_{ij} near 0 means values of x_i and x_j are unrelated (uncorrelated)].

The anthropologist would like to find good linear combinations of the characteristics that ''explain'' the variability of the data. For example, one might define the *length index L* to be

$$L = .2x_1 + .4x_3 - .3x_{11} + .4x_{17} + .5x_{18} \tag{32}$$

where x_1 might be length of forearm, x_3 length of thigh, and so on. The idea is that although we may find a certain amount of variability in individual variables from fossil to fossil, such as varying length of forearm, the ''right'' measure that gives the best way to distinguish one fossil from another is some composite index, such as the length index.

Among all possible indices formed by a linear combination of variables, the index I_1 that shows the greatest variability (i.e., largest variance) is called the **first principal component.** Among those other indices that are *uncorrelated to I_1* (covariance is 0), the index I_2 with the largest variance is called the **second principal component,** and so on. We want index I_2 uncorrelated so that it gives us new (additional) information about variability that was not contained in I_1.

In summary, the first principal component gives an index that explains the maximum variability of the data from one fossil to another. The first four principal components will typically account for over 90% of the variability in a set of 25 variables. Clearly, there are great advantages in describing each fossil with three or four numbers rather than 25 numbers. The same is true for studies in psychology, finance, quality control, and any other field where people collect large amounts of data.

So how do we find these principal components? That is, how do we determine the weights (coefficients) of the x_i, such as the weights in (32)? The answer is that we form a covariance matrix **C** of all the covariances of the fossil data, where entry (i, j) is Cov_{ij} (and $\text{Cov}_{ii} =$

V_i). Then one can show that *the vector of weights used in the first principal component is the unit-length dominant eigenvector* $\mathbf{u}_1$ *of* $\mathbf{C}$. The first simple matrix $\lambda_1(\mathbf{u}_1 * \mathbf{u}_1)$ in the eigenvalue decomposition of $\mathbf{C}$ shows how much of the variability in $\mathbf{C}$ is explained by the first principal component.

As an example, consider the following 4-by-4 covariance matrix (representing 4 of the 25 characteristics mentioned above)

$$\mathbf{C} = \begin{bmatrix} .86 & 1.19 & 2.02 & 1.45 \\ 1.19 & 1.68 & 2.86 & 2.06 \\ 2.02 & 2.86 & 5.05 & 3.50 \\ 1.45 & 2.06 & 3.50 & 2.53 \end{bmatrix} \tag{33}$$

All the covariances here happen to be positive (this is often the case), although they can be negative. We can determine the eigenvalues and associated eigenvectors by deflation, as in the previous examples, or by using some computer package. We find that

$$\lambda_1 = 10, \qquad \lambda_2 \approx .098, \qquad \lambda_3 \approx .022, \qquad \lambda_4 = .003 \tag{34}$$

The dominant (unit-length) eigenvector is

$$\mathbf{u}_1 = [.289, .408, .707, .5]$$

The simple matrix $\lambda_1 \mathbf{u}_1 * \mathbf{u}_1$ should approximate $\mathbf{C}$ well, since λ_1 is much larger than the other eigenvalues.

$$\lambda_1 \mathbf{u}_1 * \mathbf{u}_1 = \begin{bmatrix} .84 & 1.18 & 2.04 & 1.44 \\ 1.18 & 1.67 & 2.88 & 2.04 \\ 2.04 & 2.88 & 5.00 & 3.54 \\ 1.44 & 2.04 & 3.54 & 2.50 \end{bmatrix} \tag{35}$$

Upon comparing (35) with $\mathbf{C}$, it is clear that I_1 accounts for almost all the variability in the covariance matrix $\mathbf{C}$.

The first principal component index I_1 is the linear combination of variables x_1, x_2, x_3, x_4 with weights given by $\mathbf{u}_1$:

$$I_1 = .289x_1 + .408x_2 + .707x_3 + .5x_4 \tag{36} \quad \blacksquare$$

The eigenvalue decomposition of a matrix $\mathbf{A}$ and the deflation method for finding successive eigenvalues and eigenvectors depended on special properties of $\mathbf{A}$. There had to be a set of n linearly independent eigenvectors for the eigenvalue decomposition and $\mathbf{A}$ had to be symmetric for the deflation method to work. Symmetry is easy to recognize. What about linearly independent eigenvectors? The following theorem answers this question. It is a companion to Theorem 4 (which stated that an n-by-n symmetric matrix has n orthogonal eigenvectors).

Theorem 5

(i) Eigenvectors associated with different eigenvalues are linearly independent.

(ii) If the *n*-by-*n* matrix **A** has *n* distinct eigenvalues, any set of *n* eigenvectors, each associated with a different eigenvalue, will form a basis for *n*-space, and the results in Theorems 1, 2, and 3 apply.

Proof. For explicitness, assume that $n = 3$ and let $\mathbf{u}_1$, $\mathbf{u}_2$, and $\mathbf{u}_3$ be three eigenvectors of **A** associated with different eigenvalues λ_1, λ_2, and λ_3, respectively. It is easy to show that $\mathbf{u}_1$ and $\mathbf{u}_2$ are linearly independent (Exercise 14). Suppose that $\mathbf{u}_1$, $\mathbf{u}_2$, and $\mathbf{u}_3$ are not linearly independent, so that $\mathbf{u}_3$ can be expressed as a *unique* linear combination of $\mathbf{u}_1$ and $\mathbf{u}_2$: $\mathbf{u}_3 = c_1\mathbf{u}_1 + c_2\mathbf{u}_2$. Now we compute $\mathbf{Au}_3$ in two ways.

$$\mathbf{Au}_3 = \lambda_3\mathbf{u}_3 = \lambda_3(c_1\mathbf{u}_1 + c_2\mathbf{u}_2) \tag{37}$$

and

$$\mathbf{Au}_3 = \mathbf{A}(c_1\mathbf{u}_1 + c_2\mathbf{u}_2) = \lambda_1 c_1\mathbf{u}_1 + \lambda_2 c_2\mathbf{u}_2 \tag{38}$$

The representation of $\mathbf{Au}_3$ as a linear combination of $\mathbf{u}_1$ and $\mathbf{u}_2$ is unique. That is, the weights of $\mathbf{u}_1$ and $\mathbf{u}_2$ on the right sides of (37) and (38) must be equal: $\lambda_3 c_1 = \lambda_1 c_1$ and $\lambda_3 c_2 = \lambda_2 c_2$. Thus $\lambda_3 = \lambda_1$ and $\lambda_3 = \lambda_2$. This contradiction proves that $\mathbf{u}_1$, $\mathbf{u}_2$, and $\mathbf{u}_3$ must be linearly independent. ∎

The following is an example of a "defective" matrix to which Theorem 5 does not apply.

Example 7. A Matrix Without an Eigenvector Basis

The matrix

$$\mathbf{A} = \begin{bmatrix} 0 & 1 \\ 0 & 0 \end{bmatrix}$$

has the characteristic polynomial which equals $\det(\mathbf{A} - \lambda\mathbf{I}) = \lambda^2$ (check this), so its two eigenvalues are both 0: $\lambda_1 = \lambda_2 = 0$. **A** is not symmetric and does not have two different eigenvalues. Thus Theorem 5 does not apply to **A**.

Any eigenvector **u** of **A** must satisfy $(\mathbf{A} - 0\mathbf{I})\mathbf{u} = \mathbf{0}$. That is,

$$\mathbf{Au} = \mathbf{0} \quad \text{or} \quad \begin{aligned} 0u_1 + 1u_2 &= 0 \\ 0u_1 + 0u_2 &= 0 \end{aligned} \tag{39}$$

The first equation reduces to $u_2 = 0$, and the second equation is vacuous. Thus a solution **u** to (39) can have any value for u_1 while u_2

must be 0. However, all such vectors $\mathbf{u} = [u_1, 0]$ are multiples of one another, so the eigenvectors of $\mathbf{A}$ do not form a basis for 2-space. ∎

Another issue that we have skirted is complex-valued eigenvalues and eigenvectors. Complex eigenvalues were encountered in our discussion of population models in Section 4.5. In many other applications complex numbers arise naturally. (Incidentally, one can show that the eigenvalues of a symmetric matrix are always real.)

What happens when a matrix is not square (so that it has no eigenvalues or eigenvectors)? Can we find something like an eigenvalue decomposition for these matrices just as a pseudoinverse substitutes for the inverse in a nonsquare matrix? The answer is yes.

In the spirit of the development of the pseudoinverse of a nonsquare matrix, we again turn to the n-by-n matrix $\mathbf{A}^T\mathbf{A}$ which is square and symmetric [since entry (i, j) is just the scalar product of the ith and jth columns of $\mathbf{A}$]. From the eigenvalue decomposition of $\mathbf{A}^T\mathbf{A}$, one can obtain a decomposition for $\mathbf{A}$.

Theorem 6. *Singular-Value Decomposition.* For any m-by-n matrix $\mathbf{A}$ with linearly independent columns, let $|\lambda_1| \geq |\lambda_2| \geq \cdots \geq |\lambda_n|$ be the eigenvalues of $\mathbf{A}^T\mathbf{A}$ and $\mathbf{U}$ be an n-by-n matrix whose columns $\mathbf{u}_i$ are the associated (orthonormal) eigenvectors of $\mathbf{A}^T\mathbf{A}$. The $\mathbf{u}_i$ form a basis for the row space of $\mathbf{A}$.

Define the ith **singular value** s_i to be $s_i = \sqrt{\lambda_i}$. Let $\mathbf{U}'$ be an m-by-n matrix with columns $\mathbf{u}_i' = (1/s_i)\mathbf{A}\mathbf{u}_i$. The $\mathbf{u}_i'$ form an orthonormal basis for the range of $\mathbf{A}$.

Then $\mathbf{A}$ can be decomposed in the form

$$\mathbf{A} = \mathbf{U}'\mathbf{D}_s\mathbf{U}^T \tag{40}$$

where $\mathbf{D}_s$ is the diagonal matrix whose ith diagonal entry is s_i.

The proof of Theorem 6 is given in the Exercises. Recall from Theorem 4 of Section 5.3 that if $\mathbf{A}$ has linearly independent columns, then $\mathbf{A}^T\mathbf{A}$ has linearly independent columns. There is a generalized form of Theorem 6 for linearly dependent columns.

The definition of the columns of $\mathbf{U}'$ means that $\mathbf{U}'$ has the matrix formula

$$\mathbf{U}' = \mathbf{A}\mathbf{U}\mathbf{D}_s^{-1}$$

The factorization (40) leads to the following simple matrix decomposition (by the same argument that led up to Theorem 3).

Corollary A. Let $\mathbf{A}$ be as in Theorem 6, let $\mathbf{u}_i$ be the ith column of $\mathbf{U}$, and let $\mathbf{u}_i'$ be the ith column of $\mathbf{U}'$. Then

$$\mathbf{A} = s_1 \mathbf{u}_1' * \mathbf{u}_1 + s_2 \mathbf{u}_2' * \mathbf{u}_2 + \cdots + s_k \mathbf{u}_k' * \mathbf{u}_k \tag{41}$$

We also get from (40) a result like Theorem 1.

Corollary B. The computation $\mathbf{c} = \mathbf{Ab}$ becomes $\mathbf{c}^* = \mathbf{D}_s \mathbf{b}^*$ (or $c_i^* = s_i b_i^*$) when $\mathbf{b}^* = \mathbf{U}^T \mathbf{b}$ and $\mathbf{c}^* = \mathbf{U}'^T \mathbf{c}$. That is, multiplying a vector by $\mathbf{A}$ reduces to scalar multiplication of the coordinates when we express $\mathbf{b}$ in the proper row space coordinates and $\mathbf{c}$ in the proper column space (range) coordinates.

The singular-value decomposition of $\mathbf{A}$ can be "inverted" to obtain the pseudoinverse of $\mathbf{A}$.

Corollary C. Let $\mathbf{A}$ be an m-by-n matrix. Then the pseudoinverse of $\mathbf{A}$ equals

$$\mathbf{A}^+ = \mathbf{U}\mathbf{D}_s^{-1}\mathbf{U}'^T \tag{42}$$

Recall that $\mathbf{D}_s^{-1}$ is a diagonal matrix with entry $(i, i) = 1/s_i$.

We now give the singular-value decomposition for the two-refinery matrix discussed in Section 5.3.

Example 8. Singular-Value Decomposition of Two-Refinery Model

The two-refinery variant was

$$\begin{aligned} 20x_1 + 4x_2 &= b_1 \\ 10x_1 + 14x_2 &= b_2 \qquad \text{or} \qquad \mathbf{Ax} = \mathbf{b} \\ 5x_1 + 5x_2 &= b_3 \end{aligned} \tag{43}$$

Let us compute the singular-value decomposition of $\mathbf{A}$. First we form $\mathbf{A}^T\mathbf{A}$:

$$\mathbf{A}^T\mathbf{A} = \begin{bmatrix} 525 & 245 \\ 245 & 237 \end{bmatrix} \tag{44}$$

We need to find the eigenvalues λ_1, λ_2 of $\mathbf{A}^T\mathbf{A}$, since the singular values s_1, s_2 are their square roots. Further, the eigenvectors of $\mathbf{A}^T\mathbf{A}$ are the columns of the matrix $\mathbf{U}$. By some method, discussed previously or in the appendix to this section, we find that

$$\begin{aligned} \lambda_1 &\simeq 665 \qquad \text{and} \qquad \mathbf{u}_1 = [.87, .49] \\ \lambda_2 &\simeq 97 \qquad \text{and} \qquad \mathbf{u}_2 = [-.49, .87] \end{aligned} \tag{45}$$

So the singular values of $\mathbf{A}$ are $s_1 = \sqrt{665} = 25.6$, $s_2 = \sqrt{97} = 9.8$, and

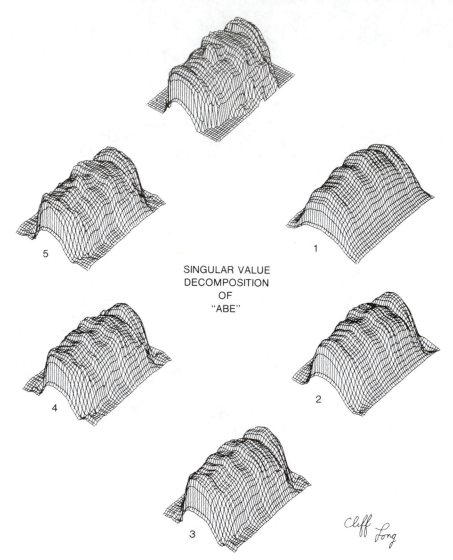

SINGULAR VALUE
DECOMPOSITION
OF
"ABE"

Figure 5.10 From Cliff Long, "Visualization of Matrix Singular Value Decomposition," *Mathematics Magazine,* Vol. 56 (1983), pp. 161–167.

$$\mathbf{U} = \begin{bmatrix} .87 & -.49 \\ .49 & .87 \end{bmatrix} \tag{46}$$

As an aside, we note that $\mathbf{U}$ performs a rotation of 20° (see Theorem 3 of Section 5.4).

Next we compute $\mathbf{U}' = [\mathbf{u}_1', \mathbf{u}_2']$, where $\mathbf{u}_i' = (1/s_i)\mathbf{A}\mathbf{u}_i$. We obtain

$$\mathbf{U}' = \begin{bmatrix} .76 & -.65 \\ .60 & .74 \\ .26 & .19 \end{bmatrix} \tag{47}$$

Then the singular-value decomposition of $\mathbf{A} = \mathbf{U}'\mathbf{D}_s\mathbf{U}^T$ is

$$
\begin{bmatrix} 20 & 4 \\ 10 & 14 \\ 5 & 5 \end{bmatrix} = \begin{bmatrix} .76 & -.65 \\ .60 & .74 \\ .26 & .19 \end{bmatrix} \begin{bmatrix} 25.6 & 0 \\ 0 & 9.8 \end{bmatrix} \begin{bmatrix} .87 & .49 \\ -.49 & .87 \end{bmatrix} \quad (48)
$$

As a sum of simple matrices [see (41)], (48) becomes

$$
\mathbf{A} = \begin{bmatrix} 16.9 & 9.5 \\ 13.4 & 7.5 \\ 5.8 & 3.3 \end{bmatrix} + \begin{bmatrix} 3.1 & -5.5 \\ -3.4 & 6.5 \\ -.8 & 1.7 \end{bmatrix} \quad (49)
$$

The decomposition in (48) can be interpreted as follows. There are two basic "input" and "output" units for the refinery model. The first input unit consists of .87 barrel of petroleum for refinery 1 and .49 barrel for refinery 2, and one such input unit yields 25.6 output units, each consisting of .76 gallon of diesel oil, .60 heating oil, and .26 gasoline. The second input unit consists of $-.49$ barrel (production is "reversed") for refinery 1 and .87 barrel for refinery 2, and it yields 9.8 output units, each consisting of $-.65$ diesel, .74 heating, and .19 gasoline. ∎

A more impressive use of the singular-value decomposition is given in Figure 5.10. Figure 5.10 (top) shows a 49-by-36 digitized image of a bust of Abe Lincoln [entry (i, j) is the height of the bust in that position]. The remaining figures in this set show the digitized image produced by the matrix $\mathbf{A}_k$, the sum of the first k simple matrices in the singular-value decomposition of $\mathbf{A}$.

Section 5.5 Exercises

Summary of Exercises
Exercises 1–5 involve the diagonalization of matrices presented in Theorem 2. Exercises 6–13 involve the eigenvalue decomposition of matrices into a sum of simple matrices (many require deflation to find the eigenvalues). Exercises 14 and 15 are about independence of eigenvectors. Exercise 16 involves defective matrices. Exercises 17–22 involve computing the singular-value decomposition. Exercises 23–25 prove the results in Theorem 6.

1. Compute the representation $\mathbf{U}\mathbf{D}_\lambda\mathbf{U}^{-1}$ of Theorem 2 for the following matrices whose eigenvalues and largest eigenvector you were asked to determine in Exercise 23 of Section 3.1.

   (a) $\begin{bmatrix} 4 & 0 \\ 2 & 2 \end{bmatrix}$ (b) $\begin{bmatrix} 1 & 2 \\ 3 & 4 \end{bmatrix}$ (c) $\begin{bmatrix} 2 & 1 \\ 2 & 3 \end{bmatrix}$ (d) $\begin{bmatrix} 4 & -1 \\ 1 & 2 \end{bmatrix}$

2. For a starting vector of $\mathbf{p} = [10, 10]$, compute $p^{(10)} = \mathbf{A}^{10}\mathbf{p}$ for each matrix $\mathbf{A}$ in Exercise 1.

3. Compute A^5 for each matrix in Exercise 1 using the formula $A^5 = UD_\lambda^5 U^{-1}$.

4. (a) Given that $A = UD_\lambda U^{-1}$, prove that $A^2 = UD_\lambda^2 U^{-1}$.
   (b) Use induction to prove that $A^k = UD_\lambda^k U^{-1}$.

5. (a) Obtain a formula for A^{-1} similar to $A = UD_\lambda U^{-1}$.
      Hint: Only the matrix D_λ will be different.
   (b) Verify your formula in part (a) for $A = \begin{bmatrix} 3 & 1 \\ 2 & 2 \end{bmatrix}$ (the matrix in Example 1).

6. Give the eigenvalue decomposition into a sum of two simple matrices (Theorem 3) for each matrix in Exercise 1.

7. Give the eigenvalue decomposition into a sum of two simple matrices for the following symmetric matrices.

   (a) $\begin{bmatrix} 1 & 4 \\ 4 & 1 \end{bmatrix}$ (b) $\begin{bmatrix} 3 & -4 \\ -4 & 3 \end{bmatrix}$ (c) $\begin{bmatrix} -1 & -5 \\ -5 & -1 \end{bmatrix}$

8. Use the deflation method to compute all eigenvalues and eigenvectors of the following symmetric matrices.

   (a) $\begin{bmatrix} 1 & 6 \\ 6 & 1 \end{bmatrix}$ (b) $\begin{bmatrix} 2 & -3 \\ -3 & 2 \end{bmatrix}$ (c) $\begin{bmatrix} 0 & 1 \\ 1 & 0 \end{bmatrix}$

   (d) $\begin{bmatrix} 1 & 2 & 3 \\ 2 & 4 & -1 \\ 3 & -1 & 0 \end{bmatrix}$ (e) $\begin{bmatrix} 1 & 0 & 1 \\ 0 & 1 & 1 \\ 1 & 1 & 1 \end{bmatrix}$ (f) $\begin{bmatrix} 2 & -1 & 0 & 1 \\ -1 & 3 & 1 & -2 \\ 0 & 1 & 0 & 1 \\ 1 & -2 & 1 & 0 \end{bmatrix}$

9. Using a software package for finding eigenvectors and eigenvalues, determine the eigenvalue decomposition into simple matrices for the following symmetric matrices.

   (a) $\begin{bmatrix} 1 & 3 & 0 \\ 3 & 2 & 1 \\ 0 & 1 & 0 \end{bmatrix}$ (b) $\begin{bmatrix} 1 & 1 & 1 \\ 1 & 1 & 1 \\ 1 & 1 & 0 \end{bmatrix}$ (c) $\begin{bmatrix} 2 & -1 & 0 & 1 \\ -1 & 3 & 1 & -2 \\ 0 & 1 & 0 & 1 \\ 1 & -2 & 1 & 0 \end{bmatrix}$

10. Approximate the following symmetric digital pictures by:
    (i) The first simple matrix in the eigenvalue decomposition.
    (ii) The sum of the first two simple matrices in the eigenvalue decomposition.
    Use deflation (or a software package) to determine the two first eigenvectors and eigenvalues.

(a) $\begin{bmatrix} .3 & .2 & .1 \\ .2 & 0 & .5 \\ .1 & .5 & .2 \end{bmatrix}$

(b) $\begin{bmatrix} .2 & .4 & .4 & .2 \\ .4 & .2 & .2 & .4 \\ .4 & .2 & .2 & .4 \\ .2 & .4 & .4 & .2 \end{bmatrix}$

(c) $\begin{bmatrix} .1 & .2 & .3 & .4 \\ .2 & .3 & .2 & .1 \\ .3 & .2 & .1 & .4 \\ .4 & .1 & .4 & .6 \end{bmatrix}$

(d) $\begin{bmatrix} .1 & .2 & .1 & 0 & 0 \\ .2 & .3 & .1 & .1 & 0 \\ .1 & .1 & .1 & .2 & .1 \\ 0 & .1 & .2 & .4 & 0 \\ 0 & 0 & .1 & 0 & .3 \end{bmatrix}$

(e) $\begin{bmatrix} .4 & .3 & .1 & .1 & .3 & .6 \\ .3 & .6 & 0 & 0 & .1 & .5 \\ .1 & 0 & 0 & 0 & .1 & .3 \\ .1 & 0 & 0 & .1 & .3 & .2 \\ .3 & .1 & .1 & .3 & .4 & 0 \\ .6 & .5 & .3 & .2 & 0 & 0 \end{bmatrix}$

11. Explain how the symmetry in the 8-by-8 digital picture (25) in Example 5 would allow one to find the eigenvalue decomposition for the 4-by-4 upper right corner submatrix $\mathbf{A}'$ and use this to get the eigenvalue decomposition for the whole matrix.

    Find the dominant eigenvalue and associated (normalized) eigenvector for $\mathbf{A}'$; approximate $\mathbf{A}'$ by the first simple matrix in the eigenvalue decomposition.

12. Verify that the eigenvalues (34) and dominant eigenvector in Example 6 are correct. Determine the second principal component in Example 6 (the normalized eigenvector for λ_2).

13. Determine the first principal component for each of the following covariance matrices. In each case, tell how well it accounts for the variability (how well does $\lambda_1 \mathbf{u}_1 * \mathbf{u}_1$ approximate

(a) $\begin{bmatrix} 3.1 & 1.1 & 0.5 \\ 1.1 & 2.0 & 1.5 \\ 0.5 & 1.5 & 4.2 \end{bmatrix}$

(b) $\begin{bmatrix} 2.4 & 0.6 & 3.1 & 1.5 \\ 0.6 & 4.1 & 0.8 & 1.2 \\ 3.1 & 0.8 & 2.7 & 5.2 \\ 1.5 & 1.2 & 5.2 & 3.2 \end{bmatrix}$

14. In the proof Theorem 4, show that $\mathbf{u}_1$ and $\mathbf{u}_2$ are linearly independent by supposing the opposite, that $\mathbf{u}_2 = r\mathbf{u}_1$, and obtaining a contradiction when $\mathbf{A}\mathbf{u}_2 \neq \mathbf{A}(r\mathbf{u}_1)$.

15. Two n-by-n matrices $\mathbf{A}$ and $\mathbf{B}$ are called **similar** if there exists an invertible n-by-n matrix $\mathbf{U}$ such that $\mathbf{A} = \mathbf{U}\mathbf{B}\mathbf{U}^{-1}$ (or equivalently, $\mathbf{B} = \mathbf{U}^{-1}\mathbf{A}\mathbf{U}$).

(a) Show that similar matrices have the same set of eigenvalues.

(b) Show that if $\mathbf{A}$ has a set of n linearly independent eigenvectors, then $\mathbf{A}$ is similar to a diagonal matrix.

16. Find the eigenvalues and as many eigenvectors as you can for the following defective matrices.

(a) $\begin{bmatrix} 1 & 0 \\ 2 & 1 \end{bmatrix}$
(b) $\begin{bmatrix} 1 & -1 \\ 1 & 3 \end{bmatrix}$
(c) $\begin{bmatrix} 2 & 1 & 0 \\ 0 & 2 & 1 \\ 0 & 0 & 2 \end{bmatrix}$

17. Find the singular-value decomposition [equation (40)] for the following matrices and use the decomposition to write the matrices as a sum of simple matrices.

(a) $\begin{bmatrix} 2 \\ 1 \\ 0 \end{bmatrix}$
(b) $\begin{bmatrix} 1 & 0 \\ 3 & 3 \\ 2 & 5 \end{bmatrix}$
(c) $\begin{bmatrix} 3 & 0 \\ 2 & 5 \\ 1 & 3 \\ 4 & 0 \end{bmatrix}$
(d) $\begin{bmatrix} 1 & 2 \\ 2 & 3 \\ 3 & 4 \end{bmatrix}$

18. For a refinery problem with p refineries and q products as in Example 8, suppose that the coefficient matrix is the matrix in each part of Exercise 17. In each case, interpret the singular-value decomposition in terms of units of "input" and "output," as was done at the end of Example 8.

19. Use the formula for the pseudoinverse in Corollary C to compute the pseudoinverse of each matrix in Exercise 17.

20. With the help of deflation or a software package to find eigenvectors of $\mathbf{A}^T\mathbf{A}$, find the singular-value decomposition for the following matrices and use the decomposition to approximate these matrices as a sum of two simple matrices.

(a) $\begin{bmatrix} 2 & 0 & -1 \\ 1 & 2 & 0 \\ 4 & -1 & 2 \\ 5 & 0 & -1 \end{bmatrix}$
(b) $\begin{bmatrix} .1 & .2 & .3 & .4 \\ .4 & 0 & .1 & .2 \\ .3 & .7 & .1 & 0 \\ .1 & .6 & .1 & 0 \\ .4 & 0 & .1 & .3 \end{bmatrix}$

21. If $\mathbf{A}$ is a symmetric matrix, show that the singular-value decomposition reduces to the standard eigenvalue decomposition in Theorem 2.

22. Verify the formula for the pseudoinverse in Corollary C.

23. Show that the n orthonormal eigenvectors of $\mathbf{A}^T\mathbf{A}$ in $\mathbf{U}$ (in the singular-value decomposition) are a basis for the row space of the m-by-n matrix $\mathbf{A}$.

Hint: What is the dimension of the row space of **A** if its columns are linearly independent?

24. Show that the columns of **U**′ (in the singular-value decomposition) must form a basis for the range of **A** (the column space of **A**).

    Hint: Use Exercise 23, and Exercise 30 of Section 5.3.

25. Verify the singular-value decomposition (40) by showing that

$$\mathbf{D}_s = \mathbf{U}'^T\mathbf{A}\mathbf{U} \tag{*}$$

(a) First show that (40) and (*) are equivalent matrix equations.
    Hint: Use the fact that (by orthonormality) $\mathbf{U}^T\mathbf{U} = \mathbf{U}'^T\mathbf{U}' = \mathbf{I}$.
(b) Prove (*) by verifying the following sequence of matrix equations:

$$\mathbf{U}'^T\mathbf{A}\mathbf{U} = (\mathbf{A}\mathbf{U}\mathbf{D}_s^{-1})^T\mathbf{A}\mathbf{U} = (\mathbf{D}_s^{-1}\mathbf{U}^T\mathbf{A}^T)\mathbf{A}\mathbf{U} = \mathbf{D}_s^{-1}\mathbf{U}^T(\mathbf{A}^T\mathbf{A}\mathbf{U})$$
$$= \mathbf{D}_s^{-1}\mathbf{U}^T(\mathbf{U}\mathbf{D}_\lambda) = \mathbf{D}_s^{-1}\mathbf{D}_\lambda = \mathbf{D}_s$$

Appendix to Section 5.5: Finding Eigenvalues and Eigenvectors

In this appendix we present two methods for finding eigenvalues and eigenvectors. The first is a way to speed up the search for an eigenvalue λ and associated eigenvector **u** once we have a rough approximation to λ, say, obtained by guesswork or by a few rounds of the iterative method $\mathbf{A}^k\mathbf{x}$. The following basic theorems about eigenvalues are needed.

Theorem 1

(i) For any nonzero integer k, **A**'s eigenvectors are eigenvectors of $\mathbf{A}^k$. If λ is an eigenvalue of **A**, then λ^k is an eigenvalue of $\mathbf{A}^k$. In particular, $1/\lambda$ is an eigenvalue for $\mathbf{A}^{-1}$. (If $k < 0$, we assume that $\mathbf{A}^{-1}$ exists.)

(ii) For any scalar r, **A** and $\mathbf{A} - r\mathbf{I}$ have the same set of eigenvectors and λ is an eigenvalue for **A** if and only if $\lambda - r$ is an eigenvalue of $\mathbf{A} - r\mathbf{I}$.

Proof. We give a proof of part (i) for $k = -1$ [positive k and Theorem 1, part (ii) are left as exercises]. Suppose that **u** is an eigenvector of **A** with associated eigenvalue λ. Then

$$\mathbf{u} = \mathbf{I}\mathbf{u} = (\mathbf{A}^{-1}\mathbf{A})\mathbf{u} = \mathbf{A}^{-1}(\mathbf{A}\mathbf{u}) \tag{1}$$
$$= \mathbf{A}^{-1}\lambda\mathbf{u}$$

Dividing both sides of (1) by λ, we have

$$\left(\frac{1}{\lambda}\right)\mathbf{u} = \mathbf{A}^{-1}\mathbf{u} \tag{2}$$

So $1/\lambda$ is an eigenvalue of $\mathbf{A}^{-1}$ with eigenvector $\mathbf{u}$, as claimed.   ■

Theorem 2. The rate at which the iterative method $\mathbf{A}^k\mathbf{x}$ converges to a dominant eigenvector λ_1 of $\mathbf{A}$ is proportional to $|\lambda_2|/|\lambda_1|$.

Proof. Any vector $\mathbf{x}$ can be written as a linear combination of the eigenvectors (assuming that the *n*-by-*n* matrix $\mathbf{A}$ has *n* linearly independent eigenvectors)

$$\mathbf{x} = c_1\mathbf{u}_1 + c_2\mathbf{u}_2 + \cdots + c_n\mathbf{u}_n$$

Then

$$\begin{aligned}
\mathbf{A}^k\mathbf{x} &= c_1\mathbf{A}^k\mathbf{u}_1 + c_2\mathbf{A}^k\mathbf{u}_2 + \cdots + c_n\mathbf{A}^k\mathbf{u}_n \\
&= c_1\lambda_1^k\mathbf{u}_1 + c_2\lambda_2^k\mathbf{u}_2 + \cdots + c_n\lambda_{nn}^{ku} \\
&= \lambda_1^k\left\{c_1\mathbf{u}_1 + c_2\left(\frac{\lambda_2}{\lambda_1}\right)^k\mathbf{u}_2 + \cdots + c_n\left(\frac{\lambda_n}{\lambda_1}\right)^k\mathbf{u}_n\right\}
\end{aligned} \tag{3}$$

The last line of (3) shows how the size (λ_2/λ_1) affects the convergence to $\lambda_1^k c_1\mathbf{u}_1$.   ■

Suppose that $\mathbf{A}$ has an inverse $\mathbf{A}^{-1}$. Since λ is an eigenvalue of $\mathbf{A}$ if and only if $1/\lambda$ is an eigenvalue of $\mathbf{A}^{-1}$ and both have the same eigenvectors, we have

Corollary. An eigenvector $\mathbf{u}_n$ associated with the smallest eigenvalue λ_n of $\mathbf{A}$ can be found (if λ_n is unique) by applying the iterative method to find the largest eigenvalue of $\mathbf{A}^{-1}$:

$$\mathbf{y}^{(k)} = \mathbf{A}^{-1}\mathbf{y}^{(k-1)} \tag{4}$$

The rate of convergence will be proportional to $|\lambda_{n-1}|/|\lambda_n|$, where λ_{n-1} is the second smallest eigenvalue of $\mathbf{A}$.

Rather than compute the inverse of $\mathbf{A}^{-1}$ and then use (4), we can write (4) as

$$\mathbf{A}\mathbf{y}^{(k)} = \mathbf{y}^{(k-1)} \tag{5}$$

In (5), we are applying the regular iterative method in reverse. To find $\mathbf{y}^{(k)}$ given $\mathbf{y}^{(k-1)}$, we solve (5) by Gaussian elimination (saving the matrices $\mathbf{L}$ and $\mathbf{U}$—see Section 3.2—to use again for each successive $\mathbf{y}^{(k)}$). We call the method for finding the smallest eigenvalue of **A** and an associated eigenvector using (5) the **inverse iterative method.**

Surprisingly, the inverse iterative method yields a faster way to compute the dominant eigenvector than the (forward) iterative method.

> ***Theorem 3. Shifted Inverse Iterative Method.*** If σ is an approximate value for an eigenvalue of a square matrix **A**, an associated eigenvector **u** can be found quickly by applying the inverse iterative method to $\mathbf{A} - \sigma \mathbf{I}$. The eigenvalue can then be obtained from **u** using the Raleigh quotient. If λ_p is the true value of the eigenvalue and λ_q is the next-closest eigenvalue of **A**, the rate of convergence is $|\lambda_q - \sigma|/|\lambda_p - \sigma|$.

This method is called the shifted inverse iterative method because of the "shift" of eigenvalues caused by $-\sigma \mathbf{I}$. Recall that from Theorem 1, part (ii), $\mathbf{A} - \sigma \mathbf{I}$ has the same eigenvectors as **A** and its eigenvalues are shifted by σ. If σ is close to λ_p, then $\lambda_p - \sigma$ will be the smallest eigenvalue of $\mathbf{A} - \sigma \mathbf{I}$ by far, and the rate of convergence $|\lambda_q - \sigma|/|\lambda_p - \sigma|$ of the inverse iterative method will be very fast (i.e., two or three iterations should suffice.

We should note that computations are very unstable with $\mathbf{A} - \sigma \mathbf{I}$, since when λ_p is close to σ, $1/(\lambda_p - \sigma)$ is an eigenvalue of $\mathbf{A}^{-1}$ of immense size. This implies that $\|\mathbf{A}^{-1}\|$, and hence the condition number of **A**, are very large. However, the only effect that this instability has on the computations of the inverse iterative method is a distortion of the total size of the $\mathbf{y}^{(k)}$ but not the direction of these vectors (the total size does not concern us, since we use scaling).

Hybrid Deflation Procedure to Find All Eigenvalues and Eigenvectors of a Symmetric Matrix

Step 1. Starting with $\mathbf{A}_1 = \mathbf{A}$, use the (forward) iterative method a few times on $\mathbf{A}_i$ to get an approximation σ to the dominant eigenvalue λ_i of $\mathbf{A}_i$ together with an approximate eigenvector **v**.

Step 2. Starting with **v**, use the shifted inverse iterative method on $\mathbf{A}_i - \sigma \mathbf{I}$ to get more accurate values for the unit-length eigenvector $\mathbf{u}_i$. Obtain λ_i from $\mathbf{u}_i$ by the Raleigh quotient $\mathbf{u}_i \cdot \mathbf{A}\mathbf{u}_i / \mathbf{u}_i \cdot \mathbf{u}_i$.

Step 3. Compute $\mathbf{A}_{i+1} = \mathbf{A}_i - \lambda_i \mathbf{u}_i {}^* \mathbf{u}_i$, set $i = i + 1$, and if $i \leq n$, go to step 1.

We now present an almost magical procedure to find all the eigenvalues at once of a square matrix **A** with distinct eigenvalues.

QR *Method for Finding All Eigenvalues of a Matrix with Distinct Eigenvalues*

Step 1. Let $A_0 = A$. For successive k,
(a) Given A_k, compute the **QR** decomposition $A_k = Q_k R_k$ (by the Gram–Schmidt orthogonalization procedure); and then
(b) Set $A_{k+1} = R_k Q_k$ and go to step (a).
Stop when the entries below the main diagonal of A_k are all almost 0 (entries above the main diagonal do not converge to 0 unless A is symmetric).

Step 2. The entries on the main diagonal of the last A_k will be approximately the eigenvalues of A (in order of decreasing absolute value). Use the shifted inverse iterative method to find the eigenvector [or if the eigenvalue λ is essentially exact, solve the homogeneous system $(A - \lambda I)u = 0$].

Example 1. **Example of QR and Shifted Inverse Iterative Method**

Let us use the **QR** method on the matrix L in the Leslie model from Example 1 of Section 4.5.

$$x' = Lx, \qquad \text{where } L = \begin{bmatrix} 0 & 4 & 1 \\ .4 & 0 & 0 \\ 0 & .6 & 0 \end{bmatrix}$$

We noted in Section 4.5 that $L^k x$ takes a long time to converge to a multiple of the dominant eigenvalue λ_1. Since convergence is slow, Theorem 2 says that the second largest eigenvalue λ_2 must be close to λ_1. The **QR** method is also slow for this L. If we run it until all below-diagonal entries are $< .001$ (then the eigenvalues are accurate to about three decimal places), it requires 60 iterations and yields

$$A_{60} = \begin{bmatrix} 1.334 & -3.586 & -1.142 \\ .000 & -1.118 & -.394 \\ .000 & .000 & -.152 \end{bmatrix} \tag{6}$$

So $\lambda_1 = 1.334$, $\lambda_2 = -1.118$, $\lambda_3 = -.152$. Solving the homogeneous system $(L - \lambda_i I) = 0$ will give a corresponding eigenvector.

Let us next try the shifted iterative method. Starting with $x = [100, 50, 30]$, we gave a table in Section 4.5 of iterates up to $L^{20}a$ (Table 4.3), at which point there was still a little cyclic behavior. Consider the ninth, tenth, and eleventh iterates:

$$x^{(9)} = [2021, 493, 279], \ x^{(10)} = [2250, 808, 295],$$
$$x^{(11)} = [3529, 900, 485]$$

When we compute the Raleigh quotients for $k = 9$ and $k = 10$, we get

$$\frac{\mathbf{x}^{(9)} \cdot \mathbf{x}^{(10)}}{\mathbf{x}^{(9)} \cdot \mathbf{x}^{(9)}} = \frac{5{,}027{,}899}{4{,}414{,}515} = 1.138$$

$$\frac{\mathbf{x}^{(10)} \cdot \mathbf{x}^{(11)}}{\mathbf{x}^{(10)} \cdot \mathbf{x}^{(10)}} = \frac{8{,}668{,}230}{5{,}802{,}389} = 1.493 \tag{7}$$

The average of our two estimates is $(1.138 + 1.493)/2 \simeq 1.32$. Presumably, we are alternating above and below the true eigenvalue, so this average value of 1.32 should be close to the true eigenvalue. Now we apply the shift step of computing $\mathbf{L} - 1.32\mathbf{I}$:

$$\mathbf{L}' = \mathbf{L} - 1.32\mathbf{I} = \begin{bmatrix} 0 & 4 & 1 \\ .4 & 0 & 0 \\ 0 & .6 & 0 \end{bmatrix} - \begin{bmatrix} 1.32 & 0 & 0 \\ 0 & 1.32 & 0 \\ 0 & 0 & 1.32 \end{bmatrix}$$

$$= \begin{bmatrix} -1.32 & 4 & 1 \\ .4 & -1.32 & 0 \\ 0 & .6 & -1.32 \end{bmatrix}$$

Let us scale $\mathbf{x}^{(11)}$ by dividing its entries by the largest entry, 3529, to obtain $\mathbf{x}' = [1, .255, .137]$. We use $\mathbf{x}'$ as the starting vector for backward iteration with $\mathbf{L}'$ (in search of an eigenvector associated with the smallest eigenvalue of $\mathbf{L}'$). After two rounds we have the vector (rounded to integers) [4658, 1396, 628], which divided by the sum of its entries (to be like a population probability distribution) yields

$$\mathbf{u}_1 = [.697, .209, .094] \tag{8}$$

Computing $\mathbf{L}\mathbf{u}_1$, we obtain

$$\mathbf{L}\mathbf{u}_1 = [.930, .278, .125] = 1.334\mathbf{u}_1 \tag{9}$$

so $\lambda_1 = 1.334$. Further, (9) confirms that $\mathbf{u}_1$ is an eigenvector.

If we wanted to get all eigenvalues and associated eigenvectors for $\mathbf{L}$, we could use 10 or 15 rounds of the QR method to get estimates for each eigenvalue and then use the shifted inverse method to home in on the associated eigenvector of each eigenvalue (any vector can be used as the starting vector for the inverse iterative method). ∎

The proof of convergence for the **QR** method is beyond the scope of this book (see G. W. Stewart, *Introduction to Matrix Computations*, Academic Press, 1973, and the classic textbook by J. H. Wilkinson, *The Algebraic Eigenvalue Problem*, Oxford University Press, 1965, for a fuller discussion of the QR method).

The convergence is not fast, especially when two eigenvalues are close

together (in absolute value). The computation of each stage is slow (propro-
tional to n^3), but schemes are available to "preprocess" **A** so that the successive **QR** decompositions can be computed quickly. Other shortcuts further
speed this procedure, including shifts (as in the shifted inverse iterative
method). This method works in some cases where there are multiple eigenvalues (of the same absolute value).

A related method called the **LU method** reverses the matrices in the
LU decomposition of a matrix the way the **QR** method reverses the matrices
in a **QR** decomposition (see the books mentioned above). The **LU** method
also converges for any matrix with distinct eigenvalues.

Section 5.5 Appendix Exercises

Summary of Exercises

Exercises 1–5 relate to Theorem 1. Exercises 6–8 illustrate the hybrid deflation procedure.

1. (a) Show that for any positive integer k, any eigenvector for the square
matrix **A** is also an eigenvector for $\mathbf{A}^k$.
   (b) Also show that if $\mathbf{A}^{-1}$ exists, part (a) is true for negative integers.

2. (a) Show that if $\mathbf{A}^{-1}$ exists, any eigenvector for $\mathbf{A}^k$ is also an eigenvector for **A**.
   (b) The existence of $\mathbf{A}^{-1}$ is essential for the result in part (a). Verify
this by showing that for $\mathbf{A} = \begin{bmatrix} 0 & 1 \\ 0 & 0 \end{bmatrix}$, **1** is an eigenvector for $\mathbf{A}^2$
but not for **A**.

3. (a) Show that for any positive integer, if λ is an eigenvalue of the
square matrix **A**, then λ^k is an eigenvalue of $\mathbf{A}^k$.
   (b) Also show that if $\mathbf{A}^{-1}$ exists, part (a) is true for negative integers.

4. Show that for any scalar r, **A** and $\mathbf{A} - r\mathbf{I}$ have the same set of eigenvectors.

5. Show that for any scalar r, λ is an eigenvalue for **A** if and only if
$\lambda - r$ is an eigenvalue for $\mathbf{A} - r\mathbf{I}$.

6. Use the inverse power method to find the dominant eigenvalue and
eigenvector for the following matrices.

   (a) $\begin{bmatrix} 4 & 0 \\ 2 & 2 \end{bmatrix}$ (b) $\begin{bmatrix} 1 & 2 \\ 3 & 4 \end{bmatrix}$ (c) $\begin{bmatrix} 1 & -1 \\ 1 & 3 \end{bmatrix}$ (d) $\begin{bmatrix} 0 & 1 \\ -1 & 0 \end{bmatrix}$

7. Use the hybrid deflation procedure to find all eigenvalues and associated
eigenvectors for the following symmetric matrices.

   (a) $\begin{bmatrix} 1 & 6 \\ 6 & 1 \end{bmatrix}$ (b) $\begin{bmatrix} 2 & -3 \\ -3 & 2 \end{bmatrix}$ (c) $\begin{bmatrix} 0 & 1 \\ 1 & 0 \end{bmatrix}$

(d) $\begin{bmatrix} 1 & 2 & 3 \\ 2 & 4 & -1 \\ 3 & -1 & 0 \end{bmatrix}$ **(e)** $\begin{bmatrix} 1 & 0 & 1 \\ 0 & 1 & 1 \\ 1 & 1 & 1 \end{bmatrix}$ **(f)** $\begin{bmatrix} 2 & -1 & 0 & 1 \\ -1 & 3 & 1 & -2 \\ 0 & 1 & 0 & 1 \\ 1 & -2 & 1 & 0 \end{bmatrix}$

8. Apply the **QR** method (using the program in Exercise 10 or a software package) to find the eigenvalues and associated eigenvectors for the matrices in Exercise 7.

Programming Projects

9. Write a program to implement the hybrid deflation procedure.

10. Write a program to implement the **QR** method.

A Brief History of Matrices and Linear Algebra

Matrices and linear algebra did not grow out of the study of coefficients of systems of linear equations, as one might guess. Arrays of coefficients led mathematicians to develop determinants, not matrices. Leibniz, coinventor of calculus, used determinants in 1693, about one hundred and fifty years before the study of matrices in their own right. Cramer presented his determinant-based formula for solving systems of linear equations in 1750, and Gauss developed Gaussian elimination around 1820. These events occurred before matrix notation even existed. As an aside, we note that Gaussian elimination was for years considered part of the development of geodesy, not mathematics; the Gauss–Jordan method, which we called elimination by pivoting, first appeared in a handbook on geodesy.

For matrix algebra to develop, one needed two things: (i) the proper notation, such as a_{ij} and $\mathbf{A}$; and (ii) the definition of matrix multiplication. It is interesting that both of these critical factors occurred at about the same time, around 1850, and in the same country, England. Except for Newton's invention of calculus, the major mathematical advances in the seventeenth, eighteenth, and early nineteenth centuries were all made by continental mathematicians, names such as Bernoulli, Cauchy, Euler, Gauss, and Laplace. But in the mid-nineteenth century, English mathematicians pioneered the study of the underlying structure of various algebraic systems. For example, Augustus DeMorgan and George Boole developed the algebra of sets (Boolean algebra) in which symbols were used for propositions and abstract elements.

The introduction of matrix notation and the invention of the word ''matrix'' were motivated by attempts to develop the right algebraic language

507

for studying determinants. In 1848, J. J. Sylvester introduced the term "matrix," the Latin word for "womb," as a name for an array of numbers. He used "womb" because he viewed a matrix as a generator of determinants. That is, every subset of k rows and k columns in a matrix generated a determinant (associated with the submatrix formed by those rows and columns).

In search of good notation for working with determinants, Sylvester in 1851 proposed writing a square matrix in the form

$$
\begin{array}{cccc}
a_1\alpha_1 & a_1\alpha_2 & \cdots & a_1\alpha_n \\
a_2\alpha_1 & a_2\alpha_2 & \cdots & a_2\alpha_n \\
\vdots & \vdots & \vdots & \vdots \\
a_n\alpha_1 & a_n\alpha_2 & \cdots & a_n\alpha_n
\end{array}
\tag{1}
$$

with each entry represented by a product of symbols. He also introduced the shorthand notation for a square matrix of

$$
\begin{pmatrix}
a_1 & a_2 & \cdots & a_n \\
\alpha_1 & \alpha_2 & \cdots & \alpha_n
\end{pmatrix}
\tag{2}
$$

He referred to the a's and α's as *umbrae,* or ideal elements. Using this umbral notation, Sylvester then wrote the determinant of (2), which involves summing the signed products of all permutations of the a's with the α's, as

$$
\begin{Bmatrix}
a_1 & a_2 & \cdots & a_n \\
\alpha_1 & \alpha_2 & \cdots & \alpha_n
\end{Bmatrix}
\tag{3}
$$

Soon after the introduction of (1), the two symbols a and α were merged into one with double subscripts—a_{ij} (Cauchy had actually used a_{ij} in 1812, but the notation was not accepted then).

Matrix algebra grew out of work by Arthur Cayley in 1855 on linear transformations. Given transformations,

$$
T_1: \quad \begin{aligned} x' &= ax + by \\ y' &= cx + dy \end{aligned} \qquad T_2: \quad \begin{aligned} x'' &= \alpha x' + \beta y' \\ y'' &= \gamma x' + \delta y' \end{aligned}
$$

he considered the transformation obtained by performing T_1 and then performing T_2.

$$
T_2 T_1: \quad \begin{aligned} x'' &= (a\alpha + b\gamma)x + (a\beta + b\delta)y \\ y' &= (c\alpha + d\gamma)x + (c\beta + d\delta)y \end{aligned}
$$

In studying ways to represent this composite transformation, he was led to define matrix multiplication: The matrix of coefficients for the composite transformation $T_2 T_1$ is the product of the matrix for T_2 times the matrix for T_1. Cayley went on to study the algebra of these compositions—matrix algebra—including matrix inverses. The use of a single symbol **A** to represent the matrix of a transformation was essential notation of this new algebra. A link between matrix algebra and determinants was quickly established with the result: $\det(\mathbf{AB}) = \det(\mathbf{A}) \det(\mathbf{B})$. But Cayley believed that matrix algebra would grow to overshadow the theory of determinants. He wrote, "There would be many things to say about this theory of matrices which should, it seems to me, precede the theory of determinants."

It is a curious sidelight to this discussion that another prominent English mathematician of this time was Charles Babbage, who built the first modern calculating machine. Abstracting the mechanics of computation as well as its algebraic structure and notation seems to have been all part of the same general intellectual development in mathematics at that time.

Mathematicians also tried to develop an algebra of vectors, but there was no natural definition for the product of two vectors. The first vector algebra, involving a noncommutative vector product, was proposed by Hermann Grassmann in 1844. Later, Grassmann introduced what we called simple matrices, formed by a column vector times a row vector.

Matrices remained closely associated with linear transformations and, from the theoretical viewpoint, were by 1900 just a finite-dimensional subcase of an emerging general theory of linear transformations. Matrices were also viewed as a powerful notation, but after an initial spurt of interest, were little studied in their own right. More attention was paid to vectors, which are basic mathematical elements of physics as well as many areas of mathematics. The modern definition of a vector space was introduced by Peano in 1888. Abstract vector spaces, whose elements were functions or even linear transformations, soon followed.

Interest in matrices, with emphasis on their numerical analysis, re-emerged after World War II with the development of modern digital computers. Von Neumann and Goldstein in 1947 introduced condition numbers in analyzing roundoff error. Alan Turing, the other giant (with von Neumann) in the development of stored-program computers, gave the **LU** decomposition of a matrix in 1948. The usefulness of the **QR** decomposition was realized a decade later.

References

Bell, E. T., *The Development of Mathematics*. McGraw-Hill, New York, 1940.

F. Cajori, *A History of Mathematical Notations,* Vol. 2. Open Court Publishing Company, Chicago, 1929.

Text and Software References

Reference Texts

Introductory Linear Algebra

Anton, H., *Elementary Linear Algebra*, 3rd ed. Wiley, New York, 1981.
Campbell, S., *Linear Algebra with Applications*. Appleton-Century-Crofts, New York, 1971.
Gewirtz, A., H. Sitomer, and A. W. Tucker, *Constructive Linear Algebra*. Prentice-Hall, Englewood Cliffs, N.J., 1974.
Grossman, S., *Elementary Linear Algebra*, 2nd ed. Wadsworth, Belmont, Calif., 1984.
Kolman, B., *Elementary Linear Algebra*, 4th ed. Macmillan, New York, 1986.
Kumpel, P., and J. Thorpe, *Linear Algebra*. W. B. Saunders, Philadelphia, 1983.
Nicholson, W. K., *Elementary Linear Algebra with Applications*. Prindle, Weber & Schmidt, Boston, 1986.
Strang, G., *Linear Algebra and Its Applications*, 2nd ed. Academic Press, New York, 1980.
Williams, G., *Linear Algebra with Applications*. Allyn and Bacon, Boston, 1984.

Freshman-Level Linear Algebra

Althoen, S., and R. Bumcrot, *Matrix Methods in Finite Mathematics*. W. W. Norton, New York, 1976.
Brown, J., and D. Sherbert, *Introductory Linear Algebra with Applications*. Prindle, Weber & Schmidt, Boston, 1984.

511

Applied Linear Algebra

Helzer, G., *Applied Linear Algebra*. Little, Brown, Boston, 1983.

Magid, A., *Applied Matrix Models*, Wiley, New York, 1985.

Noble, B., and J. Daniels, *Applied Linear Algebra*. Prentice-Hall, Englewood Cliffs, N.J., 1977.

Rorres, C., and H. Anton, *Applications of Linear Algebra*, 2nd ed. Wiley, New York, 1979.

Numerical Analysis

Cheney, W., and J. Kincaid, *Numerical Methods and Computing*. Brooks/Cole, Monterey, Calif., 1980.

Conte, S., and C. deBoor, *Elementary Numerical Analysis*. McGraw-Hill, New York, 1978.

Hildebrand, F., *Introduction to Numerical Analysis*. McGraw-Hill, New York, 1974.

James, M., G. Smith, and J. Wolford, *Applied Numerical Methods for Digit Computation*. Harper & Row, New York, 1985.

More Advanced

Golub, G., and C. VanLoan, *Matrix Computations*. Oxford University Press, New York, 1983.

Stewart, G., *Introduction to Matrix Computations*. Academic Press, New York, 1973.

Wilkinson, J., *The Algebraic Eigenvalue Problem*. Oxford University Press, New York, 1965.

Specific Applications

Graphics

Berger, M., *Computer Graphics*. Benjamin-Cummins, Menlo Park, Calif., 1986.

Magnenat-Thalman, N., and D. Thalman, *Computer Animation: Theory and Practice*. Springer-Verlag, New York, 1985.

Preparata, F., and M. Shamos, *Computational Geometry, An Introduction*. Springer-Verlag, New York, 1985.

Rogers, D., and J. Adams, *Mathematical Elements for Computer Graphics*. McGraw-Hill, New York, 1976.

Linear Models in Statistics

Dunn, O., and V. Clark, *Applied Statistics: Analysis of Variance and Regression*. Wiley, New York, 1974.

Graybill, F., *An Introduction to Linear Statistical Models*. McGraw-Hill, New York, 1961.

Mendelhall, W, *Introduction to Linear Models and the Design and Analysis of Experiments*. Duxbury Press, Belmont, Calif., 1968.

Differential Equations and Other Physical Science Applications

Braun, M., *Differential Equations and Their Applications: An Introduction to Applied Mathematics*. Springer-Verlag, New York, 1975.

Noble, B., *Applications of Undergraduate Mathematics in Engineering*. Mathematical Association of America, Washington, D.C., 1967.

Spiegel, M., *Applied Differential Equations*, 3rd ed. Prentice-Hall, Englewood Cliffs, N.J., 1981.

Strang, G., *Introduction to Applied Mathematics*. Wellesley-Cambridge Press, Wellesley, Mass., 1986.

Markov Chains

Hoel, P., S. Port, and C. Stone, *Introduction to Stochastic Processes*. Houghton Mifflin, Boston, 1972.

Kemeny, J., and L. Snell, *Finite Markov Chains*. D. Van Nostrand, New York, 1960.

Growth Models and Recurrence Relations

Goldberg, S., *Introduction to Difference Equations*. Wiley, New York, 1958.

Kemeny, J., and L. Snell, *Mathematical Models in the Social Sciences*. The MIT Press, Cambridge, Mass., 1969.

Linear Programming

Bradley, H., A Hax, and T. Magnanti, *Applied Mathematical Programming*. Addison-Wesley, Reading, Mass., 1977.

Gass, S., *Linear Programming*, 4th ed. McGraw-Hill, New York, 1975.

Hillier, F., and G. Lieberman, *Introduction to Operations Research*, 4th ed. Holden-Day, Oakland, Calif., 1986.

General Applications

Berman, A., and R. Plemmons, *Nonnegative Matrices in the Mathematical Sciences*. Academic Press, New York, 1979.

Gantmacher, F., *Applications of the Theory of Matrices*. Wiley-Interscience, New York, 1963.

For more references, see *A Basic Library List*, published by the Mathematical Association of America, Washington, D.C., 1988.

Matrix Algebra Software

General-Purpose Computer Languages and Computation Packages with Basic Matrix Operations

APL. Reference: See Heltzer's book under "Applied Linear Algebra."

MACSYMA, Symbolics, Inc.

MAPLE, University of Waterloo.

MINITAB. Reference: Ryan, T., et al., *Minitab Student Handbook*. Duxbury Press, Belmont, Calif., 1976.

muMATH (for IBM PC, Apple), The Soft Warehouse (Microsoft).

TRUE BASIC and other versions of the language BASIC that have matrix operations built-in; for example, MATRIX 100, an enhanced BASIC for IBM PCs from Stanford Business Software.

Matrix Computation Packages

GAUSS (IBM PC), Applied Technical Systems.

Linear Algebra Computer Companion (Apple), Allyn and Bacon, Boston.

LIN*KIT (for IBM PC and Apple), Wiley, New York.

MAC (MatrixAlgebraCalculator) (IBM PC, Rainbow), Professor E. Herman, Mathematics Department, Grinnell College, Grinnell, Iowa.

Matrix Calculator (Apple), CONDUIT.

MATRIX (IBM PC, Apple, Macintosh), Decision Science Software.

PC-MATLAB (for IBM PC), The Math Works, Portola Valley, Calif.

The following two packages, designed for larger computers, are the best matrix computation software in existence. PC-MATLAB and MAC use parts of these packages.

EISPACK. Public domain. Reference: Smith, B., et al., *Matrix Eigensystems Routines—EISPACK Guide*, 2nd ed. Springer-Verlag, New York, 1976.

LINPACK. Public domain. Reference: Dongarra J., J. Bunch, C. Moler, and G. Stewart, *LINPACK User's Guide*. SIAM, Philadelphia, 1979.

Solutions to Odd-Numbered Exercises

Chapter 1

Section 1.1

1. 84 feet   **3.** $\sqrt{H_0}/4$.   **5.** $A = 1500, B = 500$.   **7.** $w = 3, h = 6$.
9. $C = 7, R = 2$.   **11.** $M = 22{,}000, S = 14{,}000$.   **13.** $A \simeq 12, B \simeq 44$.
15. Slower ferry $\simeq 13.5$, faster $\simeq 18.5$.   **17.** $\lim\limits_{k \to 1} = -\infty \ (k \le 1)$.

19. $k = \frac{2}{3}, W = 12$.   **21.** Answer not reasonable: $IQ_m = 240, IQ_f = 0$.
23. $P = 5, J = -3$ (cannot be negative).

Section 1.2

1. Heating oil and diesel oil both off by 120.   **3.** $x_1 \simeq 4, x_2 \simeq 30, x_3 \simeq 40$.
5. (b) x_2 must be negative.   **7. (b)** $x_1 \simeq 45, x_2 \simeq 25, x_3 \simeq 35$.
9. Energy off by 10.   **11. (a)** $x_1 \simeq 306, x_2 \simeq 212, x_3 \simeq 160, x_4 \simeq 37$;
(b) $x_1 \simeq 332, x_2 \simeq 263, x_3 \simeq 163, x_4 \simeq 43$;   **(c)** $x_1 \simeq 378, x_2 \simeq 287$,
$x_3 \simeq 175, x_4 \simeq 46$.   **13. (b)** $x_1 = 2.3, x_2 = 28, x_3 = 63$.

Section 1.3

3. (a) $[0, .25, .5, .25, 0, 0]$;   **(b)** $[.125, .375, .375, .125, 0, 0]$;

(c) $[\frac{1}{16}, \frac{3}{16}, \frac{5}{16}, \frac{5}{16}, \frac{2}{16}, 0]$ **(d)** $[.1, .2, .2, .2, .2, .1]$.   **5. (a)** $\begin{bmatrix} \frac{2}{3} & \frac{1}{2} \\ \frac{1}{3} & \frac{1}{2} \end{bmatrix}$;

(b) $\frac{7}{12}$ **(c)** $\frac{11}{18}$

7. (a)
$$\begin{array}{c} I \\ E \\ S \\ M \end{array} \begin{bmatrix} .5 & .25 & 0 & 0 \\ .5 & .5 & .25 & 0 \\ 0 & .25 & .5 & 0 \\ 0 & 0 & .25 & 1 \end{bmatrix};$$
(b) [.375, .5, .125, 0], [.31, .47, .19, .03].

9.
$$\begin{bmatrix} 1 & .4 & 0 & 0 & 0 & 0 & 0 \\ 0 & .3 & .4 & 0 & 0 & 0 & 0 \\ 0 & .3 & .3 & .4 & 0 & 0 & 0 \\ 0 & 0 & .3 & .3 & .4 & 0 & 0 \\ 0 & 0 & 0 & .3 & .3 & .4 & 0 \\ 0 & 0 & 0 & 0 & .3 & .3 & 0 \\ 0 & 0 & 0 & 0 & 0 & .3 & 1 \end{bmatrix}$$
next round = [0, .4, .3, .3, 0, 0, 0],
second round = [.16, .24, .33, .18, .09, 0, 0].

11. (a) $[\frac{2}{3}, \frac{1}{3}]$;   **(b)** $[\frac{2}{3}, \frac{1}{3}]$;   **(c)** and   **(d)** [.1, .2, .2, .2, .2, .1];
(e) [.86, .14];   **(f)** [.6, .4];   **(g)** [.4, .2, .4];   **(h)** [.01, .01, .01, .96];
(i) [.3, .2, .2, .3];   **(j)** [.83, ~0, ~0, ~0, ~0, ~0, .17];
(k) [.70, ~0, ~0, ~0, ~0, ~0, .30];   **(l)** [.53, ~0, ~0, ~0, ~0, ~0, .47];
(m) [ABC 0, AB 0, AC 0, BC 0, A .27, B .19, C .44, none .10].
13. (a) [45, 50], [39, 50], [32, 49];   **(b)** [−20, 33].
15. (a) [30, 110], [3, 126], [−34, 151];   **(b)** [50, 130], [47, 172], [39, 231].
17. (a) [29.4, 22.8], [28.9, 21.8], [28.5, 21.0], converges to [27, 18];
(b) [8.4, 3.7], [8.7, 4.3], [8.9, 4.8], converges to [9.75, 6.5];
(c) converges to [4.5, 3]; **(d)** converges [7.5, 5].   **19. (a)** Line $F = \frac{2}{3}R$;
(b) [10, 15] converges to [−15, −10];   **(c)** [1, 2] converges to [−3, −2];
(d) convergence in one period.   **21.** After 3000 days, you are closer to starting point than before.

Section 1.4

1. $x_1 \simeq 15$, $x_2 \simeq 65$.   **3.** (18.4, 3).   **5.** Mathematics/Science.
7. $C = 0$, $W = 160$, objective function = 6400.
9. (a) Minimize $50x_1 + 40x_2$ subject to x_1, $x_2 \geq 0$, $20x_1 + 50x_2 \geq 500$,
$30x_1 + 100x_2 \geq 1000$, $10x_1 + x_2 \geq 200$, $15x_1 + 2x_2 \geq 50$;   **(c)** minimum of
1144 at $x_1 = 19.6$, $x_2 = 4.1$.   **11.** Max (i) = Min (ii) = 18.

Section 1.5

1. (a) 23;   **(b)** 8;   **(c)** 1;   **(d)** 25;   **(e)** 8.   **3. (a)** YX;   **(b)** KU;   **(c)** ZA.
5. (a) $x \equiv 7$ (mod 26);   **(b)** $x \equiv 13$ (mod 26);   **(c)** no solution;
(d) $x \equiv 15$ (mod 26).   **7. (a)** 25, 24, 27, 27, 29, . . . ;
(b) 24, 26, 26, 27, 28, . . . ;   **(c)** 30, 24, 28, 26, ...
9. (a) 2, 5, 4, 5, 4, 7, 5, 9, 5, 10, 7, 11, 12, 11, 16, 14, 18, 15, 19, weak
smoothing;   **(b)** 3, 3, 4, 5, 5, 6, 6, 7, 8, 9, 8, 10, 11, 12, 14, 16, 17, 17, 18,
good smoothing;   **(c)** 5, 4, 4, 4, 6, 4, 8, 6, 9, 6, 11, 10, 12, 14, 13, 16, 10, 18,
15, fair smoothing.   **11. (a)** $d_i'' = (d_{i-4} + 2d_{i-2} + 3d_i + 2d_{i+2} + d_{i+4})/9$;
(b) $d_i'' = (d_{i-4} + d_{i-3} + 2d_{i-2} + 2d_{i-1} + 3d_i + 2d_{i+1} + 2d_{i+2} + d_{i+3} + d_{i+4})/15$.   **15. (a)** 6;   **(b)** K;   **(c)** U.

Chapter 2

Section 2.1

1. (a) $[1 \quad 2 \quad 3 \quad 4]$; **(b)** $\begin{bmatrix} 2 \\ 4 \\ 5 \end{bmatrix}$; **(c)** $\begin{bmatrix} 3 \\ 6 \\ 7 \end{bmatrix}$; **(d)** 4; **(e)** 3.

3. (a) $\begin{matrix} A \\ H \\ S \end{matrix} \begin{bmatrix} \frac{1}{3} & \frac{1}{3} & 0 \\ \frac{1}{3} & \frac{1}{3} & 0 \\ \frac{1}{3} & \frac{1}{3} & 1 \end{bmatrix}$; **(b)** 0; $\begin{bmatrix} \frac{1}{3} \\ \frac{1}{3} \\ \frac{1}{3} \end{bmatrix}$; **(c)** rows 1 and 2, columns 1 and 2.

5. (a) a_{23} = gallons of diesel oil from 1 barrel by refinery 3,

(b) $\begin{bmatrix} 20 & 4 & 8 \\ 10 & 14 & 10 \\ 5 & 5 & 24 \end{bmatrix}$; **(c)** $\begin{bmatrix} 20 & 4 \\ 10 & 4 \\ 5 & 5 \end{bmatrix}$.

7. $\begin{bmatrix} 1 & 1 \\ 4 & 1 \\ 20 & 10 \end{bmatrix}$. **9.**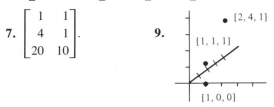

11. Operations not commutative at entry where row and column intersect, value at entry (1, 2) depends on which interchange first.

13. (a) $2\mathbf{J} + 4\mathbf{I}$, **(b)** $\mathbf{J} - \mathbf{A}$; **(c)** $3\mathbf{J} - 2\mathbf{A} + 4\mathbf{I}$.

15. $\begin{bmatrix} 5.8 & 7.6 & 8.8 \\ 7.6 & 6.0 & 8.8 \\ 7.8 & 7.2 & 8.0 \\ 5.4 & 5.4 & 6.2 \end{bmatrix}$.

17. 5 INPUT R: INPUT S
10 FOR I = 1 TO M: FOR J = 1 TO N
15 C[I, J] = R*A[I, J] + S*B[I, J]
20 NEXT J: NEXT I: END

Section 2.2

1. (a) 2; **(b)** 5; **(c)** 38; **(d)** 14. **3. (a)** $\mathbf{aA}$ = [14, 25, 36, 47];

(b) $\mathbf{bB}$ = [5, −7, 2]; **(c)** not defined; **(d)** not defined; **(e)** $\mathbf{Bb}$ = $\begin{bmatrix} 0 \\ -8 \\ 4 \end{bmatrix}$;

(f) $\mathbf{Cc}$ = $\begin{bmatrix} 38 \\ 18 \\ 24 \\ 29 \end{bmatrix}$.

5. (a) $\begin{bmatrix} .30 & .10 & .10 & .75 \\ .25 & .15 & .08 & .80 \end{bmatrix} \begin{bmatrix} 5 \\ 4 \\ 3 \\ 2 \end{bmatrix}$, **(b)** store A \$3.70, store B \$3.69.

7. (a) $A = \begin{bmatrix} 3 & 4 \\ 2 & -5 \end{bmatrix}$, $b = \begin{bmatrix} 5 \\ 3 \end{bmatrix}$, $Ax = b$; **(b)** $A = \begin{bmatrix} 2 & 1 & -2 \\ 1 & 0 & 3 \\ 3 & -1 & 0 \end{bmatrix}$,

$b = \begin{bmatrix} 0 \\ 3 \\ 5 \end{bmatrix}$, $Ax = b$; **(c)** $A = \begin{bmatrix} 2 & -1 & 0 \\ 3 & 2 & 0 \\ 4 & -3 & 0 \end{bmatrix}$, $x = Ax$.

9. $p' = p + Ap$.   **11.** $dAp = 2210$.   **13.** $\begin{bmatrix} 1 & 0 & 0 \\ 2 & 1 & 0 \\ 3 & 2 & 1 \\ 4 & 3 & 2 \\ 0 & 4 & 3 \\ 0 & 0 & 4 \end{bmatrix} \begin{bmatrix} 4 \\ -3 \\ 1 \end{bmatrix} = \begin{bmatrix} 4 \\ 5 \\ 7 \\ 9 \\ -9 \\ 4 \end{bmatrix}$.

17. (a) $\begin{bmatrix} 5 & 19 & -4 \\ 11 & 37 & -8 \end{bmatrix}$; **(b)** $\begin{bmatrix} 7 & 11 \\ 17 & 23 \end{bmatrix}$; **(c)** not possible; **(d)** $\begin{bmatrix} 6 & 10 \\ 17 & 24 \end{bmatrix}$;

(e) $\begin{bmatrix} 26 & 94 & -20 \\ 65 & 223 & -48 \end{bmatrix}$.

19. (a) $[1, -1, 0]$; **(b)** $\begin{bmatrix} 14 \\ 28 \\ 35 \end{bmatrix}$; **(c)** $\begin{bmatrix} -6 \\ -3 \\ -4 \\ -3 \end{bmatrix}$.   **21.** $[(BA)C]_{23} = -40$.

23. (a) $\begin{bmatrix} 160 & 155 \\ 182 & 169 \\ 95 & 100 \end{bmatrix}$; **(b)** $[70 \;\; 235 \;\; 95]$; **(c)** $[1835 \;\; 1765]$;
$\qquad\qquad AB \qquad\qquad\qquad\quad C^T A \qquad\qquad\qquad C^T(AB)$

(d) $[2.7, 5.3, 3.7]$; **(e)** $\begin{bmatrix} 156.5 \\ 172.9 \\ 98.5 \end{bmatrix}$.
$\qquad\quad BD \qquad\qquad\qquad\quad ABD$

25. (a) $[280 \;\; 120 \;\; 100]$; **(b)** $A^T D$; **(c)** entry $(1, 1)$ in $B^T(AC)$.

27. (a) $\begin{bmatrix} \frac{5}{9} & \frac{4}{9} \\ \frac{4}{9} & \frac{5}{9} \end{bmatrix}$, $\frac{4}{9}$; **(b)** $\begin{bmatrix} \frac{14}{27} & \frac{13}{27} \\ \frac{13}{27} & \frac{14}{27} \end{bmatrix}$, $\frac{14}{27}$; **(c)** $\begin{bmatrix} \frac{41}{81} & \frac{40}{81} \\ \frac{40}{81} & \frac{41}{81} \end{bmatrix}$, approaching $[\frac{1}{2}, \frac{1}{2}]$.

29. (a) $\begin{bmatrix} 6 & 0 & 0 \\ 0 & 2 & 0 \\ 0 & 0 & 15 \end{bmatrix}$; **(b)** $(AB)_{ii} = a_{ii}b_{ii}$.   **31. (d)** $\begin{bmatrix} 1 & 0 & 0 \\ 0 & 1 & 0 \\ 0 & 2 & 0 \end{bmatrix}$;

(e) $\begin{bmatrix} 1 & -4 & 2 \\ 0 & 1 & 0 \\ 0 & 0 & 1 \end{bmatrix}$.

Section 2.3

1. (a)

; **(b)**

; **(c)**

3. (a)
$$\begin{bmatrix} 2 & 0 & 2 & 0 \\ 0 & 2 & 0 & 2 \\ 2 & 0 & 2 & 0 \\ 0 & 2 & 0 & 2 \end{bmatrix};$$
(b)
$$\begin{bmatrix} 3 & 1 & 3 & 1 & 2 \\ 1 & 3 & 1 & 3 & 2 \\ 3 & 1 & 3 & 1 & 2 \\ 1 & 3 & 1 & 3 & 2 \\ 2 & 2 & 2 & 2 & 4 \end{bmatrix};$$
(c)
$$\begin{bmatrix} 1 & 0 & 1 & 0 & 0 \\ 0 & 2 & 0 & 1 & 0 \\ 1 & 0 & 2 & 0 & 1 \\ 0 & 1 & 0 & 2 & 0 \\ 0 & 0 & 1 & 0 & 1 \end{bmatrix};$$

(d) $A^2(G) = I$; **(e)**
$$\begin{bmatrix} 5 & 2 & 2 & 2 & 0 & 0 \\ 2 & 3 & 2 & 2 & 1 & 1 \\ 2 & 2 & 3 & 2 & 1 & 1 \\ 2 & 2 & 2 & 3 & 1 & 1 \\ 0 & 1 & 1 & 1 & 1 & 1 \\ 0 & 1 & 1 & 1 & 1 & 1 \end{bmatrix};$$
(f)
$$\begin{bmatrix} 2 & 0 & 1 & 0 & 2 & 0 \\ 0 & 3 & 0 & 2 & 0 & 2 \\ 1 & 0 & 2 & 0 & 2 & 0 \\ 0 & 2 & 0 & 2 & 0 & 1 \\ 2 & 0 & 2 & 0 & 3 & 0 \\ 0 & 2 & 0 & 1 & 0 & 2 \end{bmatrix}.$$

5. Every entry in $A(G)$ or $A^2(G)$ is positive for G_1 and G_2.
7. D_1 points vector $[5, 3, 2, 0]$, D_2 points vector $[9, 5, 4, 2, 0]$.   **11. (a)** 0;
(b) 1; **(c)** 0.   **13.** Parity will remain even if two bits are changed.

15. (a) $c = \begin{bmatrix} 0 \\ 0 \\ 1 \end{bmatrix}$, fourth bit; **(b)** $\begin{bmatrix} 0 \\ 1 \\ 1 \end{bmatrix}$, sixth bit; **(c)** $\begin{bmatrix} 0 \\ 1 \\ 0 \end{bmatrix}$, second bit.

17. New Q obtained from old Q by interchanging rows 3 and 4; $c =$
$[1, 1, 0, 1, 1, 0, 0]$

19. $M = \begin{bmatrix} 1 & 0 & 1 & 0 & 1 & 0 & 1 & 0 & 1 & 0 & 1 & 0 & 1 & 0 & 1 \\ 0 & 1 & 1 & 0 & 0 & 1 & 1 & 0 & 0 & 1 & 1 & 0 & 0 & 1 & 1 \\ 0 & 0 & 0 & 1 & 1 & 1 & 1 & 0 & 0 & 0 & 0 & 1 & 1 & 1 & 1 \\ 0 & 0 & 0 & 0 & 0 & 0 & 0 & 1 & 1 & 1 & 1 & 1 & 1 & 1 & 1 \end{bmatrix}.$

Section 2.4

1. (a) I; **(b)** n; **(c)** 1; **(d)** e_i; **(e)** 1; **(f)** 1; **(g)** 0; **(h)** n; **(i)** 0.
3. (a) $Qx = x$; **(b)** $(Q - I)x = 0$;   **5. (a)** $Ax = By + c$;
(b) $Ax - By = c$; **(c)** $x = (I - A) + By = c$.   **7. (a)** $p' = p + Ap$;
(b) $p' = (I + A)p$; **(c)** $p^{(20)} = (I + A)^{20}p$.
9. $A(x^0 + x^*) = Ax^0 + Ax^* = b + 0 = b$.

11. (a) $\begin{bmatrix} 3 & 2 & 10 \\ 4 & -5 & 8 \\ 0 & 0 & 1 \end{bmatrix};$ **(b)** $\begin{bmatrix} 1 & 2 & 5 & 20 \\ 2 & -1 & -2 & -10 \\ 3 & 4 & 6 & 30 \\ 0 & 0 & 0 & 1 \end{bmatrix};$

(c) $\begin{bmatrix} .4 & .2 & .2 & .2 & 100 \\ .3 & .2 & .2 & .1 & 50 \\ .1 & .1 & 0 & .2 & 100 \\ 0 & .1 & .1 & 0 & 0 \\ 0 & 0 & 0 & 0 & 1 \end{bmatrix}$.

13. $(1A)1 = 1 \cdot 1 = 5$.   **15. (a)** $1A^2 = 1$;
(b) $1A^2 = (1A)A = (1)A = 1$;   **(c)** $1A^3 = (1A^2)A = (1)A = 1$.

19. $\begin{bmatrix} 5 & 1 & 3 & 1 & -1 & 5 \\ 1 & 5 & -1 & 5 & -1 & 1 \\ 3 & -1 & 5 & -1 & 1 & 3 \\ 1 & 5 & -1 & 5 & -1 & 1 \\ -1 & -1 & 1 & -1 & 5 & -1 \\ 5 & 1 & 3 & 1 & -1 & 5 \end{bmatrix}$.

21. Entry (i, j) of $\mathbf{M}(G)\mathbf{M}(G)^T$ tells how many (0 or 1) entries rows i and j have in common, that is, if an edge joins i to j.   **23.** $(\mathbf{A}^2)_{ij} = \mathbf{a}_i^R \cdot \mathbf{a}_j^C = \mathbf{a}_i^C \cdot \mathbf{a}_j^R = (\mathbf{A}^2)_{ji}$.   **27.** Let $\mathbf{C}$ have at least two columns.   **33. (a)** $(\mathbf{AB})^2 = \mathbf{ABAB}$; **(b)** let $\mathbf{A}$ and $\mathbf{B}$ be diagonal matrices.

Section 2.5

3. (a) $|\mathbf{a} - \mathbf{b}|_e = \sqrt{46}$, $|\mathbf{a} - \mathbf{b}|_s = 10$, $|\mathbf{a} - \mathbf{b}|_{mx} = 6$;   **(b)** sum norm equals sum of (absolute) differences of the entries in $\mathbf{a}$ and $\mathbf{b}$;   **(c)** max norm equals largest (absolute) difference between an entry in $\mathbf{a}$ and the corresponding entry in $\mathbf{b}$.
5. (a) (a) and (b) $\mathbf{x}^* = \mathbf{e}_2$, (c) and (d) $\mathbf{x}^* = \mathbf{e}_1$;   **(b)** (a) $\mathbf{x}^* = [1, 1]$,
(b) $\mathbf{x}^* = [-1, 1]$, (c) $\mathbf{x}^* = [1, 1, 1]$, (d) $\mathbf{x}^* = [-1, 1, 1]$.   **7. (a)** 6;
(b) 24;   **(c)** $[4, 4, 4]$;   **(d)** $6^3 \cdot 6 = 1296$.
9. (a) (i) $\|\mathbf{A}\|_s = 1.2$, $\|\mathbf{A}\|_{mx} = 1.25$, (ii) $\|\mathbf{A}\|_s = 1.3$, $\|\mathbf{A}\|_{mx} = 1.7$;
(b) (i) $|\mathbf{p}'|_s = 38 \le \|\mathbf{A}\|_s |\mathbf{p}|_s = 48$, $|\mathbf{p}'|_{mx} = 19 \le \|\mathbf{A}\|_{mx} |\mathbf{p}|_{mx} = 25$,
(ii) $|\mathbf{p}'|_s = 32 \le \|\mathbf{A}\|_s |\mathbf{p}|_s = 52$, $|\mathbf{p}'|_{mx} = 18 \le \|\mathbf{A}\|_{mx} |\mathbf{p}|_{mx} = 34$;
(c) (i) 69, (ii) 88.   **11.** 375.   **13. (a)** $\frac{2}{4}, \frac{7}{6}$;   **(b)** $\frac{2}{10}, \frac{17}{2}$.
15. (a) Assuming that $|a_1|$ is largest absolute entry in $\mathbf{a}$, then

$$|\mathbf{a}|_{mx} = |a_1| = \sqrt{a_1^2} \le \sqrt{a_1^2 + a_2^2 + \cdots} = |\mathbf{a}|_e$$

17. Assuming that $|b_1|$ is largest absolute entry in $\mathbf{b}$, then

$$|\mathbf{a} \cdot \mathbf{b}| = |\Sigma_i\, a_i b_i| \le \Sigma_i\, (|a_i||b_i|) \le \Sigma_i\, |a_i||b_1| = (\Sigma_i\, |a_i|)|b_1| = |\mathbf{a}|_s \cdot |\mathbf{b}|_{mx}.$$

19. $\mathbf{A}$ symmetric means sum of ith row equals sum of ith column.
21. (a) $\|\mathbf{A} + \mathbf{B}\|_s = $ largest (absolute) column sum in

$$\mathbf{A} + \mathbf{B} = \max_i |\mathbf{a}_i^C + \mathbf{b}_i^C|_s \le \max_i |\mathbf{a}_i^C|_s + \max_i |\mathbf{b}_i^C|_s = \|\mathbf{A}\|_s + \|\mathbf{B}\|_s$$

25. (a) $\mathbf{Av} = \begin{bmatrix} 2a + 5b \\ 1a + 3b \end{bmatrix}$; both entries in $\mathbf{Av}$ increase as a and b increase, so $a = b = 1$ maximizes $|\mathbf{Av}|_{mx}$;   **(c)** If first row of $\mathbf{A}$ has larger absolute sum and, say, a_{12} is negative, then let $\mathbf{v} = [1, -1]$.
29. (a) $2^4[1, 1] + 2[1, 0] = [18, 16]$;   **(b)** $2^n[1, 1]$;   **(c)** $9\mathbf{u}_1 - 3\mathbf{u}_2$, $[285, 288]$.   **31.** $\mathbf{A}^2\mathbf{u} = \mathbf{A}(\mathbf{Au}) = \mathbf{A}(\lambda\mathbf{u}) = \lambda(\mathbf{Au}) = \lambda(\lambda\mathbf{u}) = \lambda^2\mathbf{u}$.

Section 2.6

1. (a) 1000; (b) 10^6; (c) 2000; (d) 500; (e) 2000; (f) 9000.

3. (a) If $\mathbf{R}$ is a 2×4 matrix of 1's, $\mathbf{M} = \begin{bmatrix} 2\mathbf{R} & \mathbf{R} \\ \mathbf{R} & 2\mathbf{R} \\ 2\mathbf{R} & \mathbf{R} \\ \mathbf{R} & 2\mathbf{R} \end{bmatrix}$;

(b) if $\mathbf{R} = \begin{bmatrix} 1 & 0 & 1 & 0 \\ 0 & 1 & 0 & 1 \\ 1 & 0 & 1 & 0 \\ 0 & 1 & 0 & 1 \end{bmatrix}$, $\mathbf{M} = \begin{bmatrix} \mathbf{R} & \mathbf{J} - \mathbf{R} \\ 2\mathbf{R} & \mathbf{J} - \mathbf{R} \end{bmatrix}$.

5. Additional last row of $\mathbf{D}''$ is $[0 \ \ 0 \ \ 0 \ \ 0 \ \ 1]$.

7.

$$
\begin{array}{c c} & \begin{array}{c c c c c c c c} a & b & c & d & e & f & g & h \end{array} \\
\begin{array}{c} a \\ b \\ c \\ d \\ e \\ f \\ g \\ h \end{array} &
\begin{bmatrix}
0 & \frac{1}{2} & \frac{1}{2} & \frac{1}{3} & 0 & 0 & 0 & \frac{1}{4} \\
\frac{1}{4} & 0 & 0 & \frac{1}{3} & 0 & 0 & 0 & 0 \\
\frac{1}{4} & 0 & 0 & \frac{1}{3} & 0 & 0 & 0 & 0 \\
\frac{1}{4} & \frac{1}{2} & \frac{1}{2} & 0 & 0 & 0 & 0 & 0 \\
0 & 0 & 0 & 0 & 0 & \frac{1}{2} & \frac{1}{2} & \frac{1}{4} \\
0 & 0 & 0 & 0 & \frac{1}{3} & 0 & 0 & \frac{1}{4} \\
0 & 0 & 0 & 0 & \frac{1}{3} & 0 & 0 & \frac{1}{4} \\
\frac{1}{4} & 0 & 0 & 0 & \frac{1}{3} & \frac{1}{2} & \frac{1}{2} & 0
\end{bmatrix}
\end{array}
= \begin{bmatrix} \mathbf{M} & \mathbf{N}^T \\ \mathbf{N} & \mathbf{M} \end{bmatrix}.
$$

9. $\mathbf{M}^2 = \begin{bmatrix} \mathbf{R} & 2\mathbf{J} \\ 4\mathbf{J} & \mathbf{R} \end{bmatrix}$, where $\mathbf{R} = \begin{bmatrix} 2 & 4 & 2 & 4 \\ 4 & 2 & 4 & 2 \\ 2 & 4 & 2 & 4 \\ 4 & 2 & 4 & 2 \end{bmatrix}$ and $\mathbf{J}$ is 4-by-4 matrix of 1's.

11. (a) $\begin{bmatrix} \mathbf{B} & \mathbf{O} & \mathbf{O} \\ \mathbf{O} & 2\mathbf{B} & \mathbf{O} \\ \mathbf{O} & \mathbf{O} & 3\mathbf{B} \end{bmatrix}$; (b) $\begin{bmatrix} \mathbf{B}^2 & \mathbf{O}_2 & \mathbf{O} \\ \mathbf{O} & 4\mathbf{B}^2 & \mathbf{O}_2 \\ \mathbf{O} & \mathbf{O} & 9\mathbf{B}^2 \end{bmatrix}$;

(c) 45 mults. versus 729 mults.; (d) $[2 \ \ 0 \ \ 2 \ \ 4 \ \ 8 \ \ 4 \ \ 18 \ \ 0 \ \ 18]$.

13. (a)

; (b) $\begin{bmatrix} 27\mathbf{J} & \mathbf{O} \\ \mathbf{O} & 27\mathbf{J} \end{bmatrix}$.

15. (a) If $\mathbf{A} = [\mathbf{A}_1 \ \ \mathbf{A}_2 \ \ \mathbf{A}_3]$, $\mathbf{B} = \begin{bmatrix} \mathbf{B}_1 \\ \mathbf{B}_2 \\ \mathbf{B}_3 \end{bmatrix}$, then $\mathbf{AB} =$

$[\mathbf{A}_1\mathbf{B}_1 + \mathbf{A}_1\mathbf{B}_2 + \mathbf{A}_1\mathbf{B}_3]$;

17. $\begin{bmatrix} \frac{26}{36} & \frac{1}{4} & \frac{1}{36} & 0 & \\ \frac{1}{4} & \frac{1}{2} & \frac{2}{9} & \frac{1}{36} & \\ \frac{1}{36} & \frac{2}{9} & \frac{1}{2} & \frac{2}{9} & \frac{1}{36} \\ 0 & \frac{1}{36} & \frac{2}{9} & \frac{1}{2} & \frac{2}{9} \\ & & & & \text{etc.} \end{bmatrix}$.

19. (a) Bandwidth is 2, .5 on main diagonal, .25 just off main diagonal, columns 1 and 14 same as original frog Markov chain; **(b)** bandwidth is 3, . . . , .0625, .25, .375, .25, .0625, ... **21.** Bandwidth is $2k - 1$.

23. (a) $A = \begin{bmatrix} R & Q \\ Q & R \end{bmatrix}$;

(b) $A^2 = \begin{bmatrix} A' & A'' \\ A'' & A' \end{bmatrix}$, where $A' = \begin{bmatrix} 5 & 0 & 8 & 0 & 4 & 0 \\ 0 & 5 & 0 & 8 & 0 & 4 \\ 4 & 0 & 5 & 0 & 8 & 0 \\ 0 & 4 & 0 & 5 & 0 & 8 \\ 8 & 0 & 4 & 0 & 5 & 0 \\ 0 & 8 & 0 & 4 & 0 & 5 \end{bmatrix}$,

$$A'' = \begin{bmatrix} 0 & 2 & 0 & 2 & 0 & 4 \\ 2 & 0 & 2 & 0 & 4 & 0 \\ 0 & 2 & 0 & 4 & 0 & 2 \\ 2 & 0 & 4 & 0 & 2 & 0 \\ 0 & 4 & 0 & 2 & 0 & 2 \\ 4 & 0 & 2 & 0 & 2 & 0 \end{bmatrix};$$

(c) call submatrix Q: for $n = $ even, $Q^n = I$; $n = $ odd, $Q^n = Q$.

Chapter 3

Section 3.1

1. (a) -17; **(b)** 0; **(c)** -2.   **3.** $x_1 = (2D - G)/15$, $x_2 = (10G - 5D)/15a$.   **5. (a)** Unique $x = y = 0$; **(b)** $(x, y) = r(1, 4)$; **(c)** $(x, y) = r(3, 1)$.   **7. (a)** -2; **(b)** 0; **(c)** 0; **(d)** 0; **(e)** -4; **(f)** 3.3×10^{-8}.   **9.** $x_2 = 33\frac{1}{8}$, $x_3 = 67\frac{1}{2}$.   **11.** Det $= -.002$. **13.** $x = [(2e)d - (2b)f]/[(2a)d - (2b)c]$, similarly for y. **15. (a)** Det $= a_{21}a_{12} - a_{22}a_{11}$.   **19. (a)** 24; **(b)** 0; **(c)** 24. **21.** Area$(ABC) = $ area$(ABB'A') + $ area $(BCC'B') - $ area$(ACC'A')$, where area$(ABB'A') = \frac{1}{2}(x_2 - x_1)(y_2 + y_1)$, similarly for $BCC'B'$, $ACC'A'$. **23. (i)** $\lambda = 4, 2$, $u = [1, 1]$; **(ii)** $\lambda \approx 5.3, -.37$, $u = [.46, 1]$; **(iii)** $\lambda = 4, 1$, $u = [1, 2]$; **(iv)** $\lambda = 3, 3$, $u = [1, 1]$; **(v)** $\lambda = 2, 1, -1$, $u = [1, 1, 0]$. **25.** $\lambda = 1.05 \pm .09i$ (imaginary), $u_1 = [-0.58 + i, .67i]$. **27. (a)** $\lambda = 5, 2$, $u = [1, 2]$, $v = [1, -1]$; **(b)** $p = 10/3u + 5/3v$; **(c)** $(10/3)5^8[1, 2]$.   **33.** det $(A - \lambda I) = (a - \lambda)(b - \lambda)$.   **35. (a)** 3; **(b)** 7; **(c)** 3.

Section 3.2

1. (a) $x = \frac{14}{3}$, $y = \frac{1}{3}$; **(b)** $x = \frac{23}{13}$, $y = -\frac{2}{13}$; **(c)** $x = 2$, $y = 6$. **3. (a)** $x = [30, 14, -9]$; **(b)** $x = [\frac{1}{3}, -\frac{7}{3}, 0]$; **(c)** $x = [-1, 2, \frac{3}{2}]$; **(d)** $x = (\frac{1}{29})[37, 15, 9]$; **(e)** not unique; **(f)** $x = (\frac{1}{11})[-21, -76, 164]$.

5. (a) $\begin{bmatrix} 1 & 0 & 0 \\ \frac{1}{2} & 1 & 0 \\ -\frac{1}{2} & 7 & 1 \end{bmatrix}\begin{bmatrix} 2 & -3 & 2 \\ 0 & \frac{1}{2} & 0 \\ 0 & 0 & 5 \end{bmatrix}$; **(b)** $\begin{bmatrix} 1 & 0 & 0 \\ -1 & 1 & 0 \\ -2 & 0 & 1 \end{bmatrix}\begin{bmatrix} -1 & -1 & 1 \\ 0 & -3 & 4 \\ 0 & 0 & -2 \end{bmatrix}$;

(c) $\begin{bmatrix} 1 & 0 & 0 \\ -2 & 1 & 0 \\ -5 & \frac{11}{5} & 1 \end{bmatrix} \begin{bmatrix} -1 & -3 & 2 \\ 0 & -5 & 7 \\ 0 & 0 & \frac{3}{5} \end{bmatrix}$; (d) $\begin{bmatrix} 1 & 0 & 0 \\ \frac{1}{2} & 1 & 0 \\ -1 & -\frac{3}{4} & 1 \end{bmatrix} \begin{bmatrix} 2 & 4 & -2 \\ 0 & -4 & -3 \\ 0 & 0 & -\frac{29}{4} \end{bmatrix}$;

(e) $\begin{bmatrix} 1 & 0 & 0 \\ 2 & 1 & 0 \\ 5 & 3 & 1 \end{bmatrix} \begin{bmatrix} 1 & 1 & 4 \\ 0 & -1 & -5 \\ 0 & 0 & 0 \end{bmatrix}$; (f) $\begin{bmatrix} 1 & 0 & 0 \\ \frac{3}{2} & 1 & 0 \\ \frac{9}{2} & -39 & 1 \end{bmatrix} \begin{bmatrix} 2 & -3 & -1 \\ 0 & -\frac{1}{2} & -\frac{1}{2} \\ 0 & 0 & -11 \end{bmatrix}$.

7. (a) $\mathbf{x} = [2, 0, 3]$; **(b)** $\mathbf{x} = [\frac{5}{6}, -\frac{140}{6}, -\frac{75}{6}]$; **(c)** $\mathbf{x} = [-\frac{40}{3}, \frac{20}{3}, \frac{25}{3}]$;
(d) $\mathbf{x} = (\frac{1}{29})[-55, 60, -80]$; **(e)** no solution;
(f) $\mathbf{x} = (\frac{1}{11})[-40, -205, 425]$.

9. (a) $\mathbf{x} = (\frac{1}{3})[13, 13, -16, 1]$, $\mathbf{L} = \begin{bmatrix} 1 & 0 & 0 & 0 \\ 1 & 1 & 0 & 0 \\ 2 & 4 & 1 & 0 \\ 1 & 3 & 0 & 1 \end{bmatrix}$,

$\mathbf{U} = \begin{bmatrix} 1 & 3 & 2 & -1 \\ 0 & -2 & -1 & 2 \\ 0 & 0 & 1 & -7 \\ 0 & 0 & 0 & -3 \end{bmatrix}$;

(b) $\mathbf{x} = [1, -1, 2, -2]$, $\mathbf{L} = \begin{bmatrix} 1 & 0 & 0 & 0 \\ \frac{1}{3} & 1 & 0 & 0 \\ \frac{2}{3} & -1 & 1 & 0 \\ \frac{1}{3} & 1 & -\frac{1}{2} & 1 \end{bmatrix}$,

$\mathbf{U} = \begin{bmatrix} 3 & 2 & 1 & 0 \\ 0 & \frac{1}{3} & -\frac{1}{3} & -1 \\ 0 & 0 & -2 & 0 \\ 0 & 0 & 0 & 2 \end{bmatrix}$;

(c) $\mathbf{x} = (\frac{1}{7})[19, -2, -29, 18]$, $\mathbf{L} = \begin{bmatrix} 1 & 0 & 0 & 0 \\ 2 & 1 & 0 & 0 \\ 4 & 3 & 1 & 0 \\ 3 & \frac{3}{2} & 0 & 1 \end{bmatrix}$,
(rows 3 and 4 switched $\rightarrow$
to avoid zero pivot) $\rightarrow$

$\mathbf{U} = \begin{bmatrix} 1 & 1 & -1 & -1 \\ 0 & -2 & 2 & 3 \\ 0 & 0 & -1 & -2 \\ 0 & 0 & 0 & -\frac{7}{2} \end{bmatrix}$.

11. About $n^3/2$. **13. (a)** Multiple solutions; **(b)** no solution; **(c)** $\mathbf{x} =$
$[47.8, -54.3, 160.9]$; **(d)** $\mathbf{x} = [\frac{125}{3}, \frac{125}{3}, 0]$. **15. (a)** $\mathbf{x} = [10, -2, 0]$;
(b) no solution; **(c)** $\mathbf{x} = [\frac{17}{3}, 6, \frac{5}{3}, \frac{14}{3}]$; **(d)** $\mathbf{x} = [3, 0, 0, 1]$.
17. Jello $= 5.04$, fish $= 4.89$, meat $= 1.53$. **19.** 40 micros, 200 terminals,
20 word processors.

21. (a) $\begin{bmatrix} 1 & \frac{1}{10} & 0 & | & 15 \\ 0 & -\frac{3}{2} & 0 & 450 \\ 0 & \frac{1}{2} & 1 & | & 50 \end{bmatrix}$; **(b)** $\begin{bmatrix} -3 & 0 & 0 & | & -1 \\ -5 & 1 & 0 & -4 \\ 3 & 0 & 1 & | & -1 \end{bmatrix}$.

23. (a) $\mathbf{L} = \begin{bmatrix} 1 & 0 \\ c/a & 1 \end{bmatrix}$, $\mathbf{U} = \begin{bmatrix} a & b \\ 0 & d - bc/a \end{bmatrix}$.

25. $\mathbf{L} = \begin{bmatrix} 1 & 0 & 0 \\ x/3 & 1 & 0 \\ 2/x & 9/x^2 & 1 \end{bmatrix}$, $\mathbf{U} = \begin{bmatrix} 3x & 6x^2 & e^x \\ 0 & -x^3 & \frac{2}{3}xe^x \\ 0 & 0 & -7e^x/x \end{bmatrix}$, det $= 21x^3 e^x$.

Section 3.3

3. (a) $\begin{aligned} x_1 \quad\; + 2x_3 &= 1 \\ x_2 + 3x_3 &= 0; \\ x_1 \quad\; + \; x_3 &= 0 \end{aligned}$ **(b)** $\begin{bmatrix} -1 \\ -3 \\ 1 \end{bmatrix}$.

5. (a) $\begin{bmatrix} 2 & -2 \\ -1 & 3 \end{bmatrix}$; **(c)** (i) $[0, 1]$, (ii) $[\frac{2}{3}, \frac{1}{3}]$, (iii) $[-2, 3]$-nonsense;

(d) (i) Nonsense, (ii) $[\frac{2}{3}, \frac{1}{3}]$, (iii) nonsense.

7. (a) $\begin{bmatrix} -\frac{9}{5} & \frac{22}{5} & -\frac{1}{5} \\ -1 & 2 & 0 \\ \frac{4}{5} & -\frac{7}{5} & \frac{1}{5} \end{bmatrix}$; **(b)** $\begin{bmatrix} -\frac{1}{3} & \frac{1}{6} & \frac{1}{3} \\ -\frac{5}{3} & -\frac{2}{3} & -\frac{1}{3} \\ -1 & -\frac{1}{2} & 0 \end{bmatrix}$;

(c) $\begin{bmatrix} -2 & \frac{26}{3} & -\frac{11}{3} \\ 1 & -\frac{16}{3} & \frac{7}{3} \\ 1 & -\frac{11}{3} & \frac{5}{3} \end{bmatrix}$; **(d)** $\begin{bmatrix} \frac{2}{58} & \frac{14}{58} & -\frac{20}{58} \\ \frac{11}{58} & -\frac{10}{58} & \frac{6}{58} \\ -\frac{5}{58} & -\frac{6}{58} & -\frac{8}{58} \end{bmatrix}$;

(e) no solution; **(f)** $\begin{bmatrix} -\frac{8}{11} & \frac{6}{11} & \frac{1}{11} \\ -\frac{30}{11} & \frac{17}{11} & \frac{1}{11} \\ \frac{63}{11} & -\frac{39}{11} & -\frac{1}{11} \end{bmatrix}$.

9. (a) $[\frac{22}{5}, 2, -\frac{7}{5}]$, (b) $[\frac{1}{6}, -\frac{2}{3}, -\frac{1}{2}]$, (c) $[\frac{26}{3}, -\frac{16}{3}, -\frac{11}{3}]$,
(d) $\frac{1}{58}[14, -10, 58]$, (e) no solution, (f) $[\frac{6}{11}, \frac{17}{11}, -\frac{39}{11}]$; **(b) (a)** $[\frac{2}{5}, 0, -\frac{2}{5}]$,
(b) $[-\frac{2}{3}, \frac{2}{3}, 0]$, (c) $[\frac{22}{3}, -\frac{14}{3}, -\frac{10}{3}]$, (d) $\frac{1}{58}[40, -12, 16]$, (e) no solution,
(f) $[-\frac{2}{11}, -\frac{2}{11}, \frac{2}{11}]$; **(c) (a)** $[\frac{23}{5}, 2, -\frac{8}{5}]$, (b) $-[\frac{1}{6}, \frac{1}{3}, \frac{1}{2}]$, (c) $\frac{1}{3}[37, -23, -16]$,
(d) $\frac{1}{58}[34, -16, 2]$, (e) no solution, (f) $[\frac{5}{11}, \frac{16}{11}, -\frac{38}{11}]$.

11. (a) $\begin{bmatrix} 5 & 4 & 3 & 2 & 1 \\ 4 & 4 & 3 & 2 & 1 \\ 3 & 3 & 3 & 2 & 1 \\ 2 & 2 & 2 & 2 & 1 \\ 1 & 1 & 1 & 1 & 1 \end{bmatrix}$; **(b)** $\begin{bmatrix} \frac{5}{6} & \frac{2}{3} & \frac{1}{2} & \frac{1}{3} & \frac{1}{6} \\ \frac{2}{3} & \frac{4}{3} & 1 & \frac{2}{3} & \frac{1}{3} \\ \frac{1}{2} & 1 & \frac{3}{2} & 1 & \frac{1}{2} \\ \frac{1}{3} & \frac{2}{3} & 1 & \frac{4}{3} & \frac{2}{3} \\ \frac{1}{6} & \frac{1}{3} & \frac{1}{2} & \frac{2}{3} & \frac{5}{6} \end{bmatrix}$.

13. No inverse.

15. (a) $\begin{bmatrix} .0001 & -.127 & .038 \\ -.0004 & .382 & -.013 \\ .005 & -.089 & .001 \end{bmatrix}$, $[5.04, 4.89, 1.53]$; **(b)** .51 less Jello;

(c) .013k more.

17. (a) $\begin{bmatrix} \frac{1}{2} & -1 & \frac{1}{2} \\ \frac{1}{2} & 0 & -\frac{1}{2} \\ -\frac{3}{2} & 2 & \frac{1}{2} \end{bmatrix}$, [100,000, 50,000, 100,000];

(b) change = $[-100{,}000, 0, 200{,}000]$; **(c)** change = $[-5000k, -5000k, 15{,}000k]$.
21. $(BA)C = (A^{-1}A)C = IC = C$ or $B(AC) = B(AA^{-1}) = BI = B$.
25. Entry (i, i) in inverse is $1/a_{ii}$. **27. (a)** A^{-1} is same as A except the a in entry (i, j) becomes $-a$; **(b)** part (a) is true wherever a is.
29. (a) Has inverse, **(b)** no inverse; **(c)** cannot say, **(d)** no inverse; **(e)** cannot say (A may not be square).

33. (a) $\begin{bmatrix} 1 & 0 \\ 1 & 1 \end{bmatrix}\begin{bmatrix} 4 & 0 \\ 0 & 2 \end{bmatrix}\begin{bmatrix} 1 & 0 \\ -1 & 1 \end{bmatrix}$;

(b) $\begin{bmatrix} .46 & 1 \\ 1 & -.69 \end{bmatrix}\begin{bmatrix} 5.4 & 0 \\ 0 & -.37 \end{bmatrix}\begin{bmatrix} .52 & .76 \\ .76 & -.35 \end{bmatrix}$;

(c) $\begin{bmatrix} 1 & 1 \\ 2 & -1 \end{bmatrix}\begin{bmatrix} 4 & 0 \\ 0 & 1 \end{bmatrix}\begin{bmatrix} \frac{1}{3} & \frac{1}{3} \\ \frac{2}{3} & -\frac{1}{3} \end{bmatrix}$; **(d)** not possible, only one eigenvector.

35. (a) If $A = UD_\lambda U^{-1}$, then $A^2 = AA = (UD_\lambda U^{-1})(UD_\lambda U^{-1}) = UD_\lambda(U^{-1}U)D_\lambda U^{-1} = UD_\lambda D_\lambda U^{-1} = UD_\lambda^2 U^{-1}$.
37. If 0 is an eigenvalue of A, then $Ax = 0$ has multiple solutions. So Theorem 4, part (i) is false, and A has no inverse.

Section 3.4

1. (a) $\lambda = 2$, $[1, 1]$; **(b)** $\lambda = 3$, $[-1, 1]$; **(c)** $\lambda = 4$, $[1, 1]$; **(d)** iteration does not converge. **3. (a),(b)** Cycles 45° around unit circle, iteration does not converge; **(c)** $\lambda = (\sqrt{2}/2)(1 \pm i)$. **5. (a)** $\lambda \approx 1.33$, $[.70, .21, .09]$; **(b)** first and second largest eigenvalues are close together.
7. (a) $[178, 150, 150]$; **(b)** $[202, 148, 135]$. **9.** Sum of powers fails because $\|D\| > 1$. **11. (b)** (i) $[4.11, 1.15, .04]$, (ii) $[-3.10, 11.03, .52]$.

13. (a) $D = \begin{bmatrix} 0 & -\frac{4}{9} & \frac{1}{3} \\ -\frac{1}{2} & 0 & \frac{3}{8} \\ \frac{3}{10} & -\frac{3}{10} & 0 \end{bmatrix}$; **(b)** $x = [-.60, 4.25, .54]$.

15. (a) (i) Does not apply, (ii) applies for (30) and (31), (iii) does not apply,
(b) (i) $2x_1 + x_2 = 4$, apply (30); (ii) $\frac{4}{3}x_1' - x_2 = 5$, apply (30).
$\quad\quad 3x_1 - 4x_2 = 2$ $\quad\quad\quad\quad \frac{2}{3}x_1' + x_2 = 3$
17. (a) Convergence much faster.

Section 3.5

1. (a) 8^3; **(b)** $8^3/3$; **(c)** 8^3. **3.** $k = 10$. **5.** $[x_6 + 5, x_6 + 4, x_6 + 3, x_6 + 3, x_6 + 2, x_6]$. **7.** $[\frac{1}{38}, \frac{1}{19}, \frac{1}{19}, \frac{1}{19}, \ldots, \frac{1}{19}, \frac{1}{38}]$. **9. (a)** No new nonzero entries are created during elimination; **(b)** All the upper right side of the matrix becomes nonzero. **11.** w^2 multiplications per pivot.
13. (a) $x = 0$, $y = -1$; **(b)** $x = 0$, $y = .5$; **(c)** $x = 0$, $y = -.333$.
15. (a) Any $[x, y, z]$ with $y = z + 1$ is a solution; **(b)** $x = 1$, $y = .333$, $z = .00011$. **17. (a)** $\varepsilon = [-.333, 0]$, $x^* - \varepsilon = [.333, -1]$;
(b) $\varepsilon = [1.75, 0]$, $x^* - \varepsilon = [-1.75, .5]$; **(c)** $\varepsilon = [-.333, 0]$,

$\mathbf{x}^* - \boldsymbol{\varepsilon} = [.333, -.333]$.   **19. (a)** 21;   **(b)** 5;   **(c)** not invertible;
(d) 4606.   **21. (a)** $|\mathbf{e}|_s/|\mathbf{x} + \mathbf{e}|_s \leq 1.4$;
(b) $\mathbf{x}' \approx [12, 28.4, 66.5]$, $|\mathbf{e}|_s/|\mathbf{x} + \mathbf{e}|_s \approx .13$.   **23. (a)** 525%;   **(b)** 386%;
(c) (i) $|\mathbf{e}| \leq \|\mathbf{A}^{-1}\| \cdot |\mathbf{e}'|$, (ii) $\|\mathbf{A}\| \cdot |\mathbf{x}| \geq |\mathbf{b}|$ (or $|\mathbf{x}| \geq |\mathbf{b}|/\|\mathbf{A}\|$), then dividing (i) by
(ii) yields $|\mathbf{e}|/|\mathbf{x}| \leq \|\mathbf{A}\| \cdot \|\mathbf{A}^{-1}\| \cdot |\mathbf{e}'|/|\mathbf{b}|$.   **25.** $c(\mathbf{A}) = \frac{34}{3}$, 94% error.
27. $1 = \|\mathbf{I}\|_s = \|\mathbf{A}\mathbf{A}^{-1}\|_s \leq \|\mathbf{A}\|_s\|\mathbf{A}^{-1}\|_s = c(\mathbf{A})$.

Chapter 4

Section 4.1

1. (a) $x' = -x$, $y' = -y$;   **(b)** $x' = 2x + 7$, $y' = y + 3$;   **(c)** $x' = x + y$,
$y' = y$;   **(d)** $x' = -x$, $y' = y$.   **3. (a)** $(0, 0), (-1, 0), (0, -1), (-1, -1)$;
(b) $(7, 3), (9, 3), (7, 4), (9, 4)$;   **(c)** $(0, 0), (1, 0), (1, 1), (2, 1)$;   **(d)** $(0, 0)$,
$(-1, 0), (0, 1), (-1, 1)$.   **5.** Lower left corner of grid and x-by-y size of
each square given   **(a)** $(-3, -2)$, 1-by-1;   **(b)** $(10, 1)$, 2-by-1;   **(c)** $(-1, -2)$,
1-by-1 trapezoid;   **(d)** $(-3, -2)$, 1-by-1.   **7. (a)** $x' = -y$, $y' = 2x$;
(b) $x' = -x$, $y' = -y$;   **(c)** $x' = -x - 7$, $y' = -y - 3$.   **9.** $x' = x$,
$y' = 4 - y$.   **11.** $x' = -y + 4$, $y' = x - 2$.   **13. (a)** $x' = -2y$,
$y' = -2x$;   **(b)** $x' = x$, $y' = y$.   **17. (b)** $x' = -x + y$, $y' = -2x + 2y$.
19. (a) $x' = .866x - .5z$, $y' = y$, $z' = .5x + .866z$;
(b) $x' = \cos 10°x - \sin 10°y$, $y' = \sin 10°x + \cos 10°y$, $z' = z$;
(c) $x' = .75x - .5y - .433z$, $y' = .433x + .866y - .25z$, $z' = .5x + .866z$.
21. $x' = x + .216y + .375z$, $y' = .991y - .284z$.
23. (a) $x' = x$, $y' = .866y - .5z + .134$, $z' = .5y + .866z - .5$.
25. (a) $\frac{7}{3} \cdot 2^k[1, 1]$;   **(b)** $\frac{1}{3} \cdot 2^k[1, 1]$;   **(c)** $-\frac{4}{3} \cdot 2^k[1, 1]$.
27. $4 \cdot 4^5[1, 1] + [1, -2]$, $4 \cdot 4^{20}[1, 1]$.   **29. (a)** $x = x' - y'$, $y = y'$;
(b) $x = .707x' + .707y$, $y = -.707x + .707y$;   **(c)** not invertible.

31. $T([a, b]) = T(a\mathbf{e}_1 + b\mathbf{e}_2) \to aT(\mathbf{e}_1) + bT(\mathbf{e}_2) = a\begin{bmatrix} \cos \theta° \\ \sin \theta° \end{bmatrix} +$

$b\begin{bmatrix} -\sin \theta° \\ \cos \theta° \end{bmatrix} = \mathbf{A}\begin{bmatrix} a \\ b \end{bmatrix}$.

Section 4.2

1. (a) $\hat{y} = 1.82x$, SSE $= 43.2$;   **(b)** $\hat{y} = x + 3.57$, SSE $= 15.7$;
(c) $\hat{y} = x' + 6.57$.   **3. (a)** $\hat{x} = -.86y + 12.43$;   **(b)** SSE different for
x- and y-values.   **5. (a)** $\hat{y} = .17x - 6.56$;   **(b)** $\hat{y} = .5x - 28.2$;
(c) $\hat{y}' = .5x' + 3.3$.   **7.** $q' \approx .089$, $\hat{y} \approx 11/x$.

Section 4.3

1. (a) $2N_2H_4 + N_2O_4 \to 3N_2 + 4H_2O$;
(b) $2C_6H_6 + 15O_2 \to 12CO_2 + 6H_2O$.
3. $15PbN_6 + 44CrMn_2O_8 \to 22Cr_2O_3 + 88MnO_2 + 5Pb_3O_4 + 90NO$.
5. $i_1 = 2$, $i_2 = 4$, $i_3 = 6$.   **7.** $i_1 = 8$, $i_2 = 2$, $i_3 = 1$, $i_4 = 3$, $i_5 = 11$.
9. $y'(t) = 2y(t)$.   **11. (a)** $y'(t) = x(t)$, $x'(t) = -4y(t) + 5x(t)$;
(b) $y'(t) = x(t)$, $x'(t) = -6y(t) - 5x(t)$;
(c) $y'(t) = x(t)$, $x'(t) = z(t)$, $z'(t) = -2y(t) + 3x(t) + 4z(t)$;
(d) $y'(t) = x(t)$, $x'(t) = z(t)$, $z'(t) = y(t) + 2x(t)$.

15. (a) $x(t) = y(t) = 10e^{4t}$;
(b) $x(t) = 10/3\{2e^{4t} + e^t\}$, $y(t) = 10/3\{4e^{4t} - e^t\}$; (c) $x(t) = y(t) = 10e^{5t}$.

Section 4.4

1. Cycles through $[1, 0, 0]$, $[0, 0, 1]$, $[0, 1, 0]$, $[\frac{1}{3}, \frac{1}{3}, \frac{1}{3}]$ is stable distribution.
3. $[\frac{2}{3}, \frac{1}{3}]$.   **5.** $[\frac{3}{5}, \frac{2}{5}]$.   **7.** $\mathbf{p}^* = [\frac{16}{35}, \frac{4}{35}, \frac{15}{35}]$, columns of $\mathbf{A}^{1000} \simeq \mathbf{p}^*$.
9. (b) Not regular.   **11.** (a) $1/(2n - 2)[1, 2, 2, 2, \ldots, 2, 1]$;
(b) $1/q_n[2^{2n-3}, 2^{2n-4}, 2^{2n-6}, 2^{2n-8}, \ldots, 4, 1]$, where $q_n = 2^{2n-3} + \sum_{k=0}^{n-2} 2^{2k}$;
(c) $\frac{1}{2}q_{n/2}[2^{n-3}, 2^{n-4}, 2^{n-6}, 2^{n-8}, \ldots, 4, 1, 1, 4, \ldots, 2^{n-8}, 2^{n-6}, 2^{n-4}, 2^{n-3}]$;
(d) $1/(2n - 2)[1, 2, 2, 2, \ldots, 2, 1]$.
13. If $\mathbf{A}$ begins,

$$\begin{bmatrix} 1 - a & b & \cdots \\ a & 1 - b - c & \cdots \\ 0 & c & \cdots \\ 0 & 0 & \cdots \end{bmatrix}$$

$\mathbf{A} - \mathbf{I}$ is then

$$\begin{bmatrix} -a & b & \cdots \\ a & -b - c & \cdots \\ 0 & c & \cdots \\ 0 & 0 & \cdots \end{bmatrix}$$

For the first-column elimination, add first row to second row:

$$\begin{bmatrix} -a & b & \cdots \\ 0 & -c & \cdots \\ 0 & c & \cdots \\ 0 & 0 & \cdots \end{bmatrix}$$

For second-column elimination, add second row to third row (all other middle columns like this).
15. 3.3, 14.2.   **17.** $\frac{10}{7}$, 5.   **19.** (a) $\frac{25}{16}$; (b) $\frac{35}{16}$; (c) $\frac{15}{16}$.

21. $\mathbf{N} = \begin{bmatrix} \frac{5}{2} & 2 & \frac{3}{2} & 1 & \frac{1}{2} \\ 2 & 4 & 3 & 2 & 1 \\ \frac{3}{2} & 3 & \frac{9}{2} & 3 & \frac{3}{2} \\ 1 & 2 & 3 & 4 & 2 \\ \frac{1}{2} & 1 & \frac{3}{2} & 2 & \frac{5}{2} \end{bmatrix}$, $\mathbf{1N} = [\frac{15}{2}, 12, \frac{27}{2}, 12, \frac{15}{2}]$,

$\mathbf{RN} = \begin{bmatrix} \frac{5}{6} & \frac{2}{3} & \frac{1}{2} & \frac{1}{3} & \frac{1}{6} \\ \frac{1}{6} & \frac{1}{3} & \frac{1}{2} & \frac{2}{3} & \frac{5}{6} \end{bmatrix}$.

25. 48, 8.

Section 4.5

1. (a) 52%, $[.48, .31, 21]$; (b) 0%, $[.57, .29, .14]$; (c) 17%, $[.62, .27, .11]$;
(d) 13%, $[.58, .26, .11, .05]$; (e) -16%, cyclic.

3. $\lambda = 1, -.5 \pm .866i$ (for all λ, $|\lambda| = 1$).

5. Reduced matrix $\mathbf{U} = \begin{bmatrix} -1 & 0 & 0 & 0 & 0 & 2 \\ 0 & -1 & 0 & 0 & 0 & 2 \\ 0 & 0 & -1 & 0 & 0 & 1.2 \\ 0 & 0 & 0 & -1 & 0 & 1.2 \\ 0 & 0 & 0 & 0 & -.25 & 0 \\ 0 & 0 & 0 & 0 & 0 & .47 \end{bmatrix}$.

7. For herd of constant size, we want a small number of females (we only harvest 50), but then there are not enough females to give birth to as many males as are needed to be able to harvest 100 males annually.

9. (a) [742, 742, 405, 405, 408, 408];

(b) $\begin{bmatrix} -.091 & 2.21 & 0 & 4.05 & 0 & 7.42 \\ 0 & 1.30 & 0 & 4.05 & 0 & 7.42 \\ -.5 & 1.20 & -.91 & 2.21 & 0 & 4.05 \\ 0 & .71 & 0 & 1.30 & 0 & 4.05 \\ -.085 & 2.06 & -1.56 & 3.78 & -2.86 & 6.94 \\ 0 & 1.21 & 0 & 2.23 & 0 & 4.08 \end{bmatrix}$.

11. $a_n = 0.8a_{n-1}$.   **13.** $a_n = a_{n-1} + a_{n-2}$, $a_8 = 34$.   **15.** (a) $a_{20} = 2$;
(b) $a_{20} = .5 \cdot 4^{20}$;   (c) $a_{20} = .4 \cdot 4^{20}$;   (d) $a_{20} = 1$.

17. (a) $2\begin{bmatrix} 1 \\ 1 \end{bmatrix}$;   (b) $\frac{1}{2} 4^n \begin{bmatrix} 4 \\ 1 \end{bmatrix} - \frac{1}{2} 2^n \begin{bmatrix} 2 \\ 1 \end{bmatrix}$;   (c) $\frac{2}{5} 4^n \begin{bmatrix} 4 \\ 1 \end{bmatrix} + \frac{3}{5}(-1)^n \begin{bmatrix} 1 \\ -1 \end{bmatrix}$

(d) $1\begin{bmatrix} 1 \\ 1 \end{bmatrix}$.

21. If $a_n = n\alpha^n$, then $a_{n-1} = (n-1)\alpha^{n-1}$ and $a_{n-2} = (n-2)\alpha^{n-2}$; substituting in $a_n - 2\alpha a_{n-1} + \alpha^2 a_{n-2} = 0$, we have

$$n\alpha^n - 2\alpha \cdot (n-1)\alpha^{n-1} + \alpha^2 \cdot (n-2)\alpha^{n-2} = \{n - 2(n) - 1) + n - 2)\}\alpha^n = 0$$

Section 4.6

1. Min $40x_1 + 55x_2$, $50x_1 + 100x_2 \geq 500$, $100x_1 + 100x_2 \geq 800$, $500x_1 + 700x_2 \geq 8000$, $x_1 \geq 0$, $x_2 \geq 0$.   **3.** (a) Max $40x_1 + 30x_2$, $x_1 + x_2 \leq 400$, $5x_1 + 3x_2 \leq 500$, $15x_1 + 20x_2 \leq 4000$, $x_1 \geq 0$, $x_2 \geq 0$;   (b) $x_1 = 0$, $x_2 = 166\frac{2}{3}$, income = \$5000.   **5.** $x_1 = 0$, $x_2 = 166\frac{2}{3}$, cost = \$4166.67.
7. Min $x_{11} + 2x_{12} + 3x_{13} + x_{21} + 3x_{22} + 2x_{23} + 2x_{31} + 4x_{32} + 3x_{33}$, $x_{11} + x_{12} + x_{13} = 1500$, $x_{21} + x_{22} + x_{23} = 2000$, $x_{31} + x_{32} + x_{33} = 2500$, $x_{11} + x_{21} + x_{31} = 1000$, $x_{12} + x_{22} + x_{32} = 2000$, $x_{13} + x_{23} + x_{33} = 3000$, $x_{ij} \geq 0$.
9. Min $\Sigma_i \Sigma_j a_{ij}x_{ij}$ [where a_{ij} is entry (i, j) in hours matrix and $a_{ij} = $ "$-$" means that term is not in the sum], $x_{11} + x_{12} + x_{15} = 1$, $x_{22} + x_{23} + x_{24} + x_{25} = 1$, $x_{31} + x_{32} + x_{33} = 1$, $x_{41} + x_{42} + x_{44} + x_{45} = 1$, $x_{53} + x_{54} + x_{55} = 1$, $x_{11} + x_{31} + x_{41} = 1$, $x_{12} + x_{22} + x_{32} + x_{42} = 1$, $x_{23} + x_{33} + x_{53} = 1$, $x_{24} + x_{44} + x_{54} = 1$, $x_{15} + x_{25} + x_{45} + x_{55} = 1$, $x_{ij} = 0$ or 1.
11. $X_{Ai} = $ money invested in A at start of ith year, same for x_{Bi}, $u_i = $ money not invested in ith year: Max $u_5 + 1.4x_{A4} + 1.7x_{B3} + 2x_C + 1.3x_D$, $x_{A1} + x_{B1} + u_1 = 10,000$, $x_{A2} + x_{B2} + x_C + u_2 = u_1$, $x_{A3} + x_{B3} + u_3 = u_2 + 1.4x_{A1}$, $x_{A4} + u_4 = u_3 + 1.4x_{A2} + 1.7x_{B1}$, $x_{A5} + x_D + u_5 = u_4 + 1.4x_{A3} + 1.7x_{B2}$, all variables ≥ 0.

15. **(a)** $x_1 = 0$, $x_2 = 0$, $x_3 = 6$, 24; **(b)** $x_1 = 7$, $x_2 = 0$, $x_3 = 1$, 23;
(c) $x_1 = \frac{5}{3}$, $x_2 = \frac{8}{3}$, $x_3 = 0$, $\frac{47}{3}$; **(d)** $x_1 = 3$, $x_2 = 0$, $x_3 = 0$, 6.
17. $x_1 = 0$, $x_2 = 10$, $x_3 = 30$, $x_4 = 0$, $x_5 = 20$, 1530.
19. Delete last equation and [by subtracting equations (5), (6), and (7)] change
first equation to $x_{14} - x_{21} - x_{22} - x_{23} - x_{31} - x_{32} - x_{33} - x_{41} - x_{42} -$
$x_{43} = -2$. Inequalities are $x_{21} + x_{22} + x_{23} + x_{31} + x_{32} + x_{33} + x_{41} + x_{42} +$
$x_{43} \geq 2$, $x_{21} + x_{22} + x_{23} \leq 1$, $x_{31} + x_{32} + x_{33} \leq 1$, $x_{41} + x_{42} + x_{43} \leq 1$,
$x_{21} + x_{31} + x_{41} \leq 1$, $x_{22} + x_{32} + x_{42} \leq 1$, $x_{23} + x_{33} + x_{43} \leq 1$, $x_{ij} \geq 0$.
21. **(a)** $60x_{11} + 30x_{21} + 2900$;
(b) $x_{11} = 0$, $x_{21} = 10$ ($x_{12} = x_{22} = 20$, $x_{31} = 15$, $x_{32} = 0$).
23. If 1 acre less planted: \$20 less income, 1 acre more of corn, 2 acres less of
wheat; if \$1 less used: \$2 less income, .1 acre less of corn, .3 acre less of wheat.
25. If 1 unit less of metal, \$500 less profit, $\frac{3}{4}$ more cars, $\frac{5}{8}$ less trucks; if 1 unit
less of labor, \$125 less profit, $\frac{5}{16}$ less cars, $\frac{7}{32}$ more trucks.

Section 4.7

1. **(a)** $\frac{5}{3}$; **(b)** -1; **(c)** 1; **(d)** 1; **(e)** $\pi/2$; **(f)** no solution; **(g)** 3;
(h) -1. **3.** **(a)** 4, 2; **(b)** 3, 3.45, -1.45; **(c)** -1.18, $-.15$, 1.33;
(d) $-.72$, 1.22, $-.25 \pm 1.03i$ (imaginary). **5.** **(a)** 68; **(b)** 65.
7. **(a)** 9.4; **(b)** (i) 8.4, (ii) 8.1. **9.** **(a)** (i) $\frac{4}{3}$, (ii) $\frac{26}{3}$; **(b)** 64 (exact);
(c) 53.8; **(d)** 8.4. **11.** $y_k = -.5(k^2 - 100)$, $1 \leq k \leq 99$.

Section 4.7 Appendix

1. $s(3.1) = -.98$ versus true value $-.96$, $s(3.9) = -1.18$ versus true value
-1.20. **3.** **(a)** $s(.1) = .308$ versus true value of .309, $s(.65) = .890$ versus
true value of .891; **(b)** integral $= .636$. **5.** **(b)** Spline approximation
$-.047$ versus true integral -0.45.

Chapter 5

Section 5.1

1. **(a)** Line $2x + 3y = 10$; **(b)** line $x - 2y = -4$. **3.** **(a)** No solution;
(b) one solution; **(c)** infinite solutions. **5.** **(a)** [2, 1]; **(b)** [1, 0, -2];
(c) [5, 4, -7]; **(d)** [1, -7, 5]; **(e)** [3, 1, -1]; **(f)** [0, 0, 0] (invertible
matrix); **(g)** [0, 0]. **7.** **(a)** [0, 0, 10] + r[1, 1, -1]; **(b)** [3, 3, 7].
9. **(a)** [5, 0, 5, 0, 5] + r[1, -1, -1, 1, 0] + s[1, 0, 0, 0, -1];
(b) [10, -10, -5, 10, 10]; **(c)** [10, 5, 10, -5, -5].
11. **(a)** r[2, 11, -15]; **(b)** [$\frac{56}{3}$, $\frac{38}{3}$, 10]; **(c)** [15, $-\frac{15}{2}$, $\frac{75}{2}$].
13. $3SO_2 + 2NO_3 + 2H_2O \rightarrow 4H + 3SO_4 + 2NO$. **15.** **(a)** r[1, -2, 1];
(b) r[1, 1, -2]; **(c)** r[1, 0, -1]; **(d)** r[1, -2, 2, -2, 1];
(e) r[1, -2, 1, 1, -2, 1]; **(f)** [0, 0, 0, 0, 0, 0];
(g) r[-1, 1, 1, -1, -1, 1]. **19.** **(a)** [5, -10]; **(b)** [5, 0,] + r[2, 1].
21. **(a)** [5, 2, $\frac{7}{3}$]; **(b)** [$\frac{8}{3}$, $-\frac{1}{3}$, 0, 0] + r[2, 1, -1, 0] + s[-1, 2, 0, 1].
23. For all probability vectors, $p \geq 0$ and $p_1 + p_2 + \cdots + p_n = 1$.
(a) $p_2 = 0$; **(b)** $p_3 = p_2$; **(c)** $p_3 = p_1$; **(d)** $p_1 + p_3 + p_5 = p_2 + p_4$;
(e) $2p_2 + 2p_5 = p_1 + p_3 + p_4 + p_6$; **(f)** none; **(g)** $p_1 + p_4 + p_5 = p_2 +$
$p_3 + p_6$; **25.** For all (finite) powers, $p_1 + p_3 + p_5 = p_2 + p_4 + p_6$.
27. $\mathbf{Ax}' = \mathbf{A}(c\mathbf{x}_1 + d\mathbf{x}_2) = c\mathbf{Ax}_1 + d\mathbf{Ax}_2 = c\mathbf{b} + d\mathbf{b} = (c + d)\mathbf{b} = \mathbf{b}$.

Section 5.2

1. $x_1 = \frac{850}{37}$, $x_2 = \frac{840}{37}$, $x_3 = \frac{260}{37}$.   **3. (a)** $\frac{3}{4}[2, 1] - \frac{1}{4}[2, -1]$;
(b) $8[2, -3] + \frac{13}{3}[-3, 6]$;   **(c)** $\frac{7}{9}[1, 3] - \frac{10}{9}[-2, 3]$;
(d) $[\frac{1}{2}, \frac{1}{2}, 0] + r[3, 1, -2]$.

5. (a) $\mathbf{A}^{-1} = \begin{bmatrix} \frac{1}{4} & 0 \\ 0 & \frac{1}{3} \end{bmatrix}$;   **(b)** $\mathbf{A}^{-1} = \begin{bmatrix} \frac{1}{3} & \frac{1}{3} \\ -\frac{2}{3} & \frac{1}{3} \end{bmatrix}$;   **(c)** $\mathbf{A}^{-1} = \begin{bmatrix} 8 & -3 \\ -5 & 2 \end{bmatrix}$.

7. (a) Any column, rank 1;   **(b)** first two columns, rank 2;   **(c)** any two columns, rank 2;   **(d)** first two columns, rank 2.   **9 (a)** Col: $[-1, 2]$,
Null: $[5, 3]$, rank 1;   **(b)** Col: $[2, 1, 1]$, $[1, 2, 1]$, Null: $[-3, -1, 1]$, rank 2;
(c) Col: first three cols., Null: $[-1, 0, 0, 0, 1]$, $[1, -1, -1, 1, 0]$, rank 3;
(d) Col: first two cols., Null: $[-2, 1, 1, 0, 0]$, $[1, -2, 0, 1, 0]$,
$[-1, -1, 0, 0, 1]$, rank 2;   **(e)** Col: all but fourth column,
Null: $[1, 1, -1, -1, 0, 0]$., rank 5.

11. $\mathbf{N} = \begin{bmatrix} -\mathbf{R} \\ \mathbf{I} \end{bmatrix}$, where $\mathbf{I}$ is $(n - r)$-by-$(n - r)$.

13. (a) $\mathbf{a}_3^R = \mathbf{a}_1^R + \mathbf{a}_2^R$, $\mathbf{a}_3^C = -\frac{1}{5}\mathbf{a}_1^C + 2\mathbf{a}_2^C$;   **(b)** $\mathbf{a}_3^R = \mathbf{a}_1^R + 2\mathbf{a}_2^R$,
$\mathbf{a}_3^C = 2\mathbf{a}_1^C - \mathbf{a}_2^C$;   **(c)** $\mathbf{a}_3^R = \mathbf{a}_1^R - \mathbf{a}_2^R$, $\mathbf{a}_3^C = -\mathbf{a}_1^C + 2\mathbf{a}_2^C$.
15. (a) Obtain $\mathbf{A}$ from $\mathbf{A}^*$ by reversing steps in elimination by pivot.
17. (a) Rows in final upper triangular matrix $\mathbf{U}$ are linearly independent unless
some row is all 0's (by Exercise 16(b)). Since $\mathbf{U}$ is derived from $\mathbf{A}$, if $\mathbf{U}$'s rows
are linearly independent, $\mathbf{A}$'s rows are linearly independent;   **(b)** columns linearly dependent $\leftrightarrow$ rows linearly dependent, now use part (a).   **19. (b)** The two
rows of $\mathbf{A}$;   **(c)** first four rows of $\mathbf{A}$.   **21. (a)** Rank($[\mathbf{A}, \mathbf{b}]$) = rank($\mathbf{A}$) means
$\mathbf{b}$ is linearly dependent on columns of $\mathbf{A}$, that is, $\mathbf{b}$ is in Range($\mathbf{A}$);   **(b)** if $\mathbf{b}$ not
in Range($\mathbf{A}$), then $[\mathbf{A}, \mathbf{b}]$ has one more linearly independent vector (namely, $\mathbf{b}$)
than the columns of $\mathbf{A}$.

23. $[\mathbf{x}' \quad \mathbf{x}'']$ in Null($[\mathbf{A} \quad -\mathbf{B}]$) $\leftrightarrow [\mathbf{A} \quad -\mathbf{B}]\begin{bmatrix} \mathbf{x}' \\ \mathbf{x}'' \end{bmatrix} = \mathbf{0} \leftrightarrow \mathbf{Ax}' = \mathbf{Bx}''\ (= \mathbf{d})$.

29. (a) $\begin{bmatrix} 1 & 2 \\ 2 & 3 \end{bmatrix} = \begin{bmatrix} 1 \\ 1 \end{bmatrix} * [1 \quad 2] + \begin{bmatrix} 0 \\ 1 \end{bmatrix} * [1 \quad 1]$;

(b) $\begin{bmatrix} 1 & 2 & 3 \\ 3 & 4 & 5 \\ 7 & 8 & 9 \end{bmatrix} = \begin{bmatrix} 1 \\ 1 \\ 1 \end{bmatrix} * [1 \quad 2 \quad 3] + \begin{bmatrix} 0 \\ 2 \\ 6 \end{bmatrix} * [1 \quad 1 \quad 1]$.

31. Col $(\mathbf{a} * \mathbf{b})$ = all multiples of $\mathbf{a}$, Row $(\mathbf{a} * \mathbf{b})$ = all multiples of $\mathbf{b}$.

Section 5.3

1. $c(\mathbf{A}) = 9.2$ (sum norm).   **3. (a)** $\hat{y} = -.46x + 9.53$;   **(b)** 392;
(c) $\hat{y} = -.46x + 6\frac{1}{3}$.

5. $\begin{bmatrix} .0069 & .0826 & -.0367 \\ .0162 & -.0381 & .0938 \end{bmatrix}$, $\mathbf{w} = [36.96, 69.52]$.

7. (a) $[\frac{1}{5}, \frac{2}{5}]$;   **(b)** $\frac{1}{14}[1, 2, 3]$;   **(c)** $\frac{1}{11}\begin{bmatrix} 2 & 3 & 3 \\ 1 & -4 & 7 \end{bmatrix}$;

(d) $\frac{1}{6}\begin{bmatrix} 2 & 1 & 0 & -1 \\ 0 & 1 & -2 & 1 \\ 1 & 0 & 1 & 2 \end{bmatrix}$;   **(e)** $\begin{bmatrix} .081 & .098 & .404 & -.124 \\ .098 & -.024 & .098 & .171 \\ -.097 & .258 & -.065 & .129 \end{bmatrix}$.

9. (a) $\mathbf{w} = [5.64, 13.88]$; **(b)** .13 day less; **(c)** .17 day less.

11. (a)
$$\begin{bmatrix} -.35 & -.81 & .6 & .68 & -.11 \\ -.99 & .46 & .9 & -2.4 & 2.03 \\ 2.53 & 2.34 & -2.83 & .75 & -1.8 \end{bmatrix}, \ c(\mathbf{X}^T\mathbf{X}) = 3390;$$

(b) $\hat{y} = .47\text{GPA}_{\text{col}} + 1.88\text{GPA}_{\text{hi}} - .49$; **(c)** $\mathbf{e} = [-.1, 0, -.15, .11, .14]$.
13. $\hat{y} = 3.92x^3 - 9.86x^2 - 1.13x + 8.5$, $c(\mathbf{X}^T\mathbf{X}) \approx 6850$.
15. $\mathbf{X}^+ = \{1/(\mathbf{x} \cdot \mathbf{x})\}\mathbf{x}$ and $q = \mathbf{X}^+\mathbf{y} = [\{1/(\mathbf{x} \cdot \mathbf{x})\}\mathbf{x}] \cdot \mathbf{y} = \mathbf{x} \cdot \mathbf{y}/\mathbf{x} \cdot \mathbf{x}$.
17. (a) $1/\sqrt{2}$, 45°; **(b)** .28, 74°; **(c)** $1/\sqrt{2}$, 45°; **(d)** 0, 90°; **(e)** .47, 62°;
(f) .62, 52°.   **19. (a)** $-.265$; **(b)** .080; **(c)** $-.765$.
21. (a) $[1, -2, 0, 2, -1]$; **(b)** $[-1, 3, -1, -2, 1]$;
(c) $[1, -1, 1, -1, 1, -1]$.
23. (a) $|\mathbf{a} - \mathbf{b}|^2 = |\mathbf{a}|^2 + |\mathbf{b}|^2 - 2|\mathbf{a}| \, |\mathbf{b}| \cos \theta \le |\mathbf{a}|^2 + |\mathbf{b}|^2 + 2|\mathbf{a}| \, |\mathbf{b}| =$
$(|\mathbf{a}| + |\mathbf{b}|)^2$.
25. If $\mathbf{v}^* = \Sigma \, r_i\mathbf{v}_i$, $\mathbf{w}^* = \Sigma q_j\mathbf{w}_j$, then $\mathbf{v}^* \cdot \mathbf{w}^* = \Sigma_i \Sigma_j \, r_i q_j \mathbf{v}_i \cdot \mathbf{w}_j = \Sigma_i \Sigma_j 0 = 0$.
27. (a) $\mathbf{v}_1 = [1, 2]$, $\mathbf{w}_1 = [-2, 1]$; **(b)** $\mathbf{v}_1 = [1, 2, 3]$, $\mathbf{w}_1 = [3, 0, -1]$,
$\mathbf{w}_2 = [2, -1, 0]$; **(c)** $\mathbf{v}_1 = [1, 2, 1]$, $\mathbf{v}_2 = [0, -1, 1]$, $\mathbf{w}_1 = [3, -1, -1]$;
(d) $\mathbf{v}_1 = [2, 1, 0, -1]$, $\mathbf{v}_2 = [0, 1, -2, 1]$, $\mathbf{v}_3 = [1, 0, 1, 2]$,
$\mathbf{w}_1 = [-1, 2, 1, 0]$; **(e)** $\mathbf{v}_1 = [0, 1, 2, -1]$, $\mathbf{v}_2 = [2, -1, 1, 4]$,
$\mathbf{v}_3 = [-1, 3, 0, 1]$, $\mathbf{w}_1 = [.54, .24, -.2, -.16]$.   **29.** $x = [\tfrac{1}{7}, \tfrac{3}{7}, \tfrac{2}{7}]$.

Section 5.4

1. (a) $\begin{bmatrix} .6 & -.8 \\ .8 & .6 \end{bmatrix}$; **(b)** $\frac{1}{9}\begin{bmatrix} 2 & -2 & 1 \\ 2 & 1 & -2 \\ 1 & 2 & 2 \end{bmatrix}$, $\mathbf{x} = \frac{1}{9}[1, -2, 11]$.

3. First column, $\mathbf{a}_1$, must be $\mathbf{e}_1$ (is zero below main diagonal); $\mathbf{a}_2$ is nonzero is first two positions and $\mathbf{a}_2 \cdot \mathbf{e}_1 = 0 \to \mathbf{a}_2 = \mathbf{e}_2$; etc.   **5. (a)** Weights are $\tfrac{5}{6}, \tfrac{3}{4}, \tfrac{1}{3}$;
(b) weights are $\tfrac{1}{3}, \tfrac{8}{3}, \tfrac{4}{3}$; **(c)** weights are .7755, .1633, 3265.

7. $\begin{bmatrix} \frac{3}{14} & \frac{1}{14} & \frac{2}{14} \\ \frac{4}{45} & -\frac{2}{45} & -\frac{5}{45} \end{bmatrix}$, $\mathbf{x} = [\tfrac{3}{7}, -\tfrac{1}{15}]$.

11. (a) 93° (close to orthogonal); **(b)** 37°; **(c)** 173°.
13. (a) $1/\sqrt{2}[1, 1]$, $1/\sqrt{2}[1, -1]$; **(b)** $\tfrac{1}{3}[2, 1, 2]$, $1/\sqrt{369}[14, -2, -13]$;
(c) $1/\sqrt{11}[3, 1, 1]$, $1/\sqrt{330}[-7, 16, 5]$, $1/\sqrt{177870}[-77, -154, 385]$.

15. (a) $\mathbf{A}^+ = \mathbf{R}^{-1}\mathbf{Q}^T = \begin{bmatrix} \frac{1}{3} & -1/\sqrt{2} \\ 0 & 1/\sqrt{2} \end{bmatrix}\begin{bmatrix} \frac{2}{3} & \frac{1}{3} & \frac{2}{3} \\ -1\sqrt{2} & 0 & 1/\sqrt{2} \end{bmatrix} =$

$\frac{1}{18}\begin{bmatrix} 13 & 2 & -5 \\ -9 & 0 & 9 \end{bmatrix}$.

17. (a) $\frac{1}{12}\begin{bmatrix} -3 & 0 & 3 \\ 10 & 4 & -2 \end{bmatrix}$; **(b)** $\frac{1}{10}\begin{bmatrix} -3 & -1 & 1 & 3 \\ 16 & 7 & -2 & -11 \end{bmatrix}$;

(c) $\frac{1}{12}\begin{bmatrix} -3 & 0 & 3 \\ 4 & 4 & 4 \end{bmatrix}$.

19. $\mathbf{A}^+ = (\mathbf{A}^T\mathbf{A})^{-1}\mathbf{A}^T = [(\mathbf{QR})^T\mathbf{QR}]^{-1}(\mathbf{QR})^T = [\mathbf{R}^T\mathbf{Q}^T\mathbf{QR}]^{-1}\mathbf{R}^T\mathbf{Q}^T =$
$[\mathbf{R}^T\mathbf{R}]^{-1}\mathbf{R}^T\mathbf{Q}^T = \mathbf{R}^{-1}\mathbf{R}^{T-1}\mathbf{R}^T\mathbf{Q}^T = \mathbf{R}^{-1}\mathbf{Q}^T$.
25. (a) $f(x) \approx -.085 + .856x^2$; **(b)** $f(x) \approx \tfrac{3}{16} + \tfrac{15}{16}x^2$;
(c) $f(x) \approx 2.80x - 2.17x^3$;   **27. (a)** $L_4(x) = x^4 - 6x^2/7 + \tfrac{3}{35}$;
(b) $L_5(x) = x^5 - 10x^3/9 - 5x/21$;

29. (a) $K_3(x) = x^3 - 3x^2/2 + 3x/5 - \frac{1}{20}$;
(b) $2x^3 - 1.286x^2 + .296x - .014$

31. (a) $\begin{bmatrix} 648 & -720 \\ -720 & 810 \end{bmatrix}$, 361; **(b)** $\begin{bmatrix} 900 & -2520 & 1680 \\ -2520 & 7350 & -5040 \\ 1680 & -5040 & 3528 \end{bmatrix}$, 9195;

(c) $\begin{bmatrix} 16{,}201 & -39{,}603 & 23{,}762 \\ -39{,}603 & 98{,}018 & -59{,}405 \\ 23{,}762 & -59{,}405 & 36{,}303 \end{bmatrix}$, 66,222.

Section 5.5

1. (a) $\begin{bmatrix} 1 & 0 \\ 1 & 1 \end{bmatrix}\begin{bmatrix} 4 & 0 \\ 0 & 2 \end{bmatrix}\begin{bmatrix} 1 & 0 \\ -1 & 1 \end{bmatrix}$;

(b) $\begin{bmatrix} .46 & 1 \\ 1 & -.69 \end{bmatrix}\begin{bmatrix} 5.4 & 0 \\ 0 & -.37 \end{bmatrix}\begin{bmatrix} .52 & .76 \\ .76 & -.35 \end{bmatrix}$;

(c) $\begin{bmatrix} 1 & 1 \\ 2 & -1 \end{bmatrix}\begin{bmatrix} 4 & 0 \\ 0 & 1 \end{bmatrix}\begin{bmatrix} \frac{1}{3} & \frac{1}{3} \\ \frac{2}{3} & -\frac{1}{3} \end{bmatrix}$; **(d)** defective matrix, only one eigenvector.

3. (a) $\begin{bmatrix} 1024 & 0 \\ 992 & 32 \end{bmatrix}$; **(b)** $\begin{bmatrix} 1069 & 1558 \\ 2337 & 3406 \end{bmatrix}$; **(c)** $\begin{bmatrix} 342 & 341 \\ 682 & 683 \end{bmatrix}$;

(d) $\begin{bmatrix} 648 & -405 \\ 405 & -162 \end{bmatrix}$.

7. (a) $\begin{bmatrix} 2.5 & 2.5 \\ 2.5 & 2.5 \end{bmatrix} + \begin{bmatrix} -1.5 & 1.5 \\ 1.5 & -1.5 \end{bmatrix}$; **(b)** $\begin{bmatrix} 3.5 & -3.5 \\ -3.5 & 3.5 \end{bmatrix} +$

$\begin{bmatrix} -.5 & -.5 \\ -.5 & -.5 \end{bmatrix}$; **(c)** $\begin{bmatrix} -3 & -3 \\ -3 & -3 \end{bmatrix} + \begin{bmatrix} 2 & -2 \\ -2 & 2 \end{bmatrix}$.

9. (a) $\lambda_1 \simeq 4.67$ $\mathbf{u}_1 \simeq [.82, 1, .21]$, $\lambda_2 \simeq -1.79$, $\mathbf{u}_2 \simeq [1, -.93, .52]$, $\lambda_3 \simeq .12$ $\mathbf{u}_3 \simeq [-.41, .12, 1]$,

$\begin{bmatrix} 1.83 & 2.23 & .47 \\ 2.23 & 2.72 & .58 \\ .47 & .58 & .12 \end{bmatrix} + \begin{bmatrix} -.84 & .78 & -.44 \\ .78 & -.74 & .41 \\ -.44 & .41 & -.22 \end{bmatrix} + \begin{bmatrix} .01 & -.01 & -.03 \\ -.01 & .02 & .01 \\ -.03 & .01 & .10 \end{bmatrix}$

(b) $\lambda_1 \simeq 2.73$ $\mathbf{u}_1 \simeq [1, 1, .73]$, $\lambda_2 \simeq -.73$ $\mathbf{u}_2 \simeq [-.37, -.37, 1]$, $\lambda_3 = 0$, $\mathbf{u}_3 = [-1, 1, 0]$,

$\begin{bmatrix} 1.08 & 1.08 & .79 \\ 1.08 & 1.08 & .79 \\ .79 & .79 & .57 \end{bmatrix} + \begin{bmatrix} -.08 & -.08 & .21 \\ -.08 & -.08 & .21 \\ .21 & .21 & -.57 \end{bmatrix}$

(c) $\lambda_1 \simeq 4.71$ $\mathbf{u}_1 \simeq [-.56, 1, .1, -.52]$, $\lambda_2 \simeq -1.97$ $\mathbf{u}_2 = [-.12, .54, -.78, 1]$,

$$
\begin{bmatrix}
.93 & -1.65 & -.17 & .86 \\
-1.65 & 2.95 & .29 & -1.54 \\
-.17 & .29 & .03 & -.15 \\
.86 & -.154 & -.15 & .80
\end{bmatrix}
+
\begin{bmatrix}
-.01 & .07 & -.10 & .12 \\
.07 & -.30 & .43 & -.56 \\
-.10 & .43 & -.63 & .80 \\
.12 & -.56 & .80 & -1.03
\end{bmatrix}
$$

$$
+
\begin{bmatrix}
.95 & .58 & .48 & .17 \\
.58 & .35 & .29 & .10 \\
.48 & .29 & .24 & .08 \\
.17 & .10 & .08 & .03
\end{bmatrix}
+
\begin{bmatrix}
.13 & .01 & -.21 & -.16 \\
.01 & .00 & -.02 & -.01 \\
-.21 & -.02 & .36 & .26 \\
-.16 & -.01 & .26 & .20
\end{bmatrix}
$$

11. $\lambda \simeq 1.387$ $\mathbf{u} \simeq [.11, .267, .577, .764]$, first half of $\mathbf{u}_1 = \sqrt{2}\mathbf{u}$, $\lambda \mathbf{u} * \mathbf{u}$ is upper right 4-by-4 submatrix of (28).

13. **(a)** $\begin{bmatrix} .9 & 1.0 & 1.7 \\ 1.0 & 1.2 & 2.0 \\ 1.7 & 2.0 & 3.2 \end{bmatrix}$ poor fit; **(b)** $\begin{bmatrix} 1.6 & 1.0 & 2.5 & 2.4 \\ 1.0 & .6 & 1.6 & 1.5 \\ 2.5 & 1.6 & 4.0 & 3.9 \\ 2.4 & 1.5 & 3.9 & 3.8 \end{bmatrix}$ fair fit.

17. **(a)** $\begin{bmatrix} 2/\sqrt{5} \\ 1/\sqrt{5} \\ 0 \end{bmatrix} [\sqrt{5}][1], \begin{bmatrix} 2 \\ 1 \\ 0 \end{bmatrix};$

(b) $\begin{bmatrix} .08 & .53 \\ .61 & .64 \\ .79 & -.55 \end{bmatrix} \begin{bmatrix} 6.74 & 0 \\ 0 & 1.59 \end{bmatrix} \begin{bmatrix} .52 & .86 \\ .86 & -.52 \end{bmatrix}, \begin{bmatrix} .3 & .4 \\ 2.1 & 3.5 \\ 2.7 & 4.6 \end{bmatrix} + \begin{bmatrix} .7 & -.4 \\ .9 & -.5 \\ -.7 & .4 \end{bmatrix};$

(c) $\begin{bmatrix} .29 & .52 \\ .76 & -.40 \\ .44 & -.28 \\ .39 & .70 \end{bmatrix} \begin{bmatrix} 6.72 & 0 \\ 0 & 4.34 \end{bmatrix} \begin{bmatrix} .65 & .76 \\ .76 & -.65 \end{bmatrix}, \begin{bmatrix} 1.3 & 1.5 \\ 3.3 & 3.9 \\ 1.9 & 2.2 \\ 1.7 & 2.0 \end{bmatrix} + \begin{bmatrix} 1.7 & -1.5 \\ -1.3 & 1.1 \\ -.9 & .8 \\ 2.3 & -2.0 \end{bmatrix};$

(d) $\begin{bmatrix} .34 & -.85 \\ .55 & -.17 \\ .76 & .50 \end{bmatrix} \begin{bmatrix} 6.55 & 0 \\ 0 & .37 \end{bmatrix} \begin{bmatrix} .57 & .82 \\ .82 & -.57 \end{bmatrix}, \begin{bmatrix} 1.3 & 1.8 \\ 2.1 & 3.0 \\ 2.8 & 4.1 \end{bmatrix} + \begin{bmatrix} -.3 & .2 \\ -.1 & .0 \\ .2 & -.1 \end{bmatrix}.$

19. **(a)** $[\frac{2}{3}, \frac{1}{3}, 0]$; **(b)** $\begin{bmatrix} .30 & .39 & -.23 \\ -.16 & -.13 & .28 \end{bmatrix}$;

(c) $\begin{bmatrix} .12 & .003 & -.006 & .16 \\ -.05 & .15 & .09 & -.06 \end{bmatrix}$; **(d)** $\begin{bmatrix} -\frac{11}{6} & -\frac{1}{3} & \frac{7}{6} \\ \frac{4}{3} & \frac{1}{3} & -\frac{2}{3} \end{bmatrix}.$

Section 5.5 Appendix

7. **(a)** $\lambda_1 = 7$, $\mathbf{u}_1 = [1, 1]$, $\lambda_2 = -5$, $\mathbf{u}_2 = [1, -1]$; **(b)** $\lambda_1 = 5$, $\mathbf{u}_1 = [1, -1]$, $\lambda_2 = -1$, $\mathbf{u}_2 = [1, 1]$; **(c)** $\lambda_1 = 1$, $\mathbf{u}_1 = [1, 1]$, $\lambda_2 = -1$, $\mathbf{u}_2 = [1, -1]$; **(d)** $\lambda_1 \simeq 5.06$, $\mathbf{u}_1 \simeq [.62, 1, .17]$, $\lambda_2 \simeq -3.14$, $\mathbf{u}_2 \simeq [-.92, .4, 1]$, $\lambda_3 \simeq 3.08$, $\mathbf{u}_3 = [.81, -.66, 1]$; **(e)** $\lambda_1 \simeq 2.41$, $\mathbf{u}_1 \simeq [.71, .71, 1]$, $\lambda_2 = 1$, $\mathbf{u}_2 = [-1, 1, 0]$, $\lambda_3 = -.41$, $\mathbf{u}_3 = [-.71, -.71, 1]$; **(f)** $\lambda_1 \simeq 4.71$, $\mathbf{u}_1 \simeq [-.56, 1, .1, -.52]$, $\lambda_2 \simeq -1.97$, $\mathbf{u}_2 = [-.12, .54, -.78, 1]$, $\lambda_3 \simeq 1.58$, $\mathbf{u}_3 \simeq [1, .61, .50, .18]$, $\lambda_4 \simeq 0.68$, $\mathbf{u}_4 \simeq [-.60, -.05, 1, .74]$.

Index

A

Absorbing Markov chain, 309
Absorbing state in a Markov chain, 309
Adams, J., 512
Addition of matrices, 67
Adjacency matrix for a graph, 99, 145
Affine linear transformation, 256
Alphabetic code; *see* Coding models
Althoen, S., 511
Angle between vectors, 444, 446
Anthropologist's data, 489
Anton, H., 511
Approximate (least squares) solution, 433, 438
Associative rules in matrix algebra, 120, 121

B

Babbage, Charles, 509
Band matrix, 147
Bandwidth of a matrix, 147

Basic variables in a linear program, 428
Basis of a vector space, 418
Berger, M., 512
Berman, A., 513
Binary code, 103
Boole, George, 507
Bradley, H., 513
Braun, M., 512
Brown, J., 511
Bumcrot, R., 511
Byte, 103

C

Campbell, S., 511
Canoe problem, upstream-downstream, 5
Canoe problem, with sail, 6, 161, 248, 460
Cayley, Arthur, 508
Cayley–Hamilton theorem, 174
Cauchy, A.-L., 508
Change of basis, 457